WAVE MECHANICS the first fifty years

A tribute to
Professor Louis De Broglie, Nobel Laureate
on the 50th anniversary of the discovery
of the wave nature of the electron

WAVE MECHANICS the first fifty years

Edited by

William C. Price, F.R.S.,
Seymour S. Chissick
Tom Ravensdale

University of London King's College

A HALSTED PRESS BOOK

JOHN WILEY & SONS
New York-Toronto

English edition first published in 1973 by
Butterworth & Co (Publishers) Ltd
88 Kingsway, London WC2B 6AB

Published in the U.S.A. and Canada by
Halsted Press, a Division of John Wiley & Sons, Inc.,
New York

Library of Congress Cataloging in Publication Data
Main entry under title:

Wave Mechanics.

  "A Halsted Press book."
  Contents: Robertson, A. J. The beginnings of chemical physics.—
Broglie, L. de. The beginnings of wave mechanics.—Slater, J. C.
The development of quantum mechanics in the period 1924–1926.
[etc.]
    1. Wave mechanics—Addresses, essays, lectures. 2. Broglie, Louis,
prince de, 1892– —Addresses, essays, lectures. I. Broglie, Louis,
prince de, 1892–   II. Price, William Charles, 1909– ed.
III. Chissick, Seymour S., ed. IV. Ravensdale, Tom, ed.
QC174.22.W38 1973   531'.1133   73–6652

ISBN 0 470 69731 8

Filmset by Photoprint Plates Ltd, Rayleigh, Essex
Printed and bound in England by Hazell Watson & Viney Ltd,
Aylesbury Bucks.

Dedication

PROFESSOR SIR GEORGE PORTER, F.R.S.
The Royal Institution, London

It is given to very few men to discover one of those fundamental relationships between natural phenomena which take us a step nearer to the unity of all things. In this connection we think particularly of Newton, Faraday and Maxwell, Planck, Einstein and de Broglie.

The relation between the wave and particle properties of matter was discovered half a century ago, not very long after the revolutionary concepts of relativity and quantum theory. Nothing of equal importance to physics has appeared since, though its author would be one of the last to suppose that any finality has been reached in our understanding, particularly of the wave nature of matter.

If, in some ways, progress in fundamental understanding has been disappointing over the last fifty years, and less than might have been hoped for after the rapid developments of the first quarter of this century, the applications of the new wave mechanics to the whole of physical science have been one of the great successes of the period. To take only one example, though perhaps the most important and certainly the one involving most scientists, theoretical chemistry as we know it today would not exist without quantum mechanics, whether we consider the formal calculation of molecular properties or the more qualitative description of molecular orbitals which are now the basis for even the most elementary accounts of molecular structure and behaviour.

All branches of physical science are increasingly dependent on the wave mechanics, and it is therefore most appropriate that this volume, containing articles written by many of those who have developed its concepts in various ways, should be dedicated, fifty years after his discovery of the wave nature of the electron, to Louis de Broglie.

George Porter

Foreword

THE RT. HON. EDWARD HEATH, M.P.,
Downing Street, London

At this time of enlargement of the European Community it is a special pleasure for me to welcome this book, dedicated to Professor Louis de Broglie, one of Europe's greatest scientists, on the occasion of the 50th anniversary of his discovery of the wave nature of matter, a profound and influential conception which has made possible a new unity in the physical sciences.

Edward Heath

Preface

'It is well that the reader should appreciate through personal experience the agony of the physicists of the period. They could but make the best of it, and went around with woebegone faces sadly complaining that on Mondays, Wednesdays, and Fridays they must look on light as a wave; on Tuesdays, Thursdays, and Saturdays, as a particle. On Sundays they simply prayed.'

> The Strange Story of the Quantum
> Banesh Hoffmann

It is only rarely that the 50th anniversary of a great and fundamental scientific discovery occurs during the lifetime of its discoverer. On such occasions it is customary on the part of the international scientific community to commemorate the event in various ways and traditionally one such way is to prepare a volume of essays especially written for the occasion. This volume commemorates the submission of Louis de Broglie's fundamental manuscript to the scientific press in December 1923. Although his elegant and ingenious work was at that time unrecognised, within three years Erwin Schroedinger had taken up the basic ideas and single-handed produced a theory of the atom which stands today as the basis of theoretical chemistry and physics.

In compiling this volume the editors have been fortunate in securing the help and co-operation of some of the worlds leading scientists. In addition to the chapter by Louis de Broglie himself, telling the story of how he was led to his conclusions concerning the wave nature of the electron, the volume contains articles by many of the early pioneers in the field such as John C. Slater, John H. Van Vleck, Linus Pauling and Charles Coulson. A few of the articles are historical but the majority review topics which are at the frontier of studies on the nature and function of electrons in matter and the theoretical and experimental techniques developed for their explanation.

The editors wish to record their thanks to: the University of London King's College, for the facilities provided; Sir John Hackett, Principal of King's College, for his help and encouragement; Professor V. Gold, F.R.S., Professor S. F. Mason, Professor A. J. B. Robertson and Dr. M. P. Melrose for reading many of the chapters and making helpful comments; and finally, but by no means least, the staff of Butterworths for their patience and co-operation at all stages in the preparation of the book.

Royalties from the sale of the book are being given to the Open University to establish the de Broglie Prize, which will be awarded to outstanding students in the physical sciences.

William C. Price, F.R.S.
Seymour S. Chissick
Tom Ravensdale

University of London King's College

February 1973

Contents

Prologue

The Beginnings of Chemical Physics

ANDREW J. B. ROBERTSON
Department of Chemistry, King's College, Strand, London

1. INTRODUCTION

Much of the work of such early natural philosophers as Boyle, Black and Cavendish involved the study of chemical problems with the help of the methods and concepts of physics, and could therefore be regarded as chemical physics. The view that the foundations of chemistry in its present form were laid in France by Lavoisier has, however, been expressed by Wurtz in his history of chemistry, and, if this view is accepted, Lavoisier and Laplace might be regarded as the founding fathers of chemical physics. The calorimetric methods they developed were used to study a range of problems in both physics and chemistry, and these researches, like the earlier ones of Black in Scotland, are notable for the exact and quantitative approach. The early workers in the borderland between chemistry and physics used the concepts of Newtonian mechanics; but even before Lavoisier and Laplace the discontinuous nature of the light emitted from some coloured flames had been discovered, and as further spectroscopic phenomena were encountered a body of observations arose which did not receive an explanation in fundamental terms, despite numerous attempts, such as the vortex atom theory, until the much later advent of wave mechanics.

2. THE BEGINNINGS OF OPTICAL SPECTROSCOPY

Early Observations

A paper 'Observations on Light and Colours' read in 1752 before the Medical Society of Edinburgh by the Scottish experimental philosopher, and former Glasgow divinity student, Melvill, contained a remarkable deduction on the yellow light emitted from a flame produced by burning spirits containing sodium salts. Melvill examined with a prism various coloured flames,

1

produced with salts of different metals, and in the case of the sodium flame he deduced (see Wolf, 1938) '. . . the bright yellow . . . must be of one determined degree of refrangibility; and the transition from it to the fainter colour adjoining, not gradual, but immediate'. Like Newton before him, Melvill did not use a well-defined thin beam of light with his prism, and such a beam seems to have been first subjected to prismatic analysis by Wollaston (1802), who discovered some dark lines in the solar spectrum. He noted that some of these lines could be the boundaries of colours, but two lines were not explicable in this way. Many more of these dark lines were later independently discovered and carefully investigated by Fraunhofer. The inquiry thus launched was followed by four principal branches of research: first, the study of black lines in spectra produced by absorption of light by gases and vapours; second, the study of spark and flame spectra; third, the spectroscopic investigation of light from planets, stars and nebulae; and fourth, the application of optical spectroscopy to chemical analysis.

ABSORPTION SPECTRA OF GASES AND VAPOURS

Pioneering work in this field was carried out by Brewster (1836) of Edinburgh, Scotland, and by Daniell and W. H. Miller in the newly-founded chemical laboratory at King's College, London, England. The principal object of Brewster's work was, in his words, '. . . the discovery of a general principle of chemical analysis, in which simple and compound bodies might be characterised by their action on definite parts of the spectrum, . . .' His first trial with gaseous bodies was with 'nitrous acid gas' (NO_2) which gave several hundred dark lines in the spectrum. Brewster wrote: 'The result of this experiment . . . presented me with a phænomenon so extraordinary in its aspect, — bearing so strongly on the rival theories of light, — extending so widely the resources of the practical optician, and lying so close to the root of atomical science, that I am persuaded it will open up a field of research which will exhaust the labours of philosophers for centuries to come'. Brewster announced his discovery at the Oxford meeting of the British Association in 1832, and the subject was then speedily pursued by Daniell and W. H. Miller. The latter wrote, in a report of his communication of the joint work in 1833 to the Philosophical Society of Cambridge (Miller, 1833): 'As it does not appear that Sir D. Brewster examined the effects produced by any of the other coloured gases, I beg to offer the Society a short account of some experiments which I made conjointly with Professor Daniell, in the laboratory of King's College'. They passed light through a jar of gas, and then brought the light to a focus. The line of light was viewed through a prism. Miller wrote: 'When the air in the jar was slightly coloured with the vapour of bromine, the whole of the spectrum was seen interrupted by probably more than a hundred equidistant lines; as the vapour became denser the blue end of the spectrum disappeared, and the lines in the red part grew stronger'. With iodine vapour the effect was similar. 'Euchlorine' also gave lines. At the time of this joint research, Miller was Professor of Mineralogy at Cambridge and a Fellow of St. John's College.

These remarkable discoveries were not pursued further until 1845, when W. A. Miller, who succeeded Daniell as Professor of Chemistry at King's College, initiated a notable series of researches. Miller (1845) in his first paper mentioned the first two researches and remarked: '. . . the subject appeared to have been completely laid aside by the philosophers to whom these observations are due, . . .'.

The extension of absorption measurements into the ultra-violet part of the spectrum presented experimental difficulties, but a very helpful discovery was made in the laboratory at the Royal Institution, London, by Stokes, Lucasian Professor of Mathematics at Cambridge, who was preparing experimental demonstrations for a Friday Evening Discourse 'On the Change of Refrangibility of Light, and the exhibition thereby of the Chemical Rays'. Stokes (1862) described this discovery as follows: 'In making preparations for a lecture on the subject delivered at the Royal Institution in February 1853, in which I had the benefit of the kind assistance of Mr Faraday, recourse was naturally had to electric light, on account of the extraordinary richness which it had been found to possess in rays of high refrangibility. Although fully prepared to expect rays of much higher refrangibility than were found in the solar spectrum, I was perfectly astonished, on subjecting a powerful discharge from a Leyden jar to prismatic analysis with quartz apparatus, to find a spectrum extending no less than six or eight times the length of the visible spectrum, and could not help at first suspecting that it was a mistake arising from the reflection of stray light'. Miller (1862) in the King's College laboratory used such a source of ultra-violet light and observed absorption spectra, photographically recorded, by passing light from a spark through gases, vapours, liquids and solids. This initiated many later researches. The notable spectroscopic researches of Hartley, Professor of Chemistry in Dublin, Ireland, were initiated in the King's College laboratory, where Hartley became Demonstrator in 1870.

SPARK AND FLAME EMISSION SPECTRA

That the spectrum of the electric arc is not continuous was noted by Wollaston (1802) and by Fraunhofer, but it was Wheatstone, Professor of Physics, King's College, London, in a paper 'On the Prismatic Decomposition of the Electric, Voltaic and Electro-magnetic Sparks' read at the 1835 Dublin meeting of the British Association who first drew attention to the important fact that the nature of the metals employed modifies the resultant spectrum. He found that the number, position and colours of the bright lines emitted vary from metal to metal, and that the metals may be readily distinguished by their spectra. A voltaic spark from mercury gave the same spectrum in an air-pump vacuum, in a Torricellian vacuum, and in gaseous carbon dioxide, showing that the emitted light does not arise from the combustion of the metal. A pioneering investigation of the emission spectra from wire electrodes with various salts on them was contributed by van der Willigen (1859) of Deventer, Holland, who wrote: 'Die Anbringung von Salzsäure auf die

Poldrähte brachte mich jedoch auf eine andere Weise, die charakterischen Streifen aufzuspüren, die ich hier noch in Vorbeigehen anfuhren will'.

Early experiments on flame spectra were reported by Herschel in 1822 and by Talbot, one of the pioneers of photography, in 1826, and Talbot proposed that 'a glance at the prismatic spectrum of a flame may show it to contain substances which it would otherwise require a laborious chemical analysis to effect'. The great sensitivity of this method of analysis was established by Swan in 1857, when he was a teacher in the Scottish Naval and Military Academy, Edinburgh. Other important contributions were made by Miller, starting in 1845, and by Kirchhoff and Bunsen in Heidelberg, Germany, who developed the spectroscope for qualitative analysis with flames. By that time, therefore, a considerable body of experimental data existed which was not interpreted by wave-mechanical theory until many more years elapsed.

THE INVESTIGATION OF PLANETS, STARS AND NEBULAE

The full explanation of Fraunhofer's lines in the solar spectrum given by Kirchhoff pointed the way to a knowledge of the chemical composition of the sun and the fixed stars. In Kirchhoff's words: 'The sun consists of a glowing gaseous atmosphere, surrounding a solid nucleus which possesses a still higher temperature. If we could see the spectrum of the solar atmosphere without that of the solid nucleus, we should notice in it the bright lines which are characteristic of the metals it contains. The more intense luminosity of the internal nucleus does not, however, permit the spectrum of the solar atmosphere to become apparent; it is *reversed* according to my newly discovered proposition; so that, instead of the *bright* lines which the luminous atmosphere by itself would have shown, *dark* ones appear. We do not see the spectrum of the solar atmosphere itself, but a negative image of it'. Tyndall, Professor of Natural Philosophy, Royal Institution, in a Friday Evening Discourse in June 1861 'On the Physical Basis of Solar Chemistry' gave an account of earlier workers who had been near to Kirchhoff's great discovery, and considered that Angström (1855) had come nearest to the philosophy of the subject in a paper which Tyndall himself had translated. This paper states: 'Now, as according to the fundamental principle of Euler, a body absorbs all the series of oscillations which it can itself assume, it follows from this that the same body, when heated so as to become luminous, must emit the precise rays which, at its ordinary temperature, is absorbed' (sic). Angström then shows the relation between electric spark spectra and Fraunhofer lines.

Over twenty years before Kirchhoff's generalisation the idea of the Fraunhofer lines arising from absorption in the solar atmosphere was clearly stated by Daniell (1839) who wrote in his book, after giving a lucid account of the solar spectrum and Fraunhofer lines: 'This phenomenon has been ascribed to the action of the solar atmosphere, which may have the property of absorbing particular portions of the rays of light. A similar specific action has been found in different gases and vapours, to which we shall again refer when examining the properties of heterogeneous kinds of matter'. Daniell in his book modestly does not mention that he with W. H. Miller had observed these specific actions of different gases and vapours.

Daniell also drew attention to the dark bands in the spectra of the fixed stars, discovered by Fraunhofer, and this topic was pursued after Kirchhoff's generalisation by Daniell's successor W. A. Miller, by Rutherfurd in New York and by Secchi. Miller worked with the distinguished astronomer Huggins using a spectrum apparatus attached to the latter's telescope at Upper Tulse Hill in South London, where Huggins and Miller were neighbours. Their first paper (Huggins and Miller, 1864) describes studies of some planets and nearly fifty fixed stars. The spectra showed the similarity of essential constitution which exists among the stars and the sun. A supplement to this paper (Huggins, 1864) describes observations on the light emitted from some nebulae. This light was unexpectedly found to be nearly monochromatic. Describing his first observation, Huggins wrote: 'At first I suspected some derangement of the instrument had taken place; for no spectrum was seen, but only a short line of light perpendicular to the direction of dispersion. I then found that the light of this nebula, unlike any other ex-terrestrial light which had yet been subjected by me to prismatic analysis, was not composed of light of different refrangibilities and therefore could not form a spectrum'.

3. SOME EARLY PHOTOCHEMISTRY

As with optical spectroscopy, wave mechanics provided a fundamental basis for the discussion of photochemical phenomena, and again many observations preceded their explanation by many years. Draper (1843a, 1843b), Professor of Chemistry, University of New York, investigated with the help of the prismatic analysis of light the power of green plants to decompose carbon dioxide to oxygen under the influence of sunlight. He also studied the photochemically-induced combination of hydrogen and chlorine, and devised a 'tithonometer' to measure the 'chemical force' of light rays by the amount of hydrochloric acid produced on irradiating a mixture of equal volumes of hydrogen and chlorine. This remained the first and only attempt to refer the chemical action of light to a standard measure until the researches at Heidelberg by Bunsen and Roscoe (Professor of Chemistry, Owens College, Manchester). They studied the hydrogen–chlorine reaction in a paper notable for the careful experimentation and mathematical approach (Bunsen and Roscoe, 1857), and later they devised a constant standard photographic paper and a technique for measuring light intensity from its blackening (Bunsen and Roscoe, 1863).

4. PHYSICAL AND CHEMICAL PROPERTIES AND THEIR RELATION

Early measurements of gas densities were used by Prout to deduce atomic weights, making use of the doctrine of volumes of Gay-Lussac. The two papers by Prout (1815, 1816) were published anonymously. In the first paper he wrote: 'The author of the following essay submits it to the public with the greatest diffidence; for though he has taken the utmost pains to arrive at the truth, yet he has not that confidence in his abilities as an experimentalist as to induce him to dictate to others far superior to himself in chemical acquirements and fame'. From his derived atomic weights he notes 'That all the elementary numbers, hydrogen being considered as 1, are divisible by 4, except carbon, azote and barytium, and these are divisible by 2, appearing therefore to indicate that they are modified by a higher number than that of unity or hydrogen'. In his second paper, Prout deduces that taking hydrogen as 1, the specific gravities of most, or perhaps all, elementary substances will either exactly coincide with, or be some multiple of, the weights of their atoms. The first matter of the ancients may then, he says, be considered to be realised in hydrogen. Subsequent arguments and controversy, however, remained confused until the experimental situation was clarified by the discovery of isotopes and the development of the mass spectrograph.

Important results were obtained from measurements of densities of solids and liquids by Kopp, Lecturer at Giessen in Germany, whose book in German was abstracted at length in English by Croft (1842). In that year Kopp (1842) extended his work, writing: 'I have since extended my researches to the organic combinations, and have likewise arrived at very simple results'. He noted that measurements should be made at temperatures equidistant from the boiling points. The general field of the relation between physical properties and chemical constitution was extended by many workers but the approach was of necessity largely empirical until wave mechanics began to resolve some of the problems in this field. Now wave mechanical calculations of molecular properties have gone a long way towards realising the aim of converting chemistry into a branch of applied mechanics, which, according to von Meyer in his history of chemistry, was noted as very important by the French savants Berthollet and Laplace.

The early measurements of Dulong and Petit revealed a temperature variation of the specific heat of a number of metals, and later measurements on such variation were made by Regnault, Kopp and Weber. This variation was only explained with the advent of the quantum theory, which also resolved the long-standing mystery of the failure of the Dulong and Petit law for some of the elements. Before this, Kopp had speculated, somewhat fancifully, that diamond, with the lowest atomic heat, was the only truly elementary substance known, the elements conforming with the Dulong and Petit law being possibly compounds.

5. CHEMICAL AFFINITY

The somewhat ill-defined concept of chemical affinity was used by earlier workers in considerations of the formation and stability of molecules and in connection with reversible reactions. However, as Williamson, Professor of Chemistry at University College, London, said in a Royal Institution Discourse in 1851, 'It unfortunately often occurs that names are mistaken for explanations, and people deceive themselves with the belief that, for instance, in attributing chemical decompositions to affinity, attraction, contact-force, catalysis, etc., they explain them'. He went on to remark that 'The atomic theory has hitherto been tacitly connected with an unsafe and unjustifiable hypothesis, namely, that the atoms are in a state of rest; the dynamics of chemistry will commence by the rejection of this supposition, and will study the degree and kind of motion which atoms possess, and reduce to this one fact the various phenomena of change, which are now attributed to occult forces'. One of these occult forces, the catalytic force of Berzelius, had already been characterised as an unnecessary hypothesis by Daniell (1839) in his textbook. He wrote: '. . . but there is no occasion for the hypothesis of a new force, for all the phenomena may be explained by the action of *heterogeneous adhesion*'. In the field of solutions an interesting and far-sighted idea which did not involve occult forces came from Graham in 1826, when he was working in Edinburgh. He contended that solutions of gases in liquids are mixtures of a more volatile with a less volatile liquid, and therefore obey the laws which hold for such mixtures.

6. EARLY CHEMICAL PHYSICS AS A RECOGNISED SUBJECT

The facilities available in England at the end of the eighteenth century and the beginning of the nineteenth century for the teaching of theoretical and experimental chemistry were exceedingly poor compared with those in Scotland, and in this subject the Universities of Oxford and Cambridge could not measure up to those of Edinburgh and Glasgow — indeed it has been said that Paris and Edinburgh were, at about this time, the twin intellectual capitals of Europe. Therefore the foundation of the first two London Colleges with their Chemistry Departments was a very great step forward in chemical teaching in England. At University College the chemistry course began in 1829 with Turner from Edinburgh as Professor, succeeded in 1837 by Graham from Glasgow. King's College opened in 1831 with Daniell as Professor of Chemistry, succeeded in 1845 by Miller. All four of these Professors wrote books for their students, the first being that of Turner, written in Edinburgh, and the last being the three-volume *Elements of Chemistry* by Miller (1855, 1856, 1857). Figure 1 shows the title page of the first volume of the first edition of this work, and Figure 2 shows the first page of the Advertisement, taking the place of a Preface. This book marks the emergence of chemical physics as a recognised subject.

As Miller remarks, the thoughts and mode of arrangement of his chemical physics book resemble those adopted by Daniell in his *Introduction to the Study of Chemical Philosophy*, the first book written for the King's College chemistry students. This book was dedicated to Faraday, and Daniell wrote in his Preface: 'The origin of the following work was a desire to present to students of chemistry an elementary view of the discoveries of Dr. Faraday in Electrical Science. From the very first publication of his *Experimental Researches in Electricity*, I have felt that from them Chemical Philosophy will date one of its most splendid epochs; and perceiving, at the same time,

ELEMENTS OF CHEMISTRY:

THEORETICAL AND PRACTICAL.

BY

WILLIAM ALLEN MILLER, M.D., F.R.S., F.C.S.

PROFESSOR OF CHEMISTRY IN KING'S COLLEGE, LONDON.

PART I.

CHEMICAL PHYSICS.

LONDON:

JOHN W. PARKER AND SON, WEST STRAND.

1855.

Figure 1. The title page of the first edition of Miller's book of 1855 on chemical physics

ADVERTISEMENT.

THE work, of which the First Part is now presented to the Reader, was originally designed to supply the Students who were attending the Course of Lectures on Chemistry at King's College with a text-book to guide them in their studies.

The present Part, on *Chemical Physics*, is devoted to a subject upon which no elementary work has appeared in this country since the publication of the excellent Treatise of the late Professor Daniell, and in attempting to supply what the author, in his own experience, has felt to be a want, he ventures to hope that the result of his labours may be found useful to persons beyond the circle of his own immediate Class. Much new matter, which has never yet been reduced to a systematic form, is now presented to the Student, particularly in the chapters on *Adhesion*, on *Heat*, and on *Voltaic Electricity*.

It is proposed to complete the work in Three Parts. The Second Part, which will be devoted to *Inorganic Chemistry*, is expected to be ready by the end of the present year; and the Third Part, which will embrace *Organic Chemistry*, in the spring of next year.

As the author was originally a pupil of Professor Daniell, and was subsequently, for several years, associated with him as Lecturer on Chemistry, it has happened that in some of the subjects treated of in this volume, the thoughts and mode of arrangement resemble those

Figure 2. The first page of the Advertisement for Miller's Book

that the results bear upon them the great impress of natural truths, namely, that they simplify while they extend our views, I have, from the first, availed myself of them in my instruction to my classes. I have enjoyed particular advantages in doing this from the kindness of the discoverer, for in every difficulty which arose, he has assisted me with his explanation and advice. At the same time, when consulted by my pupils upon the best mode of following up the oral instruction of lectures by the study of the subject in books, as they must do who intend to derive benefit from such instruction, I have been greatly at a loss to direct them'. Daniell therefore resolved to write his work, which became the recommended textbook for the first two years of the King's course. In 1840 this course started with 'A Preparatory View of the Forces which concur to the production of Chemical Phenomena' in the first year, inorganic and organic chemistry in the second year, and practical chemical manipulation in the third year, under the superintendence of the Demonstrator in Chemistry, in a laboratory fitted up for the purpose. On receiving a copy of Daniell's book, Faraday wrote to him from the Royal Institution on 11 April 1839, saying: 'I have received your book. I do not wait to read it, but write at once in reply to its affection, which is to me of far more value than all its philosophy. I know well the sincerity of your words and you will believe me too when I say that to have gained and retained the kindly feelings of your heart is to me matter of great self congratulation'. This letter is signed: 'Ever my kind and faithful friend, your devoted M. Faraday'.

7. SUMMARY

Early work in the borderland between physics and chemistry is reviewed. Early optical spectroscopy led to work on absorption spectra of gases, spark and flame emission spectra, and on the investigation of the spectra of stars and nebulae. Parallel developments in early photochemistry are mentioned and some early work on the relation of physical and chemical properties is reviewed. The concept of chemical affinity is considered and an account is given of the emergence in 1855 of chemical physics as a recognised subject, the development of which can be traced back directly to Faraday.

REFERENCES

Angström, A. J. (1855) 'Optical Researches', *Phil. Mag.,* **9,** 327
Brewster, D. (1836) 'Observations on the Lines of the Solar Spectrum, and on those Produced by the Earth's Atmosphere, and by the Action of Nitrous Acid Gas', *Phil. Mag.,* **8,** 384
Bunsen, R. and Roscoe, H. E. (1857) 'Photochemical Researches. Part I. Measurement of the Chemical Action of Light', *Phil. Trans. R. Soc.,* **147,** 355
Bunsen, R. and Roscoe, H. E. (1863) 'Photochemical Researches. Part V. On the Direct Measurement of the Chemical Action of Sunlight', *Phil. Trans. R. Soc.,* **153,** 139
Croft, H. (1842) 'Abstract of Dr. Hermann Kopp's Researches on the Specific Weight of Chemical Compounds', *Phil. Mag.,* **20,** 177
Daniell, J. F. (1839) *An Introduction to the Study of Chemical Philosophy,* Parker, London
Draper, J. W. (1843a) 'On the Decomposition of Carbonic Acid Gas and the Alkaline Carbonates, by the Light of the Sun; and on the Tithonotype', *Phil. Mag.,* **23,** 161

Draper, J. W. (1843b) 'Description of the Tithonometer; an Instrument for Measuring the Chemical Force of the Indigo-tithonic Rays', *Phil. Mag.*, **23**, 401

Huggins, W. (1864) 'On the Spectra of some of the Nebulae', *Phil. Trans. R. Soc.*, **154**, 437

Huggins, W. and Miller, W. A. (1864) 'On the Spectra of some of the Fixed Stars', *Phil. Trans. R. Soc.*, **154**, 413

Kopp, H. (1842) 'On a Great Regularity in the Physical Properties of Analogous Organic Compounds', *Phil. Mag.*, **20**, 187

Miller, W. A. (1845) 'Experiments and Observations on some Cases of Lines in the Prismatic Spectrum Produced by the Passage of Light through Coloured Vapours and Gases, and from certain Coloured Flames', *Phil. Mag.*, **27**, 81

Miller, W. A. (1855) *Elements of Chemistry: Theoretical and Practical. Part I. Chemical Physics*, Parker, London

Miller, W. A. (1856) *Elements of Chemistry: Theoretical and Practical. Part II. Inorganic Chemistry*, Parker, London

Miller, W. A. (1857) *Elements of Chemistry: Theoretical and Practical. Part III. Organic Chemistry*, Parker, London

Miller, W. A. (1862) 'On the Photographic Transparency of Various Bodies, and on the Photographic Effects of Metallic and other Spectra Obtained by means of the Electric Spark', *Phil. Trans. R. Soc.*, **152**, 861

Miller, W. H. (1833) 'Report of the Meeting of the Philosophical Society of Cambridge on March 11 1833', *Phil. Mag.*, **2**, 381

Prout, W. (1815) 'On the Relation between the Specific Gravities of Bodies in their Gaseous State and the Weights of their Atoms', *Ann. Philos.*, **6**, 321

Prout, W. (1816) 'Correction of a Mistake in the Essay on the Relation between the Specific Gravities of Bodies in their Gaseous State and the Weights of their Atoms', *Ann. Philos.*, **7**, 111

Stokes, G. G. (1862) 'On the Long Spectrum of Electric Light', *Phil. Trans. R. Soc.*, **152**, 599

van der Willigen, V. S. M. (1859) 'Ueber das elektrische Spectrum', *Annln Phys. Chem. Poggendorff.*, **106**, 610

Wolf, A. (1938) *A History of Science Technology and Philosophy in the Eighteenth Century*, p. 171, Allen and Unwin, London

Wollaston, W. H. (1802) 'A Method of Examining Refractive and Dispersive Powers by Prismatic Reflection', *Phil. Trans. R. Soc.*, **92**, 365

1

The Beginnings of Wave Mechanics

LOUIS DE BROGLIE
Académie des Sciences, Paris

1. BIRTH OF WAVE MECHANICS

I believe it necessary to restate the ideas which guided me when I submitted the first principles of Wave Mechanics in 1923–1924. I believe it necessary because these ideas have never been formulated in actual expositions of quantum mechanics.

In my early youth, between 1911 and 1919, I had studied with great enthusiasm all the recent results of theoretical physics of that period. The works of Poincaré, Lorentz, Langevin, ..., were familiar to me as were also those of Boltzmann and Gibbs on statistical mechanics. But my attention had been particularly seized by Planck's, Einstein's and Bohr's works on quanta. I saw in the coexistence of waves and particles in radiation, propounded in 1905 by Einstein in his quantum theory of light, a fundamental fact lying right at the heart of nature itself. Having followed the work of my brother Maurice on X-ray spectra I perceived the great importance in this field of the two-fold nature of electromagnetic radiation. Further, having studied in mechanics the Hamilton-Jacobi theory, I saw in it a sort of embryonic theory of the union of waves and particles. Finally, I had also studied the Theory of Relativity in depth and I was convinced that this had to be the basis of all new hypotheses.

Such was my state of mind when in 1919, free from the military obligations of the 1914–1918 war, I reapplied myself to research. Having a very 'realist' conception of the nature of the physical world and little given to purely abstract considerations, it was my wish to represent the union of waves and particles in a concrete manner; the particle would be a tiny localised object incorporated in the structure of a wave under propagation. Naturally, I had begun by studying the case of light and other electromagnetic rays where I strove to represent the particle today called a 'photon' as transported by the electromagnetic wave. Then suddenly, in 1923, the idea struck me that the coexistence of waves and particles was not solely confined to the case studied by Einstein but had to be extended to include all particles.

12

Applied to the electron, it seemed necessary to me to explain the strange properties of the motion of the electron in an atom; such properties were propounded by Bohr in his theory on the stationary states of atoms. In Bohr's atomic theory whole numbers are, in effect, seen; this also obtains in wave theory in the study of resonance and interference phenomena.

But I must emphasise here an idea which constantly guided me at the time and which is today never stated. As I said, I was convinced that one must constantly take as a basis for theoretical developments the ideas of the Theory of Relativity. Guided by the fine expositions of Paul Langevin of the Collège de France I made a thorough study of the properties of the relativistic representation of a wave under propagation. In addition, inspired by one of the fundamental ideas of quantum theory, I was led to define an internal rest frequency v_0 of the particle, connected with the energy m_0c^2 of the rest mass by the relation $hv_0 = m_0c^2$. This led me to think of the particle as being like a little clock in motion. I was then greatly smitten with the fact that the transformation formula of a wave according to Lorentz is

$$v = \frac{v_0}{\sqrt{(1-\beta^2)}} \qquad (v = \beta c)$$

and the transformation formula for the frequency of a clock, translating the famous 'retardation' of clocks in motion, is $v = v_0\sqrt{(1-\beta^2)}$. Intrigued by this difference I asked myself how a particle similar to a little clock should be displaced in its wave in such a manner as to remain incorporated in the wave, that is to say, in such a manner that its internal phase remains constantly equal to that of the wave. Applying this picture, albeit a little too schematically, to the simple case of a plane monochromatic wave being propagated along the x-axis I was led to write for the variation $d\phi$ of the phase of this wave

$$d\phi = 2\pi\left(v\,dt - \frac{dx}{\lambda}\right) = 2\pi\left(\frac{v_0}{\sqrt{(1-\beta^2)}}\,dt - \frac{dx}{\lambda}\right) \tag{1}$$

$$= \frac{2\pi}{h}\left(\frac{m_0c^2}{\sqrt{(1-\beta^2)}}\,dt - h\frac{dx}{\lambda}\right)$$

and for the variation in the interval of time dt of the internal phase of the particle being displaced along the x-axis with speed v

$$d\phi_i = 2\pi v_0\sqrt{(1-\beta^2)}\,dt = \frac{2\pi}{h}m_0c^2\sqrt{(1-\beta^2)}\,dt \tag{2}$$

On combining $d\phi = d\phi_i$ with $dx = v\,dt$,

$$\frac{m_0c^2}{\sqrt{(1-\beta^2)}} - m_0c^2\sqrt{(1-\beta^2)} = \frac{mv^2}{\sqrt{(1-\beta^2)}} = \frac{hv}{\lambda} \tag{3}$$

is obtained, whence for the momentum p of the particle

$$p = \frac{m_0v}{\sqrt{(1-\beta^2)}} = \frac{h}{\lambda} \tag{4}$$

Thus two fundamental relations of Wave Mechanics have been found, $W = hv$, $p = h/\lambda$ associating with them the image of a localised corpuscle which is displaced in the wave along one of its rays yet remaining constantly in phase with it. This was the concrete image I had when I had the first idea of Wave Mechanics. Perhaps I didn't explain this sufficiently thoroughly in my thesis but I emphasise that it was this which guided me.

In my notes of Autumn 1923 and in my thesis of 1924 I was able to give the first interpretation of the conditions of quanta used in Bohr's atomic theory, showing that the propagation of a wave in an atom is in accord with the approximation of geometric optics; this is not exact but it furnishes an initial and very striking interpretation of these quantum conditions.

Many other interesting considerations were explored in my thesis, notably that which concerns the resulting identity of my thoughts regarding Fermat's principle and Maupertuis' principle of minimum action. I also considered what is today called Bose–Einstein statistics, and I introduced, in order to incorporate the theory of the photon in the body of general wave mechanics of particles, the hypothesis that the rest mass of the photon, although extraordinarily small, is positively not zero. Subsequently I have constantly introduced this hypothesis in all my work on the quantum theory of light, which has today been experimentally confirmed, notably in the excellent work of M. Imbert.*

2. THE WORK OF SCHROEDINGER. THE DISCOVERY OF ELECTRON DIFFRACTION

I was wondering how to clarify and generalise my thoughts when, early in 1926, I got to know the fine memoirs of Erwin Schroedinger in the Annalen der Physik. Schroedinger, inspired by the results of my thesis and of the Hamilton–Jacobi theory, wrote in a correct (Newtonian) albeit non-relativistic form, the wave equation of Wave Mechanics which he designated by the now famous symbol ψ.

Ignoring the validity of the geometric optics approximation that he had found correcting results in my thesis, he again took up the wave propagation calculation in Bohr's atom. Then, thanks to a truly remarkable transposition, he showed that the determination of energies quantified by his method of calculating the rest values of a wave equation gave exactly the same numerical results as the more abstract method developed one year earlier by Werner Heisenberg in his Mechanics of Matrices.

Finally Schroedinger solved the problem of the wave mechanics of groups of interacting particles by writing a wave equation in configurational space formed by the co-ordinates of all the particles of the system. The abstract character of this mode of calculation is evident since the wave ψ of configurational space cannot be considered as a real wave, being propagated in physical space. Moreover, this abstract method was rapidly shown to be very effective; it soon gave very exact results of great interest.

At this time I read Schroedinger's memoirs with the liveliest admiration and I reflected a great deal on their contents. On three points, however, I did not feel myself to be in agreement with the eminent Austrian physicist. In

* *Int. J. theor. Phys.* **4**, No. 2, 125–140, (1971)

particular, the wave equation which he attributed to the wave was not relativistic and I was too convinced of the close liaison existing between the Theory of Relativity and Wave Mechanics to be satisfied with a non-relativistic wave equation; but this difficulty was soon removed, for since July 1926 several workers, myself included, have found a form of wave equation, known today as the Klein–Gordon equation, of which the Schroedinger equation is the degenerate form, having a Newtonian approximation. Another point where my view did not agree with Schroedinger's was that whilst keeping the idea that the wave ψ in physical space is a real wave it seemed to abandon completely the idea of particle localisations in the wave, and this was not in accord with my initial concepts. Finally, on recognising that the idea of a wave ψ in configurational space constituted a very useful formalism in visualising the properties of a group of interacting particles, I considered it certain that the motion of separate particles and the propagation of their waves operated in physical space during the flow of time.

During Spring 1927, whilst I was thinking how to develop my own concepts, bearing in mind the indisputable part of Schroedinger's results, I learned of the sensational discovery of electron diffraction made in the United States by Davisson and Germer. Soon repeated in different forms by G. P. Thomson in England, by Maurice Ponte in France and by yet others in other countries, these observations enabled me to verify the formula which I had given connecting the length of an electron wave with its momentum and they also gave complete confirmation of the basic ideas of Wave Mechanics. Subsequently the diffraction of other particles, such as neutrons and protons, was observed and, together with the diffraction of electrons, was shown to possess the same classical properties that had been associated with light for a very long time.

3. THE THEORY OF THE DOUBLE SOLUTION AND THE SOLVAY COUNCIL OF 1927

In the Spring of 1927 Schroedinger's work and the discovery of electron diffraction appeared to have afforded a complete confirmation of the ideas contained in my thesis. Yet the majority of theoretical physicists began to turn towards concepts which were quite different from those which had guided me. Schroedinger abandoned the idea of a localised corpuscle and only partially retained the idea of the wave having a physical reality. Going yet further, Max Born, in an important memoir where he dealt with the problem of collisions in the new mechanics, has considered the wave ψ to be a representation of probabilities, and allowed that it must be arbitrarily 'normalised'. This removed to ψ the essential character of a physical wave having a determined amplitude.

I was disturbed to see the clear and concrete physical image completely disappear, the image which had guided me in my first researches, so I made an effort to clarify my point of view. In May 1927 I published an article in the Journal de Physique under the title 'Wave Mechanics and the Atomic Structure of Matter and Radiation'. In this article, which is interesting to re-read today, I began by defining very clearly the goal I pursued and I introduced under the name 'The Theory of the Double Solution' the idea that it was necessary to distinguish between two distinct solutions; each is nevertheless intimately connected to the wave equation. I called the one wave v, being a real and non-normalised physical wave having a local significance, thereby defining the particle and represented by a singularity, and I called the other the wave ψ of Schoedinger, normalised and deprived of singularity, which could only be a representation of probabilities. This led me to generalise the formulae which I had given in my thesis for the plane monochromatic wave and to express the movement of the particle in its wave by the aid of a 'guiding formula' defining the mode by which the motion is guided by wave propagation. I was thus led to envisage that the motion of the particle in its wave is effected by the action of a force deriving from a 'quantic potential' proportional to the square of Planck's constant, the quantic potential depending on the second derivative of the wave amplitude which is zero in the case of the plane monochromatic wave. I observed that the quantic potential could be expressed by the variations of the rest mass of the particle, a result whose importance I have recently better understood; this is particularly notable in my naming of the 'states of constraint' where the free propagation of the wave was impeded by the existence of limiting conditions. I propounded also, albeit in a very incomplete fashion, the method whereby one could endeavour to justify Schroedinger's use of a statistical wave ψ defined in the configurational space of a group of particles.

I was very satisfied with the results which I thus obtained because they seemed to me to pave the way to a true interpretation of Wave Mechanics, taking account of Schroedinger's results and the success of the statistical interpretation of the wave ψ. And I still believe today I was right. I freely admit that my article only constituted a first stage and was destined to be subject to modifications and improvements, but I hoped I would be helped in this task.

In early Spring 1927 I was invited by Lorentz to take part in the Solvay Physics Council which was held at Brussels in October 1927. I there gave an exposition on my theory of the double solution, unfortunately in the somewhat simplified form of 'the pilot wave'. It received hardly any attention. Physicists such as Planck, Lorentz and Langevin, accustomed to the old methods, were hoping for an interpretation of Wave Mechanics in accordance with classical concepts but they made no pronouncement upon its nature. Schroedinger remained faithful to a pure wave interpretation. Only Einstein encouraged me somewhat on the path I wished to tread. But I was faced with redoubtable adversaries: Niels Bohr and Max Born, scientists of world renown. And there was also the group of young researchers who formed the Copenhagen School, amongst them, in particular, Pauli, Heisenberg and Dirac, who were already authors of remarkable works. They interpreted the duality of corpuscles and waves by the theory of complementarity recently proposed by Bohr and, no longer attributing to the arbitrarily normalised wave of Schroedinger any more than the role of representing the probability of certain observations being obtained, they concluded by abandoning any clear picture of a wave or a particle. I was very distressed.

I found Bohr's complementarity quite obscure and I did not like abandoning physical images which had guided me for many years. However, the probabilistic interpretation of 'quantum mechanics', developed by numerous young keen researchers possessing great facility in mathematical calculations, rapidly took the form of elegant and rigorous mathematical formalism.

Back at Paris after the Solvay Council, I hesitated for several months between following my early ideas and going over to the concepts of the Copenhagen School. However, at the end of 1928, having been nominated Professor of Theoretical Physics in the Faculty of Sciences in Paris, a post which I subsequently occupied for 34 years, and being unable to teach a theoretical interpretation which did not possess a truly satisfactory form, I decided to acquaint my students with what was beginning to be accepted in all other places and I no longer proceeded along the difficult path I had wished to walk. This enabled me to make very extensive expositions over a period of more than 20 years and I did profound work on the whole field of quantum physics as it was then being expounded. However, I never entirely lost sight of my initial concepts, and traces of these may be seen in the way I introduced basic Wave Mechanics, avoiding giving them a too abstract and axiomatic form.

4. RETURN TO ORIGINAL IDEAS SINCE 1951

Since 1951 a complete change has taken place in my spirit of interpreting Wave Mechanics and I have returned to the ideas which had directed my work when I sought to obtain a clear picture of the coexistence of waves and particles. This unforeseen change certainly originated in the studies which I pursued from 1948 to 1952. I first made some headway in the statistical interpretation of thermodynamics by insisting on the introduction of this theory in relativistic concepts. It then occurred to me that there existed a curious analogy between the relativistic heat transformation formula and the relativistic transformation of the frequency of a clock which had played such a great part in my thoughts when I submitted my Doctorate thesis; and I was also greatly smitten by the analogy, already vaguely sensed by Eddington, between the two relativistic invariants action and entropy. I was thus led to the ideas which had guided me since the discovery of Wave Mechanics, and the important role played by these thermodynamic analogies in my recent work will be yet further seen.

In succeeding years I devoted two courses to the interpretation of Wave Mechanics as expounded by the Copenhagen School and the controversies which there had been on this subject some 15 years previously between Niels Bohr on the one hand the Einstein and Schroedinger on the other. In the course of taking part in the editing of these courses I felt my ideas being modified. My confidence in the generally adopted interpretation was shaken and I asked myself if I ought not again to take up my old suggested interpretation of Wave Mechanics by the theory of the double solution. It was then that I got to know of an article by David Bohm in *The Physical Review*, where he revived the majority of the ideas contained in my article in the Journal de Physique in 1927; I decided to resume the development of my old suggestion.

Since I had many University and other obligations at this time, and was able to find only a small number of young researchers to help me, I at first made only slow progress in the work of criticism and re-interpretation which I had set myself. But for several years now I have found it possible to give a more elaborate form to the theory of the double solution and to complete it with a thermodynamics which today seems to me to open vast new perspectives.

Those who wish to delve deeper into these new ideas to which I have just alluded will be able to refer to the recent exposition which I have made in the book entitled *La Réinterprétation de la Mécanique Ondulatoire* ('The Re-Interpretation of Wave Mechanics'), Paris, Gauthier-Villars (1971).

Translated by W. V. Robinson, MSc, MIL, FRIC

2

The Development of Quantum Mechanics in the Period 1924–1926

JOHN C. SLATER

Quantum Theory Project, Williamson Hall, University of Florida, Gainesville, Florida

A few times in the history of science the pieces needed to fit together into a great theoretical synthesis have been lying around for some time, awaiting a unifying thought which would bring them together. The work of Copernicus, of Galileo, of Kepler, and of other astronomers, waited for the genius of Newton to bring them into the finished pattern of celestial mechanics. Oersted, Faraday, Henry, and many others, had already made their contributions to electromagnetism, when these pieces were brought together through the suggestion of Clerk Maxwell. In our time we have seen a synthesis as remarkable as either of these, brought into being by the inspiration of two men: Louis de Broglie, who provided the key to unlock the puzzle of wave mechanics, and Erwin Schroedinger, who put the theory together.

It may be interesting to hear at first hand from one like myself, who had the good fortune to live through the exciting days of the 1920s, nearly 50 years ago, and to get an impression of how the synthesis looked to one who was not only alive, but was expecting it. I took my Ph.D. at Harvard in 1923, working with Bridgman on an experimental study of the compressibility of the alkali halides, learning my quantum mechanics from Kemble, and from Sommerfeld's book *Atombau und Spektrallinien*. By 1923 I had convinced myself that something remarkable was in the air. I realised that the older quantum mechanics was entirely unable to explain the forces between atoms, the problem which I had met in trying to understand the alkali halide crystals.

The Compton effect had shown open-minded physicists that the photon theory of light had to be assumed, and yet the facts of physical optics showed that the wave theory must be equally true. Compton's report entitled *Secondary Radiations Produced by X-rays* had come out in October 1922, the beginning of my last year of graduate work, and though it did not give the theory of the Compton effect as completely as his paper of 1923, still he

19

stated the duality of waves and particles very clearly. I and the others at Harvard were particularly conscious of this work, since Duane, of Harvard, was engaged in a controversy with Compton over the question of whether the effect actually existed experimentally, a controversy which was not really settled until December, 1924 (Washington meeting, A.P.S., 1924). In addition to these facts, the spectroscopy of multiplets was beginning to be unravelled, with an obvious display of complicated numerical relations which needed to be understood. Bohr's correspondence principle was very suggestive of relations between classical and quantum mechanics. But the pieces did not seem to fit together.

At this point a Sheldon travelling fellowship from Harvard made it possible for me to spend a year in Europe. I chose to go to Copenhagen to work with Niels Bohr. His work on the periodic system of the elements, a couple of years before, suggested this was the direction from which we were most likely to learn about atomic structure in more detail. Bohr's absence from Copenhagen during the fall of 1923 made it desirable for me to go elsewhere for those months, and I went to the Cavendish to get an impression of that great centre of science. During the time there I worked by myself, with some contact with R. H. Fowler, on an idea which I hoped would provide some synthesis of the wave and photon pictures of light.

I proposed the existence of waves to guide the photons, a rather obvious idea which had occurred to many people, but I included the unfamiliar suggestion that the electromagnetic waves forming the light should be emitted or absorbed during the stationary states, rather than just at the time of quantum transitions. Thereby I secured something which it seemed to me clearly needed incorporation into quantum mechanics: the finite line breadth originating from a finite wave train. It is well known from the theory of Fourier analysis that the breadth of an emitted or absorbed spectral line, times the length of time over which it is emitted or absorbed, is of the order of magnitude of unity. Thus a sharp spectral line must be emitted or absorbed over a long time, comparable with the lifetime in a stationary state, and an instantaneous emission or absorption of wave-like radiation such as Bohr had assumed was not consistent with this fact. If Δv is the frequency breadth, ΔT the time over which the train is emitted, one has the relation $\Delta v \Delta T = 1$, or if we let $E = hv$, $\Delta E \Delta T = h$.

When I went to Copenhagen at the end of 1923, Bohr and Kramers were much taken by some aspects of this idea, but Bohr had not yet brought himself to the point where he would admit the existence of corpuscular photons along with the waves of light. My proposal to Bohr and Kramers had been a straightforward probability connection between the waves and photons: the intensity of the continuous radiation field, at any point of space, was taken to determine the probability of finding photons at that point. Bohr objected to the photons so strongly that it was obvious that I would have a fight on my hands if I insisted on them, though I never doubted their existence. As a result, the paper of Bohr, Kramers, and Slater (1924a, b), which resulted from our discussions, came out in an unsatisfactory hybrid form in which the energy in the radiation field was regarded as continuous, while the atomic energy changed discontinuously at the time of a transition. I published

(Slater, 1924) a note to *Nature* which to some extent indicated this difference of opinion. A short time later the work of Bothe and Geiger (1925) showed that the radiation energy really had to be considered as discrete particles, thereby preserving conservation and finally forcing Bohr to give up his opposition to the photons. At that time I sent another note to *Nature* (Slater, 1925a) pointing out that this involved essentially a return to the view which I had at the time I went to Copenhagen in 1923. Bohr himself, by 1926, wrote me regretting that he had opposed me in 1923.

Rather than spending my time in controversy with Bohr, I decided during my Copenhagen period to develop the detailed application of the ideas I was working on to the optical properties of matter. In particular, it was possible to bring into the theory the breadth of spectral lines, in a way which had not been commonly done, and this could be done in identical ways whether one believed in the photons or not. Consequently I went ahead with this, against the opposition of Bohr, but with one great element of help from Kramers. He showed me, in January, 1924, the unpublished dispersion formula which now generally goes by the name of the Kramers–Heisenberg dispersion formula, and I was able at once to understand the derivation of this formula, which he had already worked out. He published this later in the spring (Kramers, 1924a, b). This was a correspondence-theoretical derivation which led to a definite way of describing the virtual oscillators met in the Bohr–Kramers–Slater paper. Heisenberg did not enter the picture actively until some months later, when he came to Copenhagen in the summer of 1924, at the time when I left to return to America (Kramers and Heisenberg, 1925).

During the remainder of my time in Copenhagen, I worked on this theory, which was finished and published after I returned to Harvard (Slater, 1925c, d). It is an interesting fact that G. Breit (1925), J. H. Van Vleck (1925), and I (Slater, 1925b) all gave similar talks on various aspects of the subject at the same Washington meeting of the American Physical Society in December, 1924, at which the Duane–Compton controversy was finally cleared up. Breit and Van Vleck had both been at Harvard during the year 1922–1923, my last year of graduate work. Our thoughts on the finite lifetime of stationary states, and its effect on line breadths, were very similar. This work attracted very little attention at the time, though it represented one of the first appearances of the uncertainty principle as applied to energy and time. However, the work of Weisskopf and Wigner (1930) in 1930 showed that my paper (Slater, 1925c) correctly led to the relation between lifetime and line breadth which arose from the uncertainty principle and wave mechanics.

These thoughts about the relation between finite lifetime and a broadening of the energy levels suggested to me an interesting application to problems of degenerate states and their quantisation. Ehrenfest, Breit, and Tolman (Ehrenfest and Breit, 1922; Ehrenfest and Tolman, 1924) had made several contributions to a problem which they called weak quantisation, an intermediate case between the multiply periodic problem in which Sommerfeld's quantum condition led to sharp energy levels, and a non-periodic problem where there was no quantisation. Stimulated by their ideas, I showed (Slater, 1925e) the relation of these cases to finite wave trains. In particular, I considered such problems as space quantisation in a very small external magnetic field, examining the magnitude of the field, and the lifetime, which would be required to get sharp quantisation. Some of these applications look very much like the continuous wave functions found in wave mechanics.

While these things were going on at Harvard, Heisenberg, who arrived in Copenhagen about the time I left, was also intrigued by the Kramers dispersion formula. In the joint paper with Kramers (Kramers and Heisenberg, 1925), they discussed scattering of light by this formula, which I also treated as a special case in my work (Slater, 1925c). They did not, however, go as far into questions of line breadth as I had. They concentrated more on the scattered light of wavelength different from the incident wavelength, leading to the Smekal–Raman effect, a phenomenon which I had noticed in my study, but had not stressed in my paper. Heisenberg, however, was led by the existence of the virtual oscillators and their amplitudes, and in particular by their interesting relationship with the correspondence principle, to believe that a complete matrix theory of these components could be set up (Heisenberg, 1925). This paved the way for the papers of Born, Jordan, and Heisenberg (Born and Jordan, 1925; Born, Heisenberg and Jordan, 1926). Through my acquaintance with Heisenberg, whom I had met in Copenhagen before he came there for his longer stay, and through the close relationship between his ideas and mine through Kramers' dispersion formula, I naturally followed these developments very closely. I had a further chance to develop these ideas during November 1925–January 1926, when Max Born visited M.I.T., and I worked quite closely with him. During this period, I had derived the relations between Poisson's brackets and the commutators, and had written this up, before Born's departure in January 1926, but Dirac's (Dirac, 1925) first quantum-mechanical paper, in which he independently and earlier had derived these relations, came out before I had a chance to send this in for publication.

I mention these ways in which I was in touch with the developments which played their part in the explosive events of 1926, only to point out how I happened to be aware of what was going on. In the spring of 1926, I had turned my attention to complex spectra. The work of Russell and Saunders (1925), on the so-called Russell–Saunders coupling, had become possible as soon as the role of the electron spin was pointed out by Uhlenbeck and Goudsmit (1925, 1926). Saunders was an experimental spectroscopist at Harvard, Russell an astrophysicist at Princeton, and naturally these developments were much in the air at Harvard. Hund at this time was working out his theory of complex spectra, based on Pauli's (1925) interpretation of the exclusion principle. Not unnaturally I tried to incorporate some of these ideas into a treatment by the old quantum theory (Slater, 1926), carrying things as far as was possible by those methods. It was while I was finishing this up that the explosive events occurred which led to such momentous results.

There is always a certain time lag between the lighting of a fire and its bursting into a blaze. No one at Harvard, or for that matter in Copenhagen, was looking to see what a French physicist named Louis de Broglie (1924) would do for his doctor's thesis. I first learned of it by looking over the files of the *Annales de Physique* in a routine way, noticing it well after its publication (de Broglie, 1925) in the January–February 1925 number. I was at once struck by the resemblance of his treatment of light to mine, with the parallelism between waves and particles. He, like me, had been led to this idea through X-rays, in his case through his brother Maurice de Broglie. I felt that the extension of this idea to material particles, and the beautiful way in which this led to the quantum conditions, was a step of major importance. On the other hand, his insistence on relativistic methods for getting at the relation between wavelength and momentum seemed to me unnecessarily complicated, though I note that he continues to feel that this is required. After all, Einstein and later Compton had freely used the relation that the momentum of a photon had to be set equal to h/λ, and I was quite ready to use a non-relativistic version of the theory. But while I realised the beauty of de Broglie's use of the simultaneous existence of waves and corpuscles for material particles as well as photons, my preoccupation with complex spectra prevented me from carrying the idea further at the time it was proposed. It is only natural that de Broglie's ideas did not spread rapidly in the Copenhagen school, considering Bohr's prejudice against the photon, a prejudice which had not been overcome in 1924.

The rest of the story is known to every physicist (see Jammer, 1966 for an account). The news of de Broglie's work reached Schroedinger through a round-about route, and he (Schroedinger, 1926) proceeded with immense speed to convert de Broglie's general idea of the existence of waves to accompany the electrons and other material particles, into a mathematical form, by setting up his wave equation. Immediately the various features of quantum theory which had been lying around began to fall into place. Schroedinger not only showed how to compute his wave functions, but showed that Heisenberg's matrices followed from his formulation in a very simple way. He was not willing at first to accept the statistical relation between waves and particles which was at the basis of de Broglie's idea, but this hesitation was soon removed by Born's (1926a, b) espousal of the statistical interpretation of the theory, which was soon accepted by almost all physicists. Heisenberg (1926a, b, 1927) quickly brought out his interpretation of the relation of symmetry in the many-body problem to the exclusion principle and to the concept of exchange, laying the foundation for the theories of complex spectra and of ferromagnetism. Hartree (1928) showed how wave mechanics naturally led to the central-field model of the atom, which he (Hartree 1923, 1924a, b), Schroedinger (1921), and others, following Bohr's lead, had been working on rather unsuccessfully before 1926. Dirac (1927) showed that the paradoxes connected with the radiation theory could be overcome if one treated the radiation, as well as the atoms, by wave mechanics. In this way, by 1927 or 1928, we had an essentially complete theory of quantum mechanics, and were ready to proceed with the numberless applications to the properties of atoms, molecules, solids, and radiation, which have followed ever since.

All this happened with almost unbelievable speed after de Broglie supplied the necessary idea in his thesis in 1924, and Schroedinger followed this up with his monumental mathematical development in 1926. It is of course a familiar fact that the discoveries in nuclear physics in the period following

24 Wave Mechanics

1930 seemed so exciting to many theorists that they turned their interest in that direction, rather than keeping on with the problems of atoms, molecules, and radiation which had been the major interest before 1930. I did not follow their lead, but preferred to go into the application of the theory to these problems which are closer to everyday life than is nuclear physics. In doing this, it has been constantly obvious to me what an advantage I had over those who came along later, in that I had seen with my eyes the first stages of quantum mechanics, instead of having to be taught them at second hand. To have lived through the 1920s and the development of quantum mechanics was an experience which one can never forget.

SUMMARY

The development of quantum mechanics in the years from the discovery of the Compton effect to the work of Schroedinger on wave mechanics is described. There is particular emphasis on those aspects which the author observed personally, as related to the work discussed in the paper of Bohr, Kramers, and Slater in 1924.

REFERENCES

Bohr, N., Kramers, H. A. and Slater, J. C. (1924a) 'The Quantum Theory of Radiation', *Phil. Mag.*, **47**, 785–802
Bohr, N., Kramers, H. A. and Slater, J. C. (1924b) 'Über die Quantentheorie der Strahlung', *Z. Phys.*, **24**, 69–87
Born, M. (1926a, b) 'Zur Quantenmechanik der Stossvorgänge', *Z. Phys.*, **37**, 863–867; **38**, 803–827
Born, M., Heisenberg, W. and Jordan, P. (1926) 'Zur Quantenmechanik. II', *Z. Phys.*, **35**, 557–615
Born, M. and Jordan, P. (1925) 'Zur Quantenmechanik', *Z. Phys.*, **34**, 858–888
Bothe, W. and Geiger, H. (1925) Über das Wesen des Comptoneffekts; ein Experimenteller Beitrag zur Theorie der Strahlung', *Z. Phys.*, **32**, 639–663
Breit, G. (1925) 'Polarization of Resonance Radiation and the Quantum Theory of Dispersion', *Phys. Rev.*, **25**, 242 (Abs.)
de Broglie, L. (1924) 'Recherches sur la Théorie des Quanta', *Thesis*, Masson et Cie, Paris
de Broglie, L. (1925) 'Recherches sur la Théorie des Quanta', *Annln Phys., Paris*, **3**, 22–128
Compton, A. H. (1922) 'Secondary Radiations Produced by X-rays', *Bull. Natn. Res. Coun., Wash.*, **4**, Part 2, No. 20, 1–56
Compton, A. H. (1923) 'A Quantum Theory of the Scattering of X-rays by Light Elements', *Phys. Rev.*, **21**, 483–502
Dirac, P. A. M. (1925) 'Fundamental Equations of Quantum Mechanics', *Proc. R. Soc.*, **109**, 642–653
Dirac, P. A. M. (1927) 'Quantum Theory of Emission and Absorption of Radiation', *Proc. R. Soc.*, **114**, 243–265
Ehrenfest, P. and Breit, G. (1922) 'Ein bemerkenswerter Fall von Quantisierung', *Z. Phys.*, **9**, 207–210
Ehrenfest, P. and Tolman, R. C. (1924) 'Weak Quantization', *Phys. Rev.*, **24**, 287–295
Hartree, D. R. (1923) 'Some Approximate Numerical Applications of Bohr's Theory of Spectra', *Proc. Camb. phil. Soc.*, **21**, 625–641
Hartree, D. R. (1924a) 'The Spectra of Some Lithium-like and Sodium-like Atoms', *Proc. Camb. phil. Soc.*, **22**, 409–425
Hartree, D. R. (1924b) 'Calculation of Successive Ionization Potentials', *Proc. Camb. phil. Soc.*, **22**, 464–474

Slater 25

Hartree, D. R. (1928) 'The Wave Mechanics of an Atom with a Non-Coulomb Central Field. Part I. Theory and Methods. Part II. Some Results and Discussions', *Proc. Camb. phil. Soc.*, **24**, 89–132

Heisenberg, W. (1925) 'Über quantentheoretische Umdeutung Kinetischer und Mechanischer Beziehungen', *Z. Phys.*, **33**, 879–893

Heisenberg, W. (1926a) 'Mehrkörperproblem und Resonanz in der Quantentheorie', *Z. Phys.*, **38**, 411–426

Heisenberg, W. (1926b) 'Über die Spektra von Atomsytemen mit zwei Elektronen', *Z. Phys.*, **39**, 499–518

Heisenberg, W. (1927) 'Mehrkörperproblem und Resonanz in der Quantenmechanik II', *Z. Phys.*, **41**, 239–267

Jammer, M. (1966) *The Conceptual Development of Quantum Mechanics*, McGraw-Hill, New York

Kramers, H. A. (1924a) 'The Law of Dispersion and Bohr's Theory of Spectra', *Nature, Lond.*, **113**, 673–674

Kramers, H. A. (1924b) 'The Quantum Theory of Dispersion', *Nature, London.*, **114**, 310–311

Kramers, H. A. and Heisenberg, W. (1925) 'Über die Streung von Strahlung durch Atome', *Z. Phys.*, **31**, 681–708

Pauli, W., Jr. (1925) 'Über den Zusammenhang des Abschlusses der Elektronengruppen im Atom mit der Komplexstruktur der Spektren', *Z. Phys.*, **31**, 765–783

Russell, H. N. and Saunders, F. A. (1925) 'New Regularities in the Spectra of the Alkaline Earths', *Astrophys. J.*, **61**, 38–69

Schrödinger, E. (1921) 'Versuch zur Modelmässigen Deutung des Terms der Scharfen Nebenserien', *Z. Phys.*, **4**, 347–354

Schrödinger, E. (1926a) 'Quantisierung als Eigenwertproblem, I', *Annln Phys., Lpz.*, **79**, 361–376

Schrödinger, E. (1926b) 'Quantisierung als Eigenwertproblem, II', *Annln Phys., Lpz.*, **79**, 489–527

Schrödinger, E. (1926c) 'Über das Verhältnis der Heisenberg-Born-Jordanschen Quantenmechanik zu der Meinen', *Annln Phys., Lpz.*, **79**, 734–756

Schrödinger, E. (1926d) 'Quantisierung als Eigenwertproblem, III', *Annln Phys., Lpz.*, **80**, 437–490

Schrödinger, E. (1926e) 'Quantisierung als Eigenwertproblem, IV', *Annln Phys., Lpz.*, **81**, 109–139

Slater, J. C. (1924) 'Radiation and Atoms', *Nature, Lond.*, **113**, 307–308

Slater, J. C. (1925a) 'The Nature of Radiation', *Nature, Lond.*, **116**, 278–279

Slater, J. C. (1925b) 'The Nature of Resonance Radiation', *Phys. Rev.*, **25**, 242 (Abs.)

Slater, J. C. (1925c) 'A Quantum Theory of Optical Phenomena', *Phys. Rev.*, **25**, 395–428

Slater, J. C. (1925d) 'Methods for Determining Transition Probabilities from Line Absorption', *Phys. Rev.*, **25**, 783–790

Slater, J. C. (1925e) 'Physically Degenerate Systems and Quantum Dynamics', *Phys. Rev.*, **26**, 419–430

Slater, J. C. (1926) 'A Dynamical Model for Complex Atoms', *Phys. Rev.*, **28**, 291–317

Uhlenbeck, G. E. and Goudsmit, S. (1925) 'Ersetzung der Hypothese vom Unmechanischen Zwang durch eine Forderung bezüglich des Inneren Verhaltens jedes einzelnen Elektrons', *Naturwissenschaften*, **13**, 953–954

Uhlenbeck, G. E. and Goudsmit, S. (1926) 'Spinning Electrons and the Structure of Spectra', *Nature, Lond.*, **117**, 264–265

Van Vleck, J. H. (1925) 'Virtual Oscillators and Scattering in the Quantum Theory', *Phys. Rev.*, **25**, 242 (Abs.)

Washington meeting, American Physical Society, December 29–31, 1924, *Phys. Rev.*, **25**, 233–258; see pages 235, 236

Weisskopf, V. F. and Wigner, E. (1930) 'Berechnung der Natürlichen Linienbreite auf Grund der Diracschen Lichttheorie', *Z. Phys.*, **63**, 54–73

3

Central Fields in Two Vis-a-Vis Three Dimensions: an Historical Divertissement

JOHN H. VAN VLECK
Lyman Laboratory of Physics, Harvard University,
Cambridge, Mass.

1. THE DIFFERENCE BETWEEN RESULTS IN TWO AND THREE DIMENSIONS

The eigenvalues of the energy and other properties of a particle subject to a central force field are different when the problem is treated two- and three-dimensionally. This result, which came to light in 1926 after the true quantum mechanics had replaced the old quantum theory, was indeed startling to anyone reared in the classical Keplerian tradition.

To demonstrate the dependence on dimensionality, relatively little mathematical analysis is necessary. The Schroedinger equation satisfied by the radial wave function R is

$$[\mathcal{H}_r - E]R = 0 \tag{1}$$

where the radial Hamiltonian operator is

$$\mathcal{H}_r = \frac{1}{2m}\left[p_r^2 + \frac{2}{r}p_r + \frac{h^2}{4\pi^2 r^2}l(l+1)\right] + V(r) \left(p_r = \frac{h}{2\pi i}\frac{\partial}{\partial r}\right) \tag{2}$$

This familiar result is obtained in a standard way by expressing the three-dimensional Laplacian operator ∇^2 in polar co-ordinates. The complete wave function then can be taken to be

$$\psi = R_{nl}(r)\gamma_l^{m_l}(\theta,\phi) \tag{3}$$

where $\gamma_l^{m_l}$ is a Tesseral harmonic, and one finds that R must satisfy (1). As customary, l denotes the azimuthal quantum number such that the square of the angular momentum is $l(l+1)h^2/4\pi^2$ and $m_l h/2\pi$ is its component along the direction of spatial quantisation.

26

The wave functions R are orthogonal in the sense that

$$\int_0^\infty R_{nl}^* R_{n'l}\rho \, dr = 0 \qquad (n' \neq n) \tag{4}$$

with $\rho = r^2$. This is a reflection of the fact that the complete wave functions are orthogonal with the volume element $dV = r^2 \, d\omega$. For many purposes it is convenient to have a radial function normalised to dr rather than $r^2 \, dr$. This is done by absorbing a factor r in the definition of R, i.e. $Rr \to R$. One finds that the new R satisfies (1) provided that now the radial operator is

$$\mathcal{H}_r = \frac{1}{2m}\left[p_r^2 + \frac{h^2}{4\pi^2 r^2} l(l+1) \right] + V(r) \qquad (\rho = 1) \tag{5}$$

The orthogonality relation (4) now has $\rho = 1$.

For the two-dimensional, i.e. plane problem, the separation of variables takes the form

$$\psi = R_{n|k|}(r)e^{ik\phi} \tag{6}$$

with $kh/2\pi$ the angular momentum.

The radial wave function satisfies a differential equation similar to (1) provided the Hamiltonian operator be taken as

$$\mathcal{H}_r = \frac{1}{2m}\left[p_r^2 + \frac{1}{r}p_r + \frac{h^2 k^2}{4\pi^2 r^2} \right] + V(r) \qquad (\rho = r) \tag{7}$$

The orthogonality relation (4) is valid if l is replaced by $|k|$ and if one takes $\rho = r$ corresponding to the fact that in two dimensions the volume element is $dV = 2\pi r \, dr$.

To achieve orthogonality with $\rho = 1$, one absorbs a factor $r^{\frac{1}{2}}$ in R, i.e. $r^{\frac{1}{2}}R \to R$, and then the Hamiltonian operator becomes

$$\mathcal{H}_r = \frac{1}{2m}\left[p_r^2 + \frac{h^2}{4\pi^2 r^2}(k-\tfrac{1}{2})(k+\tfrac{1}{2}) \right] + V(r) \qquad (\rho = 1) \tag{8}$$

Note that the index involved in the radial wave funcion in (6) is $|k|$ rather than k since clearly the operators (7) and (8) do not involve the sign of k. The three-dimensional quantum number l is non-negative whereas k, unlike l, can have either sign. The quantum number k is necessarily an integer (as are l and m_l). In the radial differential equation the quantum number l or $|k|$ enters only as a parameter.

Solution of this equation brings in the radial quantum number n_r which specifies the number of nodes of the radial function and is necessarily an integer in both cases. Comparison of (8) and (5) shows that the eigenvalues of the three-dimensional problem are exactly like those of the two-dimensional one if we artificially give the quantum number k half integral rather than integral values. Conversely the eigenvalues for the two-dimensional problem can be obtained from those of the three by taking l to be half integer. For the particular case of a Coulomb field the energy is a function only of $n_r + l$ or $n_r + |k|$. Thus for a two-dimensional Coulomb field the Balmer formula

$$E = -\frac{R_\infty Z^2}{n^2} \qquad \left(R_\infty = \frac{2\pi^2 m e^4}{h^2} \right) \tag{9}$$

is still valid except that the n in the denominator is a half rather than whole integer. In other words the principal quantum number n has to be defined as $n = n_r + |k| + \frac{1}{2}$ rather than $n_r + l + 1$. Consequently if one tries to solve the eigenvalue problem of the hydrogen atom two- rather than three-dimensionally, one gets a completely fallacious answer. Another example is provided by the isotropic harmonic oscillator. Treated two-dimensionally, the eigenvalues can be shown to be

$$E = h\nu_0[2n_r + |k| + 1]$$

So one sees immediately that in the three-dimensional problem the energy levels are $h\nu_0(n + \frac{3}{2})$, while in the two-dimensional one they are $h\nu_0(n + 1)$, where n is an integer in either case. These results are also apparent in more obvious fashion if one uses Cartesian rather than polar co-ordinates, since then the harmonic oscillator factorises into three one-dimensional oscillators whose eigenvalues are each of the form $h\nu_0(n_i + \frac{1}{2})$.

The expectation or mean value of any power of the radius is given by a formula of the form

$$\langle r^s \rangle = \int_0^\infty R^* r^s R \, \mathrm{d}r \tag{10}$$

where R is R_{nl} or $R_{n|k|}$ depending on whether one is dealing with the three- or two-dimensional problem, and provided in both cases one uses the form appropriate to $\rho = 1$, and provided the radial functions are normalised to unity.

In the Coulomb case it is customary to express the results in terms of the principal quantum number n, rather than of n_r. From the preceding discussion, it follows immediately that the mean values of r^s can be obtained for the two- from the three-dimensional and vice versa, by appropriately replacing integers by half integers. The ability to obtain mean values for two dimensions from those for three dimensions and vice versa by this simple replacement is not confined to the Coulomb case and applies to any function of the radius alone rather than just a power as in (10). In the non-Coulomb case, n_r and $|k|$ or l no longer enter only additively in the energy so that it is better to label the states by n_r, l or n_r, $|k|$ rather than n, l or n, $|k|$. The correlation then is

$$n_r, l+1 \leftrightarrow n_r, |k| + \frac{1}{2} \tag{11}$$

For an arbitrary central field there is of course no simple functional relation expressing E in terms of quantum numbers, and n_r becomes simply an integer which designates successive eigenvalues of the energy for given $|k|$ or l.

2. MOTIVATION FOR THE PRESENT PAPER

Probably theoretical physicists were more cognisant in the late 1920s of the difference between the two- and three-dimensional treatments which we have described than they are today, when quantum mechanics has become commonplace. Recently I was with a group of young physicists, and I asked them of they knew of any reference where central fields were examined in two dimensions. None of them did, and I failed to find an appropriate reference in a rapid perusal of several well known texts on quantum mechanics. Then I chanced to look at Sommerfeld's *Wellenmechanischer Erganzungsband*, written in 1929, and there I found the two- and three-dimensional Keplerian problem discussed in considerable detail, more than here. Sommerfeld even integrated the radial equation separately for the two cases.

He did not call attention explicitly to the general behaviour in an arbitrary central field, or compare the formulas for $\langle r^s \rangle$. This relation has received scant, if any, attention in the literature, but is an obvious consequence of the similarity of the structure of equations (5) and (8) and is essential for some of the discussion later in the present paper.

If the results of the preceding section were known over forty years ago, why am I writing the present article? There are three reasons which we amplify in Sections 3, 4 and 5 respectively.

The first reason is a purely historical one. The distinction between the two- and three-dimensional results was not adequately recognised in the brief period between the appearance of matrix mechanics and that of the Schroedinger equation (almost a delta function of time in the history of physics). As a result, during this period prominent physicists published papers in which they purportedly treated the hydrogen atom two-dimensionally, but nevertheless by a happy combination of empiricism and intuition which we discuss later managed to come up instead with answers appropriate to our real world, apparently not realising the breach in rigor in passing from two to three dimensions.

The second reason is that the two three dimensional contrast furnishes a prime example of a situation where the wave-mechanical differential equation approach is much simpler and more intuitive than the matrix one. It seems particularly appropriate to stress this feature in a volume written in honour of de Broglie.

The third reason is that in preparing the present paper the writer went back and re-examined the two-dimensional problem, which has been somewhat neglected because it is in a certain sense academic. As a result, we present in Section 5 a comparatively simple two-dimensional matrix treatment of the hydrogen atom.

3. HISTORICAL OBSERVATIONS ON PAPERS STARTING WITH THE TWO-DIMENSIONAL MODEL

Heisenberg in his epoch-making paper on the 'Quantum Theoretical Reinterpretation of Kinematic and Mechanical Relations' (1925), written just before the dawn of the quantum mechanics which it motivated, treats among other things, the two-dimensional rotator. He derives the energy levels $E = h^2 k^2 / 8\pi^2 I$ and concludes that k is a half integer rather than integer. The argument which he uses to justify this choice is a strange one, viz. that there is no radiation between the two lowest states $k = \pm\frac{1}{2}$ as they have the same energy.

This radiation argument is unconvincing, as when k is an integer as it really is two-dimensionally, $k = 0$ is the lowest state and so can have no radiation. This remark should not be regarded as a criticism of his wonderful paper, for in those days there was no real quantum theory of radiation. Perhaps he was influenced by the fact that half quantum numbers worked better than whole ones in interpreting band spectra of diatomic molecules without internal angular momentum. Later in the same paper he derives in three-dimensional space the Goudsmit–Kronig–Honl intensity formulas for the rotator. Had he calculated the corresponding formula for the energy by his multiplication scheme (essentially a matrix one), he would have obtained $E = h^2 l(l+1)/8\pi^2 I$ the correct formula for the diatomic molecule in three dimensions.

Dirac's paper on 'Quantum Mechanics and a Preliminary Investigation of the Hydrogen Atom' (1926) treated the hydrogen atom two-dimensionally. He was able to derive the formula $E = -R_\infty Z^2/(n-\Delta)^2$ where n is an integer and Δ is a constant. He hoped that $\Delta = 0$ but was unable to prove it. Actually, carrying on his type of calculation for a two-dimensional model to full fruition yields $\Delta = \frac{1}{2}$. This fact is apparent from what we have said in Section 1, and we verify this conclusion explicitly with matrix algebra in Section 5. We there use the Heisenberg matrix language, but it can immediately be reworded into that of Dirac's q-numbers (or vice versa) by noting that in an expansion of the type

$$q = \Sigma_\tau q_\tau(J)e^{i\tau w} \quad \text{or} \quad q = \Sigma_\tau e^{i\tau w} q_\tau'(J)$$

in Dirac's q-number, angle and action variables symbolism the coefficients $q_\tau(J)$ and $q_\tau'(J)$ are the same as the Heisenberg matrices $\langle n | q | n-\tau \rangle$ and $\langle n+\tau | q | n \rangle$ with $n = J/h$.

The relativity corrections to the hydrogen spectrum, through terms in $1/c^2$, were first calculated by Heisenberg and Jordan (1926). First-order perturbation theory (cf. for instance, Kemble, 1937) shows that to this approximation, the shift in energy due to the combined effect of relativity and spin-orbit coupling is equivalent to adding to the energy a term

$$\Delta E = -\frac{1}{2mc^2}\left\langle E_0 + \frac{Ze^2}{r}\right\rangle^2 + \frac{Ze^2h^2}{8\pi^2m^2c^2}\, \boldsymbol{l}\cdot\boldsymbol{s}\left\langle\frac{1}{r^3}\right\rangle \tag{12}$$

where $2\boldsymbol{l}\cdot\boldsymbol{s} = j(j+1) - l(l+1) - \frac{3}{4}$, and E_0 is the energy without the relativity and spin connections. Evaluation of (12) requires knowledge of the mean values of r^{-1}, r^{-2}, r^{-3}, which in three dimensions are

$$\left\langle\frac{1}{r}\right\rangle = \frac{Z}{n^2a_0}, \quad \left\langle\frac{1}{r^2}\right\rangle = \frac{Z^2}{a_0^2 n^3(l+\frac{1}{2})}, \quad \left\langle\frac{1}{r^3}\right\rangle = \frac{Z^3}{a_0^3 n^3 l(l+\frac{1}{2})(l+1)} \tag{13}$$

while in two they are

$$\left\langle\frac{1}{r}\right\rangle = \frac{Z}{n^2a_0}, \quad \left\langle\frac{1}{r^2}\right\rangle = \frac{Z^2}{a_0^2 n^3|k|}, \quad \left\langle\frac{1}{r^3}\right\rangle = \frac{Z^3}{a_0^3 n^3|k|(k-\frac{1}{2})(k+\frac{1}{2})} \tag{14}$$

with n a whole integer in (13) and a half integer in (14). As customary the symbol a_0 denotes the Bohr radius $h^2/4\pi^2 e^2 m$. (We will discuss in Section 4 how these values are most easily obtained.) Heisenberg and Jordan started with a two-dimensional model, and if they had made a calculation rigorously for this model, they would have employed (14) and obtained a formula for ΔE that in no way agreed with the experiment. Instead, they utilised the values in (13): thus ending up with the celebrated right answer, viz. that the relativity corrections with spin give in quantum mechanics exactly the same energy values as in the old quantum theory without spin, but with a different interpretation of one of the integers (viz. it is $j + \frac{1}{2}$ rather than k).

The last days of the old quantum theory were the golden age of empiricism, where physicists often obtained correct answers by appropriate doctoring of formulas based on questionable theory, and some of this empiricism still survived in the very early days of quantum mechanics. In a footnote Heisenberg and Jordan state that Pauli in some work in course of publication (his matrix treatment of the hydrogen atom) had obtained results which are tantamount to our Hamiltonian (5), thus essentially justifying the use of (14) with half integral k and integral n. However, it is not clear whether Heisenberg and Jordan were aware at the time of the difference between the results with the two- and three-dimensional treatments.

I have described elsewhere, in another 'memoir' type of paper, (Van Vleck, 1968), how, guided by Dirac's paper on 'Quantum Mechanics and a Preliminary Investigation of the Hydrogen Atom', I had in June 1926 calculated with his q-number algebra the expectation value of (12), only to find when I reached Copenhagen that I had been anticipated by Heisenberg and Jordan. In a first draft of still another manuscript which I prepared for a volume in honour of Dirac's seventieth birthday (Van Vleck, 1972), I originally stated that I felt the calculation based on Dirac's formula was more rigorous than Heisenberg and Jordan's. Then at the last minute before the paper was sent in, I looked again at Dirac's paper, and discovered to my consternation that he used a two-dimensional model, and I was just as guilty as Heisenberg and Jordan in replacing (14) by (13). Of course the answers came out right, as is clear from the conversion relations discussed in Section 1. The same is true of the paper that I wrote in 1934 in which I calculated $\langle r^{-5} \rangle$ and $\langle r^{-6} \rangle$ by formulas from Dirac's paper. The choice between half and whole integral quantum numbers was a perfectly obvious one as far as the right answer was concerned, but not until 1972 did I notice (nor apparently has any of the readers!) this choice involved a switch in dimensionality, a rather inexcusable oversight, as 1934 was not 1926.

4. MATRIX v. WAVE-MECHANICAL TREATMENT OF CENTRAL FIELDS

It is not our intent to discuss generally the relative merits of the Heisenberg matrix and the wave-mechanical interpretations of quantum mechanics. This is a subject on which there is a vast literature by physicists, mathematicians and philosophers, with emphasis ranging from the practical or computational to the aesthetic. In the early days of quantum mechanics, the differential equation and algebraic approaches were regarded as more or less competitive ways of getting the same correct answer, whereas nowadays they are regarded as simply different aspects of the general transformation theory formalism of quantum mechanics. Much of my own research work has been in the field of magnetism, and elsewhere (Van Vleck, 1928, 1972), I have stressed the fact that apart from co-operative effects most of the computations can most easily be formulated in the language of matrix theory. This is because angular momentum matrices lend themselves particularly well to treatment by non-commutative algebra. However, for discussion of selection rules and factorisation of the secular equation the use of group theory for the crystallographic point group is essential. Here the transformation properties of the solution of the Schroedinger equation are involved, but the basis functions of group theory can be so abstract that one loses sight of the de Broglie wavelength.

   On the other hand, the wave-mechanical approach is certainly the most facile one for studying central fields. As Sommerfeld remarks in his book, comparison of Schroedinger's and Pauli's calculations for the hydrogen atom shows that here the wave-mechanical method is superior to the matrix one in 'mathematical ease and lucidity', Dirac himself, in his classic volume on *The Principles of Quantum Mechanics* (1930) abandons his earlier algebraic approach we have mentioned, and treats the non-relativistic Keplerian problem in the usual way involving terminating series development of the radial factor in the solution of the Schroedinger equation.

   Schroedinger reputedly once said he wished he had never discovered his wave functions if they were used simply as computational tools for calculating matrix elements. Nevertheless, it is true that the use of wave functions greatly facilitates the understanding of matrix mechanics. This is presumably what Pauli had in mind in the opening sentences of his review (1930) of Born and Jordan's book *Elementare Quantanmechnik*. This sentence reads 'This book is the second volume of a series, in the steady aim and sense that the nth volume is made comprehensible by the virtual existence of an $(n+1)$th volume'. The book he reviewed was the second of the series he caricatured. The first volume written in the days of the old quantum theory, obviously could not come up with the correct answers. The choice of the word 'Elementare' in the title of the second volume was a particularly inept one, as the word was used to connote the purely algebraic or matrix approach without any introduction of the Schroedinger equation and its wave functions. A third volume which was to encompass wave mechanics, was promised when the authors had the 'time and energy', but was never written. Pauli's review is probably the most caustic in history of a volume where one of the authors and the reviewer ultimately were Nobel laureates. In Born and Jordan's defence it is to be noted that at the time there were already several texts on the wave-mechanical approach, and none on the purely matrix one. The book therefore filled a special, but not 'elementare' place. I am afraid I was inadvertently responsible for the final sentence of Pauli's review. In the course of a conversation with him in my broken German in 1930, I happened to remark about how an American review of a book on another subject by other authors began by saying 'the faults of the book are distributed equally between the writer, the translator, and the printer' and ended by saying 'the paper is adequate'. (I have tried recently without success to locate and so be able to cite this review explicitly; I suspect that the tactful Professor Tate, then Editor-in-Chief of the Physical Review insisted that the reviewer tone it down and that he showed me the original manuscript which both amused and worried him.) In any event in our conversation Pauli instantly reacted and said that he could now supply a missing finale for his review of the Born and Jordan book. As a result he added at the end the following incongruous sentence. 'The printing and paper of the book are outstanding'.

The comparison of the behaviour of central fields in two and three dimensions is an excellent illustration of the Pauli theme that oftentimes the results of matrix mechanics are more lucid in the light of wave mechanics. Although a considerable part of the present paper is concerned with a historical critique of early articles written with the purely algebraic approach, we have introduced differential equations in Section 1 as the basis on which to develop this critique. For instance, the appearance of $(k+\frac{1}{2})(k-\frac{1}{2})$ rather than k^2 as the coefficient of r^{-2} in (8) arises from normalisation to dr rather than $r\,dr$ as in equation 7. It is thus no justification for making k half-integral. Of course all the various contrasts between the two and three dimensions could be reworded, though less conveniently, by using matrix rather than wave-mechanical language. The comparison of the two- and three-dimensional versions of the radial problem can, for instance, be treated as a problem in standard quantum-mechanical matrix algebra in a one-dimensional space with the Hamiltonian function (5) or (8), rather than presented as a differential equation.

Heisenberg and Jordan employed (8) as their radial Hamiltonian function. They do not explain explicitly how they got it, but presumably they obtained it by making the transformation from Cartesian to polar co-ordinates and momenta such as to make P_r a Hermitean matrix [corresponding to $\rho = 1$ in our orthogonality relation (4)]. The transformation of this character is

$$x \pm iy = re^{\pm i\phi}$$
$$p_x \pm ip_y = p_r e^{\pm i\phi} \pm i\bar{r}\left[\tfrac{1}{2}p_\phi e^{\pm i\phi} + \tfrac{1}{2}e^{\pm i\phi}p_\phi\right] \tag{15}$$

If one does not Hermiteanise and instead replaces the bracketed factor of (15) by $e^{\pm i\phi}p_\phi$, a procedure equivalent to setting up the Laplacian in polar co-ordinates, one obtains (7) rather than (8) and the adaptation to three dimensions is less obvious. The easiest way of calculating the averages of r^{-1}, r^{-2}, r^{-3} needed in connection with (12) and given in (13) or (14) is not by means of the quadrature (10) involving the wave functions, but by use of some simple tricks. The virial theorem yields $\langle 1/r \rangle$ immediately once the energy is known. Differentiation of the energy with respect to the parameter l in connection with first-order perturbation theory yields $\langle 1/r^2 \rangle$, as explained, for example in Kemble's book (1937). [Heisenberg and Jordan used a more or less equivalent procedure. They employed the relation $\langle \dot{\phi} \rangle = \partial E/h\partial n$ and $r^{-2} = m\dot{\phi}/p_\phi$ and took $\rho_\phi = (k+\frac{1}{2})h/2\pi$ instead of $kh/2\pi$.]

This procedure does not, however, lend itself to three dimensions, as unlike classical mechanics, the problem is no longer a planar one and there is no total angle Ψ such that $(l+\frac{1}{2})h/2\pi = mr^2\dot{\Psi}$). Once $\langle 1/r^2 \rangle$ has been found one can immediately obtain $\langle 1/r^3 \rangle$ as with either (5) or (8) the equations of motion immediately express $m\ddot{r}$ as a linear function of r^{-2} and r^{-3} with constant coefficients and the expectation value of $\ddot{r}$ vanishes. This method of obtaining inter-relationship between $\langle r^{-2} \rangle$ and $\langle r^{-3} \rangle$ works in either two or three dimensions. It was utilised by Heisenberg and Jordan (1926), and is described in greater detail by Kemble (1937).

One nice feature of wave as compared with matrix mechanics is that it shows, or at least gives an indication why in the two-dimensional central field the angular momentum is an integral multiple of $h/2\pi$. Namely, if we require the wave function $R(r)e^{ik\phi}$ to be single-valued then k is necessarily integral. A similar conclusion is reached if one requires that the wave functions be classifiable as regards parity, i.e. be even or odd under the substitution $x, y \rightarrow -x, -y$. The ladder arguments of matrix theory show that successive eigenvalues of k differ by integers. If left and right handed senses of rotation are equivalent then the sequence of allowed k values must be one of whole or half integers but the distinction between the two alternatives is less obvious without resorting to wave functions. (It is thus small wonder that Heisenberg in his initial paper could not decide unambiguously between the two choices, nor could Dirac.)

5. THE TWO-DIMENSIONAL HYDROGEN ATOM TREATED BY MATRIX METHODS

The derivation of the Balmer formula for the hydrogen atom in three dimensions by matrix rather than wave-mechanical methods requires a rather involved calculation (Pauli, 1926). For this reason in his review of the Born and Jordan book which we have already mentioned, Pauli criticises the authors for including it. Their doing so, he said, was not a case of the 'grapes being sour because they were too high', inasmuch as the original calculation was by the reviewer and so not beyond his intellectual grasp! In two dimensions the algebra is considerably simpler, and so we give here a matrix calculation similar to Pauli's but for the two-dimensional case.

In the treatment of Coulomb fields by matrix mechanics the vector

$$a = \frac{1}{2Ze^2m}[P \times p - p \times P] + \frac{r}{r}$$

plays a fundamental role, as it commutes with the Hamiltonian function $\mathcal{H}$. For the two-dimensional case one has

$$\mathcal{H} = \frac{1}{2m}(p_x^2 + p_y^2) - \frac{Ze^2}{r} \tag{16}$$

$$a_x \pm ia_y = \frac{\pm i}{2me^2Z}[p_\phi(p_x \pm ip_y) + (p_x \pm ip_y)p_\phi] + \frac{x \pm iy}{r} \tag{17}$$

Here $p_\phi = xp_y - yp_x$ is the angular momentum, a scalar, which replaces the angular momentum vector P entering in the three-dimensional analysis. The position and momentum co-ordinates, and also the components of the two-dimensional vector a all satisfy commutation rules of the form

$$p_\phi(q_x \pm iq_y) - (q_x \pm iq_y)p_\phi = \pm \frac{h}{2\pi}(q_x \pm iq_y) \tag{18}$$

In other words, (18) is valid for $q_x, q_y = x, y$, also p_x, p_y and a_x, a_y. With the aid of (18) one verifies by elementary but rather tedious calculation that (16) commutes with (17). Also one similarly finds that

$$(a_x \pm ia_y)(a_x \pm ia_y) = 1 + \frac{\mathcal{H}}{2mZ^2e^4}\left[4p_\phi^2 \pm 4p_\phi\left(\frac{h}{2\pi}\right) + \frac{h^2}{4\pi^2}\right] \tag{19}$$

At this point one must pause to assess and interpret the physics of the situation. From (18) it follows that

$$\langle n_r k \mid a_x \pm a_y \mid n_r' k' \rangle \tag{20}$$

vanishes unless $k' = k \mp 1$, where $p_\phi = kh/2\pi$. In writing down the expression (20) we have utilised the fact that there are two quantum numbers entering in our problem, i.e. there is the radial quantum number n_r in addition to the azimuthal one. We note that (20) is not diagonal in k and that the energy is augmented by increased rotation and so is an increasing function of $|k|$ for fixed n_r. The fact that $a_x \pm a_y$ commutes with $\mathscr{H}$ implies that the change of energy due to change in k just offsets that due to that in n_r. (The compensating effect of changes in two quantum numbers is not too clear on reading Pauli's paper. His index k, not to be confused with our k, is really a double index involving the pair of quantum numbers n_r, l which enter in the three-dimensional problem in addition to m_l.) Hence among the states belonging to the same E, an eigenvalue of $\mathscr{H}$, there is one of maximum k, with $n_r' = 0$, and there can be no matrix element of $a_x - ia_y$ taking one to the non-existent state $k_{\max} + 1$, n_r'. Hence we conclude that (19) must vanish when $k = k_{\max}$ and so the corresponding eigenvalue of $\mathscr{H}$ is given by

$$E = -\frac{R_\infty Z^2}{(k_{\max} + \frac{1}{2})^2} \qquad \left(R_\infty = \frac{2\pi^2 m e^4}{h^2} \right) \tag{21}$$

In general there are $2k_{\max} + 1$ states belonging to a given E. First let us suppose that the allowed values of k are integers. The non-vanishing matrix elements of (20) are of the form $(n_r k \to n_r' \pm 1, k \mp 1$ if $k \geq 0$ and of the type $\to n_r \mp 1$, $k \mp 1$ if $k < 0$. In other words we must have $E = f(|k| + n_r)$ and we conclude from (21) that generally

$$E = -\frac{R_\infty Z^2}{(n_2 + \frac{1}{2})^2} \tag{22}$$

where n_2 is an integer.

To complete our derivation of (22) we must verify the correctness of our assumption that k takes on only integral values. This obviously is the case if one grants that the wave function is to be single-valued (cf. equation 6), but the integral character of k can also be demonstrated as follows by using purely matrix algebra. The usual ladder type argument and the cut-off condition on $a_x + ia_y$ at $k = k_{\min}$ shows immediately that the sequence of allowed values of k is either one of integers or half integers. To eliminate the half integral case one proceeds as follows. One first notes that the matrix elements (20) vanish when $k = -k' = \pm\frac{1}{2}$ unless $n_r = n_r'$. Otherwise a would connect states of different energy inasmuch as E is independent of the sign of k. Furthermore one has

$$\langle n_r, \pm\tfrac{1}{2} \mid p_\phi(p_x \pm ip_y) + (p_x \pm ip_y)p_\phi \mid n_r', \mp\tfrac{1}{2} \rangle = 0$$

because of the different sign of p_ϕ fore and aft. By using these facts and by setting $p_\phi = \mp\frac{1}{2}(h/2\pi)$ in (19) one concludes from (17) that

$$|\langle n_r, \pm\tfrac{1}{2} \mid (x \pm iy)r^{-1} \mid n_r', \pm\tfrac{1}{2} \rangle| = \delta_{n_r'}^{n_r}$$

and so

$$|\langle n_r, \pm\tfrac{1}{2} \mid (x \pm iy)r^{-3} \mid n_r, \pm\tfrac{1}{2} \rangle| = \langle n_r, \tfrac{1}{2} \mid r^{-2} \mid n_r, \tfrac{1}{2} \rangle > 0 \tag{23}$$

On the other hand one has

$$\left\langle n_r, \pm\tfrac{1}{2} \left| \frac{\mathrm{d}p_x}{\mathrm{d}t} \pm \frac{\mathrm{i}}{\mathrm{d}t} \frac{\mathrm{d}p_y}{\mathrm{d}t} \right| n_r, \pm\tfrac{1}{2} \right\rangle = 0 \tag{24}$$

inasmuch as the frequency factor thrown down by time differentiation vanishes. Comparison of (23) and (24) shows that the equations of motion

$$\frac{\mathrm{d}(p_x \pm \mathrm{i}p_y)}{\mathrm{d}t} = -Ze^2 \frac{(x \pm \mathrm{i}y)}{r^3}$$

are violated. The assumption of half integral k is hence not tenable.

Equation 22 is equivalent to a Balmer formula with half rather than whole integers. This is to be expected in the light of the discussion in Section 1. The conversion relations of Section 1 show that if there are half integers in two dimensions there must be whole ones in the actual three-dimensional world. When this trick is used, our procedure probably provides the simplest derivation of the Balmer formula by matrix algebra. Even so, most readers will feel it is easier to grasp the grapes of the wave-mechanical derivation whose seeds were sown by de Broglie in 1923.

3.6 SUMMARY

One of the surprises of quantum mechanics, now seldom noticed, is that the energy levels and the expectation values of functions of the radius for a particle subject to a central field are different when a planar rather than true three-dimensional model is used. It is shown that there is, however, a simple general rule which enables one to convert the results obtained in two dimensions to those for three; viz. keep the radial quantum number integral but artificially give the azimuthal quantum number half integral rather than integral values. Some historic papers in the early days of quantum mechanics started with two-dimensional models but nevertheless curiously came up with right answers because the authors used the conversion rules apparently without realising what they were doing. The dependence on dimensionality is more evident in the differential equation than the matrix approach, and this fact explains some of the confusion in the early matrix days. The derivation of the eigenvalues of the energy for a Coulomb field by matrix methods is much easier in two than in three dimensions, and we therefore give a derivation for the plane case by appropriate adaptation and simplification of Pauli's calculation. The result is a Balmer type formula, but with half rather than whole integers in the denominator. This is to be expected and the conversion rule restores the whole integers when the dimensionality is switched to that of the real three-dimensional world.

REFERENCES

Born, M. and Jordan, P. (1930) *Elementare Quantenmechanik*, Springer, Berlin
Dirac, P. A. M. (1926) 'Quantum Mechanics and a Preliminary Investigation of the Hydrogen Atom', *Proc. R. Soc.*, **A110**, 561–579

Dirac, P. A. M. (1930) *The Principles of Quantum Mechanics*, Oxford University Press

Heisenberg, W. (1925) 'Über quantentheoretische Umdeutung kinematischer und mechanischer Bezeihungen', *Z. Phys.*, **33**, 879–893

Heisenberg, W. and Jordan, P. (1926) 'Anwendung der Quantenmechanik auf das Problem der Anomalen Zeemaneffekte', *Z. Phys.*, **36**, 263–277

Kemble, E. C. (1937) *The Fundamental Principles of Quantum Mechanics*, pp. 503–509, McGraw-Hill, New York

Pauli, W. (1926) 'Uber das Wasserstoffspektrum von Standpunkt der Neuen Quantenmechanik', *Z. Phys.*, **36**, 336–363

Pauli, W. (1930) 'Review of Born and Jordan's "Elementare Quantenmechanik"', *Naturwissenschaften*, **18**, 602

Sommerfeld, A. (1929) *Atombau und Spektrallinien, Wellenmechanischer Ergänzungband*, Section 7B, Vieweg, Braunschweig

Van Vleck, J. H. (1928) 'On Magnetic Susceptibilities in the New Quantum Mechanics', *Phys. Rev.*, **32**, 587–613

Van Vleck, J. H. (1934) 'A New Method of Calculating the Mean Value of $1/r^s$ for Keplerian Systems in Quantum Mechanics', *Proc. R. Soc.*, **A143**, 679–681. There are some purely numerical errors in the tabulated values of $\langle r^{-6} \rangle$ for specific values of 1 at the bottom of p. 580 see Rockanten, K. (1956) *Phys. R.*, **102**, 729

Van Vleck, J. H. (1968) 'My Swiss Visits of 1906, 1926 and 1930', *Helv. Physica Acta*, **41**, 1234–37

Van Vleck, J. H. (1972) 'Travels with Dirac in the Rockies', *Some Aspects of Quantum Theory*, 7–17 (edited by Salam, A. and Wigner, E. P.) Cambridge University Press

4

Mathematical Foundations of a Quantum Theory of Valence Concepts

KEITH R. ROBY

Research School of Chemistry, Institute of Advanced Studies,
The Australian National University, Canberra, A.C.T.

1. INTRODUCTION

Professor Louis de Broglie's discovery (1923a, b) of the wave nature of the electron has had a profound influence on the way chemists think about molecules in terms of atoms and bonds. At the same time, the wave nature of the electron has associated with it the properties of probability, uncertainty, and the indistinguishability of the electrons in a system. In addition there is the problem of properly accounting for the correlation of electron motions (see Löwdin, 1959; Sinanoğlu, 1964). The difficulties that these properties pose for chemical valence theory have yet to be resolved, and it is a great honour to be able to dedicate this assessment of the present situation to Professor de Broglie.

Many of the concepts with which we are concerned — the electron pair chemical bond itself, lone pair and inner shell electrons, ionic and covalent bonding — appear in G. N. Lewis' famous paper of 1916. After de Broglie's discovery and the development of the full quantum theory, Lewis' ideas received confirmation in the work of Heitler and London, Slater, Pauling, Mulliken, Hund and others, and in addition new concepts such as hybridisation, electronegativity and resonance were introduced (see especially Pauling, 1931, 1960). In all of this work, various approximate forms of the quantum theory were drawn upon, and so long as we are prepared to remain within these approximations, the relation between quantum theory and valence concepts is fairly well understood.

From the point of view of rigorous theory, however, the status of such basic concepts is not very clear at all. In fact it seems at first sight as though 'the simple bond has got lost' (Coulson, 1970). Yet we know that a more rigorous theory including electron correlation is necessary if molecular properties are to be calculated at a level of accuracy equivalent to that

38

obtained in their experimental measurement. In sum, we are faced with a situation that is best described as the 'conceptual dilemma of quantum chemistry': agreement with experiment demands more rigorous theory, yet in more rigorous theory the conceptual structure of chemistry is not readily apparent.

There is thus a real and pressing need to consider very closely the definition, validity and role of fundamental valence concepts within rigorous quantum theory. We will attempt here to set down certain basic ideas and to outline a mathematical framework within which some progress may be made. A complete exposition of some of the mathematical detail and also of resulting numerical methods will be left to subsequent papers.

We propose to seek a way through the conceptual dilemma by beginning with the fundamental connection between physical reality and quantum theory. Key references include the works of von Neumann (1955), Dirac (1958), Mackey (1963), Jauch (1968) and Messiah (1968). By drawing upon the mathematical structure formulated in these works, one may ask whether valence concepts have any status in the physical interpretation of molecular quantum theory.

Propositions made or questions asked about a physical system enable the generation of a 'propositional calculus' (von Neumann, 1955; Mackey, 1963; Jauch, 1968). The propositional calculus in turn is fundamentally related to the mathematics of abstract Hilbert space in such a way that statements about the physical system can be translated into quantities and relations in Hilbert space. Key notions include those of 'states', which are related to vectors and subspaces of the Hilbert space, and 'observables', which physically are measurable properties and mathematically are self-adjoint operators in Hilbert space. In our analysis of valence concepts, we shall argue that the particular observables, projection and density operators, by means of which the states of a system are defined, are of basic importance for valence theory.

In Section 2, the meaning of the term 'Hilbert space' is briefly explained, and some of the relevant properties are given. We are especially concerned with the way in which, from the Hilbert space for an N-electron molecular system, Hilbert spaces for smaller subsystems of electrons may be defined, and the way in which any Hilbert space may be partitioned geometrically into subspaces. In regard to systems and subsystems, various density operators have central roles, and we concentrate particularly on the density operator that describes the behaviour of a single electron within the system (Section 3). In regard to subspaces, projection operators are the essential quantities, and we show such operators may be associated with atoms in molecules and chemical bonds (Section 4). Combining the concepts of subsystems and subspaces by bringing together the density and projection operators, we obtain a partitioning of the molecular electron density such that the notions of the charge of an atom in a molecule and the sharing of electron density receive a mathematical definition (Section 5). Application of the same ideas to the orbital basis of rigorous quantum theory leads to a procedure for the classification of molecular orbitals and to some conclusions regarding their localisation and the hybridisation of atomic orbitals (Section 6).

2. THE MOLECULAR HILBERT SPACE

In the formulation of von Neumann (1955), quantum theory proceeds from the notion of the abstract Hilbert space. Mathematically, any molecular system in any state may be defined by a Hilbert space and the observables that operate within it.

An abstract Hilbert space H is defined as a set of vectors that obey certain axioms and for which we shall use the Dirac ket notation $|f\rangle, |g\rangle, \ldots$ The space of Schroedinger wave functions is a particular 'realisation' of this more general abstract space, and obeys the mathematics of the abstract space. Manipulations with the vectors $|f\rangle, |g\rangle, \ldots$ bear a one-to-one correspondence to manipulations of wave functions.

The abstract Hilbert space is a linear vector space, which means that if $|f\rangle$ and $|g\rangle$ belong to H, then so does the sum $|f\rangle + |g\rangle$, and if α is any real or complex number, then so does $\alpha|f\rangle$. Consequently, any linear combination of any number of vectors in the space also belongs to the space.

We need to be familiar with the related notions of 'linear independence', 'basis', 'scalar product' and 'orthonormality'. A set of vectors $|f_1\rangle, |f_2\rangle, \ldots$ is said to be linearly independent if the relation

$$\sum_i \alpha_i |f_i\rangle = 0 \tag{1}$$

implies that all coefficients $\alpha_i = 0$. The maximum number of linearly independent vectors that can be found in a vector space is the dimension of that space, and the vectors then are said to form a basis for the space. The quantum mechanical Hilbert space is infinite dimensional, and this is the source of some of our trouble.

Given a basis for a vector space, one may expand any other vector $|x\rangle$ in the space in terms of the basis:

$$|x\rangle = \sum_i \beta_i |f_i\rangle \tag{2}$$

and the basis set is said to be complete. Thus the question of finding a suitable basis in the Hilbert space is a central one.

One of the axioms of Hilbert space relates to the definition of an inner or scalar product $\langle f|g\rangle$ for all pairs of vectors $|f\rangle, |g\rangle$. The scalar product is in general a complex number and has the properties:

$$\langle f|g\rangle = (\langle g|f\rangle)^* \tag{3}$$

$$\langle f|g+h\rangle = \langle f|g\rangle + \langle f|h\rangle \tag{4}$$

$$\langle f|\lambda g\rangle = \lambda\langle f|g\rangle, \lambda \text{ any complex number} \tag{5}$$

$$\langle f|f\rangle \geq 0 \text{ and } = 0 \text{ if and only if } |f\rangle = |0\rangle \tag{6}$$

In the wave function realisation, the scalar product of any two wave functions is simply the overlap integral over electron co-ordinates:

$$\langle \Psi|\Phi\rangle = \int\int\ldots\int \Psi^*(r_1, r_2 \ldots r_N)\Phi(r_1, r_2 \ldots r_N)\, dr_1\, dr_2 \ldots dr_N \tag{7}$$

A set of vectors $|f_1\rangle, |f_2\rangle, \ldots$ is said to be orthonormal if the relation

$$\langle f_i|f_j\rangle = \delta_{ij} \tag{8}$$

δ_{ij} being the Kronecker delta, is obeyed for all $|f_i\rangle$, $|f_j\rangle$. Such a system cannot be linearly dependent, and so one way of generating a basis is to find the set of all vectors orthogonal to a given vector.

Operators appear in the theory as quantities that act on one vector to give another. Let A be an operator, then in general

$$A|f\rangle = |g\rangle \tag{9}$$

while if $|f\rangle$ is an eigenvector of A, the equation becomes

$$A|f\rangle = \alpha|f\rangle \tag{10}$$

α being the corresponding eigenvalue. It may be shown that all eigenvectors of an observable form a complete orthonormal set, and thus a basis in the Hilbert space. This will be an important consideration in the following discussion, but we need also to keep in mind that a basis need not be orthonormal.

Having defined the scalar product, we may begin to speak of the geometry of Hilbert space, leading ultimately to the concept of projection (Section 4). Here we are content to define the norm or magnitude $\|f\|$ of any element $|f\rangle$ according to:

$$\|f\| = \sqrt{\langle f|f\rangle} \tag{11}$$

and having the property, from equation 6,

$$\|f\| \geq 0 \tag{12}$$

Then the distance $D(f, g)$ between any two elements $|f\rangle$ and $|g\rangle$ is given by

$$D(f, g) = \|f - g\| \tag{13}$$

and has the properties of distance in three-dimensional space:

$$D(f,g) \geq 0 \tag{14}$$

$$D(f, g) = D(g,f) \tag{15}$$

$$D(f, g) + D(g, h) \geq D(f, h) \tag{16}$$

Many other properties of abstract Hilbert space have been given by von Neumann, Mackey, and Jauch, and we have mentioned only those important in the present context. With regard to molecular electronic structure, the abstract vectors $|f\rangle$, $|g\rangle$, ... may be taken as representing N-electron wave functions, and we may refer to the 'N-electron molecular Hilbert space', H_N. The symbols $|\Psi_1\rangle$, $|\Psi_2\rangle$, ... will be used instead of $|f\rangle$, $|g\rangle$ when we wish to refer specifically to this space, and the Schroedinger equation becomes

$$H|\Psi_i\rangle = E_i|\Psi_i\rangle \tag{17}$$

But the properties of abstract Hilbert space apply equally well to the description of subsystems of the N electrons, and this is an essential feature of the approach to valence concepts. It becomes appropriate to speak of the one-, two-, ..., $(N-1)$-electron Hilbert spaces H_1, H_2, ..., H_{N-1}. The N-electron Hilbert space H_N is given in terms of the appropriate subsystem spaces by a procedure known as the tensor or direct product. For example:

$$H_N = H_1 \otimes H_{N-1} \tag{18}$$

or

$$H_N = H_2 \otimes H_{N-2} \tag{19}$$

or

$$H_N = H_1 \otimes H_1 \otimes \ldots \otimes H_1, N \text{ factors} \tag{20}$$

and so on. The first of these equations, for example, means that if $|f\rangle$ is a vector in H_1 (that is, $|f\rangle$ corresponds to a one-electron wave function or orbital) and $|g\rangle$ is a vector in H_{N-1} [corresponding to an $(N-1)$-electron wave function], then the product $|fg\rangle$ is a vector in H_N. Furthermore an observable for a subsystem becomes an observable for the total system upon taking the tensor product with the identity operator for its related subsystem. For example a one-electron observable A_1 in H_1 becomes an N-electron observable A_N in H_N through the relation

$$A_N = A_1 \otimes I_{N-1} \tag{21}$$

I_{N-1} being the identity operator in H_{N-1}. In this way the idea of subsystems of the total system is given meaning, and we will consider especially the electron density operators for subsystems (Section 3).

The space H_1 is given special significance in the next section. When we wish to refer specifically to H_1, we replace the general symbols $|f\rangle$, $|g\rangle$, ... with symbols $|i\rangle$, $|j\rangle$, ... to refer to molecular spin orbitals, or the symbols $|\mu_A\rangle$, $|v_B\rangle$, ... to refer to atomic orbitals centred on the various atoms A, B, ... in the molecule. Equation 20 already points to the importance of H_1 and is the basis for all forms of the 'cluster expansion' of many-electron wave functions in terms of orbitals (Primas, 1965).

3. MOLECULAR ELECTRON DENSITY OPERATORS

Electron density operators are vital quantities in the physical interpretation of quantum theory. Apart from their obvious intrinsic physical significance and the fact that they are observables, they supply a definition of the pure or mixed states of a system, a means of calculating the expectation values of molecular properties, and a characterisation of subsystems. Various aspects of their properties are treated in von Neumann (1955) and Dirac (1958) and in the papers and reviews by Husimi (1940), Löwdin (1955), Fano (1957), McWeeny (1960), ter Haar (1961), Coleman (1963), Ando (1963) and Smith (1968).

The N-electron density is well known, being simply the square of the molecular wave function and giving, by integration over a particular volume, the probability of finding the electrons in that volume. In operator terminology, the N-density operator D_N for a system in a normalised pure state $|\Psi\rangle$ is:

$$D_N = |\Psi\rangle\langle\Psi| \tag{22}$$

If the state is not pure, then

$$D_N = \sum_i |\Psi_i\rangle \alpha_i \langle\Psi_i|, \quad 0 \leq \alpha_i \leq 1 \tag{23}$$

where α_i is the probability of finding the system in state $|\Psi_i\rangle$.

The concept of the trace of an operator will be used often in this and later sections. In general terms, if the set $|f_1\rangle, |f_2\rangle, \dots$ is orthonormal and complete, then the trace of any operator A for which all $A|f_i\rangle$ are defined is:

$$\text{Tr}\, A = \sum_i \langle f_i | A | f_i \rangle \tag{24}$$

It is readily shown that:

$$\text{Tr}\, \alpha A = \alpha \,\text{Tr}\, A \tag{25}$$

$$\text{Tr}\,(A+B) = \text{Tr}\, A + \text{Tr}\, B \tag{26}$$

$$\text{Tr}\, AB = \text{Tr}\, BA \tag{27}$$

$$\text{Tr}\, ABC = \text{Tr}\, CAB = \text{Tr}\, BCA \tag{28}$$

Here, A, B and C are operators and α is any real or complex number.

D_N being an observable, its eigenvectors (belonging to zero as well as non-zero eigenvalues) constitute a complete orthonormal set in H_N. Taking the trace of D_N with respect to its eigenvectors, we obtain the normalisation condition:

$$\text{Tr}\, D_N = \sum_i \langle\Psi_i|\Psi_i\rangle \alpha_i \langle\Psi_i|\Psi_i\rangle$$

$$= \sum_i \alpha_i$$

$$= 1 \tag{29}$$

According to von Neumann, the N-density operator D_N determines uniquely the density operators for any subsystem (the 'reduced density operators'). The relation is best expressed by introducing a partial trace. For example, the density operator D_{N-1} for $(N-1)$ electrons is given by

$$D_{N-1} = \text{Tr}_1 D_N \tag{30}$$

where Tr_1 indicates the trace with respect to a complete set of vectors in the one-electron Hilbert space H_1. Similarly,

$$D_{N-2} = \text{Tr}_2 D_N$$
$$\vdots$$
$$D_2 \quad = \text{Tr}_{N-2} D_N$$
$$D_1 \quad \equiv \rho = \text{Tr}_{N-1} D_N \tag{31}$$

We have maintained the normalisation:

$$\text{Tr}\, D_{N-1} = \text{Tr}\, D_{N-2} = \dots = \text{Tr}\, D_2 = \text{Tr}\, \rho = 1 \tag{32}$$

each trace being understood as involving the Hilbert space to which the density operator refers. We shall use the more usual symbol ρ for the one-electron density operator in H_1, instead of D_1.

In the calculation of molecular properties, the expectation value of any observable A is given by:

$$\langle A \rangle = \text{Tr } AD_N \tag{33}$$

Most commonly, A is a sum of one-electron or two-electron operators. In the former case, if

$$A = \sum_{i=1}^{N} A_1(r_i) \tag{34}$$

then

$$\langle A \rangle = N \text{ Tr } A_1 \rho \tag{35}$$

while in the latter, if

$$A = \sum_{i=1}^{N} \sum_{j>i}^{N} A_2(r_i, r_j) \tag{36}$$

then

$$\langle A \rangle = \frac{N(N-1)}{2} \text{Tr } A_2 D_2 \tag{37}$$

The total non-relativistic energy in particular has this type of dependence on D_2 and ρ:

$$E = N\text{Tr } h\rho + \frac{N(N-1)}{2} \text{Tr } gD_2 \tag{38}$$

Here h is the operator for the kinetic energy of one electron and its attraction to the nuclei, while g is the operator for electron–electron repulsion. Equations 35, 37 and 38 highlight the importance of the one- and two-electron density operators.

There is, however, a theorem due to Hohenberg and Kohn (1964) in which this reduction process is carried even further. Hohenberg and Kohn found that the ground state energy and molecular properties depend in theory only on the one-density ρ. In fact the energy is given by:

$$E = N\text{Tr } v\rho + F[\rho(r)] \tag{39}$$

where $F[\rho(r)]$ is a universal but unknown functional of the representation $\rho(r)$ of the operator ρ in configuration space, and v is any external potential acting on one electron. In our case, v is the nuclear-electron attraction potential.

By virtue of the theorem of Hohenberg and Kohn, the one-density operator ρ becomes a fundamental quantity in the theory. Indeed, an attempt has already been made to base a new representation of the quantum theory on ρ (Primas, 1967). Practical difficulties that arise, concerning the nature of the functional $F[\rho(r)]$ and the conditions to be placed on ρ so that it actually represents the state of the total system (the 'N-representability problem', see Coleman, 1963) need not concern us at this stage. We need only know that a unique one-density may be generated from the wave function for the system via D_N, that ρ is an observable and so provides a direct link between theory and experiment, and that it is fundamental, so that any analysis of ρ leads at least in principle to an analysis of other subsystem density operators and molecular properties.

The orbital basis of our analysis arises from the expression of ρ in terms of its eigenvalues λ_i and eigenvectors $|i\rangle$, the latter being the so-called 'natural spin orbitals' (Löwdin, 1955):

$$\rho = \sum_i |i\rangle\lambda_i\langle i|, \quad 0 \le \lambda_i \le 1/N \tag{40}$$

and

$$\mathrm{Tr}\,\rho = \sum_i \lambda_i = 1 \tag{41}$$

The quantities $n_i = N\lambda_i$ are interpreted as the occupation numbers of the natural spin orbitals $|i\rangle$. It is convenient also to introduce a matrix notation such that equation 40 may be written instead as:

$$\rho = |\Phi\rangle\lambda\langle\Phi| \tag{42}$$

where $|\Phi\rangle$ is a row matrix of the kets $|i\rangle$, λ is a diagonal matrix whose elements are the λ_i, and $\langle\Phi|$ a column matrix of the bras $\langle i|$.

Another useful version of equation 40 results if we define 'orbital density operators' ρ_i by:

$$\rho_i = |i\rangle\lambda_i\langle i| \tag{43}$$

such that

$$\rho = \sum_i \rho_i \tag{44}$$

Then

$$N\mathrm{Tr}\,\rho_i = n_i \tag{45}$$

and

$$N\mathrm{Tr}\,\rho = \sum_i n_i = N \tag{46}$$

We note that equation 44 together with equation 35 for the expectation value of a one-electron operator gives an 'orbital additivity' of one-electron molecular properties:

$$\langle A \rangle = \sum_i N\,\mathrm{Tr}\,A_1\rho_i \tag{47}$$

If equations such as this are to explain the observed near-additivity of molecular properties, much more needs to be known about the degree of localisation of natural spin orbitals on atoms and in bonds. Some aspects of this question are considered in Section 6.

Davidson (1972) has reviewed other properties of natural spin orbitals, and spin and symmetry have been discussed by McWeeny (1960) and Bingel and Kutzelnigg (1970). When dealing with spin-free molecular properties, it is convenient to introduce with these authors a spin-averaged version of the operator ρ by taking the trace of ρ with respect to the spin variables. The eigenvectors of the spin-averaged one-density operator are often called 'natural orbitals' and in general differ from the space parts of the natural spin orbitals. Because our methods apply with only minor adjustments to either natural spin orbitals or natural orbitals, we are able to use these terms interchangeably in the following sections.

Of course, the importance of the molecular one-density for valence theory has been recognised before, and there exists a number of ways of partitioning the one-density into physically meaningful regions. Especially useful are the ideas of binding and anti-binding regions (Berlin, 1951; Bader, 1964); of 'loges' (Daudel, 1953; Daudel, Gallais and Smet, 1967); of density difference maps (Roux, Besnainou and Daudel, 1956; Roux, Cornille and Burnelle, 1962); and of orbital and total density maps (Wahl, 1966). Bader, Henneker and Cade (1967) and Ransil and Sinai (1967) have used binding and anti-binding regions, density difference maps and density maps in conjunction in elaborate analyses of molecular Hartree–Fock wave functions, and their work is capable of application to exact wave functions. Ruedenberg (1962) has devised a conceptual partitioning of the molecular energy based on an expansion of the one- and two-densities in terms of atomic orbitals, and this partitioning has recently been extended and thoroughly studied in the case of the hydrogen molecule-ion (Feinberg, Ruedenberg and Mehler, 1970).

Since it is also based on density operators, our approach is quite complementary to the above methods. It differs from them in that our partitioning of the density results from projection techniques.

4. THE CONCEPT OF PROJECTION

The concept of projection is a fundamental one in the mathematics of Hilbert space, and one of the main points of this paper is that it is equally fundamental in the quantum theory of valence concepts.

Very simply, the concept of projection is the generalisation to many-dimensional vector spaces of the resolution of a vector in three-dimensional space. In fact, the components of a vector along the x, y and z axes may be called the projections of the vector on those axes.

In many-dimensional vector spaces, projections are associated with subspaces of the total space. A subspace M is defined by any subset of the basis vectors in the total space, together with all possible linear combinations of the vectors in the subset and all vectors which are limits of any sequence of vectors in the subset. A subspace is itself a linear vector space of finite or infinite dimension, and possesses most of the properties of the total Hilbert space. Those vectors remaining in the total space that are orthogonal to all vectors in the subspace themselves form a subspace which is called the 'orthogonal complement' to M.

Alternatively, one may define the subspaces of a Hilbert space by projection operators (projectors). Any operator P that is idempotent and self-adjoint:

$$P^2 = P, P^+ = P \tag{48}$$

is a projector, and every projector defines a subspace.

The importance of subspaces and projectors arises largely from the following 'projection theorem'. Any vector $|f\rangle$ of the Hilbert space has a unique decomposition with respect to any subspace:

$$|f\rangle = |f_1\rangle + |f_2\rangle \tag{49}$$

where $|f_1\rangle$ belongs to the subspace and $|f_2\rangle$ to its orthogonal complement.

This theorem may be written equally well in terms of projectors. If P is the projector defining the subspace, and Q the projector on its orthogonal complement, then

$$P+Q = 1, \ PQ = QP = 0 \tag{50}$$

and

$$|f_1\rangle = P|f\rangle \tag{51}$$

while

$$|f_2\rangle = Q|f\rangle \tag{52}$$

It follows that

$$\langle f|f\rangle = \langle f|P|f\rangle + \langle f|Q|f\rangle \tag{53}$$

Since all quantities here are positive, for example,

$$\langle f|P|f\rangle = \langle f|P^+P|f\rangle \geq 0 \tag{54}$$

we obtain the limits

$$0 \leq \langle f|P|f\rangle \leq \langle f|f\rangle \tag{55}$$

or, in terms of norm;

$$0 \leq \|Pf\| \leq \|f\| \tag{56}$$

The norm $\|Pf\|$ measures the 'magnitude' of the component of $|f\rangle$ on the subspace, and the limits have the following meaning:

$\|Pf\| = 0$ if and only if $|f\rangle$ belongs wholly to the orthogonal complement.

$\|Pf\| = \|f\|$ if and only if $|f\rangle$ belongs wholly to the subspace.

Because of these limits and their meaning, we propose in the scheme that follows that the projection norm is a suitable quantity for measuring the degree of hybridisation of atomic orbitals and localisation of molecular orbitals, when the projectors have been appropriately defined.

In passing, we mention an approximation theorem closely related to the projection theorem. If $|f\rangle$ is any vector in the Hilbert space, and $|g\rangle$ a vector confined to a particular subspace, then the distance $D(f, g)$ is least when

$$|g\rangle = P|f\rangle \tag{57}$$

That is, the best least squares approximation to a vector $|f\rangle$ in a particular subspace is given by $P|f\rangle$. If the operator P is expanded in terms of the basis vectors of the subspace, then $P|f\rangle$ becomes the best approximation to $|f\rangle$ as a linear combination of the basis vectors.

Let us then consider the definition of a projector by the basis vectors of its subspace. Consider first an orthonormal set of basis vectors $|1\rangle$, $|2\rangle$, ..., which may, for example, be the set of occupied natural molecular spin orbitals. If there are m members of the set, then

$$P = \sum_{i=1}^{m} |i\rangle \langle i| = \sum_i P_i \tag{58}$$

where $P_i = |i\rangle \langle i|$ is itself a projector on the subspace defined by a single orbital. Alternatively, P may be written in matrix notation if we gather the individual kets $|i\rangle$ into a row matrix designated $|\Phi\rangle$ as before:

$$P = |\Phi\rangle \langle \Phi| \tag{59}$$

Consider, on the other hand, a non-orthogonal basis set, for example, atomic orbitals $|\mu\rangle$, $|\nu\rangle$, ... centred on different atoms in a molecule. Such orbitals have non-zero scalar products (overlap integrals)·

$$S_{\mu\nu} = \langle \mu | \nu \rangle \tag{60}$$

If $\mathbf{S}$ is the overlap matrix whose elements are the $S_{\mu\nu}$, and $|\chi\rangle$ is a row vector of the kets $|\mu\rangle$, $|\nu\rangle$, ..., then as Löwdin (1964, 1965) has shown,

$$P = |\chi\rangle S^{-1} \langle \chi|$$

$$= \sum_{\mu=1}^{m} \sum_{\nu=1}^{m} |\mu\rangle (S^{-1})_{\mu\nu} \langle \nu| \tag{61}$$

Within the framework of projection, we begin to see how the molecular Hilbert space may be partitioned into quantities associated with atoms and bonds. The argument may proceed at different levels. Thus, if A is an atom in a molecule and N_A is the number of electrons associated with the neutral atom, then the various states of the neutral atom define a subspace of the N_A-electron molecular Hilbert space H_{N_A}. The projector on these states may be formed, and used to analyse the reduced density operator D_{N_A}. However, we have argued in Section 3 that the one-electron molecular space H_1 is fundamental, and will therefore confine our attention to this case.

Then we may assign to each atom A, B, C, ... its orthonormal set of natural atomic (spin) orbitals $|\chi_A\rangle$, $|\chi_B\rangle$, $|\chi_C\rangle$ ..., these orbitals being the eigenvectors of the isolated neutral atom one-density operators, and therefore being sufficient to define the states of the atoms. When the atom becomes part of a molecule, the related set of atomic orbitals defines a subspace of the one-electron molecular Hilbert space H_1, and such subspaces have projectors P_A, P_B, P_C, etc. Here:

$$P_A = |\chi_A\rangle \langle \chi_A|$$

$$= \sum_{\mu \in A} |\mu_A\rangle \langle \mu_A|$$

$$= \sum_{\mu \in A} P_\mu, \quad P_\mu = |\mu_A\rangle \langle \mu_A| \tag{62}$$

When two subspaces are combined in such a way that each vector of the two is counted only once, the resulting larger set of vectors is called the union of the subspaces, and is itself a subspace (provided the larger set is linearly independent). The projector on the union is the sum of the projectors for the individual subspaces only if the subspaces are orthogonal. Otherwise the overlap matrix has to be found and its inverse taken to give the projector for the union in the general form equation 61.

This procedure may be used to generate projectors for pairs of atoms, atoms taken in threes, and so on. Consider, for example, atoms A and B, whose orbitals $|\chi_A\rangle$ and $|\chi_B\rangle$ are combined in the total set $|\chi_{AB}\rangle$. If A and B are sufficiently far apart that there is no overlap of their atomic orbitals, then the projector P_{AB} is given by

$$P_{AB} = |\chi_{AB}\rangle\langle\chi_{AB}|$$
$$= P_A + P_B, \quad P_A P_B = 0 \tag{63}$$

If there is non-negligible overlap of the atomic orbitals, then

$$P_{AB} = |\chi_{AB}\rangle S_{AB}^{-1}\langle\chi_{AB}|, \quad S_{AB} = \langle\chi_{AB}|\chi_{AB}\rangle \tag{64}$$

Similarly one may find projectors P_{ABC} over atoms in threes, P_{ABCD} over atoms in fours and so on, and finally the projector P on the subspace which is the union of all the individual atom subspaces. Let the set $|\chi\rangle$ be the combination of the sets $|\chi_A\rangle$, $|\chi_B\rangle$,... over all atoms, and let the total overlap matrix S be given,

$$S = \langle\chi|\chi\rangle \tag{65}$$

then, as before,

$$P = |\chi\rangle S^{-1}\langle\chi|$$
$$\neq \sum_A P_A \quad \text{in general} \tag{66}$$

Any problems that arise in the combination process due to linear dependencies amongst the combined sets may be readily overcome by the method of Berrondo and Löwdin (1969).

Our brief analysis of projection leads, then, to the following statement: An atom in a molecule may be given a mathematical definition as a subspace of an appropriate molecular Hilbert space pertaining to a subsystem of the electrons, and may be represented by a projector. The projector in turn is related to the states of the isolated atom and retains its identity whatever the molecule in question. If we concentrate our attention on the one-electron molecular Hilbert space, then the union of atomic subspaces produces subspaces for atoms in pairs, threes, etc., which will become important when considering the chemical bond.

We now proceed to use the projection techniques devised in this section to analyse the molecular one-density operator.

5. ANALYSIS OF THE MOLECULAR ELECTRON DENSITY BY PROJECTION TECHNIQUES

A combination of the density operator description of molecular structure and the idea of atomic subspaces of the molecular Hilbert space leads directly to a physically meaningful and mathematically fundamental definition of atomic charges and overlap regions. Again we confine our attention to the Hilbert space for one electron, although much of the discussion is capable of more general application.

Gleason (1957) derived a most important theorem which has an immediate consequence here. According to Gleason's theorem (see also Mackey, 1963; Jauch, 1968), the probability of occupancy of a given subspace is $\mathrm{Tr}DP$, where D is a density operator and P the projector on the subspace. In the one-electron Hilbert space, $N\mathrm{Tr}\rho P$ may thus be regarded as the occupation number of the subspace represented by P.

Alternatively, we may argue from the theory of 'superoperators' (Fano, 1957; Crawford, 1958; Schatten, 1960; Banwell and Primas, 1963), thereby obtaining some new insights into Gleason's theorem. The operators on an M-dimensional linear vector space themselves form a linear vector space of dimension M^2, in which the scalar product of two operators A and B is defined as the trace, $\mathrm{Tr}A^+B$. A superoperator in the operator space is a quantity that acts on an operator to give another operator. One such is the 'projection superoperator' p that gives the representation of any other operator on a subspace defined by a basis set of orthonormal operators E_i:

$$pA = \sum_i (\mathrm{Tr}E_i^+ A)E_i, \quad \mathrm{Tr}E_i^+ E_j = \delta_{ij} \tag{67}$$

If a subspace of the ordinary Hilbert space has an orthonormal basis $|f_1\rangle, |f_2\rangle, \ldots$, then the corresponding operator subspace takes a basis consisting of all operators of the form $|f_i\rangle \langle f_j|$. Then

$$pA = \sum_i \sum_j \mathrm{Tr}\,(\,|f_j\rangle \langle f_i|\,A)\,|f_i\rangle \langle f_j|$$

$$= \sum_i \sum_j |f_i\rangle \langle f_i|\,A\,|f_j\rangle \langle f_j|$$

$$= PAP, \quad P = \sum_i |f_i\rangle \langle f_i| \tag{68}$$

Operators such as PAP have been called 'outer projections' by Löwdin (1965).

In our case $P\rho P$ will be the operator representation of the density operator ρ confined to a particular subspace, and in this sense $P\rho P$ becomes the density operator for the subspace. The arguments of Section 3 lead us to define the occupation number n of the subspace by taking the trace:

$$n = N\mathrm{Tr}\,P\rho P$$

$$= N\mathrm{Tr}\,\rho P \tag{69}$$

where we have used the cyclic property of the trace, equation 28. The result is in complete agreement with Gleason's theorem.

Now the projector P above may refer to the subspace defined by a single atomic orbital on an atom in a molecule, or to any of the other atomic or bond subspaces defined in Section 4. Consequently, the occupation number of a particular atomic orbital $|\mu\rangle$ is given by

$$n_\mu = NTr\,\rho P_\mu, \quad P_\mu = |\mu\rangle\langle\mu| \tag{70}$$

and, as we show elsewhere, this has the limits:

$$0 \le n_\mu \le 1 \tag{71}$$

Furthermore, the operator $P_A \rho P_A$ may be regarded as a 'density operator' defining a particular atom A in a molecule. Then the number of electrons n_A that may be associated with that atom becomes

$$\begin{aligned} n_A &= NTr\,\rho P_A \\ &= N\sum_{\mu\varepsilon A} Tr\,\rho P_\mu \\ &= \sum_{\mu\varepsilon A} n_\mu \end{aligned} \tag{72}$$

It follows that the charge Q_A on an atom A in a molecule is given by:

$$Q_A = N_A - NTr\,\rho P_A \tag{73}$$

where N_A is the number of electrons on the isolated atom.

By a similar procedure, the hierarchy of projectors for pairs of atoms, atoms in threes, and so on, provides a ready description of the chemical bonding in a particular molecule. For instance,

$$n_{AB} = NTr\,\rho P_{AB} \tag{74}$$

becomes the occupation of the subspace which is the union of the subspaces associated with atoms A and B. From the properties of the union of projectors, we obtain the inequalities:

$$n_{AB} \ge n_A \quad \text{and} \quad n_{AB} \ge n_B \tag{75}$$

and

$$n_{AB} \le n_A + n_B \le 2n_{AB} \tag{76}$$

We find that $n_{AB} = n_A + n_B$ if and only if the subspaces do not overlap, for then P_{AB} is simply the sum of the projectors P_A and P_B as in equation 63.

The sum $(n_A + n_B)$ includes electron density that belongs solely to the individual atoms and electron density shared between them, the latter being counted twice. Since n_{AB} on the other hand counts shared density only once, the amount of sharing is measured by the difference

$$n_A + n_B - n_{AB} = NTr\,\rho(P_A + P_B - P_{AB}) \tag{77}$$

From the above limits, equation 76, this quantity is zero if there is no sharing, and reaches a maximum of n_{AB} when there is complete sharing of all electron density in the union. It is thus related to the degree of covalent bonding.

Similarly, one may find the occupancy of the union of the subspaces associated with three, four, etc., atoms by using the appropriate projector. Again, the amount of shared electron density is derived by considering how the particular subspace depends on the subspaces of lower order. For example, the electron density genuinely shared by three atoms, A, B, C at once is:

$$n_{ABC} - (n_{AB} + n_{BC} + n_{AC}) + (n_A + n_B + n_C)$$

where

$$n_{ABC} = N\mathrm{Tr}\,\rho P_{ABC} \tag{78}$$

A central feature of the process we are developing here is that shared electron density is automatically allotted to each of the atoms that participate in the sharing. Physically this is more realistic than other methods in which some arbitrary division of the shared density is made. One consequence is that the condition of the conservation of charge usually assumed, that is

$$\sum_A n_A = N \tag{79}$$

is no longer obeyed. It is replaced by a condition on the subspace of the total set of atomic orbitals, namely that

$$N\mathrm{Tr}\,\rho P \le N \tag{80}$$

Equality obtains if and only if the total set of atomic orbitals is completely sufficient to describe the molecular structure. Then each molecular natural orbital may be expanded exactly in terms of the atomic orbitals, and

$$P\rho = \rho P = \rho \tag{81}$$

Thus the difference $(N - N\mathrm{Tr}\,\rho P)$ measures the inadequacy of the atomic orbital basis for describing the structure of the molecule.

The approach we have derived is essentially a generalisation of a method due to Davidson (1967). Our derivation is, however, quite different, and the use of projection operators at once allows the generalisation, simplifies the formulation, and brings out the fundamental nature of the results.

The method also overcomes all of the problems that have become apparent in the most common previous approach, the population analysis of Mulliken (1955): its lack of generality, inability to cope with extended atomic orbital basis sets, arbitrary division of overlap density among the participating atoms, and lack of limits on the atomic orbital occupations. By contrast, our method is completely general and may be used in conjunction with calculations at any level of accuracy. Because the projectors are defined by states of the isolated atoms, it would give similar results for different calculations on the same molecule, provided only that the calculations themselves gave similar densities. As mentioned, it realistically allots genuinely shared electron density to all the participating atoms, and has definite limits, equation 71, on atomic orbital occupation numbers.

6. CLASSIFICATION OF NATURAL MOLECULAR ORBITALS. LOCALISATION AND HYBRIDISATION

In Section 5, we carried out an analysis of the complete one-density operator by projection techniques. Using the same techniques, we now look more closely at the structure of this operator and particularly at the properties of the natural (spin) orbitals of which it is composed.

We are not able to answer the intriguing question of whether natural molecular orbitals in general are intrinsically localised on atoms and in bonds. However, the localisation properties of any particular set of natural orbitals may be determined by means of the concept of the projection norm (Section 4).

Suppose that we have a particular natural orbital $|i\rangle$, the orbital density operator ρ_i (Section 3), and a subspace having the projector P. Then if $|i\rangle$ is localised in the subspace,

$$P|i\rangle = |i\rangle, \quad \|Pi\| = \|i\| = 1 \tag{82}$$

and therefore the operator representation of ρ_i in the subspace satisfies

$$P\rho_i P = \rho_i \tag{83}$$

That is, the density operator ρ_i is also localised in the subspace. More generally, the degree of localisation of a normalised natural orbital and its associated density is measured by $\|Pi\|$, approaching unity for complete localisation.

The projector P in the above may be any of the projectors for atomic subspaces or unions of atomic subspaces as defined in Section 4. For example, if $\|P_A i\|$ approaches unity, then $|i\rangle$ may be regarded as localised on atom A. If $\|P_A i\|$ and $\|P_B i\|$ take comparable values and $\|P_{AB} i\|$ approaches unity, then $|i\rangle$ may be regarded as a bonding orbital for atoms A and B. With similar conditions, three-, four-, ...centre orbitals are defined. This classification of molecular orbitals can, of course, be generated automatically by computer.

There are two theorems which enable this type of relationship between the atomic orbital set and the natural molecular orbitals to be extended. The first theorem is due to Löwdin (1965) and concerns the outer projection PAP of any self-adjoint operator A bounded from below (that is, $\langle f|A|f\rangle$ for arbitrary $|f\rangle$ is always greater than some constant). Löwdin found that the extreme values of the integral $\langle g|A|g\rangle$, where the $|g\rangle$ are constrained to belong to the subspace defined by the projector P, are given by the eigenvalues of PAP. This 'outer projection theorem' immediately provides various general criteria for the localisation of molecular orbitals and the hybridisation of atomic orbitals.

Equally important in showing the correspondence between localisation and hybridisation is the 'pairing theorem', found independently by Amos and Hall (1961) and Löwdin (1962, 1964) in connection with the method of different orbitals for different spins. We state here a general version:

'If B is some operator which is not self-adjoint, and B^+ is its adjoint, then the self-adjoint operators BB^+ and B^+B have non-zero eigenvalues in common. Furthermore the eigenvectors correspond one-to-one, such that if $|f_i\rangle$ is an eigenvector of BB^+ having eigenvalue α_i, and $|g_i\rangle$ is the eigenvector of B^+B having the same eigenvalue, then

$$B^+|f_i\rangle = \alpha_i^{\frac{1}{2}}|g_i\rangle \tag{84}$$

and

$$B\,|\,g_i\rangle = \alpha_i^{\frac{1}{2}}\,|\,f_i\rangle \tag{85}$$

$\alpha_i^{\frac{1}{2}}$ being the positive square root.'

We take as an example the outer projection $P_A\rho P_A$, which we have argued (Section 5) is a density operator for an atom A in the molecule. It has the expansion:

$$P_A\rho P_A = |\,\chi_A\rangle\langle\chi_A\,|\,\rho\,|\,\chi_A\rangle\langle\chi_A\,|$$

$$= \sum_{\mu\in A}\sum_{\nu\in A}|\,\mu'_A\rangle\langle\mu_A\,|\,\rho\,|\,\nu_A\rangle\langle\nu_A\,| \tag{86}$$

Now the outer projection theorem means that it is possible to find a new set of atomic orbitals $|\,\chi'_A\rangle$ related to $|\,\chi_A\rangle$ by a unitary transformation U_A,

$$|\,\chi'_A\rangle = |\,\chi_A\rangle U_A \tag{87}$$

where

$$U_A^+ U_A = U_A U_A^+ = 1 \tag{88}$$

and such that the $|\,\chi'_A\rangle$ are eigenvectors of $P_A\rho P_A$,

$$P_A\rho P_A\,|\,\mu'_A\rangle = \alpha_\mu\,|\,\mu'_A\rangle \tag{89}$$

The eigenvalues α_μ are extreme values of the integral

$$\langle\mu'_A\,|\,\rho\,|\,\mu'_A\rangle = \mathrm{Tr}\,\rho P_{\mu'},\quad P_{\mu'} = |\,\mu'_A\rangle\langle\mu'_A\,| \tag{90}$$

But this is simply the probability of occupancy of the orbital $|\,\mu'_A\rangle$, and $N\alpha_\mu$ is its occupation number. Therefore the set $|\,\chi'_A\rangle$ is distinguished from any other orthonormal set in the subspace of A by having extreme values of the atomic orbital occupation numbers. Those members of the set with maximum occupations are then the orbitals on A most closely related to the molecular density. They are thus a form of hybrid atomic orbital determined according to a 'criterion of maximum occupation'.

Multiplying equation 89 on the left by a different member of the set, we find also that

$$\langle\nu'_A\,|\,P_A\rho P_A\,|\,\mu'_A\rangle = \langle\nu'_A\,|\,\rho\,|\,\mu'_A\rangle,\quad \text{since } P_A\,|\,\mu'_A\rangle = |\,\mu'_A\rangle, P_A\,|\,\nu'_A\rangle = |\,\nu'_A\rangle$$

$$= \alpha_\mu\langle\nu'_A\,|\,\mu'_A\rangle$$

$$= 0,\quad |\,\mu'_A\rangle \neq |\,\nu'_A\rangle \tag{91}$$

That is, the set $|\,\chi'_A\rangle$ has the additional property that it casts the operator $P_A\rho P_A$ into the following diagonal form:

$$P_A\rho P_A = \sum_{\mu'\in A}|\,\mu'_A\rangle\,(\mathrm{Tr}\,\rho P_{\mu'})\,\langle\mu'_A\,| \tag{92}$$

The obvious resemblance in form and interpretation to the natural orbital expansion of ρ itself allows us to regard the set $|\,\chi'_A\rangle$ as the 'natural atomic orbitals for atom A in the molecule'.

Any of the projectors we have considered before may be substituted for P_A in the above. Diagonalisation of the outer projection $P_{AB}\rho P_{AB}$ is especially interesting, for the result is an orthonormal set of orbitals which are linear combinations of the atomic orbitals on A and B. This time, among those having maximum occupation, will be optimum 'two-centre natural bonding orbitals' relative to the molecular density.

In every case, the resemblance of the maximum occupation orbitals to particular natural molecular orbitals may be measured by the appropriate projection norm (for example, $\| P_\mu i \|$ approaches unity if the hybrid atomic orbital $| \mu'_A \rangle$ resembles the natural orbital $| i \rangle$). There is also the possibility of finding corresponding molecular orbitals by means of the pairing theorem. However, the adjoint operators involved in obtaining the molecular orbitals corresponding to the set $| \chi'_A \rangle$, say, are $P_A \rho^{\frac{1}{2}}$ and $\rho^{\frac{1}{2}} P_A$, where

$$\rho^{\frac{1}{2}} = \sum_i | i \rangle \lambda_i^{\frac{1}{2}} \langle i | \tag{93}$$

The self-adjoint product operators are then the outer projection $P_A \rho P_A$ as required, and the 'inner projection' (Löwdin, 1965) $\rho^{\frac{1}{2}} P_A \rho^{\frac{1}{2}}$. It is interesting in the light of Löwdin's results that by the pairing theorem the inner and outer projections have the same eigenvalues. But no useful criterion of localisation on the molecular orbital eigenvectors of $\rho^{\frac{1}{2}} P_A \rho^{\frac{1}{2}}$ appears, and it seems more appropriate to look instead in other directions.

In fact a criterion for localisation as well as a different criterion for hybridisation emerges when the outer projection and pairing theorems are applied to projectors alone. Let P and Q be projectors that define distinct and non-orthogonal subspaces, that is, such that

$$PQ \neq 0, \quad PQ \neq Q, \quad PQ \neq P \tag{94}$$

Then the outer projection theorem applies separately to the operators PQP and QPQ. The results are connected by the pairing theorem for the operator PQ and its adjoint QP, so that the self-adjoint product operators are,

$$PQ \cdot QP = PQ^2 P = PQP \tag{95}$$

and

$$QP \cdot PQ = QP^2 Q = QPQ \tag{96}$$

By the outer projection theorem, the eigenvectors of PQP are the orthonormal vectors $| f_i \rangle$ of the subspace of P whose eigenvalues are extreme values of the integral $\langle f_i | Q | f_i \rangle$. But this integral is simply the square of the projection norm, and hence the eigenvectors of PQP have extreme projection norms $\| Q f_i \|$ with respect to Q. Similarly the eigenvectors of QPQ are the orthonormal vectors $| g_i \rangle$ of the subspace of Q having extreme projection norms $\| P g_i \|$ with respect to P.

By the pairing theorem, these eigenvectors are in an interesting one-to-one correspondence. If $| f_i \rangle$ and $| g_i \rangle$ are corresponding eigenvectors of PQP and QPQ respectively, then it follows that

$$\| Q f_i \| = \| P g_i \| \tag{97}$$

and

$$QP | f_i \rangle = Q | f_i \rangle = (\| Q f_i \|) | g_i \rangle \tag{98}$$

while

$$PQ|g_i\rangle = P|g_i\rangle = (\||Qf_i\||)|f_i\rangle \tag{99}$$

Multiplication of equation 99 on the left by $\langle f_i|$ gives

$$\||Qf_i\|| = |\langle f_i|PQ|g_i\rangle| = |\langle f_i|g_i\rangle| \tag{100}$$

Therefore our derivation shows that the corresponding eigenvectors having the maximum eigenvalue, or maximum projection norm, also have the property of maximum overlap (Mulliken, 1950; Lykos and Schmeising, 1961), and our conclusions are related in this way to the results of King *et al.* (1967).

We see that the criterion of maximum overlap is one facet of a more general 'criterion of maximum (or minimum) projection', depending on the outer projection theorem. This in turn is equivalent to statements about the distance $D(f_i, g_i)$ between the vectors of the two subspaces, for one may show that $D(f_i, g_i)$ is a minimum for those corresponding vectors having maximum projection, and conversely is a maximum for those corresponding vectors having minimum projection.

A connection between hybridisation and localisation results at once from this criterion. Let Q in the above be the projector on the subspace of occupied natural molecular orbitals:

$$Q = |\Phi\rangle\langle\Phi|$$
$$= \sum_i |i\rangle\langle i| \tag{101}$$

and take for P the projector P_A, say. Then new sets $|\Phi''\rangle$ and $|\chi_A''\rangle$ may be found that are related to the sets $|\Phi\rangle$ and $|\chi_A\rangle$ by unitary transformations, that bring the operators QP_AQ and P_AQP_A, respectively, to diagonal form, and that correspond in the sense of the pairing theorem. Among the $|\Phi''\rangle$ will be molecular orbitals having maximum projection on the subspace of P_A, and among the $|\chi_A''\rangle$ will be atomic orbitals having maximum projection on the natural orbital subspace. These will be optimum molecular orbitals localised on A and optimum hybrid atomic orbitals on A according to the criterion of maximum projection.

Again one may substitute for P_A other atomic projectors P_B, P_C, etc., and bond projectors P_{AB}, P_{ABC}, etc. In the latter case, optimum bonding combinations of atomic orbitals and optimum localised bonding molecular orbitals are obtained. From all the localised molecular orbitals found in this way, it should be possible to choose a non-orthogonal set that spans the same subspace as the occupied natural orbitals, and so exhibits the localisation properties of this subspace in an optimum way.

We note in passing that the same general scheme provides hybridisation criteria when the molecular density is not known. In this way it comes closer to the original formulation of hybridisation and directed valence by Slater (1931) and Pauling (1931), and to the many other studies of maximum overlap hybrid orbitals in the literature (see the review by Randić and Maksić, 1972). Given the molecular geometry and the atomic projectors P_A and P_B, say, one may find from the operators $P_AP_BP_A$ and $P_BP_AP_B$ that hybrid orbital (or orbitals) on A having maximum projection on the subspace of B, and the corresponding hybrid orbital (or orbitals) on B having maximum projection on A. These orbitals have the equivalent properties of maximum overlap and minimum distance apart and so will be optimum directed hybrids between atoms A and B.

Alternatively, one may take the projector P_A and the projector $P_{BCD}\ldots$ on the subspace that is the union of the subspaces associated with all atoms other than A. Now among the set of orbitals on A that bring the operator $P_A P_{BCD}\ldots P_A$ to diagonal form will be orbitals having minimum projection on the remainder subspace, and therefore being a maximum distance from it. These may be interpreted as optimum inner shell and lone pair atomic orbitals on A relative to the remainder subspace and the molecular geometry.

There is a need to explore in much greater detail the relation between hybrid atomic orbitals determined with and without a knowledge of the molecular density, and the relation in practice of the criterion of maximum occupation to the criterion of maximum projection. Nevertheless it is clear that when the outer projection and pairing theorems are added to the projection operator–density operator framework developed in this paper, a unified and general treatment of the concepts of hybridisation and localisation is possible. These concepts can play an important role in the interpretation of rigorous wave functions.

7. CONCLUSIONS

We have seen that certain theorems and properties of abstract Hilbert space are very relevant to the quantum theory of valence concepts. The notions of subsystems and subspaces in particular have been central to our argument. Other notions, such as norm and distance, superoperators and outer projections, together with the projection, outer projection and pairing theorems and Gleason's theorem, have helped to provide the essential links between abstract mathematics and particular valence concepts.

The main new idea of this paper is contained in the proposition that atoms in molecules and chemical bonds receive a mathematical definition as subspaces of an appropriate molecular Hilbert space. In the one-electron molecular Hilbert space, outer projection 'density operators' for atoms in molecules and chemical bonds result from the description of molecular structure by the one-density operator. Then the definitions of atomic orbital, atom and bond occupation numbers, and of atomic charge and shared density, follow readily. The use of the projection norm as a measure of localisation or hybridisation and the criteria of maximum occupation and maximum or minimum projection are natural outgrowths of the approach.

Given a molecular wave function of any degree of complexity, the relevant one-density operator may be found, and given prior solutions for the isolated atoms, the required atomic and bond projectors may be generated. We have concentrated on exact quantum theory, but the projection-density approach is also applicable to approximate wave functions. Indeed, there are some interesting consequences for approximate LCAO MO theory which will be reported elsewhere.

While the 'conceptual dilemma' is still far from resolved, it appears that the projection-density approach is a useful addition to existing schemes for the interpretation of molecular wave functions.

6.8 SUMMARY

The meaning that simple chemical valence concepts have within rigorous quantum theory is, in some respects, obscure. An investigation of this situation is made beginning with basic principles of the physical interpretation of quantum theory. Some relevant aspects of the mathematics of abstract Hilbert space are reviewed; in particular it is shown how molecular density operators are central to the description of electron subsystems, and projection operators are central to the partitioning of any Hilbert space into subspaces.

By means of projection operators, an atom in a molecule may be given a mathematical representation, and projection operators that describe chemical bonds are also possible. When such projection techniques are applied to the molecular density operator for one electron, definitions of the charge on an atom in a molecule and of shared electron density result. When the orbital basis of the projection operator–density operator method is considered, a unified approach to the formation of hybrid atomic orbitals and localised molecular orbitals follows. The criteria for hybridisation or localisation include those of 'maximum occupation' and 'maximum or minimum projection'.

REFERENCES

Amos, A. T. and Hall, G. G. (1961) 'Single Determinant Wave Functions', *Proc. R. Soc.*, **A263**, 483–493

Ando, T. (1963) 'Properties of Fermion Density Matrices', *Rev. mod. Phys.*, **35**, 690–702

Bader, R. F. W. (1964) 'Binding Regions in Polyatomic Molecules and Electron Density Distributions', *J. Am. chem. Soc.*, **86**, 5070–5075

Bader, R. F. W., Henneker, W. H. and Cade, P. E. (1967) 'Molecular Charge Distributions and Chemical Binding', *J. chem. Phys.*, **46**, 3341–3363

Banwell, C. N. and Primas, H. (1963) 'On the Analysis of High-resolution Nuclear Magnetic Resonance Spectra. I' (see Appendix 1), *Molec. Phys.*, **6**, 225–256

Berlin, T. (1951) 'Binding Regions in Diatomic Molecules', *J. chem. Phys.*, **19**, 208–213

Berrondo, M. and Löwdin, P. O. (1969) 'The Projection Operator for a Space Spanned by a Linearly Dependent Set', *Int. J. quantum Chem.*, **3**, 767–780

Bingel, W. A. and Kutzelnigg, W. (1970) 'Symmetry Properties of Reduced Density Matrices and Natural p-states', *Adv. quantum Chem.*, **5**, 201–218

Coleman, A. J. (1963) 'Structure of Fermion Density Matrices', *Rev. mod. Phys.*, **35**, 668–687

Coulson, C. A. (1970) 'Recent Developments in Valence Theory', *Pure appl. Chem.*, **24**, 257–287

Crawford, J. A. (1958) 'An Alternative Method of Quantization: the Existence of Classical Fields', *Nuovo Cim.*, **10**, 698–713

Daudel, R. (1953) 'Sur la Localisabilité des Corpuscules dans les Noyaux et les Cortèges électroniques des Atomes et des Molécules', *C. r. hebd. Séanc. Acad. Sci., Paris*, **237**, 601–603

Daudel, R., Gallais, F. and Smet, P. (1967) 'On the General Theory of Molecular Additivity Rules: the Particular Case of the Faraday Effect', *Int. J. quantum Chem.*, **1**, 873–891

Davidson, E. R. (1967) 'Electronic Population Analysis of Molecular Wave Functions', *J. chem. Phys.*, **46**, 3320–3324

Davidson, E. R. (1972) 'Properties and Uses of Natural Orbitals', *Rev. mod. Phys.*, **44**, 451–464

de Broglie, L. (1923a) 'Ondes et Quanta', *C. r. hebd. Séanc. Acad. Sci., Paris*, **177**, 507–510

de Broglie, L. (1923b) 'Waves and Quanta', *Nature*, **112**, 540

Dirac, P. A. M. (1958) *The Principles of Quantum Mechanics*, 4th Edn., Oxford University Press

Fano, U. (1957) 'Description of States in Quantum Mechanics by Density Matrix and Operator Techniques', *Rev. mod. Phys.*, **29**, 74–93

Feinberg, M. J., Ruedenberg, K. and Mehler, E. L. (1970) 'The Origin of Binding and Anti-binding in the Hydrogen Molecule-ion', *Adv. quantum Chem.*, **5**, 27–98

Gleason, A. M. (1957) 'Measures on the Closed Subspaces of a Hilbert Space', *J. Math. Mech.*, **6**, 885–893

Hohenberg, P. and Kohn, W. (1964) 'Inhomogeneous Electron Gas', *Phys. Rev.*, **B136**, 864–871

Husimi, K. (1940) 'Some Formal Properties of the Density Matrix', *Proc. phys.-math. Soc., Japan*, **22**, 264–314

Jauch, J. M. (1968) *Foundations of Quantum Mechanics*, Addison-Wesley, Reading, Mass.

King, H. F., Stanton, R. E., Kim, H., Wyatt, R. E. and Parr, R. G. (1967) 'Corresponding Orbitals and the Nonorthogonality Problem in Molecular Quantum Mechanics', *J. chem. Phys.*, **47**, 1936–1941

Lewis, G. N. (1916) 'The Atom and the Molecule', *J. Am. chem. Soc.*, **38**, 762–785

Löwdin, P. O. (1955) 'Quantum Theory of Many-particle Systems. I. Physical Interpretations by Means of Density Matrices, Natural Spin-orbitals, and Convergence Problems in the Method of Configurational Interaction', *Phys. Rev.*, **97**, 1474–1489

Löwdin, P. O. (1959) 'Correlation Problem in Many-electron Quantum Mechanics. I. Review of Different Approaches and Discussion of some Current Ideas', *Adv. chem. Phys.*, **2**, 207–322

Löwdin, P. O. (1962) 'Quantum Theory of Nonmetallic Crystals', *J. appl. Phys.*, **33** (Suppl.), 251–280

Löwdin, P. O. (1964) *Linear Algebra and the Fundaments of Quantum Theory*, Preprint No. 125, Quantum Chemistry Group, Uppsala, Sweden

Löwdin, P. O. (1965) 'Studies in Perturbation Theory. X. Lower Bounds to Energy Eigenvalues in Perturbation-theory Ground State', *Phys. Rev.*, **A139**, 357–372

Lykos, P. G. and Schmeising, H. N. (1961) 'Maximum Overlap Atomic and Molecular Orbitals', *J. chem. Phys.*, **35**, 288–293

Mackey, G. W. (1963) *The Mathematical Foundations of Quantum Mechanics*, Benjamin, New York

McWeeny, R. (1960) 'Some Recent Advances in Density Matrix Theory', *Rev. mod. Phys.*, **32**, 335–369

Messiah, A. (1968) *Quantum Mechanics*, Volume I, North-Holland, Amsterdam

Mulliken, R. S. (1950) 'Overlap Integrals and Chemical Binding', *J. Am. chem. Soc.*, **72**, 4493–4503

Mulliken, R. S. (1955) 'Electronic Population Analysis on LCAO-MO Molecular Wave Functions. I', *J. chem. Phys.*, **23**, 1833–1840

Pauling, L. (1931) 'The Nature of the Chemical Bond. Application of Results Obtained from the Quantum Mechanics and from a Theory of Paramagnetic Susceptibility to the Structure of Molecules', *J. Am. chem. Soc.*, **53**, 1367–1400

Pauling, L. (1960) *The Nature of the Chemical Bond*, Cornell University Press, New York

Primas, H. (1965) 'Separability in Many-electron Systems', *Modern Quantum Chemistry*, Part II (edited by O. Sinanoğlu), 45–74, Academic Press, New York

Primas, H. (1967) 'A Density Functional Representation of Quantum Chemistry. I. Motivation and General Formalism', *Int. J. quantum Chem.*, **1**, 493–519

Randić, M. and Maksić, Z. B. (1972) 'Hybridization by the Maximum Overlap Method', *Chem. Rev.*, **72**, 43–53

Ransil, B. J. and Sinai, J. J. (1967) 'Toward a Charge-density Analysis of the Chemical Bond; the Charge-density Bond Model', *J. chem. Phys.*, **46**, 4050–4074

Roux, M., Besnainou, S. and Daudel, R. (1956) 'Recherches sur la Répartition de la Densité Électronique dans les Molecules. I. Effet de la Liaison Chimique', *J. Chim. phys.*, **54**, 218–221

Roux, M., Cornille, M. and Burnelle, L. (1962) 'Study of the Electron Density Distribution in the Simpler Hydrocarbons', *J. chem. Phys.*, **37**, 933–936

Ruedenberg, K. (1962) 'The Physical Nature of the Chemical Bond', *Rev. mod. Phys.*, **34**, 326–376

60 Wave Mechanics

Schatten, R. (1960) *Norm Ideals of Completely Continuous Operators,* Springer-Verlag, Berlin

Sinanoğlu, O. (1964) 'Many-electron Theory of Atoms, Molecules and their Interactions', *Adv. chem. Phys.,* **6,** 315–412

Slater, J. C. (1931) 'Directed Valence in Polyatomic Molecules,' *Phys. Rev.,* **37,** 481–489

Smith, D. W. (1968) 'Applications of Density Matrix Methods to the Electronic Structure of Atoms and Molecules', *Reduced Density Matrices with Applications to Physical and Chemical Systems* (edited by Coleman, A. J. and Erdahl, R. M.) 169-254, Queen's University, Kingston, Ontario

ter Haar, D. (1961) 'Theory and Applications of the Density Matrix', *Rep. Prog. Phys.,* **24,** 304–362

von Neumann, J. (1955) *Mathematical Foundations of Quantum Mechanics,* English Edn, Princeton University Press

Wahl, A. C. (1966) 'Molecular Orbital Densities: Pictorial Studies', *Science,* **151,** 961–967

5

The Localisation of Electrons and the Concept of the Chemical Bond

RAYMOND DAUDEL

President of the International Academy and Quantum Molecular Science, Paris

INTRODUCTION

When taking account of the Perrin–Rutherford–Bohr planetary model it became customary to assume that the electrons of an atom follow various orbits and to speak of K, L, M, N ... electrons. The extension of such a model to molecules led one to associate electrons and bonds and to distinguish between core electrons and bond electrons.

Such an approach is not consistent with the wave-mechanical viewpoint. In the framework of that theory it is not possible to describe the trajectory of an electron. Furthermore, as a consequence of the indistinguishability principle, the average value of any property associated with an electron in an atom or a molecule is the same. Each electron plays the same role in such a system. Therefore, it is forbidden to distinguish between K and L electrons, valence and core electrons. The concept of bond is no longer well defined as in a molecule each electron participates in the interaction between any pair of nuclei.

The concept of orbital provided the first approach to establish a bridge between wave mechanics and chemical intuition as it is possible to consider K orbitals, L orbitals, bond orbitals. But it is important to point out that orbitals are only mathematical functions which may be used to express approximate wave functions. They have no direct physical meaning. It is not possible to associate an electron with an orbital. Furthermore as the usual *approximate* wave functions (independent electron model) are determinants built on orbitals the wave functions are invariant with respect to any unitary transformation of the orbitals. There are an infinite number of equivalent basis sets of orbitals to represent a given system. This is why, without change in the total wave function, we can very often represent a given molecule in terms of localised *or* delocalised orbitals. It is a question of choice.

As a consequence, there is a need for a concept able to establish a bridge between wave mechanics and chemical intuition, convenient to analyse very elaborate wave functions (as those produced by modern electronic computers) and *possessing* a definite physical meaning. The concept of 'loge' provides for that need.

THE LOGE THEORY

Let us consider a two-electron system. The helium atom could be that system. Let $\Psi(M_1, \omega_1, M_2, \omega_2, t)$ be the exact wave function describing the system (or any kind of approximate wave function if the exact one is unknown).

Consider* any arbitrary partition of the space into two volumes v_A and v_B. It can be written:

$$\exists\{v_i\}, \quad \mathscr{R}^3 = U v_i$$

$$\forall\, i, j,\ \text{mes}\,(v_i \cap v_j) = \tfrac{1}{2}\delta_{ij}\,(\text{mes}\ v_i + \text{mes}\ v_j)$$

Imagine an apparatus which makes it possible to measure the positions of the two electrons at a given time. Three *events* are possible: (1) the two electrons are found in v_A, (2) they are found in v_B, (3) one of them is found in v_A, the other in v_B.

The average value $\bar{\Omega}$ of a property of the system described by the operator Ω can be divided into contributions associated with each event because:

$$\bar{\Omega} = \int_{v_A + v_B} dv_1 \int_{v_A + v_B} dv_2 \Psi * \Omega \Psi = \int_{v_A} dv_1 \int_{v_A} dv_2\, \Psi * \Omega \Psi$$

$$+ \int_{v_B} dv_1 \int_{v_B} dv_2\, \Psi * \Omega \Psi + 2 \int_{v_A} dv_1 \int_{v_B} dv_2\, \Psi * \Omega \Psi \qquad (1)$$

The first term in the last member of equation 1 can be associated with the first event, the second term with the second event, the third term with the third event.

If $\bar{\Omega}_\lambda$ denotes the contribution of the event λ we have:

$$\bar{\Omega} = \sum_\lambda \bar{\Omega}_\lambda \qquad (2)$$

If the operator Ω is simply the unity 1, $\bar{\Omega}_\lambda$ defines the probability of occurrence p_λ of the event λ and equation 2 becomes:

$$1 = \sum_\lambda p_\lambda \qquad (3)$$

The *relative missing information function*[†] associated with that distribution of probability is:

$$I_r = \frac{\sum_\lambda p_\lambda \log_2 p_\lambda^{-1}}{\log_2 v} \qquad (4)$$

if v denotes the number of events.

*We are following a process which is described in detail in: Aslangul *et al.* (in press).
[†]Shannon (1948). It would be perhaps better to introduce explicitly the *a priori* weight of each event due to the indistinguishability of the electrons.

If I_r reaches its smallest value for a given partition of the space into two particular volumes that partition gives us the *maximum amount of information* about the localisability of the electrons.

That partition is called the *best partition of the space into two loges** *for the system.*

For a given partition the quantity:

$$L = 1 - I_r \qquad (5)$$

is called the *localisation* of the electrons. For the best partition it becomes the *localisability* of the electrons.

If we go back to the helium case in its first excited state it is tempting to consider the partition of the space into two volumes produced by a sphere of radius R centred at the nucleus. Figure 1 shows the variation of the

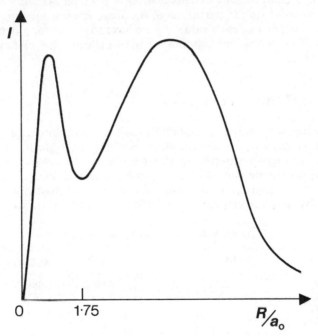

Figure 1   Variation of the missing information I as a function of the radius R for the helium atom in its first excited state

missing information I as a function of the radius R. A minimum is reached for

$$R = 1.75\, a_0$$

It corresponds to the best partition into spherical loges. [The other minima ($R = 0$ and $R \to \infty$) correspond to only one volume (the total space) and must not be considered.]

*The concept of loge is due to Daudel (1953). The information theory has been introduced in the loge theory by Aslangul (1971).

The probabilities of the various events are:

$$p_1 = 0.052, \quad p_2 = 0.028 \quad \text{and} \quad p_3 = 0.920$$

When I_r reaches a minimum value there is very often a *leading event*, that is to say an event λ for which the probability p_λ is high in comparison with the probabilities of the other ones. This is the case of the third event $(p_3 \gg p_1, p_2)$. For the first excited state of the helium atom there is a high probability of finding simultaneously one electron in the spherical loge and one outside that loge (92 per cent). Therefore it is tempting to call K loge that central spherical loge and L loge the remaining part of the space. Then, a bridge is established between the old chemical language and the wave-mechanical one, but without contradiction because *now we distinguish between parts of the space, not between electrons*.

The theory can be easily extended to an n-electron system.* It becomes necessary to compare the partitions of the space into various numbers p of volumes ($1 \leq p \leq n$). For each value of p we have to search for a best partition into loges. That which corresponds to the smallest value of I_r is the best among the best.

ATOMIC AND MOLECULAR LOGES

In practice it is very difficult to obtain the best partition into loges associated with a given system in a certain state, but it is possible to obtain good partitions into loges, i.e. partitions which correspond to minimum values of I_r for an approximate wave function and for a certain kind of partition of the space. For atoms it is possible to obtain good partition into loges separated by spheres centred at the nucleus. Table 1 shows the values

Table 1. Atomic Radii (Atomic Units)

Elements	Be	F$^-$	Ca^{+2}	Rb$^+$	Hg
R_K	1.12	0.37	0.13	0.06	0.025
R_L			0.64	0.26	0.10
R_M				1	0.28
R_N					0.93

obtained for the radius of the corresponding spheres if Hartree–Fock wave functions are used.†

By dividing the volume of a loge by the number of electrons which it contains during the leading event one obtains an idea of the space v which an electron tends to occupy when it 'visits' the loge. If now the mean value p of the electric potential which acts on the electron in the same loge is calculated the following relation is obtained (Odiot and Daudel, 1954).

$$p^{\frac{3}{2}}v = \text{constant} \tag{6}$$

*For more details see Aslangul *et al.* (1972).
†Such radii have been calculated for other kinds of wave function (Sperber, 1971).

for all atoms and loges. There is a kind of Boyle–Mariotte law between the 'size' of an electron and the electric 'pressure' which acts on it. That procedure does not apply for the superficial loges as their volumes are infinite. In that case, starting from an idea of Robb, Haines and Csizmadia (in press) we can calculate the mean value $\overline{(r_J^2)}$ of the square of the distance of an electron in the loge J from the centre of gravity of the loge and define the 'size' of the electron as:

$$v_J = \tfrac{4}{3}(\overline{r_J^2})^{\tfrac{3}{2}} \tag{7}$$

Figure 2 shows a good partition into loges of the space in the lithium molecule Li_2 (ground state) and corresponds to a leading event. Surrounding each nucleus there is a spherical loge very similar to the K loge of the lithium atom. Such a loge can be called a *core loge*. The remaining part of the space will be called the *bond loge*. It is seen that during the leading event there are

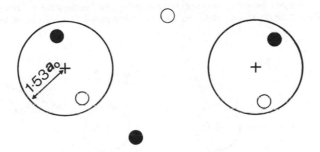

Figure 2   *Illustration of the partition into loges of the space of the Li₂ molecule in its ground state*

two electrons of opposite spins in each loge. The probability of that event is very high: 0.95. *This image derived from wave mechanics is the modern translation of Lewis' ideas.* It shows that *with the classical single bond it is possible to associate a two-electron loge, that is to say a part of the molecular space between two cores in which there is a high probability of finding two electrons (and two only) with opposite spins.*

It is possible to show (Aslangul *et al.*, 1972) that Figure 3 illustrates a typical good partition into loges for a large molecule. It consists of: (1) four *core loges* (one surrounding each nucleus); (2) a two-electron loge adjacent to one core only, i.e. a *lone-pair loge*; (3) a two-electron bond loge adjacent to two cores: it corresponds to a *localised bond*; (4) a four-electron bond extended over the three cores B, C, D: which corresponds to a *delocalised bond*.

The loge theory has been recently used to propose a criterion to distinguish between covalent and dative bonds (Daudel and Veillard, 1970).

66

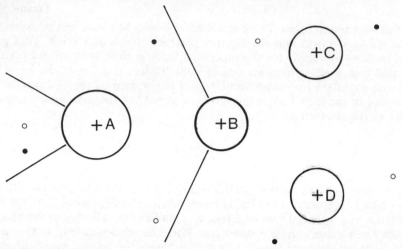

Figure 3 Partition into loges for a tetra-atomic molecule illustrating: (a) four core loges—one surrounding each nucleus; (b) a two-electron loge adjacent to one core only, i.e. a lone-pair loge; (c) a two-electron loge adjacent to two cores corresponding to a localised bond; (d) a four-electron loge extended over the three cores B, C, and D which corresponds to a delocalised bond

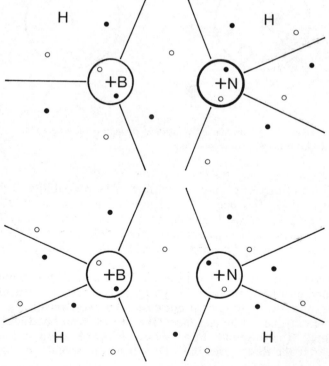

Figure 4 Partition into loges for amino borane (upper) and borazane (lower). The bond loge BN lies between two groups of loges having charges $+e$ and $+e$ for the covalent bond and $+2e$ and 0 in the case of the dative bond

Figure 4 illustrates a good partition into loges for amino borane and borazane and the corresponding leading events. The BN bond is classically considered to be a covalent bond in the first molecule, a dative bond in the second one. It is seen that during the leading event the bond loge BN lies between two groups of loges having respectively total charges of $+e$ and $+e$ for the covalent bond, of $+2e$ and 0 in the case of the dative bond (the total charge is simple the sum of the charges of the nuclei and the charges of the electrons found in the groups of loges during the leading event).

ADDITIVE PROPERTIES OF MOLECULES

It is possible to express any average value associated with a given molecule as a partition between the contributions of loges and contributions of loge pairs. Let us go back to the helium case. The Hamiltonian operator can be divided into three parts: h_1, h_2 (the monoelectronic parts) and $h_{1,2}$ (the bielectronic term).

The average energy $\overline{E}_3$ associated with the third event can be written as:

$$\overline{E}_3 = 2 \int_{v_A} dv_1 \int_{v_B} dv_2 \, \Psi^*(h_1+h_2+h_{1,2})\Psi$$

$$= 2 \int_{v_A} dv_1 \int_{v_B} dv_2 \, \Psi^* h_1 \Psi + 2 \int_{v_A} dv_1 \int_{v_B} dv_2 \, \Psi^* h_2 \Psi$$

$$+ 2 \int_{v_A} dv_1 \int_{v_B} dv_2 \, \Psi^* h_{1,2}\Psi \tag{8}$$

In the last part of equation 8 the first term is the contribution $\overline{E}_3^A$ of the loge v_A, the second the contribution $\overline{E}_3^B$ of the loge v_B and the third term the contribution $\overline{E}_3^{A,B}$ of the loge pair v_A and v_B:

$$\overline{E}_3 = \overline{E}_3^A + \overline{E}_3^B + \overline{E}_3^{A,B} \tag{9}$$

In general:

$$\overline{\Omega}_\lambda = \sum_J \overline{\Omega}_\lambda^J + \sum_{J<J'} \overline{\Omega}_\lambda^{J,J'} \tag{10}$$

Therefore

$$\overline{\Omega} = \sum_\lambda \overline{\Omega}_\lambda = \sum_\lambda \sum_J \overline{\Omega}_\lambda^J + \sum_\lambda \sum_{J<J'} \overline{\Omega}_\lambda^{J,J'} \tag{11}$$

Putting

$$\overline{\Omega}^J = \sum_\lambda \overline{\Omega}_\lambda^J \quad \overline{\Omega}^{J,J'} = \sum_\lambda \overline{\Omega}_\lambda^{J,J'} \tag{12}$$

it is readily seen that:

$$\overline{\Omega} = \sum_J \overline{\Omega}^J + \sum_{J,J'} \overline{\Omega}^{J,J'} \tag{13}$$

If Ω is a monoelectronic operator, equation 13 reduces to:

$$\overline{\Omega} = \sum_J \overline{\Omega}^J \tag{14}$$

the average value is simply the sum of the loge contributions. If furthermore in a family of molecules good partitions into loges are obtained in terms of a set of invariant loges *molecular additivity rules will be observed.* By following that way an interpretation of the Faraday effect additive properties has been given (Daudel, Gallais and Smet, 1967).

When Ω is a bielectronic operator the term associated with the loge pairs gives rise to deviations from the additivity.

In both cases *it is possible to compare the loge contributions with experimental moduli. This is a very interesting new process, a possibility of parametrising the exact wave function, a starting point for an empirical ab-initio method.* Recently (Aslangul *et al.*, 1972) an analogous procedure has been proposed to establish relationship between isomerisation energies by starting from an approximate expression of the wave function in terms of loge functions. The formalism we have just described shows that this approximation is not necessary.

EVENT FUNCTIONS AND LOGE FUNCTIONS

Consider again the helium example, $\Psi(M_1, \omega_1, M_2, \omega_2, t)$ being the wave function. With event 1 we shall associate a function Ψ_1 defined as follows: follows:

$$\Psi_1 = \Psi \text{ when } M_1 \in v_A, M_2 \in v_A$$

$$\Psi_1 = 0 \text{ otherwise.}$$

In the same way:

$$\Psi_2 = \Psi \text{ when } M_1 \in v_B, M_2 \in v_B$$

$$\Psi_2 = 0 \text{ otherwise}$$

and

$$\Psi_3 = \Psi \text{ when } M_1 \in v_A, M_2 \in v_B$$
$$\text{or } M_1 \in v_B, M_2 \in v_A$$

$$\Psi_3 = 0 \text{ otherwise}$$

We can obviously write:

$$\Psi = \sum_\lambda \Psi_\lambda \tag{15}$$

Therefore any wave function can be written as a sum of event functions Ψ_λ.

The event functions introduced until now are not normalised. If we like we can consider normalised event functions ϕ_λ. Equation 15 becomes:

$$\psi = \sum_\lambda c_\lambda \phi_\lambda \tag{16}$$

where the c's are directly related to the normalisation factors.

To go farther it is interesting to introduce (Ludeña and Amzel, 1970) *n-electron completely localised loge functions* $L_j^{(n)}$. They are functions which vanish when one point is outside of the loge J. Keeping the helium example we can write:

$$\phi_1(M_1, \omega_1, M_2, \omega_2, t) = L_A^{(2)}(M_1, \omega_1, M_2, \omega_2, t)$$

$$\phi_2(M_1, \omega_1, M_2, \omega_2, t) = L_B^{(2)}(M_1, \omega_1, M_2, \omega_2, t)$$

$$\phi_3(M_1, \omega_1, M_2, \omega_2, t) =$$

$$L_A^{(1)}(M_1, \omega_1, t)L_B^{(1)}(M_2, \omega_2, t)f(M_1, \omega_1, M_2, \omega_2, t)$$

$$+ L_A^{(1)}(M_2, \omega_2, t)L_B^{(1)}(M_1, \omega_1, t)f(M_2, \omega_2, M_1, \omega, t) \qquad (17)$$

The third event function is therefore expressed (equation 17) as a sum of products of *correlated loge functions*.

THE LOGE THEORY AS A STARTING POINT TO CALCULATE ELABORATE WAVE FUNCTIONS

Equations 16 and 17 give expressions for any kind of wave functions: the exact one or any approximate one. Therefore they provide a very powerful tool *for suggesting trial functions*. Almost all actually known methods of calculating wave functions can be derived from equations similar to equations 16 and 17. Furthermore new methods can be also produced. Let us remark that any wave function can be written as:

$$\Psi = \sum_P (-1)^P P \psi \sigma \qquad (18)$$

in the framework of the non-relativistic wave-mechanics if ψ denotes a space function, σ a convenient spin function and $\sum_P (-1)^P P$ an antisymmetrisor. To simplify the discussion we will discuss only the space function problem. Let us assume that a trial expression is obtained for ψ and we can calculate the effective shape of that function by solving the variational equation:

$$\delta \langle \sum_P (-1)^P P \psi \sigma \,|\, H - E \,|\, \sum_P (-1)^P P \psi \sigma \rangle \qquad (19)$$

The problem reduces to the choice of expressions for ψ.

Equations 16 and 17 suggest the building of trial functions in terms of completely localised loge functions and of correlation functions f. Such a procedure has never been used until now. It would be very tedious. Ludeña and Amzel (loc. cit.) used a trial function in the case of beryllium which amounts to only considering the leading event and to neglecting the correlation function. Therefore ψ is written as:

$$\psi = L_K^{(2)}(M_1, M_2)L_L^{(2)}(M_3, M_4)$$

To go further into the calculation it is necessary to know the frontier between the two loges. The frontier has been calculated from a Hartree–Fock wave function. The shape of the loge function is then calculated following a variational procedure. The energy so obtained is slightly better than the

Hartree–Fock one. To improve the wave function it would be possible to add to the localised function ψ a delocalised correction (Aslangul *et al.* in press).

To avoid a previous determination of the frontiers of the loges it is possible to introduce non-completely localised loge functions.* Only a rough idea about the topology of a good partition into loges remains necessary to use that formalism. Calculations on the simplest molecule of a family of compounds can give such an idea for all the molecules of the family. Experimental data, chemical intuition can be sufficient: a certain interatomic distance can be for example the proof of the presence of a single bond and therefore of a two-electron bond loge.

As now the loge functions slightly overlap it is not so necessary to use a correlation function. To control the overlapping we can introduce a strong or a soft orthogonality condition between the loge functions. The variational procedure will do the rest. It has been shown (Aslangul *et al.*) that the very powerful PCILO method (Diner, Malrieu and Claverie, 1969; Malrieu, Claverie and Diner, 1969) amounts to using such a process with other approximations. The leading event is taken as the leading term in a perturbation procedure. The other important events are introduced by perturbation. Furthermore the loge functions are expressed in term of monoelectronic functions.

The McWeeny group functions method (Klessinger and McWeeny, 1965) amounts to considering only the leading event. With McWeeny and Valdemoro we are working on a program for an electronic computer in which the loge functions are expressed as configuration interaction functions extended in terms of gaussian functions.

The method amounts to calculating local configuration interactions in each loge region and using a self-consistent field treatment between the various configuration interaction functions. An advantage of the procedure is that the wave function is directly furnished in terms of rather well localised loge functions. Furthermore it can be expected that some loge functions obtained at the end of the interaction procedure for a given molecule are a good starting point for other molecules of the same family. Finally, as only a local configuration interaction is needed the number of necessary configurations is reduced by a large factor.

REFERENCES

Aslangul, C. (1971) *C. r. hebd. Séanc. Acad. Sci., Paris,* Ser. B., **272,** 1
Aslangul, C., Constanciel, R., Daudel, R., Esnault, L. and Ludena, E. (1972) (in press)
Aslangul, C., Constanciel, R., Daudel, R. and Kottis, P. (1972) *Adv. Quantum Chem.,* **6,** 93
Daudel, R. (1953) *C. r. hebd. Séanc. Acad. Sci., Paris,* **237,** 601
Daudel, R. (1956) *Fondements de la Chimie Théorique,* Gauthier Villars, Paris (English version, *The Fundamentals of Theoretical Chemistry,* Pergamon Press, Oxford, 1968)
Daudel, R. and Veillard, A. (1970) *Nature et Propriétés des Liaisons de Coordination,* 21, C.N.R.S., Paris
Daudel, R., Gallais, F. and Smet, F. (1967) *Int. J. Quantum Chem.,* **1,** 873
Diner, S., Malrieu, J. P. and Claverie, P. (1969) *Theor. Chim. Acta,* **13,** 1

*Such were the loge functions introduced by Daudel (1956).

Klessinger, M. and McWeeny, R. (1965) *J. chem. Phys.,* **42,** 3343
Ludeña, E. and Amzel, V. (1970) *J. chem. Phys.,* **52,** 5923
Malrieu, J. P., Claverie, P. and Diner, S. (1969) *Theor. Chim. Acta,* **13,** 18
Odiot, S. and Daudel, R. (1954) *C. r. hebd. Séanc. Acad. Sci., Paris,* **238,** 1384
Robb, M. A., Haines, W. J. and Csizmadia, I. G. C. (1973) *J. Am. Chem. Soc.,* **95,** 42
Shannon, E. (1948) *Bell Syst. Tech. J.,* **27,** 379
Sperber, G. (1971) *Int. J. Quantum Chem.,* **5,** 189

The Wave Mechanical Treatment of Hydrogen Bonded Systems

MARK A. RATNER

Department of Chemistry, New York University, New York

and

JOHN R. SABIN

Quantum Theory Project, Williamson Hall, University of Florida, Gainesville, Florida

1. INTRODUCTION

Hydrogen bonds are liaisons dangereuses for theoreticians. Latimer and Rodebush's (1920) statement 'the hydrogen nucleus held between two octets constitutes a weak "bond"' is really a theoretical statement, and the intervening fifty years have seen all manner of theoretical effort expended in our attempt to understand hydrogen bonding. The game is clearly worth the candle: hydrogen bonds are responsible for holding together the polar ice caps and keeping them floating above the sea, which these same bonds prevent from boiling. They pass on our eye colour and help us to digest; they perform delicate measurements and complicate the smog problem; and their characterisation and description continues to challenge physical science.

The importance and uniqueness of hydrogen bonds in nature stems from their omnipresence and their weakness. For present purposes let us define a hydrogen bond as a proton bonded simultaneously to two different heavier atoms; the bonds may be of equal or different length and strength. Generally, we will picture a hydrogen bond as A—H ... B, and call A the proton donor and B the proton acceptor (this is essentially the Lowry–Brønsted Acid–Base definition); we imply here that the A—H distance is less than half the A—B separation. The bond dissociation energy is the energy for the reaction A—H ... B → A—H + B. It is generally quite small, *c.* 5–10 kcal/mol, but it may be larger in charged systems (see below).

In the early investigations (Werner, 1903; Moore and Winmill, 1912), the A and B ions were highly electronegative first-row atoms (O, F, N). Recent work has extended beyond those, but even just O, F and N are so common in the ecosphere as to render hydrogen bonding (or association) so prevalent that it is found in most naturally occurring systems. Consequently, the early workers, although they treated only a few of the many possible hydrogen-bonded systems, contributed greatly to the understanding of this important phenomenon.

Experimentally, the identification and study of hydrogen bonding has been carried out by a variety of methods. The early investigations were actually thermodynamic, and much important thermodynamics is still being done. Since the measurements of Wulf and Liddel (1935), Liddel and Wulf (1933), and Freymann (1932), the use of spectroscopic techniques, first infra-red and Raman and more recently nuclear magnetic resonance (nmr), has become the most common means of study. For condensed phases, X-ray and neutron diffraction (e.g. Hamilton and Ibers, 1968) have proven very useful. The theoretical interpretation of those experiments is providing an understanding of the dynamic processes involved in hydrogen bonding.

Hydrogen bonding has been the subject of many fine reviews. Pimentel and McClellan (1960, 1971) have provided the most useful overall discussion. Murthy and Rao (1968) have reviewed spectroscopic properties; Hofacker and Hofacker (1968) have covered nmr. Theoretical developments have been reviewed by Sokolov (1964, 1965), Bratož (1967) and Kollman and Allen (1972). Solid-state aspects are reviewed by Hamilton and Ibers (1968) and ferroelectrics by Shirane, Jona and Pepinsky (1955), Jona and Shirane (1962), and Blinc (1968). Eisenberg and Kauzman (1969) have considered water substance, and Dickerson and Geis (1969) have discussed biological applications. In addition, there are published conference proceedings from Ljubljana (Hadži, 1959) and Munich (Riehl, Bullemer and Engelhardt, 1969).

The present review focuses largely on the subject of theoretical approaches to the electronic structure problem associated with hydrogen bonding. This was first discussed using point-charge models, but these had many failings. Our knowledge of hydrogen bond energetics is far from complete; most of what progress has been made comes from quantum mechanical calculation of electronic structures. These would be unthinkable without a development based on the wave nature of the electron, a conception whose introduction by de Broglie (1925) revolutionised modern physics.

2. EARLY THEORIES

After the identification and definition of the H bond by Latimer and Rodebush (1920), many studies were conducted which observed association in a wide variety of substances in solid and liquid phases. Several attempts were made to correlate these observations, attempting to characterise the effect of H bonding in terms of various phenomenological parameters, such as parachor, dipole moment, acid constant, bond length, and infra-red or ultra-violet spectral shift. The most widely used of these correlative schemes was that of Badger and Bauer (1937), who showed that 'there appears to be a relation between the energy of a hydrogen bond and the shift in frequency of the O–H bands' (Badger, 1940). This was successful in ordering a series of molecules, and can be justified by a simple model in which the proton acceptor group lowers the force constant, and hence the frequency of the infra-red fundamental, by charge transfer to the proton [this is one of the valence-bond structures proposed later by Tsubomura (1954); see below].

The first real theories of hydrogen bond formation were electrostatic, and are all grandchildren of the descriptions of Pauling (1928, 1960). Pauling used the Pauli principle and the electrostatic model to understand the formation of hydrogen bonds. Essentially, his argument rests on the conception that chemical bonds involving hydrogen utilise only the 1s hydrogen orbital. Since this orbital is already used for the covalent link to A in A—H ... B, any additional interaction with the electrons on B can only come about by electrostatic forces—that is, by point-charge interactions. Pauling has used this model to explain many of the properties of associated systems. Later work using the electrostatic model relaxed the point-charge requirement. Lennard-Jones and Pople (1951) attempted to describe the entire charge distribution by optimising the local bond geometry. These calculations (and many others, e.g. Schneider, 1955 and, earlier, Bauer and Magat, 1938) are predicated on the ability to properly describe the bond geometry in terms of electrostatics. Practically, the most commonly cited drawback of the electrostatic model is the lack of correlation of bond strength with dipole moment. The standard example is the fact that aceto-nitrile (Pimentel and McClellan, 1960) with a dipole moment of 3.44 D forms weaker hydrogen bonds with phenol than does dioxane, with a dipole moment of 0.3 D. One should be careful of such comparison, however, as the strengths of hydrogen bonds of various proton acceptors with a common donor are more closely related to the local dipole at the acceptor site, than to the gross dipole of the acceptor molecule. There is also a variety of other hydrogen bond properties which do not correspond to an electrostatic picture. One of the most serious of these deficiencies is the inability of the electrostatic model to properly predict infra-red intensity changes of the A–H stretching mode on hydrogen bond formation. More fundamentally, the electrostatic model is essentially classical in conception, and is therefore basically ill-suited to the description of a system in which quantum effects play a dominant role. Although the electrostatic theories greatly increased our understanding of the association phenomenon, their essentially *ad hoc* and classical interpretation seems insufficiently general and accurate; these drawbacks make them inappropriate for quantitative purposes, although some interesting calculations have been carried out, such as Baur's treatment of hydrated sulphate crystals (1964, 1965).

3. QUANTUM MECHANICAL DESCRIPTION

The first true quantum mechanical theories of hydrogen bonding were based on the valence-bond scheme. The first attempt in this direction seems to have been made by Gillette and Sherman (1936). Sokolov (1947, 1959) carried out extensive valence-bond calculations. Coulson and Danielsson

(1954a, b) attempted to estimate the percentage of covalent character in what was originally considered an entirely ionic bond. This work was continued by Tsubomura (1954); the structures he considered are indicated in Figure 1. The first three wave functions are covalent, the last two ionic. Coulson and Danielsson did not include the functions ψ_c and ψ_e, which imply a negative net charge on the proton, in their calculations; they also restricted the possible values of the coefficients a, b, d in $\psi_{\text{coulson}} = a\psi_a + b\psi_b + d\psi_d$ by some bond order assumptions, although in their later work these restrictions were dropped. Tsubomura included all five structures and allowed their coefficients to vary freely, although some of the electron repulsion integrals had to be estimated. Tsubomura finds that the structures ψ_c and ψ_e are important contributors to the ground state wave function, in contrast to the intuitive belief that the proton is somewhat acidic. This finding is also in sharp disagreement with observed downfield nmr shifts of protons in hydrogen bonds (Packard and Arnold, 1951. For reviews see Hofacker and Hofacker,

A ————— H		B	ψ_a
A⁻	H ————— B⁺		ψ_b
A	H⁻	B⁺	ψ_c
A⁻	H'	B	ψ_d
A⁺	H⁻	B	ψ_e

Figure 1 Valence bond canonical structure from
Tsubomura (1954)

1968; Alexandrov, 1966). Burnelle and Coulson (1957) have also given a treatment of water in which they propose 'bent' hydrogen bonds.

Perhaps the most important point gleaned from these valence-bond calculations is that, particularly at short (2.5 Å) distances, the 'covalent contribution' to hydrogen bonding becomes extremely important. Thus, the pure electrostatic picture, as stated above, is not a useful concept for describing these bonds. However, the large changes which occur in the valence-bond description upon the addition of new canonical structures, and the near-cancellation of large contributions to the energy, as well as the necessity for including enormous numbers of canonical bond structures and the associated non-orthogonality problem for the description of such systems as water dimer or (HF)$_6$, militate against the valence-bond picture, and in fact, such calculations are now seldom performed.

Roothaan's (1951) convenient formulation of the Hartree–Fock equations for a molecular orbital wave function has led to the widespread application of the LCAO-MO-SCF procedure to a variety of molecular systems including hydrogen bonding. The first MO treatment of the A-H---B system was reported by Pimentel (1951), who qualitatively considered the HF$_2^-$ system using only p orbitals on the fluorines and a 1s orbital on the proton. Hofacker (1957, 1958, 1959) reported the first calculations on associated systems. Unlike Tsubomura, he specifically decouples the lone-pair and bond-pair electron orbitals from the remainder of the complex. He then calculates bond energies and force constant parameters as a function of distance for both water and methanol. The basis set for water was taken from the work of Ellison and Shull (1955). Despite the approximations necessitated by the use of a small basis set and approximate integrals, this calculation is extremely significant, as it was the first full calculation using the MO method. Sokolov (1959, 1965) and Nukasawa, Tanaka and Nagakura (1953) have also prioneered the use of the MO theory to describe hydrogen bonding.

Hofacker's calculation (1958) used a minimum basis set of Slater-type orbitals (STO). Paoloni (1959) soon after suggested the inclusion of p-type orbitals on the hydrogen to better describe the polarisation phenomenon associated with bond formation. It is now recognised that these polarisation functions are quite important contributors to the wave function. In the succeeding thirteen years, there has been a plethora of calculations on hydrogen-bonded systems, and it is fair to say that we now well understand aspects of this bond. Recent reviews have been given by Sokolov (1964, 1965), Bratoz (1967), Hadzi (1965), Pimentel and McClellan (1960, 1971), and Kollman and Allen (1972).

Semi-empirical MO calculations have achieved marked success. In particular, the $CNDO/2$–$INDO$ methods (Pople and Beveridge, 1970) have been employed for the all valence-electron calculations of first-row hydride dimers and mixed dimers as well as some larger systems (cf. e.g. Kollman and Allen, 1970; Sabin, 1972). These methods were originally parametrised to predict bonding properties, and encouraging results have been obtained for overall bond energies. On the other hand, they do not account very well for atomic repulsions, and tend to overstress bonding, leading to relatively bad absolute values for force constants and charge distributions. For applications to larger systems, in particular pi-electron molecules, additional approximations have been used. Ladik pioneered the use of electronic structure calculations on large molecules of biological interest. Ladik and Rein (1964), Rein and Harris (1964a, b; 1965) and Lunnell and Sperber (1967) calculated potential surfaces for proton motion in cytosine–guanine and adenine–thymine base pairs. Sabin (1968a, b) calculated the pyridine–pyridinium and pyridine–pyrrole complexes. The relative ease and approachability of these methods will probably result in their continued widespread application, especially for large molecules.

Calculations using the *ab initio* LCAO MO SCF all electron method have yielded the most satisfying results; in addition, this method is the most general and aesthetic of those used so far. The first calculations of this type were those of Clementi (1961), who analysed the bifluoride ion, which was the subject both of Pimentel's (1951) original MO investigation and of earlier work by Bessis and Bratož (1960), who did a configuration-interaction calculation in the belief that properties of negatively charged systems could not be adequately described by single-determinant wave functions. Bessis and Bratož were (unfortunately) forced to use a variety of integral approximations which make the quality of their results difficult to evaluate.

The LCAO MO SCF method has been used to find binding energies, geometries, and one-electron properties of many associated systems. Among the more interesting results are those of Kollman (1972), who discusses problems of directionality of hydrogen bonding in the light of molecular orbital theory. Sabin (1971b) has pointed out that while the hydrogen-bond energy for neutral systems is quite small (5–10 kcal/mol), the charged systems have much stronger hydrogen bonds (calculated energies as great as 32 kcal/mol for $H_5O_2^+$ (Kraemer and Diercksen, 1970) or 31 kcal/mol for $H_3S_2^-$ (Sabin, 1971a)). Clementi (1967a, 1967b) and Clementi and Gayles (1967) in very lovely work, have studied the energy of the $HCl–NH_3$ system as a function of geometry, and have concluded that the association reaction $NH_3 + HCl \rightarrow NH_4Cl$ begins by electron polarisation and charge transfer, and that '[there is] no indication of the existence of an activation energy barrier in the range of the distances between HCl and NH_3 we have studied'. They also report a variety of potential surface and thermodynamic data.

In a calculational tour-de-force, Clementi, Mehl and von Niessen (1971) have calculated the potential surface for proton motion in the guanine–cytosine base pair by an all-electron calculation. While the labour and expense of this calculation (7×10^{10} two-electron integrals were calculated) effectively preclude such calculations on a day-to-day basis, this work does give a basis for confidence that this method can indeed provide useful information for very large molecules. In addition, this work extends the earlier calculations of Rein and Harris (1964a, b) and provides insight concerning the role of the integral approximations used in the earlier work.

The question of non-additivity of hydrogen bond energies in associated systems extending over more than two free molecules has been studied by Kollman and Allen (1970), Del Bene and Pople (1969, 1970, 1971) and Hankins, Moskowitz and Stillinger (1970, cf. Hankins, 1970). These last workers found that 'three-molecule non-additivities are large in magnitude and vary in size according to the hydrogen-bond pattern involved'. This is clearly a crucial issue in determining the structure of hydrogen-bonded networks, including, in particular, the structure of condensed phases, water and ice.

4. OTHER TREATMENTS OF HYDROGEN BONDING

Proton dynamics in hydrogen bonds have been the subject of extensive investigation. Lippincott and Schroeder (1955, 1956 cf. also Lippincott, Finch and Schroeder, 1959) and Reid (1959) have developed parametrised potential functions for the proton between two electronegative groups. Several groups (Somorjai and Hornig, 1962; Brickmann and Zimmermann, 1967, 1969) have calculated the vibrational wave functions and infra-red spectral properties of associated protons. Ibers (1964) has introduced quartic terms into the potential for proton motion in HF_2^-; in his detailed and careful analysis of chromous and cobaltous acids (Delaplane et al., 1969; Ibers, Holm and Adams, 1961), Ibers proposed that the protium and deuterium acids correspond to different potential functions (Snyder and Ibers, 1962), one with and one without a double minimum for proton motion. The whole question of potential functions is reviewed by Hamilton and Ibers (1968). While these studies do help to understand the spectroscopy of the hydrogen bond, there is a fundamental question about the adiabatic separation of proton and ion (A or B atoms) that complicates any simple interpretation. The potential models do provide interesting insights into the tunnelling probabilities of protons, a subject whose original impetus was certainly de Broglie's (1925) concept of a wave-length for matter.

Ben-Naim and Stillinger (1972) have developed an effective pair potential for water, in which the non-pair (co-operative) interactions are treated on an averaged basis. While this model does deal with the co-operation problem, the non-adiabatic difficulty persists (see below).

Mulliken (1952) has suggested that the hydrogen bond may be conceived of as a charge-transfer species, and several calculations have estimated the contribution of charge-transfer to the bonding phenomenon. This model contains a great deal of chemical insight, and is extremely useful in correlating trends among analogous systems. It is essentially a valence-bond description and suffers, when adapted to quantitative purposes, the disadvantages of the valence-bond method. Nevertheless, it has been successfully utilised by Puranik and Kumar (1963a, b), Bratož (1966) and Szczepaniak and Tramer (1965) as well as Mulliken (Person and Mulliken, 1969). The method is not designed for quantitative application to energetics.

5. CURRENT PROBLEMS

Theoretical understanding of the association phenomenon has increased a great deal in the last 25 years. Yet there remains a host of unanswered questions, both fundamental and applied. The most obvious concerns the role of correlation energy. For an average neutral small associated system such as $(H_2O)_2$, the total electronic energy in the Born–Oppenheimer approximation is of the order 10^2 Hartrees, the total bonding energy is of the order 10^0 Hartree, but the hydrogen-bonding energy is only 10^{-2} Hartree. Thus an error of 0.01 per cent in the total energy could conceivably give an error in the hydrogen-bond energy of 100 per cent. It is certainly true that the correlation energy would not be expected to change drastically on formation of hydrogen bonds. Nevertheless, Clementi (1967a) has pointed out that we can scarcely expect the correlation energy to be fully constant, since, at large $A-B$ separation, the $A-H$ bond dissociation process begins to resemble the $A-H \rightarrow A+H$ process, for which the $HF-SCF$ method has long been known to give incorrect results (e.g. Coulson and Fischer, 1949). The correlation energy problem in hydrogen bonding has, to our knowledge, not been treated explicitly. While the attractive results of SCF calculations seem to suggest that correlation effects are not overly important at least for small systems, this remains as a possible difficulty.

Electronically excited states are a problem in $HF-SCF$ theories, since the ground determinantal function is obtained and only virtual (non-renormalised) orbitals are left to describe excited states. For associated systems most work on excited states has been semi-empirical, concentrating on pi-electron molecules. Experimentally, acid constants and spectral shifts, as well as changes in lifetime, can be measured; the interpretation of these is really more concerned with the pi system than with the hydrogen bond. Adam *et al.* (1968) have looked at this problem using the extended Hückel method; Rao and Murthy (1971) and de Jeu (1970) have used the CNDØ scheme. The latter paper treats the formaldehyde–water system in both ground and excited states and attempts to find the change in hydrogen bond strength and length upon electronic excitation. The CNDØ method is generally quite bad for excited states, and one should not generally expect to obtain any quantitative information about them. Bratož (1967) has reviewed the problem of excited-state hydrogen bonds, but despite his sanguine expectations, little new theoretical information seems to have been acquired about the electronically excited states of hydrogen bonds. Indeed, the original qualitative ideas of Brealey and Kasha (1955), rooted in the valence-bond picture, are probably more useful than any later work. More progress on this point for sigma-bonded associated systems awaits general progress on the excited-state problem. The recently-developed multiple scattering Xα (MSXα) scheme (Slater, 1972; Slater and Johnson, 1972; Schwarz and Connolly, 1971) has the facility to treat excited states; although no work using this method for hydrogen bonds has yet been reported, calculations are presently under way which hopefully will lead to some insight on this problem (Connolly and Sabin, 1972).

The entire idea of adiabatic separation of proton motion from ionic (A or B) motion has often been questioned (Batuev, 1950; Fischer *et al.*; Ratner, 1969; Marechal and Witkowski, 1968; Singh and Wood, 1968; Salthouse and Waddington, 1968; Bournay and Marechal, 1971). Indeed, Fischer, Hofacker and Sabin (1969) have calculated the coupling constant between protonic and ionic motion in a linear ice model, and find it to be of order unity. This is not surprising, since a Born–Oppenheimer (1927) like separation would not be expected to hold between particles of similar (different by a factor of 16) mass. Strong-coupling models have been successful both for model systems (Ratner, 1969) and for such specific systems as imidazole (Marechal and Witkowski, 1968) or acrylic acid dimer (Bournay and Marechal, 1971). This approach seems to hold great promise for the analysis of many problems concerned with the association phenomenon; conversely, its neglect can render any well-intentioned calculation suspect.

One's naive expectation that the energy of two hydrogen bonds to the same molecule should be nearly the sum of the two independent bond energies is not borne out by calculation (Hankins, Moskowitz and Stillinger, 1970). This has serious implications for the simple potential-trough calculations such as those of Somorjai (1963). It seems now that the energies of hydrogen bonds are no more additive than normal chemical bonds; indeed, on a percentage basis, the non-additivity correction to the hydrogen-bond energy is much larger than in any other system except conjugated pi-systems. The charge-transfer model of Mulliken (1952) may be used to understand this in terms of variations in charge as each succeeding unit is added to such a complex. In the HF system, for instance, the formation of the first bond H-F---H-F renders the left-hand hydrogen more electropositive than the proton in H-F, due to charge-transfer within the extant hydrogen bond (Sabin, 1972). This problem becomes important in the analysis of large systems, notably in the structure of condensed phases. One possible approach is that of an effective pair potential including the non-additive contributions in the average (Ben-Naim and Stillinger, 1972). This method is presently being applied in a molecular-dynamics study of liquid water (Stillinger and Rahman, 1972). Other possible avenues are a self-consistent treatment of proton dynamics (Fang, 1970), treatment of clusters (Connolly and Sabin, 1972), treatment as a disordered chain (Whalley and Bertie, 1967) or, for periodic solids, a CNDØ cell method such as that used by Bacon and Santry (1971). These all leave much to be desired: the structure of water in the condensed phase remains terribly difficult.

Another fascinating aspect of hydrogen-bonded solids is their abnormally high protonic conductivity which has led Onsager and Dupuis (1962) to call some of them, notably ice, 'protonic semiconductors'. The high conductivity of ice as measured, for instance, by the Munich group (Riehl, Bullemer and Engelhardt, 1969) has been explained by Onsager and Dupuis (1962) in terms of Bjerrum defects. Detailed dynamical discussions have been given by Gosar (1963, 1969) and by Fischer and Hofacker (1969a). Fascinating phenomena in these systems, such as Tyndall star formation (Jaccard and Hofacker, 1972; Riehl, Bullemer and Engelhardt, 1969) await microscopic theoretical explanation. The protonic conductivity itself depends in an essential way on non-adiabatic proton–proton interaction.

Hydrogen bonding plays a crucial role in the structure and function of many biological macromolecules. As Pauling (1960) has pointed out, the weakness of hydrogen bonds makes them ideal regulatory devices for biological reactions, which occur at temperatures corresponding to 0.8 kcal/mol. Theoretical developments here are still in a qualitative stage with the notable exception of the work of Clementi, Mehl and Von Niessen, (1971). The concepts of hydrophilic and hydrophobic (Nemethy, 1967) interaction await quantitative discussion, and the role of proton dynamics in such systems as DNA replication (Löwdin, 1965) and energy transfer in enzymes remains a fascinating and largely unexplored area. Semi-empirical work on biological systems has been extensive (e.g. Pullman and Pullman, 1963) but a caveat is provided by a comparison of the Rein–Harris (1964a) and Clementi, Mehl and Von Niessen (1971) calculations on the G–C base pair: this is a touchy area for semi-empirical methods.

In the calculations on molecular stability, it must always be borne in mind that, particularly for molecules in the condensed phase, the free energy function, not the energy, determines the equilibrium state. Most studies, particularly those involving MO calculations, have ignored the contribution of entropy to the behaviour of these systems. Notable exceptions are provided in the work of Pullman and co-workers (Perahia and Pullman, 1971), Scheraga and co-workers (Gô, Gô and Scheraga, 1968) and Froimowitz (1972) and, for a smaller system, Clementi and Gayles (1967). This issue has been well documented by workers studying the thermodynamics of such systems (Klotz and Franzen, 1962; Kresheck and Klotz, 1969), but has been insufficiently appreciated in most theoretical approaches.

The free energy function is of course also the natural context for the study of phase transitions in hydrogen-bonded systems. The most important of these are the ferroelectric and antiferroelectric crystals. Early theoretical work here was concerned with the statistical thermodynamics of the transition (Slater, 1941; Takagi, 1948; Devonshire, 1964; Jona and Shirane, 1962). As more experimental work has become available (Pepinsky, 1972), the dynamics of the transition has been subjected to closer theoretical investigation (de Gennes, 1963; Brout, Muller and Thomas, 1966; Blinc and Svetina, 1966; Fischer, 1967; Kobayashi, 1968; Nettleton, 1971). The classic examples are Rochelle salt, KH_2PO_4 and close relatives (KD_2PO_4, $NH_4H_2PO_4$, KH_2AsO_4). The last have become the focus of renewed recent attention, both for technical application as storage devices (Miller, 1969) and because the ferroelectric transition does not seem to follow the scaling hypothesis (Widom, 1965; Kadanoff, et al., 1967; Stanley, 1971; Cummins, 1972). Potential functions in all of this work have been assumed and there is need here for some electronic studies to determine a numerical potential (although here the phonon coupling is clearly essential for the transition).

The concept of the hydrogen bond, in terms of its classical definition, has recently been changing in an operational sense, as more systems are discovered which contain a hydrogen-like bond but do not fit the normal concept of this bond. Bonds of this type can be characterised by the observation that hydrogen bridges of the form A-H--B will apparently form between atoms which are not highly electronegative. The standard example of this type of structure is the boron hydrides, B_2H_6 being the simplest (cf. e.g. Lipscomb, 1963). In this case, there is the well-known two electron three centre B-H-B bridge, or a hydrogen bridge between atoms with a Pauling electronegativity of only 2.0, while hydrogen has an electronegativity of 2.1 (Pauling, 1960). Although such bonds are not standard hydrogen bonds, they must certainly be considered hydrogen bonds of some sort, as they have a finite bond energy and a doubly bonded bridging proton.

One would expect, then, that other atoms such as second and third row non-metals with low electronegativity [S(E.N. = 2.5), P(2.1), Se(2.4), Te(2.1), As(2.0)] and carbon (2.5) should form weak hydrogen bonds. Although little theoretical work has been done on this problem, it appears that this is indeed the case (cf. e.g. Sabin, 1971a, b) and in fact a number of such cases have been found experimentally (e.g. Sutor, 1962; Dulmage and Lipscomb, 1951; McDaniel and Evans, 1966). There are also known hydrogen bonds of the type Cr-H-Cr (Handy et al., 1966) and He-H-He$^+$ (e.g. Preuss and Janoschek, 1969) where the electronegativities are considerably less than that of hydrogen.

It is interesting to note that semi-empirical studies using localised orbital methods (Lindner and Sabin, 1973) seem to indicate that the three centre type of bond that exists in the boron hydrides is not unique to these compounds, but becomes more and more predominant as the electronegativity of the proton donor and acceptor atoms decreases.

Another type of a hydrogen bond occurs when a pi-electron system acts as proton acceptor. Systems employing aromatic systems as proton acceptors (Huggins and Pimentel, 1955, 1956) are well known experimentally but have received no theoretical treatment to date except as charge-transfer systems (Person and Mulliken, 1969).

As mentioned above, very little of a theoretical nature has been done with the less common hydrogen bonded systems, and a fuller understanding of the nature, energetics, and possible three centre-bonding awaits a theoretical description.

On a molecular level, our knowledge of the formation and properties of hydrogen bonds is still at a rather primitive stage. Some of the areas discussed above are in the active research stage, and these investigations are certain to lead to increased understanding of this important and enigmatic phenomenon.

ACKNOWLEDGEMENTS

One of us (MAR) is grateful to Professor P. O. Löwdin for his hospitality at the Quantum Theory Project during part of the period when this review was being written. Acknowledgement is made to the Donors of the Petroleum Research Fund, Administered by the American Chemical Society for partial support (MAR and JRS) of this work.

6. SUMMARY

The hydrogen bond was originally conceived of as a totally electrostatic phenomenon, but it is now well known that a wave mechanical picture must be employed to explain the many interesting properties of this entity. Hydrogen bonds can now be well described using a simple wave mechanical picture. Bonding moments, strengths, polarisabilities and distances can be obtained quite accurately using an appropriate molecular orbital scheme. In part, the remarkable increase in strength of charged as opposed to uncharged systems and the high mutual polarisability of hydrogen-bonded networks can be explained nicely using wave mechanics, as can some aspects of the abnormally high protonic conductivity in some protonic semiconductors. Certain aspects of hydrogen-bonded systems have still not been satisfactorily explained, notably the bonding structure of their electronically excited states. Overall, however, the wave mechanical description of this extremely important and sensitive system is remarkably successful and concise.

REFERENCES

Adam, W., Grimison, A., Hoffman, R. and de Ortiz, C. Z. (1968) 'Hydrogen Bonding in Pyridine', *J. Am. chem. Soc.,* **90,** 1509–1516

Alexandrov, I. V. (1966) *The Theory of Magnetic Resonance,* Academic Press, New York

Bacon, J. and Santry, D. P. (1971) 'Molecular Orbital Theory for Infinite Systems: Self-Consistent Field Perturbation Treatment of Hydrogen-Bonded Molecules', *J. chem. Phys.,* **55,** 3743–3751

Badger, R. M. (1940) 'The Relation Between the Energy of the Hydrogen Bond and the Frequencies of the O-H Bands', *J. chem. Phys.,* **8,** 288–289

Badger, R. M. and Bauer, S. H. (1937) 'Spectroscopic Studies of the Hydrogen Bond. II. The Shift of the O-H Vibrational Frequency in the Formation of the Hydrogen Bond', *J. chem. Phys.*, **5**, 839–851

Batuev, M. I. (1950) 'Frequency Modulation and Predissociation Theories of the Hydrogen Bond', *Izvest. Akad. Nauk SSSR. Ser. Fiz.*, **14**, 429–434

Bauer, E. and Magat, M. (1938) 'Sur la Déformation des Molécules en Phase Condensée et la "Liaison Hydrogéne"', *J. Phys. Radium*, **9**, 319–330

Baur, W. (1964) 'On the Crystal Chemistry of Salt Hydrates. II. A Neutron Diffraction Study of Magnesium Sulfate Tetrahydrate', *Acta Cryst.*, **17**, 863–869

Baur, W. (1965) 'On Hydrogen Bonds in Crystalline Hydrates', *Acta Cryst.*, **19**, 909–916

Ben-Naim, A. and Stillinger, F. H. (1972) 'Aspects of the Statistical Mechanical Theory of Water', *Water and Aqueous Solutions* (edited by Horne, R. A.), 295–331, Wiley, New York

Bessis, G. and Bratož, S. (1960) 'Etude de la Structure Electronique de l'Ion FHF⁻ a l'Aide de la Méchanique Ondulatoire', *J. Chim. Phys.*, **57**, 769–773

Blinc, R. (1968) 'Magnetic Resonance in Hydrogen Bonded Ferroelectrics', *Advances in Magnetic Resonance*, **3**, 141–204

Blinc, R. and Svetina, S. (1966) 'Cluster Approximations for Order-Disorder Type Hydrogen-Bonded Ferroelectrics, II. Application to KH_2PO_4', *Phys. Rev.*, **147**, 430–443

Born, M. and Oppenheimer, J. R. (1927) 'Zur Quantentheorie der Molekeln', *Annln Phys.*, **84**, 457–884

Bournay, J. and Marechal, Y. (1971) 'Dynamics of Protons in Hydrogen-Bonded Systems', *J. chem. Phys.*, **55**, 1230–1235

Bratož, S. (1966) 'Interactions Moleculaires', *Colloques Natn. Cent. natn. Rech. Scient.*, 29–57

Bratož, S. (1967) 'Electronic Theories of Hydrogen Bonding', *Adv. quantum Chem.*, **3**, 209–237

Brealey, G. D. and Kasha, M. (1955) 'The Rôle of Hydrogen Bonding in the $n \to \pi^*$ Blue-Shift Phenomenon', *J. Am. chem. Soc.*, **77**, 4462–4468

Brickmann, J. and Zimmermann, H. (1967) 'Über den Tunneleffekt des Protons im Doppelminimumpotential von Wasserstoffbrückenbindung', *Ber. Bunsen. Phys. Chem.*, **71**, 160–164

Brickmann, J. and Zimmermann, H. (1969) 'On the Lingering Time of the Proton in the Wells of Double Minimum Potential of Hydrogen Bonds', *J. chem. Phys.*, **50**, 1608–1618

Brout, R., Muller, K. A. and Thomas, H. (1966) 'Tunnelling and Collective Excitations in a Microscopic Model of Ferroelectricity', *Solid St. Comm.*, **4**, 507–510

Burnelle, L. and Coulson, C. A. (1957) 'Bond Dipole Moments in Water and Ammonia', *Trans. Faraday Soc.*, **53**, 403–405

Clementi, E. (1961) 'Ground State Wave Functions for Linear Molecules', *J. chem. Phys.*, **34**, 1468–1469

Clementi, E. (1967a) 'Study of the Electronic Structure of Molecules. II. Wavefunctions for the $NH_3 + HCl \to NH_4Cl$ Reaction', *J. chem. Phys.*, **46**, 3851–3880

Clementi, E. (1967b) 'Study of the Electronic Structure of Molecules. VI. Charge-Transfer Mechanism for the $NH_3 + HCl \rightleftharpoons NH_4Cl$ Reaction', *J. chem. Phys.*, **47**, 2323–2334

Clementi, E. and Gayles, J. N. (1967) 'Study of the Electronic Structure of Molecules. VII. Inner and Outer Complex in the NH_4Cl Formation from NH_3 and HCl', *J. chem. Phys.*, **47**, 3837–3841

Clementi, E., Mehl, J. and von Niessen, W. (1971) 'Study of the Electronic Structures of Molecules. VII. Hydrogen Bridges in the Guanine–Cytosine Pair and in the Dimeric Form of Formic Acid', *J. chem. Phys.*, **54**, 508–520

Connolly, J. W. D. and Sabin, J. R. (1972) 'Total Energy in the Multiple Scattering Formalism: Application to the Water Molecule', *J. chem. Phys.*, **56**, 5529–5533

Coulson, C. A. and Danielsson, U. (1954a) 'Ionic and Covalent Contributions to the Hydrogen Bond. I, *Ark. Fys.*, **8**, 239–244

Coulson, C. A. and Danielsson, U. (1954b) 'Ionic and Covalent Contribution to the Hydrogen Bond, II', *Ark. Fys.*, **8**, 245–255

Coulson, C. A. and Fischer, I. (1949) 'Notes on the Molecular Orbital Treatment of the Hydrogen Molecule', *Phil. Mag.*, **40**, 386–393

Cummins, H. Z. (1972) Personal communication

de Broglie, L. (1925) 'Recherches sur la Théorie des Quanta', *Annls Phys., Paris*, **3**, 22–128

de Gennes, P. G. (1963) 'Collective Motions of Hydrogen Bonds', *Solid St. Comm.*, **1**, 132–137

de Jeu, W. H. (1970) 'A CNDO Calculation of the Hydrogen Bond and the Blue Shift in Formaldehyde–Water', *Chem. Phys. Letters*, **7**, 153–156

Delaplane, R. G., Ibers, J. A., Ferraro, J. R. and Rush, J. J. (1969) 'Diffraction and Spectroscopic Studies of the Cobaltic Acid System $HCoO_2$–$DCoO_2$', *J. chem. Phys.*, **50**, 1920–1927

Del Bene, J. and Pople, J. A. (1969) 'Intermolecular Energies of Small Water Polymers', *Chem. Phys. Letters*, **4**, 426–428

Del Bene, J. and Pople, J. A. (1970) 'Theory of Molecular Interactions. I. Molecular Orbital Studies of Water Polymers Using a Minimal Slater-Type Basis', *J. chem. Phys.*, **52**, 4858–4866

Del Bene, J. and Pople, J. A. (1971) 'Theory of Molecular Interactions. II. Molecular Orbital Studies of HF Polymers Using a Minimal Slater-Type Basis', *J. chem. Phys.*, **55**, 2296–2299

Devonshire, A. F. (1964) 'Some Recent Work on Ferroelectrics', *Rep. Prog. Phys.*, **27**, 1–22

Dickerson, R. E. and Geis, I. (1969) *The Structure and Actions of Proteins*, Harper and Row, New York

Dulmage, W. J. and Lipscomb, W. N. (1951) 'Crystal Structures of Hydrogen Cyanide', *Acta Cryst.*, **4**, 330–334

Eisenberg, P. and Kauzmann, W. (1969) *The Structure and Properties of Water*, Oxford University Press

Ellison, F. O. and Shull, H. (1955) 'Molecular Calculations I. LCAO MO Self-Consistent Field Treatment of the Ground State of H_2O, *J. chem. Phys.*, **23**, 2348–2357

Fang, J. K. (1970) *Thesis*, Northwestern University

Fischer, S. F. (1967) 'Collective Excitations of Hydrogen-bonded Ferroelectrics', *Int. J. quantum Chem.*, **S1**, 745–753

Fischer, S. F. and Hofacker, G. L. (1969a) 'Theory of the Mobility of Structural Defects in Ice', *Physics of Ice* (edited by Riehl, N. *et al.*). Plenum Press, New York

Fischer, S. F., Hofacker, G. L. and Ratner, M. A. (1969b) 'Spectral Behavior of Hydrogen Bonded Systems: Quasiparticle Model', *J. chem. Phys.*, **52**, 1934–1947

Fischer, S. F., Hofacker, G. L. and Sabin, J. R. (1969) 'Proton–Phonon Coupling in a Hydrogen Bonded System', *Phys. kondens, Materie*, **8**, 268–278

Freymann, M. (1932) 'Effect de la Dilution et de la Temperature sur les Bandes d'Absorption Infrarouges: Associations Intermoleculaires', *C. r. hebd. Séanc., Acad. Sci., Paris*, **195**, 39–41

Froimowitz, M. (1972) 'Computer Simulation and Conformational Thermodynamics of the Acetylcholine Molecule', *Ph. D. Thesis*, New York University

Gillette, R. H. and Sherman, A. (1936) 'The Nature of the Hydrogen Bond. I. Association of Carboxylic Acids', *J. Am. chem. Soc.*, **58**, 1135–1139

Gô, N., Gô, M. and Scheraga, H. (1968) 'Molecular Theory of the Helix-coil Transition in Polyamino Acids. I. Formulation', *Proc. natn. Acad. Sci.*, **59**, 1030–1037

Gosar, P. (1963) 'Mobility of H_3O^+ in Ice Crystals', *Nuovo Cim.*, **30**, 931–946

Gosar, P. (1969) 'Proton–proton and Proton–lattice Interactions in Ice', *Physics of Ice* (edited by Riehl, N. *et. al.*). 401–415, Plenum Press, New York

Hadži, D. (1965) 'Infrared Spectra of Strongly Hydrogen Bonded Systems', *Pure appl. Chem.*, **11**, 435–53

Hadži, D. (1959) *Hydrogen Bonding*, Pergamon Press, Oxford

Hamilton, W. C. and Ibers, J. A. (1968) *Hydrogen Bonding in Solids*, Benjamin, New York

Handy, L. B., Treichel, P. M., Dahl, L. F. and Hayter, R. G. (1966) 'Structure and Bonding in $HCr_2(CO)_{10}^-$. The First Known Linear Electron–Deficient X–H–X Molecular System Stabilized by a Three-Center, One-Electron-Pair Bond', *J. Am. chem. Soc.*, **88**, 366–367

Hankins, D. (1970) 'Quantum Mechanical Investigations of Water Molecule Interactions', *Ph.D. Thesis*, New York University

Hankins, D., Moskowitz, J. W. and Stillinger, F. (1970) 'Water Molecule Interactions', *J. chem. Phys.*, **53**, 4544–4554

Hofacker, G. L. (1957) 'Zum Bindungsmechanismus von O–H---O Wasserstoffbrücken', *Z. Electrochem.*, **61**, 1048–1053

Hofacker, G. L. (1958) 'Zur Theorie der Wasserstoffbrückenbindung', *Z. Naturforsch.*, **13a**, 1044–1057

Hofacker, G. L. (1959) 'Die Behandlung der Wasserstoffbrückenbindung nach der MO-Theorie', *Hydrogen Bonding* (edited by Hadzi, D.), Pergamon Press, Oxford, 375–384

Hofacker, G. L. and Hofacker, U. A. H. (1968) 'Hydrogen Bonds in the Light of NMR Measurements', *Magnetic Resonance and Relaxation* (edited by Blinc, R.), 502–525, North-Holland, Amsterdam

Huggins, C. M. and Pimentel, G. C. (1955), 'Infrared Intensity of the C–D Stretch of Chloroform-d in Various Solvents', *J. chem. Phys.*, **23**, 896–898

Huggins, C. M. and Pimentel, G. C. (1956) 'Systematics of the Infrared Spectral Properties of Hydrogen Bonding Systems: Frequency Shift, Halfwidth, and Intensity', *J. phys. Chem.*, **60**, 1615–1619

Ibers, J. A. (1964) 'Potential Function for the Stretching Region in Potassium Acid Fluoride', *J. chem. Phys.*, **41**, 25–28

Ibers, J. A., Holm, C. H. and Adams, C. R. (1961) 'Proton Magnetic Resonance Study of Polycrystalline HCrO₂', *Phys. Rev.*, **121**, 1620–1629

Jaccard, C. and Hofacker, G. L. (1972) Lecture Delivered at Meeting on Hydrogen Bonding, Ottawa, 1972

Jona, F. and Shirane, G. (1962) *Ferroelectric Crystals*, Pergamon Press, New York

Kadanoff, L. *et al.* (1967) 'Static Phenomena Near Critical Points; Theory and Experiment', *Rev. mod. Phys.*, **39**, 395–451

Klotz, I. M. and Franzen, J. S. (1962) 'Hydrogen Bonds Between Model Peptide Groups in Solution', *J. Am. chem. Soc.*, **84**, 3461–3466

Kobayashi, K. K. (1968) 'Dynamical Theory of the Phase Transition in KH₂PO₄ Type Ferroelectric Crystals', *J. phys. Soc. Japan*, **24**, 497–508

Kollman, P. A. (1972) 'A Theory of Hydrogen Bond Directionality', *J. Am. chem. Soc.*, **94**, 1837–1842

Kollman, P. A. and Allen, L. C. (1970) 'Hydrogen Bonded Dimers and Polymers Involving Hydrogen Fluoride, Water, and Ammonia', *J. Am. chem. Soc.*, **92**, 753–759

Kollman, P. A. and Allen, L. C. (1972) 'The Theory of the Hydrogen Bond', *Chem. Rev.*, **72**, 283–303

Kraemer, W. P. and Diercksen, G. H. F. (1970) SCF–MO–LCAO Studies on Hydrogen Bonding'. The System (H₂OHOH₂)⁺, *Chem. Phys. Letters*, **5**, 463–465

Krescheck, G. C. and Klotz, I. M. (1969) 'The Thermodynamics of Transfer of Amides from an Apolar to an Aqueous Solution', *Biochemistry*, **8**, 8–12

Ladik, J. and Rein, R. (1964) 'Semiempirical SCF–LCAO–MO Calculation of the Electronic Structure of the Guanine–Cytosine Base Pair', *J. chem. Phys.*, **40**, 2466–2473

Latimer, W. M. and Rodebush, W. H. (1920) 'Polarity and Ionization from the Standpoint of the Lewis Theory of Valence', *J. Am. chem. Soc.*, **42**, 1419–1433

Lennard-Jones, J. and Pople, J. A. (1951) 'Molecular Association in Liquids. I. Molecular Association Due to Lone Pair Electrons', *Proc. R. Soc.*, **A205**, 155–162

Liddell, U. and Wulf, O. R. (1933) 'The Character of the Absorption of Some Amines in the Near Infra-red', *J. Am. chem. Soc.*, **55**, 3574–3583

Lindner, P. and Sabin, J. R. (1973) 'Localized Orbitals in Hydrogen Bonded Systems', Submitted to *J. Am. chem. Soc.*

Lippincott, E. R., Finch, J. N. and Schroeder, R. (1959) 'Potential Function Model of Hydrogen-Bond Systems', *Hydrogen Bonding* (edited by Hadzi, D.), 361–372, Pergamon Press, Oxford

Lippincott, E. R. and Schroeder, R. (1955) 'General Relation Between Potential Energy and Internuclear Distance for Diatomic and Polyatomic Molecules', *J. chem. Phys.*, **23**, 1131–1141

Lippincott, E. R. and Schroeder, R. (1956) 'General Relation Between Potential Energy and Internuclear Distance. II. Polyatomic Molecules', *J. Am. chem. Soc.*, **28**, 5171–5178

Lipscomb, W. N. (1963) *Boron Hydrides*, Benjamin, New York

Löwdin, P. O. (1965) 'Quantum Genetics and the Aperiodic Solid. Some Aspects on the Biological Problems of Heredity, Mutations, Aging, and Tumors in View of the Quantum Theory of the DNA Molecule', *Adv. quantum Chem.*, **2**, 216–360

Lunnell, S. and Sperber, G. (1967) 'Study of the Hydrogen Bonding in the Adenine–Thymine, Adenine–Cytosine, and Guanine–Thymine Base Pairs', *J. chem. Phys.*, **46**, 2119–2124

Marechal, Y. and Witkowski, A. (1968) 'Infrared Spectra of H-Bonded Systems', *J. chem. Phys.*, **48**, 3697–3705

McDaniel, D. H. and Evans, W. O. (1966) 'Strong Hydrogen Bonds. III. Hydrogen Sulfide-Hydrosulfide Anion Interactions', *Inorg. Chem.*, **5**, 2180–2182

Miller, G. L. (1969) Personal communication

Moore, T. S. and Winmill, T. F. (1912) 'The State of Amines in Aqueous Solution', *J. chem. Soc.*, **101**, 1635–1676

Mulliken, R. S. (1952) 'Molecular Compounds and Their Spectra. III. The Interaction of Electron Donors and Acceptors', *J. phys. Chem.*, **56**, 801–822

Murthy, A. S. N. and Rao, C. N. R. (1968) 'Spectroscopic Studies of the Hydrogen Bond', *App. Spect. Rev.*, **2**, 69–191

Nemethy, G. (1967) 'Hydrophobic Interactions' *Angew. Chem. Int. Edn.*, **6**, 195–206

Nettleton, R. E. (1971) 'Ferroelectric Phase Transitions: A Review of Theory and Experiment—Part II—Conclusions', *Ferroelectrics*, **2**, 93–96 (See also other papers in this series)

Nukasawa, K., Tanaka, J. and Nagakura, S. (1953) 'Note on the Hydrogen Bond', *J. phys. Soc. Japan.*, **8**, 792–793

Onsager, L. and Dupuis, M. (1962) 'The Electrical Properties of Ice', *Electrolytes* (edited by Pesce, B.), 27–46, Pergamon Press, New York

Packard, M. E. and Arnold, J. T. (1951) 'A Fine Structure in Nuclear Induction Signals for Ethyl Alcohol', *Phys. Rev.*, **83**, 210–211

Paoloni, L. (1959) 'Nature of the Hydrogen Bond', *J. chem. Phys.*, **30**, 1045–1058

Pauling, L. (1928) 'The Shared-electron Chemical Linking', *Proc. Natn. Acad. Sci.*, **14**, 349–362

Pauling, L. (1960) *'The Nature of the Chemical Bond'*, Cornell University Press

Pepinsky, R. (1972) 'Some Structural Mechanisms in Ferroelectricity', *Physics of Electronic Ceramics* (edited by Hench, L. L. and Dove, D. B.), Dekker, New York

Perahia, D. and Pullman, B. (1971) 'The Conformational Energy Map for the Disulphide Bridge in Proteins', *Biochem. Biophys. Res. Comm.*, **43**, 65–68

Person, W. B. and Mulliken, R. S. (1969) *Molecular Complexes*, Wiley Interscience, New York

Pimentel, G. C. (1951) 'The Bonding of Trihalide and Bifluoride Ions by the Molecular Orbital Method', *J. chem. Phys.*, **19**, 446–448

Pimentel, G. C. and McClellan, A. L. (1960) *The Hydrogen Bond*, Freeman, San Francisco

Pimentel, G. C. and McClellan, A. L. (1971) 'Hydrogen Bonding', *A. Rev. phys. Chem.*, **22**, 347–385

Pople, J. A. and Beveridge, D. L. (1970) *Approximate Molecular Orbital Theory*, McGraw–Hill, New York

Preuss, H. W. and Janoschek, R. (1969) 'Wave Mechanical Calculations on Molecules Taking all Electrons into Account', *J. mol. Struct.*, **3**, 423–428

Pullman, A. and Pullman, B. (1963) *Quantum Biochemistry*, Wiley Interscience, New York

Puranik, P. G. and Kumar, V. (1963a) 'Charge Transfer Theory of the H–Bond', *Proc. Indian Acad. Sci.*, **58**, 29–37

Puranik, P. G. and Kumar, V. (1963b) 'Charge Transfer Theory of the H–Bond. II. Experimental SH Band in Thiphenol (Integrated Infrared Absorption Intensities)', *Proc. Indian Acad. Sci.*, **58**, 327–335

Rao, C. N. R. and Murthy, A. S. N. (1971) 'Hydrogen Bonding in Electronically Excited States of Molecules', *Theor. Chim. Acta*, **22**, 392–395

Ratner, M. A. (1969) 'Strong Coupling Model for Analysis of Spectral Properties of Hydrogen Bonded Systems, *Thesis*, Northwestern University

Reid, C. (1959) 'Semiempirical Treatment of the Hydrogen Bond, *J. chem. Phys.*, **30**, 182–190

Rein, R. and Harris, F. E. (1964a). 'Studies of Hydrogen-Bonded Systems. I. The Electronic Structure and the Double-well Potential of the N–H——N Hydrogen Bond of the Guanine–Cytosine Base Pair, *J. chem. Phys.*, **41**, 3393–3401

Rein, R. and Harris, F. E. (1964b) 'Proton Tunneling in Radiation Induced Mutation', *Science*, **146**, 649–650

Rein, R. and Harris, F. E. (1965) 'Studies of Hydrogen-Bonded Systems. II. Tunneling and Tautomeric Equilibrium in the N–H——N Hydrogen Bond of the Guanine–Cytosine Base Pair', *J. chem. Phys.*, **42**, 2177–2180

Riehl, N., Bullemer, B. and Engelhardt, H. (1969) *Physics of Ice*, Plenum Press, New York

Roothaan, C. C. J. (1951) 'New Developments in Molecular Orbital Theory', *Rev. mod. Phys.*, **23**, 69–89

Sabin, J. R. (1968a) 'Hydrogen Bonding in Simple π-Electron Systems. I. Pyridine–Pyrrol, *Int. J. quantum Chem.*, **2**, 31–36

Sabin, J. R. (1968b) 'Hydrogen Bonding in Simple π-Electron Systems. II. Pyridinium–Pyridine, *Int. J. quantum Chem.*, **2**, 23–30

Sabin, J. R. (1971a) 'Hydrogen Bonds involving Sulfur. II. The Hydrogen Sulfide-Hydrosulfide Complex', *J. chem. Phys.*, **54**, 4675–4680

Sabin, J. R. (1971b) 'A Comment Concerning S–H–S Type Hydrogen Bonds', *Int. J. quantum Chem.*, **5**, 133–136

Sabin, J. R. (1972) 'CNDO Study of the Properties of Ionic Defect Structure in a Model One-dimensional Hydrogen Bonded Chain', *J. chem. Phys.*, **56**, 45–51

Salthouse, J. A. and Waddington, T. C. (1968) 'New Mechanisms for the Broadening of the Hydrogen Stretching Peak on Hydrogen Bonding', *J. chem. Phys.*, **48**, 5274–5275

Schneider, W. G. (1955) 'Properties of the Hydrogen Bond. The Role of Lone Pair Electrons', *J. chem. Phys.*, **23**, 26–30

Schwartz, K. and Connolly, J. W. D. (1971) 'Approximate Numerical Hartree–Fock Method for Molecular Calculations', *J. chem. Phys.*, **55**, 4710–4714

Shirane, G., Jona, F. and Pepinsky, R. (1955) 'Some Aspects of Ferroelectricity', *Proc. Inst. Radio Engrs.*, **43**, 1738–1793

Singh, J. and Wood, A. (1968) 'Two-Dimensional Double-Minimum Model of Hydrogen Bonding: The Symmetric Case', *J. chem. Phys.*, **48**, 4567–4581

Slater, J. C. (1941) 'Theory of the Transition in KH_2PO_4', *J. chem. Phys.*, **9**, 16–33

Slater, J. C. (1972) 'Statistical Exchange-Correlation in the Self-Consistent Field', *Adv. quantum Chemistry*, **6**, 1–93

Slater, J. C. and Johnson, K. H. (1972) 'Self-Consistent Field $X\alpha$ Cluster Method for Polyatomic Molecules and Solids', *Phys. Rev. B.*, **5**, 844–853

Snyder, R. G. and Ibers, J. A. (1962) 'O–H–O and O–D–O Potential Energy Curves for Chromous Acid', *J. chem. Phys.*, **36**, 1356–1360

Sokolov, N. D. (1947) *Dokl. Akad. Nauk. SSSR*, **58**, 611

Sokolov, N. D. (1959) 'Quantum Theory of the Hydrogen Bond', *Hydrogen Bonding* (edited by Hadzi, D.), 385–389, Pergamon Press, Oxford.

Sokolov, N. D. (1964) *Vodorodnaya Svyaz*, 7–39

Sokolov, N. D. (1965) 'De la Théorie de la Liaison Hydrogène', *Annls Chim. Phys.*, **10**, 497–515

Somorjai, R. L. (1963) 'On Double Minimum Potentials in Hydrogen Bonded Systems', *Thesis*, Princeton University

Somorjai, R. and Hornig, D. F. (1962) 'Double-Minimum Potentials in Hydrogen-Bonded Solids', *J. chem. Phys.*, **36**, 1980–1987

Stanley, H. E. (1971) *Introduction to Phase Transition and Critical Phenomena*, Oxford University Press

Stillinger, F. H. and Rahman, A. (1972) *J. chem. Phys.*, **57**, 1281–1292

Sutor, D. J. (1962) 'The C–H——O Hydrogen Bond in Crystals', *Nature*, **195**, 68–69

Szczepaniak, K. and Tramer, A. (1965) 'Charge-Transfer Theory of Hydrogen Bond and Infrared Spectra of Chloroform Complexes', *Bull. Acad. pol. Sci., Ser. Sci. math. astr. phys.*, **13**, 79–83

Takagi, Y. (1948) 'Theory of the Transition in KH_2PO_4', *J. phys. Soc. Japan*, **3**, 271–274

Tsubomura, H. (1954) 'The Nature of the Hydrogen Bond. I. The Delocalization Energy in the Hydrogen Bond as Calculated by the Atomic Orbital Method', *Bull. chem. Soc. Japan*, **27**, 445–450

Werner, A. (1903) 'Die Ammoniumsalze als einfachste Metallammoniake', *Ber. dt. chem. Ges.*, **36**, 147–159

Whalley, E. and Bertie, J. E. (1967) 'Optical Spectra of Orientationally Disordered Crystals', *J. chem. Phys.*, **46**, 1264–1270

Widom, B. (1965) 'Equation of State in the Neighborhood of the Critical Point', *J. chem. Phys.*, **43**, 3898–3905

Wulf, O. R. and Liddel, U. (1935) 'Quantitative Studies of the Infra-Red Absorption of Organic Compounds Containing NH and OH Groups', *J. Am. chem. Soc.*, **57**, 1464–1473

7

Hybrid Bond Orbitals

LINUS PAULING and IAN KEAVENY
Stanford University, Stanford, California

The discovery of the wave character of the electron by de Broglie (1924) and the subsequent development of quantum mechanics have had a tremendous impact on chemistry during the last half century. Many of the fundamental questions about the nature of the chemical bond that had puzzled chemists during the period before 1924 had been answered by 1934. It was shown that detailed quantum mechanical calculations for the simpler molecules give values of the energy levels and some other properties in complete agreement with the observed values. For larger molecules the calculations become so complicated that it is impractical to carry them out, and chemists have found it convenient to develop a simple semi-empirical method of discussing the structure of molecules, in which classical chemical structure theory has been greatly extended by the incorporation of new ideas based upon quantum mechanics. One of the most important and useful of these ideas is the concept of hybridisation of bond orbitals.

1. THE CHEMICAL BOND

The first satisfactory quantum mechanical calculation of the energy of a chemical bond was made in 1927 by Burrau. The simplest of all chemical bonds is the one-electron bond in the hydrogen molecule-ion, H_2^+, which has bond energy $254.6 \, \text{kJ mol}^{-1}$ (energy of dissociation to H and H^+). Burrau showed that the Schroedinger wave equation for the electron in the field of the two protons could be separated in confocal ellipsoidal co-ordinates, and he solved the equation for the normal state by numerical methods. Condon (1927) then pointed out that by placing two electrons, with opposed spins, in the normal-state molecular orbital given by Burrau for H_2^+ and estimating the repulsion energy of the two electrons one obtains a rough value for the bond energy in the hydrogen molecule, H_2. These two papers provided the beginning of the molecular-orbital method of treating the chemical bond, which was greatly expanded in later years by F. Hund (1931) and R. S. Mulliken (1933).

The valence-bond method was originated by Heitler and London in 1927. They assigned to the normal state of the hydrogen-molecule the wave function A(1) B(2) + A(2) B(1), in which A is a 1s atomic orbital about one proton and B is a similar orbital about the other, and 1 and 2 are the two electrons. The Pauli exclusion principle requires that the spins of the two electrons be opposed. With this function the normal state of the hydrogen molecule can be described as involving an electron in the normal 1s atomic orbital about each of the two nuclei, with the two electrons resonating synchronously between the two orbitals (changing places).

The calculated bond energy (not corrected for vibration of the nuclei) is 303 kJ mol^{-1}, which is 67% of the experimental value, 455.4 kJ mol^{-1}. A similar calculation for H_2^+ (Pauling, 1928a) leads to a value 64% of the correct value.

2. CONCENTRATION OF BOND ORBITALS

It was pointed out by Wang (1928) that an improved value of the bond energy for the hydrogen molecule, 80% of the correct value, is obtained by using the Heitler–London wave function with the bond orbitals A and B taken as 1s orbitals with effective nuclear charge ζ somewhat greater than 1; that is, the orbitals are concentrated closer to the nuclei. A similar treatment for H_2^+ (Finkelstein and Horowitz, 1928) leads to a bond energy 81% of the correct value. The value of ζ for the shrunken orbital in H_2^+ is 1.228, and for that in H_2 is 1.166.

In 1931 Rosen introduced a term in the bond orbital to polarise it in the direction of the other atom, thus increasing its magnitude in the region of low potential energy for the electron, between the two nuclei. He wrote for the bond orbital A the expression $a_0 1s(\zeta) + a_1 2p(2\zeta)$, with a similar expression for B. By taking 2ζ as the effective nuclear charge for the 2p function its radial extent is made nearly the same as that of $1s(\zeta)$. The effect of the angular function $\cos \theta$ in 2p is to increase A on one side of the atom and decrease it on the other side. With $a = 0.103$ and $\zeta = 1.19$, the calculated value of the bond energy (not corrected for vibrational energy) increases from 303 to 388 kJ mol^{-1}. A similar treatment of H_2^+ (Dickinson, 1933) gave $a_1 = 0.145$, with the bond energy increasing from 217 to 263 kJ mol^{-1}.

3. TETRAHEDRAL BOND ORBITALS

An apparent conflict between the equivalence of the four valence bonds formed (in the directions towards the corners of a regular tetrahedron) by the carbon atom and the lack of equivalence of the 2s orbital and the other three orbitals of the L shell ($2p_x$, $2p_y$ and $2p_z$) was resolved by the recognition that resonance could occur among the four structures corresponding to an s bond with the four respective ligands, giving each bond 25%

s character and 75% p character (Pauling, 1928b). It was then pointed out by Slater (1931a) that the p_x, p_y and p_z orbitals have their maxima along lines at right angles to one another, and that the bond angles for p bonds should be 90°, in approximate agreement with many observed values in molecules and crystals. Slater also suggested that the four equivalent sp^3 hybrid orbitals would have their maxima in tetrahedral directions. Pauling (1931) made the simplifying assumption that the bond-forming power of an orbital could be taken to be proportional to the magnitude of the angular part of the orbital wave function in the direction toward the atom with which the bond is formed. He showed that with this criterion the best sp bond orbital that can be formed is $\frac{1}{2}s + (3\frac{1}{2}/2)p_z$, or an equivalent orbital obtained from this one by rotation to a different orientation in space, that any two such best orbitals that are mutually orthogonal must have a bond angle equal to 109.47° (the tetrahedral angle), and that a complete set of four such best sp^3 bond orbitals can be formed. He also showed that the best set of four equivalent sp^2d orbitals lie in a plane, with bonds directed towards the corners of a square, and that the best set of six equivalent sp^3d^2 bond orbitals direct their bonds towards the corners of a regular octahedron.

The best hybrid orbitals, by the angular strength criterion, have cylindrical symmetry about the axis in the direction along which the value of the angular wave function has its maximum. Hultgren (1932) discussed all possible sets of equivalent orthogonal cylindrical spd hybrid bond orbitals, with evaluation of the bond strength. A general discussion of bond orbitals in relation to symmetry was presented by Kimball (1940), and various aspects of bond-orbital theory have been described by other investigators (five equivalent d orbitals by Powell, 1968, Pauling and McClure, 1970, and Keaveny and Pauling, 1972; spd tetrahedral orbitals by Kuhn, 1948; icosahedral spdf orbitals by Macek and Duffey, 1960; and best nine sp^3d^5 orbitals by McClure, 1972).

4. JUSTIFICATION OF THE BOND-STRENGTH CRITERION

The assumptions upon which the bond-strength criterion was based (Pauling, 1931; Slater, 1931a) are that the maximum bond energy results from the maximum overlapping of two orbitals, one for each atom, in the region between the two bonded atoms, and from orthogonality of the orbitals of one atom involved in forming different bonds. Slater (1931b) quickly developed a quantum mechanical theory of the electronic structure of molecules that provided a sound justification of these assumptions. He showed how to formulate a wave function corresponding to a valence-bond description of any molecule. With this function the resonance integrals corresponding to interchange of two electrons involved in a single bond between two atoms occur in the expression for the energy of the molecule with the coefficient $+1$.

Since these integrals are negative, an increase in their absolute value, which would result from increasing overlapping, would stabilise the molecule, thus increasing the bond energy. The single exchange integrals of electrons in two orbitals not forming a bond with one another occur with coefficient

−1/2. Accordingly stability would result from minimum overlap of these orbitals, which can be achieved by having them orthogonal. Slater's theory has been the basis for most of the later work on the quantum mechanical valence-bond theory of molecular structure.

. 5. A DISCUSSION OF THE BOND-STRENGTH CRITERION

The original bond-strength function was assumed to be the value of the angular wave function for the orbital, normalised to 4π, along the direction of the bond ($\theta = 0°$). The cylindrical s, p, d, f, g and h functions are given in

Table 1. Cylindrical Bond Orbitals

$$
\begin{aligned}
s &= 1 \\
p &= 3^{\frac{1}{2}} \cos^2 \theta \\
d &= 5^{\frac{1}{2}}(3/2 \cos^2 \theta - \tfrac{1}{2}) \\
f &= 7^{\frac{1}{2}}(5/2 \cos^3 \theta - 3/2 \cos \theta) \\
g &= 9^{\frac{1}{2}}(35/8 \cos^4 \theta - 15/4 \cos^2 \theta + 3/8) \\
h &= 11^{\frac{1}{2}}(63/8 \cos^5 \theta - 35/4 \cos^3 \theta + 15/8 \cos \theta)
\end{aligned}
$$

Table 1. It is seen that the values of $S(0°)$ are $(2l+1)^{\frac{1}{2}}$; that is, 1, 1.7321, 2.2361, 2.6458, and so on.

The energy of a bond involving an orbital

$$\Phi = a_0 s + a_1 p + \ldots = \sum_l a_l \phi_l \tag{1}$$

can be taken to be

$$E = E_0 S - \sum_l a_l P_l \tag{2}$$

Here S is the bond-strength function $\sum_l a_l S_l$ and P_l is the promotion energy of ϕ_l, equal to the difference in atomic energy for ϕ_l and ϕ_0. Equation 2 applies when the bond orbital on the second atom remains constant. If both bond orbitals have the form given in equation 1 the bond energy becomes

$$E = E_{00} S^2 - n \sum_l a_l P_l \tag{3}$$

Here n is 1 for a one-electron bond and 2 for an electron-pair bond. The calculations made for H_2^+ by Dickinson and for H_2 by Rosen, mentioned above, permit a simple test to be made of the reliability of the bond-strength criterion, as expressed in equation 3. Dickinson and Rosen used atomic bond orbitals $a_0 1s(\zeta) + a_1 2p(2\zeta)$, and found a_1 to have the values 0.145 for H_2^+ and 0.103 for H_2. The promotion energy P_1 can be easily calculated by use of the virial-theorem values for the potential energy and kinetic energy; it is found to be ζW_0, with W_0 the ionisation energy of the hydrogen atom, 1312.3 kJ mol^{-1}. The value of E_{00}, the calculated energy of the s–s bond, is 217 kJ mol^{-1} for H_2^+ ($\zeta = 1.228$) and 363 kJ mol^{-1} for H_2($\zeta = 1.166$), and the values of E are 263 kJ mol^{-1} and 388 kJ mol^{-1}, respectively. When these values are introduced in equation 3, to obtain S, the values of S are found to correspond to 1.25 for H_2^+ and 1.78 for H_2 for the strength of the 2p orbital, in place of the maximum value $3\frac{1}{2} = 1.7321$ of the angular wave function. There is accordingly a rough substantiation of the bond-strength criterion.

On the other hand, it is found that equation 3 cannot be reliably used to evaluate the coefficient a_1 such as to make E a maximum; the calculated values of a_1 are over 50% too large. Moreover, inclusion of further terms [3d(3ζ), 4f(4ζ), etc.] leads to ever increasing values of E. It is, of course, expected that the values of $S(0°)$ are too large, inasmuch as the region of effective overlapping of the bond orbitals extends over a considerable volume centred about the internuclear axis, and an average value of the angular wave function over this region should be used, rather than the maximum value, on the axis itself.

(a) THE 2s2p STATE OF H_2^+

Another test of the bond-strength criterion can be made by evaluating the bond energy for the 2s2p excited state of H_2^+ as a function of the ratio of the coefficients of 2s and 2p in the hybrid bond orbitals. An early calculation by Pauling and Sherman (1937) led to the conclusion that for values of the internuclear distance greater than $4a_0$ the hybrid orbital that minimises the energy is very close to a tetrahedral orbital, in agreement with the bond-strength criterion. In this calculation approximations were made, including the neglect of the lack of orthogonality with the ground-state wave function. Somewhat better calculations, including, however, some errors, were made by Pritchard and Skinner (1951), Coulson and Lester (1955), and Gray, Pritchard, and Sumner (1956). The last authors pointed out that errors had been made in the earlier calculations, but they did not correct them. Reliable calculations were then carried out by Blethen, Keaveny, and Pauling (1973), who found that the best 2s2p bond orbital for the excited state of H_2^+ is in fact very close to the tetrahedral orbital, and that for this hybrid orbital the simple bond-strength criterion is reliable.

6. EXPANDED AND CONTRACTED d ORBITALS

One of the most valuable results of the theory of hybrid bond orbitals is its explanation of the stereochemistry and magnetic properties of co-ordination complexes of the transition metals (Pauling, 1931). Octahedral complexes of essentially covalent type were assigned d^2sp^3 bond orbitals, and square complexes of this type (diamagnetic for nickel, palladium, and platinum) were assigned dsp^2 orbitals. A serious difficulty became evident (see Coulson, 1969). This difficulty, that the d orbitals sometimes were ascribed a size (average value of r, the distance from the nucleus) such as to make them unsuitable for bond formation, has been resolved only recently.

Let us consider, for example, the hexamminocobalt (III) cation,

$[Co(NH_3)_6]^{+++}$. The six hybrid orbitals of cobalt involved in forming bonds with the six nitrogen atoms are considered to be hybrids of two 3d orbitals, one 4s orbital, and three 4p orbitals. The values of $\langle r \rangle$, the average distance from the nucleus, are 54 pm for 3d, 163 pm for 4s, and 210 pm for 4p, as calculated by the screening-constant method (Pauling and Sherman, 1932). Other methods of calculating $\langle r \rangle$, such as the Hartree–Fock method, give closely similar results. From the values for the hydrogen atom, $\langle r \rangle = 80$ pm, and the hydrogen molecule, internuclear distance 74 pm, we expect $\langle r \rangle$ for the cobalt bond orbitals to be equal to or slightly greater than the Co–N internuclear distance, 200 pm. The 4s and 4p orbitals of cobalt accordingly have about the right sizes, but the 3d orbital seems to be far too small to serve as a bond orbital.

This argument has, in fact, caused some investigators to reject a considerable part of the theory of chemical bonding. For example, Jorgensen (1971), in a paper entitled 'Breakdown of the Hybridization Theory and the Ligand Field Problem', mentions that hybridisation of 3d, 4s, and 4p is impossible for the iron-group elements because their values of $\langle r \rangle$ are in the ratios $1:3:4$, and suggests that the theory of hybrid bond orbitals has been discredited since 1955.

The resolution of this difficulty has, in fact, been achieved only recently. It has been recognised that the orbital occupied by a bonding electron may be either expanded or contracted considerably with respect to the orbital in the isolated atom. This expansion or contraction is accompanied by a contraction or expansion of other orbitals occupied by unshared electrons. Expansion or contraction of one orbital decreases or increases its shielding of the nuclear charge for another orbital. Study of the screening-constant expressions (Pauling and Sherman, 1932) shows that this effect is greatest when the two orbitals are in the same sub-shell. We conclude accordingly that the three unshared electron pairs occupying three of the 3d orbitals of the cobalt atom in the hexamminocobalt (III) complex are in contracted orbitals, with $\langle r \rangle$ less than 53 pm. These six 3d electrons shield the nucleus more effectively for the two electrons in the two bonding 3d orbitals so as to increase $\langle r \rangle$ for them to a value suitable for their service as bond orbitals. The cobalt atom in this complex can accordingly be described as having three unshared electron pairs in three contracted 3d orbitals and two bonding electrons in two expanded 3d orbitals, which are hybridised with 4s and 4p to produce the set of six d^2sp^3 octahedral bond orbitals.

Long ago it was pointed out by one of us (Pauling, 1938) that an empirical analysis of some of the properties of the first series of transition metals and their alloys, especially the saturation ferromagnetic moments, leads to the conclusion that about half of the five 3d orbitals are involved in bond formation and the other half are available for occupation by unshared electrons. The latter 3d orbitals were described as showing small interatomic overlapping. We now recognise that they are the contracted 3d orbitals, with the other 3d orbitals expanded to serve as bond orbitals.

This splitting of the d orbitals in metals into diffuse and localised bands was later confirmed, through detailed calculations, by Wood (1960) for the iron d band. He states that '. . . a significant fraction of the d electrons in solids do not "look" like atomic d electrons at all'. We now reaffirm that the expanded 3d orbitals serve as the bond orbitals while the remaining 3d orbitals are contracted into localised (non-bonding) atomic orbitals, with little overlap with the orbitals of adjacent atoms.

The quantitative discussion in the next section shows that there is no obvious reason for the number of expanded 3d orbitals to be equal to the number of contracted 3d orbitals. The number of expanded 3d orbitals was originally taken to be 2.56 (Pauling, 1948). Later (Pauling, 1961) it was taken to be 2.72, and the six bonding orbitals for the metals iron, cobalt, and nickel were taken to be hybrids of one 4s, 2.28 4p, and 2.72 4d orbitals. Because the number of expanded 3d orbitals is not otherwise limited, it may be determined by the bond-forming power of the hybrid orbitals, which may be considered to approximate the best spd bond orbital. The hybrid composition corresponds to bond strength S equal to 1.51 for p and 1.65 for d.

The discussion of co-operative expansion and contraction of 3d orbitals provides an explanation of a previously puzzling fact. Whereas iron, cobalt, and nickel crystallise in cubic or hexagonal close packing or the cubic body-centred structure, with metallic valence 6, manganese crystallises with three different structures, in which all or most of the atoms have metallic valence 4. With valence 6 the manganese atom would have only one un-shared electron in a contracted 3d orbital to stabilise the expanded 3d bonding electrons, whereas with valence 4 it would have three, which would serve better to stabilise the expanded bond orbitals. In β-manganese the unit cube contains 8 atoms with valence 6 and 12 with valence 4, and in α-manganese it contains 24 with valence 6 and 34 with valence 4. The smaller 6-valent atoms have icosahedral ligation, and the larger 4-valent atoms have ligancies greater than 12; the structures are accordingly denser than close-packed structures of equivalent spheres. Chromium also forms crystals with the α-manganese structure, presumably stabilised by the same factors.

7. ORBITAL SIZES FOR IRON-GROUP ATOMS

In order to check their reliability, we have calculated values of the sizes of 3d and 4s orbitals of the atoms from scandium to zinc by several methods, with the results given in Table 2. The values headed C1, C2, and R are based on atomic wave functions that are linear combinations of Slater orbitals. It is seen that the calculated values of the radius at the outer maximum of the radial distribution function are in moderately good agreement with one another. The values of $\langle r \rangle$ calculated with one of these wave functions agree very well with those calculated by the screening-constant method. This very simple calculation is made by inserting in the expression for r for a hydrogen-like orbit, $n^2 a_0 [3/2 - l(l+1)/2n^2]/(Z - S_s)$ the appropriate values of the size screening constant S_s, as tabulated by Pauling and Sherman.

It is seen that for all these transition elements the calculated radii of the

3d orbitals are far too small to permit their effective hybridisation with 4s and 4p to form good bond orbitals, and that a threefold expansion in size would be required to permit them to function in this way.

8. THE ENERGY REQUIREMENTS FOR EXPANDED AND CONTRACTED 3d ORBITALS

We have carried out an extensive set of calculations of the energy of the nickel atom in the state $3d^{10}\,^1S_0$ as a function of the relative sizes of expanded and contracted 3d orbitals, the wave function used was a minimal-basis function of the Slater type, with the orbital functions as given by Slater's

Table 2. Calculated Sizes of 3d and 4s Orbitals (pm)

			r at maximum of $\Psi^2 r^2$						Values of $\langle r \rangle$				
	Atom $(3d^n\,4s^2)$		3d				4s		3d			4s	
n			C1	C2	R	C1	C2	R	C1	PS		C1	PS
1	Sc	^{2}D	59	67	—	172	184	—	89	88		209	203
2	Ti	^{3}F	52	59	44	163	176	177	77	78		200	195
3	V	^{4}F	48	53	41	155	171	170	70	72		192	188
4	Cr	^{5}D	44	49	38	149	166	164	65	66		185	181
5	Mn	^{6}S	41	45	37	143	161	157	60	62		179	175
6	Fe	^{5}D	38	43	35	137	156	152	57	58		172	169
7	Co	^{4}F	36	40	33	132	152	146	54	54		167	163
8	Ni	^{3}F	34	38	32	128	149	142	51	51		162	159
9	Cu	^{2}D	32	36	—	124	145	137	49	48		157	154
10	Zn	^{1}S	30	34	—	120	142	—	46	45		153	149

Cl: Clementi, 1965; C2: Clementi and Raimondi, 1963; R: Richardson, Niewpoort, Powell, and Edgell, 1962; PS: Pauling and Sherman, 1932.

rules (Slater, 1930). With the argon core invariant, the optimum orbital exponent in the expression

$$\langle \psi \rangle = \left[\frac{(2\zeta)^{2n+1}}{(2n)!} \right]^{\frac{1}{2}} r^{n-1} \exp(-\zeta r)$$

for ten electrons in equivalent 3d orbitals was found to be 3.405, corresponding to total energy -1496.165 atomic units (1 A.U. = 2625.64 kJ mol^{-1}). For a single Slater function the value of $\langle r \rangle$ is $(n+\frac{1}{2})a_0/\zeta$. Accordingly the value for the 3d electrons in Ni $3d^{10}$ is calculated to be 54.5 pm, in good agreement with value 51 pm for Ni $3d^8 4s^2$ (Table 2).

The value of the energy, -1496.165 A.U. may be compared with that found for a minimal-basis set with the sum of two Slater functions for 3d, 1503.183 A.U. (Richardson *et al.*, 1962), and that for a Hartree–Fock function, -1506.643 A.U. (Watson, 1960a, b). The energy value -1499.132 A.U. is obtained if we use the inner core of Richardson *et al.* (1962) and optimise the energy with respect to a single 3d Slater function.

With the argon-core function invariant, calculations of the energy were made for various values of two size parameters, ζ_e and ζ_c, for expanded and contracted 3d orbitals, respectively, with 2, 4, 6, and 8 electrons in expanded orbitals and 8, 6, 4, and 2 in contracted orbitals. We had expected that the energy of the atom would have its minimum for ten equivalent 3d electrons, and would rise by only a small amount when the electrons were separated into an expanded and a contracted group. We found, however, that the atom is made more stable by separation of the 3d electrons into an expanded and a contracted group (Table 3). This additional stability can be attributed to radial correlation, amounting to 0.03 A.U. per electron–electron interaction for the 2, 8 system. The effect was recognised forty years ago (Eckart, 1930). The wave function for the normal helium atom with one electron in an expanded 1s orbital and one in a contracted 1s orbital gives a calculated value of the energy 0.0278 A.U. less than that with the two electrons equivalent. Moreover, a significant improvement in energy (by -4.051 A.U.) for nickel with ten equivalent 3d electrons is achieved by taking the 3d orbital as the sum of an expanded and a contracted orbital (Richardson *et al.*, 1962),

Table 3. Properties of Expanded and Contracted 3d Orbitals

Electrons in expanded orbitals	0	2	4	6	8	10
Electrons in contracted orbitals	10	8	6	4	2	0
Extra binding energy (A.U.)	0	1.427*	1.375	0.385	0.037	0
ζ (expanded)		3.405	0.56	1.13	2.07	3.12	3.405
r (expanded)		54.5	331	164	90	59	54.5
ζ (contracted)		3.405	3.88	4.36	4.60	4.26	3.405
r (contracted)		54.5	48	43	40	44	54.5

*A similar value, 1.403 A. U., is obtained by using the description of Richardson *et al.* (1962) for the inner argon core and optimising the energy with respect to the 8 contracted and 2 expanded 3d electrons.

with $\zeta = 2.00$ and 5.75, respectively, and Richardson and his co-workers point out that many pairs of values of ζ_e and ζ_c give nearly the same value of the energy.

In our calculation the greatest stability of the nickel atom is found for about 3 electrons in expanded orbitals and 7 in contracted orbitals. The lack of symmetry may be attributed to increased repulsion between the inner shells and the contracted orbitals as they become smaller. We conclude that for the iron-group elements maximum stability of hexacovalent complexes, which involve two electrons in expanded 3d orbitals, would occur with about 5 unshared 3d electrons, as, for example, with iron (III). The value of $\langle r \rangle$ for this case (Table 3) is also just right. Other ratios of electrons in expanded to those in contracted may also be compatible with the formation of good dsp bond orbitals, inasmuch as the energy changes very little when the value of ζ_e is moved to the desired value, about 0.9. We find, for example, that for four electrons in expanded orbitals the change of ζ from the value 1.13 (Table 3) to 0.9, with the best value of ζ_c, is accompanied by a change in energy of only 0.025 A.U. per bonding electron, which is small in comparison with the total bond energy.

A somewhat different problem from that of the 3d 4s 4p hybrid orbitals of the transition metals is that of the 3s 3p 3d hybrid orbitals, such as those of the phosphorus atom in its quinquecovalent compounds (PCl_5, for example), or those of the sulphur atom in its sexicovalent compounds (SF_6, etc.). Craig, *et al.* (1954) mentioned that the values at which the Slater orbitals have their maxima for the sulphur atom are 77 pm for 3s and 3p and 289 pm for 3d. The values of $\langle r \rangle$ calculated by the screening-constant method (Pauling and Sherman, 1932) are 93 pm for 3s, 111 pm for 3p, and 232 pm for 3d. These values are quite incompatible with hybridisation to an effective spd bond orbital. Craig *et al.* (1954) suggested that interaction with the ligands leads to a suitable contraction of the 3d orbitals, and this was supported by the results of calculations carried out by Craig and Magnusson (1956), Craig and Zauli (1962a, b), and Bendazzoli and Zauli (1965). The role of 3d orbitals in multiple bonds (π-bonds) has been discussed by many authors, including Schomaker and Pauling (1939), Wolfsberg and Helmholz (1952), Jaffe (1953), Cruickshank (1961), Archibald and Perkins (1970, 1971), and Perkins (1972).

We agree with the conclusion reached by Coulson (1969) that the apparent unsuitability in size of d orbitals to use in hybrid bond orbitals is in many molecules and complexes overcome by the expansion of the orbitals that are too small or contraction of those that are too large. The rejection of the theory of hybridisation by Jorgensen (1971) on the basis of the incompatibility in size of the d orbitals with the s and p orbitals is unjustified. The theory of hybridisation continues to have great value as a part of the structure theory of molecules, contributing to the explanation of the properties of substances in terms of their molecular structure.

9. SUMMARY

Calculations have been made of the energy of a nickel atom with 0, 2, 4, 6, or 8 of the 10 3d electrons in expanded 3d orbitals, suitable to bond formation, and the others in contracted 3d orbitals. The results of these calculations show that the size of 3d orbitals may vary in such a way as to make them available for forming spd hybrid bond orbitals; this answers one criticism that has been made of the theory of hybrid orbitals. Other aspects of the theory are reviewed.

REFERENCES

Archibald, R. M. and Perkins, P. G. (1970) *Chem. Comm.*, 569
Archibald, R. M., and Perkins, P. G. (1971) *Rev. roum. Chim.*, **16**, 1137
Bendazzoli, G. L. and Zauli, C. (1965) *J. chem. Soc.*, 6827
Blethen, M., Keaveny, I. T. and Pauling, L. (1973) To be published
Burrau, Ø. (1927) *Det. Kgl. Danske Vid. Selskab.*, **7**, 1
Clementi, E. (1965) 'Tables of Atomic Functions', A supplement to the paper; *I.B.M. J. Res. Develop.*, **9**, 2
Clementi, E. and Raimondi, D. L. (1963) *J. chem. Phys.*, **38**, 2686
Condon, E. U. (1927). *Proc. natn. Acad. Sci.*, **13**, 466
Coulson, C. A. (1969) *Nature, Lond.*, **221**, 1106
Coulson, C. A. and Lester, G. R. (1955) *Trans. Faraday Soc.*, **51**, 1605
Craig, D. P., Maccoll, A., Nyholm, R. S., Orgel, L. E. and Sutton, L. E. (1954) *J. chem. Soc.*, 332

98 Wave Mechanics

Craig, D. P. and Magnusson, E. A. (1956) *J. chem. Soc.*, 4895
Craig, D. P. and Zauli, C. Z. (1962a) *J. chem. Phys.*, **37**, 601
Craig, D. P. and Zauli, C. Z. (1962b) *J. chem. Phys.*, **37**, 609
Cruickshank, D. W. J. (1961) *J. chem. Soc.*, 5486
De Broglie, L. (1924) *Thesis*, Paris
De Broglie, L. (1925) *Annln Phys.*, **3**, 22
Dickinson, B. N. (1933) *J. chem. Phys.*, **1**, 317
Eckart, C. E. (1930) *Phys. Rev.*, **36**, 878
Finkelstein, B. N. and Horowitz, G. E. (1928) *Z. Phys.*, **48**, 118
Gray, B. F., Pritchard, H. O. and Sumner, F. H. (1956) *J. chem. Soc.*, 2631
Heitler, W. and London, F. (1927) *Z. Phys.*, **44**, 455
Hultgren, R. (1932) *Phys. Rev.*, **40**, 891
Hund, F. (1931) *Z. Phys.*, **73**, (1), 565
Jaffé, H. H. (1953), *J. phys. Chem.*, **58**, 185
Jorgensen, C. K. (1971) *Chimia*, **25**, 109
Keaveny, I. T. and Pauling, L. (1972) *Israel J. Chem.*, **10**, 211
Kimball, G. E. (1940) *J. chem. Phys.*, **8**, 188
Kuhn, H. (1948) *J. chem. Phys.*, **16**, 727
Macek, J. H. and Duffey, G. H. (1960) *J. chem. Phys.*, **34**, 288
McClure, V. (1972) *Thesis*, University of California, San Diego
Mulliken, R. S. (1933) *J. chem. Phys.*, **1**, 492
Pauling, L. (1928a) *Chem. Rev.*, **5**, 173
Pauling, L. (1928b) *Proc. natn. Acad. Sci.*, **14**, 359
Pauling, L. (1931) *J. Am. chem. Soc.*, **53**, 1367
Pauling, L. (1938) *Phys. Rev.*, **54**, 899
Pauling, L. (1948) *Nature, Lond.*, **161**, 1019
Pauling, L. (1961) *J. Indian chem. Soc.*, **38**, 435
Pauling, L. and McClure, V. (1970) *J. chem. Educ.*, **47**, 15
Pauling, L. and Sherman, J. (1932) *Z. Krist.*, **81**, 1
Pauling, L. and Sherman, J. (1937) *J. Am. chem. Soc.*, **59**, 1450
Perkins, P. G. (1972) *Rev. roum. Chim.*, **17**, 773
Powell, R. E. (1968) *J. chem. Educ.*, **45**, 45
Pritchard, H. O. and Skinner, H. A. (1951) *J. chem. Soc.*, 945
Richardson, J. W., Niewpoort, W. C., Powell, R. R. and Edgell, W. F. (1962) *J. chem. Phys.*, **36**, 1057
Rosen, N. (1931) *Phys. Rev.*, **38**, 2099
Schomaker, V. and Pauling, L. (1939) *J. Am. chem. Soc.*, **61**, 1769
Slater, J. C. (1930) *Phys. Rev.*, **36**, 57
Slater, J. C. (1931a) *Phys. Rev.*, **37**, 481
Slater, J. C. (1931b) *Phys. Rev.*, **38**, 1109
Wang, S. C. (1928) *Phys. Rev.*, **31**, 579
Watson, R. E. (1960a) *Phys. Rev.*, **118**, 1036
Watson, R. E. (1960b) *Phys. Rev.*, **119**, 1934
Wolfsberg, M. and Helmholtz, L. (1952) *J. chem. Phys.*, **20**, 837
Wood, J. H. (1960) *Phys. Rev.*, **117**, 714

8

Application of Wave Mechanics to the Electronic Structure of Molecules through Configuration Interaction

P. S. BAGUS, B. LIU, A. D. McLEAN and M. YOSHIMINE
IBM Research Laboratory, San Jose, California

1. INTRODUCTION

The advent of wave mechanics, fifty years ago, provided the basic principles governing all chemical phenomena. The field of quantum chemistry deals with the application of these principles to the detailed understanding and prediction of chemical phenomena. The calculations on the hydrogen molecule by Heitler and London (1927), and later by James and Coolidge (1933), ushered in a branch of quantum chemistry, generally referred to as *ab initio* molecular quantum mechanics, which seeks solutions of the Schroedinger equation to gain detailed knowledge of molecular properties and processes. The efforts of two generations of research workers in this field, assisted by large scale electronic computing machinery, have made it possible to computationally obtain reliable information on properties of ground and excited states of small molecules. Work in this area has recently been reviewed by Schaefer (1971a).

The problem addressed in this chapter is that of finding solutions of the Schroedinger equation with an accuracy beyond that of the independent-particle Hartree–Fock (HF) approximation. Computational and numerical problems have been so thoroughly investigated for the HF procedure that tools now exist which allow research workers, not necessarily specialists in computation, to determine HF wave functions in a routine fashion. This is not at all the case for *ab initio* methods, which strive to resolve the correlation errors (Löwdin, 1959) inherent in the HF model. These errors, though small in most cases, are often too large to allow useful prediction of chemical processes.

The most general method for determining these correlation effects, and with which most calculations have been made, is the method of configuration

99

interaction (CI). In what follows we discuss our current knowledge of how to quantitatively apply the method of configuration interaction. In an attempt to give an up-to-date account, we include references to yet unpublished results by various research groups. We gratefully acknowledge the co-operation of the scientists who made this possible.

2. THE CONFIGURATION INTERACTION METHOD

We consider the determination of the eigenfunctions and eigenvalues of the n-electron clamped nuclei Hamiltonian

$$H = -\tfrac{1}{2}\sum_i \nabla_i^2 - \sum_{A,i} \frac{Z_A}{r_{Ai}} + \sum_{i<j} \frac{1}{r_{ij}} + \sum_{A<B} \frac{Z_A Z_B}{r_{AB}} \tag{1}$$

In equation 1 the terms, in order, correspond to kinetic energy, nuclear attraction, electronic repulsion, and nuclear repulsion, with i, j referring to electrons, A, B, to nuclei. Terms responsible for spectral fine structure and hyperfine structure, such as magnetic interactions, as well as those which couple electronic and nuclear motions are not included. These effects are normally small and can be accurately treated through perturbation theory.

In the configuration interaction (CI) method, approximations to the true eigenstates of equation 1 are determined by diagonalising the matrix representation of the Hamiltonian in an orthonormal basis set of n-particle functions. The n-particle functions are themselves constructed from an orthonormal basis set of one-particle functions. We therefore seek roots of

$$(\mathbf{H} - E\mathbf{S})\mathbf{C} = 0 \tag{2}$$

in which $\mathbf{H}$ is the Hamiltonian matrix in the selected basis, and $\mathbf{S}$ is the unit matrix. E is an eigenvalue of H in this basis, with C the associated eigenvector. If the n-particle basis functions are Φ_K, then elements of the $\mathbf{H}$ and $\mathbf{S}$ matrices are

$$H_{KL} = \int dx_1 \ldots dx_n \Phi_K^*(x_1,\ldots,x_n) H \Phi_L(x_1,\ldots,x_n) \tag{3}$$

$$S_{KL} = \int dx_1 \ldots dx_n \Phi_K^*(x_1,\ldots,x_n) \Phi_L(x_1,\ldots,x_n) = \delta_{KL} \tag{4}$$

where, in equations 3 and 4, integration variables x include space and spin co-ordinates of the suffixed electron. An eigenvector of equation 2 is an expansion in Φ_K

$$\Psi = \sum_K C_K \Phi_K \tag{5}$$

If, in equation 5, the Φ_K include all n-particle functions that can be constructed from the one-particle basis for the problem, then the eigenvalues of equation 2 and the wave function Ψ of equation 5 are invariant under a linear transformation of the one-particle basis. This would be a *complete CI*, yielding eigenvalues and eigenvectors which are at the one-particle basis set limit. However, complete CI calculations for systems with more than four electrons are not feasible at present, or in the near future. We must therefore seek truncated CI expansions which approximate the complete CI limits within tolerable error bounds. The results of such calculations depend on the actual choice of the one-particle and n-particle basis functions. The practical problem discussed in this chapter is the determination of one- and n-particle basis functions which achieve the best possible results from a severely truncated CI expansion.

In the limit of complete one- and n-particle basis sets, diagonalisation of the Hamiltonian yields the exact spectrum of the Hamiltonian. In practice one can think of approaching an exact eigenstate of equation 1 by following a sequence of one-particle basis limits. In these terms the questions to be asked are: (1) For a given one-particle basis how closely do we approach the one-particle basis limit, when we are using less than the full set of possible n-particle functions? and (2) how closely does the one-particle basis limit approach the exact eigenstate to be determined? It is apparent that there must be some appropriate balance between the two basis sets. The one-particle basis limit should be close to the exact result, and a sufficiently large n-particle basis set must be used so that this limit may be approached. An unbalanced calculation, for example one with an n-particle basis set sufficiently large to approach an inadequate one-particle basis limit, can be justified in terms of insight gained into the nature of CI expansions, but not in terms of the value or accuracy of predicted properties.

3. ELEMENTARY FUNCTIONS AND THE ONE-PARTICLE BASIS

The space of the one-particle basis is spanned by a set of *elementary functions*, $\{\chi\}$, chosen to provide a flexible starting point for expanding orthogonal functions, $\{\phi\}$, from which the n-particle basis functions are to be constructed (Roothaan, 1951). The orthogonal orbital functions are the one-particle basis; the expansion in $\{\chi\}$ is according to

$$\phi_{i\alpha} = \sum_{p=1}^{N} c_{i\alpha, p} \chi_p \qquad (6)$$

where in equation 6, i is an orbital serial number, α counts over functions forming a basis for an irreducible representation of the ith orbital, p is a serial number of an elementary function, and $c_{i\alpha, p}$ is a linear expansion coefficient.

The elementary functions are normally a non-orthogonal set of functions centred on the various nuclei in the molecule, chosen largely for their ability to describe the electronic structure of the separated atoms in the states to which the desired molecular state or states dissociate. In addition to these functions, which can be determined in simpler atomic calculations, the elementary functions must include functions which efficiently describe electronic polarisation effects in the molecular field, functions spanning the space of molecular Rydberg states which may correlate with separated atom occupied orbitals, and functions required to describe many-body molecular correlation effects, particularly those that are geometry dependent. Different regions of the space spanned by the elementary functions may get different treatments, depending on the information required from the calculation. For example, where the space of the valence shells is of greatest importance a much more approximate description of inner shells may be satisfactory. One set of properties where particular care does have to be taken with inner shells is the electric field and its gradients at the molecular nuclei. For these properties, polarisation effects involving shell interactions, of the anti-shielding type for example, through the entire shell structure can be important. Thus, appropriate one-particle basis functions for polarising all shells will be necessary in computing these properties.

It is a common practice to choose the elementary functions by supplementing well optimised atomic basis sets with polarisation functions, chosen by symmetry type, with exponents based on principles of maximum overlap or from previous experience (Bagus, Gilbert and Roothaan, 1972; Huzinaga, 1971; Clementi, 1965; McLean and Yoshimine, 1968). These elementary functions, particularly the polarisation ones, can be optimised in the system under consideration; a classical case study of optimisation of elementary functions for a small molecule is given by Cade, Sales and Wahl (1966). However, such processes are expensive and time consuming. We propose here a method for choosing appropriate polarisation functions by carrying out modified atomic calculations. Atomic basis sets supplemented by polarisation functions are optimised to reproduce atomic properties in external electric fields. We stress that these atomic calculations may be of HF or CI type. If the elementary functions used in such atomic calculations are capable of describing the distortions in the electric field, as indicated by the accuracy of computed atomic polarisabilities and atomic shielding factors, then they must be appropriate for describing the atomic distortions in a molecule. In fact, functions required to reproduce atomic polarisabilities should be suitable for describing molecular valence shell polarisation effects; functions required to reproduce atomic shielding factors should be suitable for describing polarisation phenomena contributing to electronic properties at the atomic nuclei. A systematic study of atoms in external electric fields along these lines could be of great value in providing a starting point for molecular calculations. The method has been applied with considerable success to the determination of the quadrupole coupling constant in LiH (McLean and Yoshimine, 1972), and to the determination of potential curves for the low-lying electronic states of hydrides of first-row atoms (Billingsley and Krauss, 1972).

Functional forms of the elementary functions most used are:

(1) Slater-type, $r^{n-1}e^{-\zeta r}Y_{l,m}(\theta, \phi)$, with $n > l \geq |m|$ integers, and ζ referred to as the elementary function exponent. These have proven superior for use in diatomic and linear molecular systems. They have achieved more limited use for systems of more complicated geometry in which matrix element evaluation of the electronic repulsions has proved a severe obstacle.

(2) Gaussian-type in either spherical polar co-ordinates, $r^{n-1}e^{-\alpha r^2}Y_{l,m}$ (θ, ϕ), or more commonly in cartesians, $x^{n_1}y^{n_2}z^{n_3}r^{n_4}e^{-\alpha r^2}$, with certain constraints on the integral powers of co-ordinates, related to the methods of matrix element evaluation.

(3) Numerical-type, $R(r)Y_{l,m}(\theta, \phi)$, where the radial dependence of the single-centre function has been determined numerically in some previous calculation. As a starting point for molecular calculations care must be taken to see that elementary functions of this type are sufficiently flexible to allow expansion of near optimal molecular orbitals. This could be done using numerically tabulated atomic orbitals, and their derivatives, in the molecular basis set.

So far the discussion has been on the elementary functions rather than the one-particle basis. This emphasis is not misplaced, since the results of molecular computation are ultimately dependent upon these functions; no inadequacy in the elementary function can be made up by further computation. The actual choice of the one-particle basis functions is strongly coupled to the choice of the n-particle basis and will therefore be deferred until after the description of the n-particle basis in the next section.

4. n-PARTICLE BASIS

The simplest n-particle functions which can be constructed from the one-particle basis, and satisfy the physical requirements of antisymmetry with respect to electron exchange, are Slater determinants. In a Slater determinant, electrons occupy a subset of the spin orbitals, $\psi_j = \phi_{i\alpha}\{^\alpha_\beta\}$, which can be constructed from an orbital member of the one-particle basis multiplied by either spin α or spin β. Each Slater determinant

$$\frac{1}{\sqrt{n!}} \begin{vmatrix} \psi_1(1) \, \psi_2(1) \dots \psi_n(1) \\ \psi_1(2) \, \psi_2(2) \dots \psi_n(2) \\ \vdots \qquad\qquad \vdots \\ \psi_1(n) \, \psi_2(n) \dots \psi_n(n) \end{vmatrix}$$

is normalised, and the set of all Slater determinants is orthonormal, the orthogonality following from spin and orbital orthogonality. Slater determinants are, in general, not eigenfunctions of the symmetry operators for the system. A preferable n-particle basis is formed by taking linear combinations of Slater determinants which are symmetry eigenfunctions, configuration state functions (CSF's), and in particular we will be interested in CSF's that belong to certain n-particle subspaces which we classify below.

To construct the desired n-particle functions it is necessary to start with electronic configurations, which are assignments of electrons to orbitals, rather than an assignment to spin orbitals as in a single Slater determinant. We define the *electronic configuration*, C, as

$$C = C(n_1, n_2, n_3, \dots, n_I, n_{I+1}, \dots, n_{I+E}) \tag{7}$$

in which occupation number n_i is the number of electrons in orbital ϕ_i, up to a maximum $2\lambda_i$ where λ_i is the degeneracy of the irreducible representation of ϕ_i. The notation of equation 7 shows a division in the orbital set between ϕ_I and ϕ_{I+1}. The first n_I orbitals are called *internal orbitals*, and the remaining n_E, spanning the remainder of the one-particle space, are called *external orbitals*. The dividing point between the internal and external set is not unique, and depends to some extent on the desired accuracy in the calculation, but the internal orbital set must contain all orbitals which are used in the construction of what will turn out to be the leading terms in the final CI expansion. For any given problem it will be necessary to select a set of 'reference' configurations, denoted $\{0,0\}$, which are the most important in describing the electronic state of interest. This calls on chemical knowledge, intuition, and previous experience, and is subject to confirmation in the course of calculation. Any other configuration can then be assigned to the set $\{n, m\}$ where the integers n, m describe the type of orbital excitation which relates any configuration to the members of $\{0, 0\}$. Suppose that the members of $\{0, 0\}$ are $C_q[n_1^q, n_2^q, ..., n_I^q, 0, 0, ..., 0]$ in which we note that, by definition, all external orbital occupation numbers are zero. Then, any configuration $C_a[n_1^a, n_2^a, ..., n_I^a, n_{I+1}^a, ..., n_{I+E}^a]$ belongs to the set $\{n, m\}$ with

$$n = \left[\frac{1}{2} \sum_{i=1}^{I+E} | n_i^a - n_i^q | - m \right]_{min} \tag{8}$$

and

$$m = \sum_{i=I+1}^{I+E} n_i^a \tag{9}$$

and where, in equation 8, min refers to the smallest value the expression inside parentheses can take as a function of q. The number n is the minimum number of electrons excited into internal orbitals in C_a compared with all C_q, and m is the number of electrons excited into external orbitals.

The total number of Slater determinants with spin orbital assignments consistent with the orbital occupation of a configuration C is

$$N_C = \prod_{i=1}^{I+E} \frac{(2\lambda_i)!}{n_i!(2\lambda_i - n_i)!} \tag{10}$$

These Slater determinants span the n-particle space of configuration C, and can be transformed to form bases of irreducible representations of the symmetry group for the n-electron system, called *configuration state functions* (*CSF's*). Methods for deriving CSF's are well known and will not be described here. It suffices to say that a comprehensive review has been given by Slater (1963). Where configurations are assigned to the sets $\{n, m\}$, CSF's, or linear combinations of them, are assigned to sets $\{n, m; i\}$ in which the first two integers correspond to the configuration from which the n-particle function is derived, and the third is assigned on the basis of perturbation theory in the following way. From the configurations in $\{0, 0\}$ we select CSF's, supposed to be the most important in describing the electronic state under consideration, and assign these CSF's to $\{0, 0; 0\}$. A zeroth-order approximation to this desired state will be an eigenstate of the Hamiltonian in the n-particle basis $\{0,0;0\}$. We now go through all configurations in the problem, and for each configuration determine the CSF's, or linear combinations of them, that would make a first-order correction to this zeroth-order function in Rayleigh–Schroedinger perturbation theory. These functions are members of $\{n, m; 1\}$. The contributions to $\{n,m; 1\}$ from some given configuration are such that there exists no linear combination of members of $\{n, m; 1\}$ which have a zero matrix element through the Hamiltonian with any function in $\{0, 0; 0\}$, independent of the numerical value of electronic matrix elements between orbitals. In determining the first-order subspace only configurations with orbital excitation $n+m \leq 2$ need be considered. As an example let us suppose that the CSF of $^2\Pi$ symmetry with $M_L = 1$ arising from a configuration $\sigma^2\pi^3$ is the only member of $\{0, 0; 0\}$. Now consider a configuration $\sigma\pi^2\sigma'\pi'$ in which, relative to $\sigma^2\pi^3$, one electron out of both σ and π shells is excited, one into σ' and one into π'. There are 96 Slater determinants spanning the n-particle space of this configuration which under a linear transformation become CSF's forming bases of $7^2\Pi$, $5^4\Pi$, $1^6\Pi$, $2^2\Phi$, and $1^4\Phi$ irreducible representations of the axial symmetry group. Of these CSF's there are seven with $^2\Pi$ symmetry and $M_L = 1$. In general, all seven will have a non-zero matrix element through the Hamiltonian with $\{0, 0; 0\}$ and a further linear transformation is necessary before we find the n-particle functions which contribute to the first-order space. In this particular case the number of functions contributing to the first-order space is 2; that is, out of the seven CSF's of appropriate symmetry it is possible to find five linearly independent combinations which have zero matrix element through the Hamiltonian with $\{0, 0; 0\}$ independent of the numerical value of integrals over molecular orbitals. An algorithm for determining a basis for n-particle functions which contribute in the first-order space has been developed (McLean and Liu, 1973).

We can now determine the second-order subspace by going through all the configurations in the problem, and from the n-particle space of each configuration, not contained in the zeroth- and first-order spaces, find n-particle functions which span the space of those, and only those, functions which have a non-zero matrix element through the Hamiltonian with some member of the first-order subspace. The same algorithm (McLean and Liu, 1973) used for determining the first-order subspace can be applied. The second-order subspace is the space of all functions which, in Rayleigh–Schroedinger perturbation theory, contribute first to the second-order correction to the zeroth-order wave function.

This partitioning of the n-particle space, based on perturbation theory, is obviously generalisable. The ith-order subspace, orthogonal to all other subspaces, contains those, and only those, functions which have a non-zero Hamiltonian matrix element with at least one function in the $(i-1)$th-order subspace.

The space spanned by $\{n, m; i\}$ for fixed values of n, m, i, which includes all possible functions of this type that can be generated from the starting one-particle basis, is invariant to a unitary transformation (rotation) of the external orbitals. Thus, the first-order subspace is invariant to a rotation of the external orbitals. It follows that a CI wave function expanded in an

n-particle basis made up of the zeroth- and first-order subspaces will be invariant under a rotation of the external orbitals since it is variationally determined in an invariant n-particle space. However, the Rayleigh–Schroedinger first-order perturbed wave function is not invariant to rotation of the external orbitals since, even though it is an expansion in the same space as the CI calculation, it is not variationally determined.

The dimension of these subspaces, and their importance, depends on the choice of the zeroth-order subspace. With an adequate choice, the zeroth- and first-order subspaces can alone account for over 90% of the correlation energy in the problem, even though they are only a small part of the full n-particle space. Selection of the zeroth-order subspace will be discussed at length in a later section.

A final remark is to emphasise that, even though n-particle functions are classified on the basis of perturbation theory, we are not committed to perturbation theory wave function expansions. In fact, if possible, a variational (CI) expansion is preferable; perturbation theory has simply been used to classify functions according to qualitative importance, not to evaluate their quantitative contribution.

5. ONE-PARTICLE BASIS AND NATURAL ORBITALS

The one-particle space is spanned by a set of elementary functions. As discussed earlier, this space must be complete enough to give desired accuracy for physical properties of interest. To construct CSF's, this set of elementary functions is transformed into an orthonormal set of basis orbitals of the same or smaller dimension. In a CI calculation, this orbital basis determines the dimension of the n-particle space, the importance of individual CSF's, and ultimately the extent to which the CI expansion may be truncated. In a good set of basis orbitals we seek the following qualities: (1) it is easy to construct; (2) it is compact in the sense that it packs the most important part of the one-particle space, spanned by the elementary functions, into a small number of orbitals; (3) it leads to a compact n-particle basis, in the sense that it packs the most important part of the n-particle space into a small number of CSF's; (4) it should facilitate the selection of these important CSF's.

The determination of orbital bases for CI calculations has been the subject of extensive study. The use of natural orbitals, underlying some of the most successful methods of orbital determination, deserves special attention. The natural orbitals (NO's) of a complete CI wave function are obtained by diagonalising the first-order density matrix (Löwdin, 1955). The eigenvectors are the NO's, and the eigenvalues are the occupation numbers. We restrict our consideration to what Davidson (1972) calls the symmetry-constrained natural orbitals. These NO's are constructed subject to symmetry and equivalence constraints, namely that orbitals span the basis of irreducible representations of the spatial symmetry group of the molecule and that the same orbitals are used for degenerate representations and for both α and β spins. A complete CI wave function constructed from a one-particle basis of NO's has a diagonal first-order density matrix. The NO's, as a one-particle basis, satisfy certain optimal convergence properties. While none of these properties has any real value in practice, computational experience has shown that the NO's satisfy most of the criteria given above for a desirable one-particle basis. The properties and uses of NO's have been reviewed by Davidson (1972). In practice we usually deal with approximations to the NO's obtained by diagonalising the density matrix of a truncated CI expansion. These approximate NO's are referred to as ANO's.

With this background we can now offer an adequate description of a path that we believe holds promise of developing accurate CI expansions with practicable effort and with a minimal ingredient of arbitrariness in the selection of the n-particle basis. Three steps are involved: (1) The determination of internal orbitals and the associated choice of $\{0, 0; 0\}$; (2) the determination of external orbitals and CI wave functions expanded in CSF's of the first-order subspace, and (3) the inclusion of higher order subspaces in a final CI wave function.

6. INTERNAL ORBITALS AND ZEROTH-ORDER SUBSPACE

Partitioning of the n-particle space is done in the hope that it will be sufficient to consider only the zeroth- and first-order subspaces; and that the higher order subspaces, if necessary, can be included to a considerably lesser degree of completeness. The success of such a scheme depends on the choice of a zeroth-order subspace; a CI wave function determined in this subspace must have a significant overlap with the complete CI wave function for all nuclear geometries of interest. The choice of the zeroth-order CSF's, $\{0, 0; 0\}$, is closely coupled to the choice of a set of internal orbitals from which they are constructed. An optimally chosen set of internal orbitals permits a compact zeroth-order subspace, one spanned by a small number of CSF's, which is desirable because a small zeroth-order subspace means a small first-order subspace.

A convenient approach to the selection of internal orbitals and zeroth-order CSF's is through the use of NO's. Consider a complete CI wave function, constructed from NO's. The internal orbitals are those NO's with large occupation numbers and the zeroth-order CSF's are the dominant terms, as measured by their expansion coefficients or estimated energy contributions, in the wave function. This appears to be a situation where the answer is needed before the calculation begins, and to some extent this is true. However, for ground states and some excited states, the most important orbitals can usually be deduced from elementary molecular orbital theory, combined with considerations of separated, and possibly united, atom limits, depending on the nuclear geometries considered in the study. In most cases it would be sufficient to include all occupied molecular orbitals, and those unoccupied orbitals nearly degenerate with the occupied ones. Alternatively the internal orbital set may include all molecular orbitals that correlate, at the united and separated atom limits, with the occupied atomic orbitals and those nearly degenerate with them. There will undoubtedly be cases where the selection of internal orbitals is less obvious, particularly for excited states.

In these cases, it is advisable to begin with a larger set of internal orbitals than is believed necessary, and base the selection on results of actual calculations. At any rate, the initial choice of internal orbitals can always be checked for consistency by comparing them to the leading ANO's extracted from more accurate wave functions obtained in more extensive CI expansions.

The determination of internal ANO's relies on the identification of a set of reference CSF's which are dominant terms in the complete CI expansion. This set of CSF's will be such that each internal orbital is occupied in at least one member of the set. Members of the set are therefore selected from the subspaces $\{n, 0; i\}$ spanning the complete set of CSF's that can be generated from the internal orbitals. The results of the internal orbital calculations will be used to select the zeroth-order space $\{0, 0; 0\}$; the space spanned by $\{n, 0; i\}$ is independent of $\{0, 0; 0\}$. A safe starting point for internal orbital determination is to use all members of $\{n, 0; i\}$. Internal orbitals resulting from a starting point which uses a subset of $\{n, 0; i\}$ must be essentially the same as would have been obtained with the full set. Given a starting set of CSF's for internal orbital determination there are two methods available for orbital determination: (1) to find solutions of self-consistent field (SCF) equations or many configuration–self-consistent field (MCSCF) equations depending on whether there are one or many CSF's in the chosen starting set, and (2) to follow an iterative natural orbital procedure (INO) using as an n-particle basis the chosen starting set of CSF's, together with members of $\{n, 0; i\}$ not in the starting set and members of $\{n, 1; i\}$.

In the framework of the expansion methods discussed in this paper, the first approach involves the determination of orbital expansions, in a given one-particle basis, which minimises the energy expectation value of a wave function set up as an expansion in the chosen starting set of CSF's. Both n-particle expansion coefficients and one-particle expansion coefficients are variationally determined. In the limit of a complete one-particle basis the orbitals determined are the Hartree–Fock (HF) orbitals or the multi-configuration Hartree–Fock (MCHF) orbitals depending on whether there are one or many starting CSF's, respectively. (Even with less than a complete one-particle basis the calculations are commonly called Hartree–Fock calculations.) Detailed discussions are available in the literature for both single CSF (Roothaan and Bagus, 1963) and many CSF (Hinze and Roothaan, 1967; Das and Wahl, 1972a) starting points.

In some cases, depending on the selection of reference CSF's, the MCSCF orbitals are only determined to within an arbitrary unitary transformation. Thus the full set of MCSCF internal orbitals may span approximately the same space as the internal NO's, while each MCSCF internal orbital bears no resemblance to the NO's they are trying to approximate. This annoying feature can be avoided, in a way that aids the selection of CSF's to be included in $\{0,0;0\}$, by using the MCSCF orbitals to construct a CI in the complete internal n-particle space $\{n, 0; i\}$. The natural orbitals from this CI wave function (eigenvectors of the first-order density matrix) are chosen as the internal orbitals, rather than the MCSCF orbitals. In fact, whenever a MCSCF calculation is done to obtain internal orbitals, we recommend that the MCSCF orbitals be transformed through a density matrix diagonalisation as described above. This guarantees that no significant change occurs in the internal orbitals obtained through later natural orbital transformations on more extended wave functions. From this it follows that the relative importance of CSF's will remain unchanged.

In the second approach to internal ANO's, the iterative natural orbital method (Bender and Davidson, 1966), a trial set of orbitals spanning the full one-particle space is selected, and a CI calculation carried out in an n-particle space spanned by members of $\{n, 0; i\}$ and $\{n, 1; i\}$. The natural orbitals of this wave function will be used as trial orbitals in the next iteration which carries out a CI calculation again in the n-particle space spanned by members of $\{n, 0; i\}$ and $\{n, 1; 1\}$. The n-particle space of the CI calculations is *not* invariant from iteration to iteration, and in fact the process may not even converge. Normally, one would select an n-particle space spanned by *all* members of $\{n, 0; i\}$ and $\{n, 1; i\}$, although this is not critically important. If all members are not chosen, it is important to select CSF's such that all orbitals, both internal and external, are occupied in at least one. The iterations should be monitored for CI eigenvalues and eigenvectors of interest, the sum of the internal orbital occupation numbers, and the degree to which the density matrix becomes diagonal. There is no point in carrying iterations beyond the stage where eigenvalues do not sensibly decrease, or sum of internal orbital occupation numbers sensibly increase. Examination of the wave functions at the final iteration allows selection of the important CSF's to be included in $\{0, 0; 0\}$.

To illustrate the selection of internal orbitals and the zeroth-order CSF's, we consider a CI study of the potential curves of the two lowest electronic states of CH, $X^2\Pi$ and $a^4\Sigma^-$ (Lie, Hinze and Liu, 1972). Both of these states have the separated atom limit $H(1s, \,^1S) + C(1s^2 2s^2 2p^2, \,^3P)$. The internal orbitals are the five molecular orbitals that correlate with occupied separated-atom orbitals: $1\sigma(C, 1s)$, $2\sigma(C, 2s)$, $3\sigma(H, 1s)$, $4\sigma(C, 2p_\sigma)$, and $1\pi(C, 2p_\pi)$. For the $X^2\Pi$, the electronic configurations $1\sigma^2 2\sigma^2 3\sigma^2 1\pi$, $1\sigma^2 2\sigma^2 4\sigma^2 1\pi$, and $1\sigma^2 2\sigma^2 3\sigma 4\sigma 1\pi$, from which four $^2\Pi$ CSF's with some designated axial angular momentum can be constructed, are necessary to give a Hartree–Fock description of the wave function at the separated atom limit. For the $a^4\Sigma^-$ case, only a single CSF from the configuration $1\sigma^2 2\sigma^2 3\sigma 1\pi^2$ is necessary for the Hartree–Fock separated atom limit description. Addition of four more CSF's to the $X^2\Pi$ calculation, derived from configurations $1\sigma^2 3\sigma^2 1\pi^3$, $1\sigma^2 4\sigma^2 1\pi^3$, and $1\sigma^2 3\sigma 4\sigma 1\pi^3$, would allow dissociation to a separated atom limit in which the carbon atom is represented by the two configurations $C_1 1s^2 2s^2 2p^2 + C_2 1s^2 2p^4$. For the $a^4\Sigma^-$ this improved dissociation limit requires inclusion of the one CSF obtainable from $1\sigma^2 3\sigma 4\sigma^2 1\pi^2$. In determining internal orbitals for these two states we would perform either MCSCF or INO calculations based on either the eight CSF's described above for $X^2\Pi$, or the two for $a^4\Sigma^-$, as members of $\{n, 0; i\}$. Members of $\{0, 0; 0\}$ will be chosen by inspecting the results of the internal orbital calculation which, in this case, will restrict members of $\{0, 0; 0\}$ to the four CSF's $(X^2\Pi)$ or one CSF $(a^4\Sigma^-)$ required for Hartree–Fock dissociation. Note, in the case of the $a^4\Sigma^-$, that while 4σ is included in the internal orbital set, it is not occupied in the single member of $\{0, 0; 0\}$.

FIRST-ORDER SUBSPACE AND EXTERNAL ORBITALS

If the computation of an electronic state or states has been properly designed, with an appropriate zeroth-order space and internal orbitals, it is expected that extending a CI calculation to include the first-order subspace will account for over 90% of the correlation effects. In addition, it is expected that analysis of CI wave functions expanded in the zeroth- and first-order subspaces through natural orbital transformations will be a necessary step towards a final wave function that includes contributions from higher order subspaces.

In what follows, we discuss methods for the systematic investigation of the first-order subspace. We shall assume, unless otherwise stated, a $\{0, 0; 0\}$ set of CSF's consistent with the chosen internal orbitals, and a starting set of external orbitals. The external orbitals are normally chosen to span the orthogonal complement of the internal orbitals in the one-particle space spanned by the elementary functions. However, one could treat the space of the elementary functions as simply a space from which internal orbitals are taken, and the space of the external orbitals selected in some other way. For example, the space of the external orbitals might be derived directly from the internal orbitals, by constructing functions from component parts of the internal orbitals. These derived functions, transformed into an orthonormal basis, would be the starting external orbitals. Many possibilities for deriving such functions can be suggested, such as multiplication of component parts of internal orbitals and their derivatives by polynomials, or through scaling transformations on component parts of the internal orbitals. Systematic studies of different methods of constructing the external orbital space are needed; our main point in introducing these possibilities into the current discussion is to indicate that the usual choice of an external orbital space left over from internal orbital determination may not be the best, and may even be far from best. It should be further noted that derivation of an external orbital space directly from the internal orbitals gives a systematic way of augmenting one-particle basis sets. If it is possible to carry out CI calculations in a space, constructed from a given one-particle basis set, in such a way that the one-particle basis limit is either computed or extrapolated from computation, then a sequence of one-particle basis set limits corresponding to systematically augmented one-particle basis sets can be extrapolated to the clamped nuclei non-relativistic limit. In other words, systematic augmenting makes convergence studies possible.

There are three basic computational steps which, arranged in various sequences, allow a systematic exploration of the first-order subspace. The actual sequence of steps used in an application will depend on basis set sizes, both one-particle and n-particle, and computer program characteristics. The ability of a program to execute the three basic steps allows great flexibility in the exploration of the CI method; at this stage in the application of a large scale CI, flexibility is essential because optimum sequences which achieve a desired result with minimal effort have yet to be determined. After defining these three basic steps we will suggest potentially useful sequences and will describe calculations in the literature in the framework of the present discussion.

The three basic steps are:

(1) From a given orbital set, generate the full n-particle zeroth- and first-order subspaces possible with internal orbital excitations restricted to those appropriate to the correlation effects being studied. In this n-particle space determine a wave function as one of:

(a) a CI expansion, in which the wave function is an eigenstate of the Hamiltonian;

(b) a Rayleigh–Schroedinger perturbation theory wave function, through first-order, with the zeroth-order function being an eigenstate of the Hamiltonian expanded in $\{0, 0; 0\}$;

(c) an approximate CI expansion, in which the wave function is an approximate eigenstate of the Hamiltonian, blocks of whose matrix elements have been set to zero.

(2) From a given wave function expansion in an n-particle space, generate a truncated n-particle space which includes all members of $\{0, 0; 0\}$, $\{1, 0; 1\}$, $\{0, 1; 1\}$, and those members of $\{2, 0; 1\}$, $\{1, 1; 1\}$, $\{0, 2; 1\}$ which make a significant contribution to the wave function as measured either by an energy criterion or by expansion coefficient. (In potential curve studies, this truncation may be made after inspecting a given wave function at several nuclear geometries to derive a common set of CSF's suitable for all geometries.) In the truncated n-particle space so generated, a CI wave function is determined.

(3) Transform the orbital set used to construct the n-particle space, in which a given wave function is expanded, in order to obtain an improved orbital set which will allow more compact expansions in subsequent calculation. This is done by evaluating, from the given wave function, the first-order density matrix followed by one of four transformations:

(a) rotate the entire orbital set using the matrix which is a collection of the density matrix eigenvectors (this is the natural orbital transformation);

(b) follow (3a) by a truncation of the external orbital set, in which orbitals are dropped from the one-particle space by some criterion, normally a threshold on occupation number;

(c) rotate only the external orbitals by using the matrix which is a collection of eigenvectors of the external orbital sub-block of the density matrix;

(d) follow (3c) by a truncation of the external orbital set.

Before proceeding to potentially useful sequences of steps, some notes on the steps themselves may be illuminating.

Calculations involved in (1b) involve only diagonal matrix elements of the Hamiltonian and those connecting functions in the first-order subspace with functions in the zeroth-order subspace; that is, only a small fraction of the full Hamiltonian matrix. Thus (1b) can be used to explore n-particle spaces too large to study with (1a). The (1b) wave function is not variationally determined and the perturbation theory energy is not an upper bound.

Calculations involved in (1c) are also to investigate larger n-particle spaces than are accessible to (1a). When the approximate Hamiltonian matrix sets all off diagonal matrix elements involving only functions in the first-order subspace to zero, calculation (1c) is essentially the same as (1b), and involves the same matrix elements. It can be made more powerful than (1b), without substantially increasing the number of matrix elements used, by accurately computing additional sub-blocks of the Hamiltonian matrix. Gershgorn and Shavitt (1968) have investigated various ways of approximating the Hamiltonian matrix. The eigenvalue of the chosen eigenvector of this matrix is not an energy upper bound.

In step (2) there are two ways that have been used to estimate the importance of a CSF in a wave function expansion. The first, appropriate to (1b) type wave functions, is the energy contribution of CSF Φ_K in second order of Rayleigh–Schroedinger perturbation theory

$$\Delta E_K = H_{K0}^2/(H_{KK}-H_{00}) \tag{11}$$

where H_{KK} is the diagonal matrix element of CSF Φ_K, and H_{K0} is its matrix element with the zeroth-order wave function. The second, which is used for (1a) and (1c) type wave functions where a zeroth-order function is not defined, is the difference in the energy expectation values of the wave function including Φ_K and the related wave function derived by dropping Φ_K and renormalising. This difference is

$$\Delta E_K = |C_K|^2(E-H_{KK})/(1-|C_K|^2) \tag{12}$$

where C_K is the expansion coefficient of Φ_K in the full wave function whose energy expectation value is E. Deletion on the basis of equation 12 has been used extensively in our laboratory, by Shavitt and co-workers, and by the theoretical chemistry group at Indiana University.

Steps (3a) and (3c) will be used when it is not necessary to limit the n-particle space in a following step (1) calculation. It is often desirable to finish a sequence of calculations with step (1a), in which case the full n-particle space of step (1) must be limited, and this is done through external orbital truncation in preceding (3b) or (3d) steps.

In step (3) when good internal orbitals, of Hartree–Fock or multiconfiguration Hartree–Fock quality for example, are available as input to a sequence of calculations it makes little difference whether orbital transformation is restricted to just the external or to the entire orbital set. It appears to be marginally advantageous to keep the internal orbitals fixed, otherwise an iterative process involving repeated application of steps (1) and (3) sometimes causes CI energies to drift upwards, although not significantly (Green et al., 1972). If the internal orbitals used in constructing the wave function being processed in step (3) are not near optimal, it is essential to transform the entire orbital set so that externals of the old orbital set become components of the new internal orbitals, thereby improving them relative to the old.

In the computational sequences of the three basic steps enumerated below we assume an input set of orbitals and a corresponding $\{0, 0; 0\}$ which will be used in the first step. Input to succeeding steps will be from the last previous step that produced data of the required type. Thus, for example, in the sequence $(1c)(3d)(1a)$ input orbitals are used to construct the n-particle space of wave function $(1c)$; this wave function will be processed in step $(3d)$ which produces a rotated, truncated orbital set used to construct an n-particle space in which a $(1a)$ type wave function will be determined.

With the preceding background we can illustrate the type of calculations that might be, and have been, done to investigate the first-order subspace by describing the following procedures.

In the case that the full n-particle space appropriate to the problem generated from the starting orbital set is small enough to permit a calculation of the type $(1a)$ we might use *Procedure A*: $(1a)(3_c^a)(1a)(2)$. Discussing these steps in order, the first $(1a)$ determines a wave function whose energy expectation value cannot be improved if the input internal orbitals are near optimal. (3_c^a) transforms to ANO's and $(1a)$ produces a wave function essentially equivalent to the previous one but in an improved orbital representation which will permit optimum truncation in the final step (2). In this case the three final steps have not improved the wave function; they have simply made it more compact, thus reducing the computational labour in extending the wave function expansion into higher order spaces and hopefully facilitating interpretation. This procedure is usually restricted to small systems. It is the procedure basically followed by Liu in calculations on He_2^+ (Liu, 1971) and on H_3 (Liu, 1973). Orbital transformation cannot improve the initially computed wave function in this procedure, since the near optimal internal orbitals are essentially unchanged under a natural orbital transformation and the wave function is invariant to a rotation of the external orbitals. If the internal orbitals are not close to NO's a wave function of the same quality can be produced by *Procedure A'*: $\overline{(1a)(3_c^a)}(1a)(2)$ where the bar over the first two steps indicate that they are iterated. Iteration is expensive, and it is usually better to obtain good internal orbitals before investigating the first-order space.

If the full n-particle space generated from the input orbitals does not permit step $(1a)$ then there are basically two procedures that can be recommended, either *Procedure B*: $(1_c^b)(3_d^a)(1a)(3_c^a)(1a)(2)$ or *Procedure C*: $(1_c^b)(3_d^a)\overline{(1_c^b)(2)(3_c^a)}$ $(1_c^b)(2)$. Procedure B will be recognised simply as Procedure A prefaced by two steps $(1_c^b)(3_d^b)$ which are supposed to transform the external orbitals so that they can be effectively truncated, bringing a calculation $(1a)$ into range. After truncation, as in Procedure A, iteration cannot improve the wave function. Procedure B, in the $(1c)(3b)(1a)(3a)(1a)(2)$ variant, has been essentially used in a recent calculation of low lying states of CH (Lie, Hinge and Liu, 1972). Procedure C starts in the same way as Procedure B where the first two steps are designed to improve the external orbitals, but instead of truncating the external orbital set it proceeds to step (1_c^b) which generates an n-particle space of the same dimension used in the previous wave function calculation. The iterated sequence $\overline{(1_c^b)(2)(3_c^a)}$ improves the external orbitals, making them closer to NO's than the orbitals obtained in the first application of (3_c^a). Procedure C, then, should result in a somewhat better wave function than Procedure B although whether the improvement is significant is not yet clear. The last two steps of these procedures serve the same role as in Procedure A; that of preparing for investigation of higher order subspaces. Another, possibly important, comparison of Procedures B and C stems from the fact that in C truncation of the n-particle space is only ever done through study of actual wave functions; after truncation, orbitals spanning the entire input orbital space may still be used. In Procedure B, however, truncation of the n-particle space is through orbital truncation. Procedure C, especially if the middle sequence is iterated, is more expensive to implement than B. The calculations of Bender and Davidson (1966) which pioneered the use of NO's in CI essentially use the iterated sequence $(1b)(2)(3a)$ of Procedure C, as do their later comprehensive calculations on the ground states of first-row hydrides (Bender and Davidson, 1969).

If the n-particle space generated in step (1) restricts the double internal orbital excitations to those in which two electrons are excited out of a single internal orbital, then the natural orbitals produced in the sequence $(1a)(3a)$ are the so-called pseudonatural orbitals (PSNO's) appropriate for describing intra shell correlation effects in the orbital from which excitations are made. If double orbital excitations are restricted to those in which one is excited from one orbital and a second electron from a second orbital the sequence $(1a)(3a)$ produces PSNO's which may, but not necessarily, be appropriate for all correlation effects involving the two orbitals. Use of PSNO's for finding a compact representation of correlation effects in wave function expansions was introduced by Edmiston and Krauss (1966) in calculations on He_2^+. They used the same technique later on H_3 (1968). PSNO calculations can be performed to set up an orbital basis for a CI calculation which treats several different types of pair excitations together. This is done by selecting the most important PSNO's from calculations which individually look at single pair excitations, and then merging them to form an orthonormal external orbital set as input to Procedure A.

The individual PSNO calculations whose results are merged should be done using the step sequence $(1a)(3c)$ so that internal orbitals are unchanged; otherwise the merging procedure is unworkable. The PSNO method can be extended to work with several types of double excitations simultaneously, followed by a merging procedure to produce external orbitals for input to Procedure A. PSNO methods have been recently used in valence shell correlation studies of LiO (Yoshimine, 1972), AlO (Liu, McLean and Yoshimine, 1972), and FeO (Bagus and Preston, 1972).

If the n-particle space generated in step (1) is restricted to functions involving an excitation of at most one electron into an external orbital, that is, the class of excitations $\{0, 2; 1\}$ is excluded, the procedure $\overline{(1a)(3a)}$ is that used extensively by Schaefer and co-workers (see for example: Schaefer, 1971a, b). This procedure may generate external orbitals which approximate NO's, although in some cases ($\{0, 0; 0\}$ contains only a single closed shell CSF, for example) it will be recognised as the equivalent of a Hartree–Fock calculation, as discussed in a previous section, and no information on external NO's is obtainable. In a recent calculation on BeH_2, Shavitt, Hsu and Hosteny (1972) reported that ANO's obtained in this manner closely resemble the exact NO's. This was not the case in calculations on CH (Lie,

Hinze and Liu, 1972). The disparity of these results may be caused by the difference in the sizes of elementary basis set employed; the BeH_2 calculation was done with a double zeta basis whereas the CH calculation was done with an extended basis including polarisation functions. It is safe to say that more experience is needed to properly evaluate this method of external orbital determination.

So far we have dealt with methods which require the use of an extended set of external orbitals in the construction of the n-particle space. A different approach is the MCSCF method discussed earlier in connection with internal orbital determination. Here, the CSF's to appear in the final CI expansion are first chosen, and then the occupied orbitals together with the CI expansion coefficients are determined variationally. The orbitals thus determined are the best orbitals for the corresponding wave function in the sense of lowest energy. However, the simultaneous determination of the orbitals and CI coefficients is a difficult numerical problem, especially in cases involving many CSF's and orbitals. In spite of recent advances (Das and Wahl, 1972a), the MCSCF method does not appear to be practical in calculations intended to approach the one-particle basis set limit. Das and Wahl in their optimised valence configuration (OVC) method combined MCSCF with a systematic method of selecting CSF's responsible for the geometry sensitive part of the correlation energy in a molecule. The OVC method has been applied to many molecular systems; examples are the van der Waal well depth of He_2 (Bertoncini and Wahl, 1970) and the binding energy of F_2 (Das and Wahl, 1972b). The MSCF method can also be used in a manner similar to the PSNO method to obtain orbitals for later CI calculations. However, this method has not been studied extensively.

8. SECOND AND HIGHER ORDER SUBSPACES

In our discussion of higher order subspaces we assume a one-particle basis closely approximating the NO's of the complete CI expansion, and a CI wave function including the full zeroth-order subspace and all the important CSF's in the first-order subspace. Using this wave function as a zeroth-order wave function, the contribution of the second-order space may be estimated, and a CI calculation including the most important CSF's can be done to obtain an energy upper bound. The resulting wave function can be used, in turn, to obtain an estimate of the contribution from the third-order subspace, and so on. The difficulty with this procedure is that the higher order subspaces may be so large that even a perturbation calculation is not practical.

Liu (1971), in his calculation on He_2^+, studied the contribution of the second-order subspace using CSF's constructed from ANO's determined in the full zeroth- and first-order subspaces. He found that the contribution from the second-order subspace was accounted for by using a severely truncated one-particle basis including only a few of the leading ANO's. In this calculation the second-order subspace, containing 15 482 CSF's, contributed 0.000 50 A.U. to the total energy, 98% of which was reproduced by the inclusion of 371 second-order CSF's constructed from the leading

ANO's. Similar results were found in calculation on H_3 (Liu, 1972). Recently, Shavitt, Hsu and Hosteny (1972) reported results confirming Liu's conclusion for systems involving more electrons. Therefore, it appears that the most efficient way of studying the higher order subspaces is to combine truncation of the one-particle basis with perturbation selection of CSF's.

9. SUMMARY

The systematic application of the method of configuration interaction in the determination of molecular bound electronic states is reviewed. Wave functions are approximated by eigenvectors of the clamped nuclei non-relativistic Hamiltonian in an n-particle space spanned by an orthonormal set of configuration state functions (CSF's). CSF's are functions, antisymmetric to particle interchange, constructed from orthonormal one-particle functions (orbitals), which form bases for irreducible representations of the symmetry group of the nuclear frame. The n-particle space is divided into a hierarchy of orthogonal subspaces defined relative to a zeroth-order subspace spanned by the dominant CSF's in the electronic states being approximated. The ith-order space is the space of those, and only those, functions which have a non-zero matrix element through the Hamiltonian with a member of the $(i-1)$th-order subspace. Basis functions spanning these subspaces are linear combinations of CSF's, whose expansion coefficients are orbital independent. Functions in the ith-order subspace contribute first in the ith-order Rayleigh–Schroedinger perturbation theory correction to a zeroth-order function which is an eigenstate of the Hamiltonian in the zeroth-order subspace. The orbitals are expanded in a basis of elementary functions which are chosen by the investigator. They are divided into two subsets, internal and external. The division is between more and less important, respectively, with the internal subset at least containing all orbitals used to construct CSF's contained in the zeroth-order n-particle subspace. A calculation can be divided into the following sequence of steps: (1) choice of elementary functions, (2) choice of the zeroth-order subspace and determination of internal orbitals, (3) inclusion of the first-order subspace and determination of external orbitals, (4) inclusion of higher order subspaces. These steps are discussed in some depth, with examples and with an enumeration of types of calculation which have been done and which might be done.

REFERENCES

Bagus, P. S., Gilbert, T. L. and Roothaan, C. C. J. (1972) 'Hartree–Fock Wave Functions of Nominal Accuracy for He through Rb^+ Calculated by the Expansion Method', *J. chem. Phys.*, **56**, 5195–5196

Bagus, P. S. and Preston, H. J. T. (1972) *The lowest $^5\Sigma^+$ State of FeO: an ab initio Investigation*, unpublished

Bender, C. F. and Davidson, E. R. (1966) 'A Natural Orbital Based Energy Calculation for Helium Hydride and Lithium Hydride', *J. phys. Chem.*, **70**, 2675–2685

Bender, C. F. and Davidson, E. R. (1969) 'Studies in Configuration Interaction: the First-row Diatomic Hydrides', *Phys. Rev.*, **183**, 23–30

Bertoncini, P. and Wahl, A. C. (1970) 'Ab initio Calculation of the Helium–Helium $^1\Sigma_g^+$ Potential at Intermediate and Large Separations', *Phys. Rev. Lett.*, **25**, 991–994

Billingsley, F. P. and Krauss, M. (1972) 'Coupled Multi-configurational Self-consistent Field Method for Atomic Dipole Polarizabilities. I. Theory and Application to Carbon', *Phys. Rev.*, **A6**, 855–865

Cade, P. E., Sales, A. D. and Wahl, A. C. (1966) 'Electronic Structure of Diatomic Molecules. III. A. Hartree–Fock Wave Functions and Energy Quantities for $N_2(X^1\Sigma_g^+)$ and $N_2^+(X^2\Sigma_g^+, A^2\Pi_u, B^2\Sigma_u^+)$ Molecular Ions', *J. chem. Phys.*, **44**, 1973–2003

Clementi, E. (1965) 'Tables of Atomic Wave Functions', *Supplement to IBM J. Res. Develop.*, **9**, 2–19

Das, G. and Wahl, A. C. (1972a) 'New Techniques for the Computation of Multi-configuration Self-consistent Field (MCSCF) Wave Functions', *J. chem. Phys.*, **56**, 1769–1775

Das, G. and Wahl, A. C. (1972b) 'Theoretical Study of the F_2 Molecule using the method of Optimized Valence Configurations', *J. chem. Phys.*, **56**, 3532–3540

Davidson, E. R. (1972) 'Properties and Uses of Natural Orbitals', *Advances in Quantum Chemistry*, Vol. 6, Academic Press, New York

Edmiston, C. E. and Krauss, M. (1966) 'Pseudonatural Orbitals as a Basis for the Superposition of Configurations. I. He_2^+'. *J. chem. Phys.*, **45**, 1833–1839

Edmiston, C. E. and Krauss, M. (1968) 'Pseudonatural Orbitals as a Basis for the Superposition of Configurations. II. Energy Surface for Linear H_3', *J. chem. Phys.*, **49**, 192–205

Gershgorn, Z. and Shavitt, I. (1968) 'An Application of Perturbation Theory Ideas in Configuration Interaction Calculations', *Int. J. quantum Chem.*, **2**, 751–759

Green, S., Bagus, P. S., Liu, B., McLean, A. D. and Yoshimine, M. (1972) 'Calculated Potential-energy Curves for CH^+', *Phys. Rev.*, **A5**, 1614–1618

Heitler, W. and London, F. (1927) 'Wechselwirkung Neutraler Atome and Homöpolare Bindung nach der Quantenmechanik', *Z. Physik*, **44**, 455–472

Hinze, J. and Roothaan, C. C. J. (1967) 'Multi-configuration Self-consistent Field Theory', *Prog. Theor. Phys.*, Kyoto, **40**, 37–51

Huzinaga, S. (1971) 'Approximate Atomic Functions I & II' *Research Report of the Dept. of Chemistry of the University of Alberta*, Unpublished

James, H. M. and Coolidge, A. S. (1933) 'Ground State of the Hydrogen Molecule', *J. chem. Phys.*, **1**, 825–835

Lie, G. C., Hinze, J. and Liu, B. (1972) *Valence Excited States of* CH, unpublished

Liu, B. (1971) 'Dissociation Energies of $He_2^+(X^2\Sigma_u^+)$ and $He_2(A^1\Sigma_u^+)$', *Phys. Rev. Lett.*, **27**, 1251–1253

Liu, B. (1973) 'Ab Initio Potential Energy Surface for Linear H_3', *J. chem. Phys.*, **58** (in press)

Liu, B., McLean, A. D. and Yoshimine, M. (1973) Band Strengths for Electric Dipole Transitions from Ab Initio Computation: LiO $(X^2\Pi - X^2\Pi)$, $(A^2\Sigma^+ - A^2\Sigma^+)$, $(X^2\Pi - A^2\Sigma^+)$; AlO $(X^2\Sigma^+ - X^2\Sigma^+)$, $(A^2\Pi - A^2\Pi)$, $(X^2\Sigma^+ - A^2\Pi)$, $(B^2\Sigma^+ - B^2\Sigma^+)$, $(X^2\Sigma^+ - B^2\Sigma^+)$, *J. chem. Phys.*, **58** (in press)

Löwdin, P. O. (1955) 'Quantum Theory of Many-particle Systems. I. Physical Interpretations by Means of Density Matrices, Natural Spin-orbitals, and Convergence Problems in the Method of Configurational Interaction', *Phys. Rev.*, **97**, 1474–1489

Löwdin, P. O. (1959) 'Correlation Problem in Many-electron Quantum Mechanics. I & II', *Advances in Chemical Physics*, Vol. II, 207–322 Interscience, New York

McLean, A. D. and Liu, B. (1973) 'Classification of Configurations and the Determination of Interacting and Non-interacting Spaces in Configuration Interaction', *J. chem. Phys.*, **58**, 1066–1078

McLean, A. D. and Yoshimine, M. (1968) 'Tables of Linear Molecule Wave Functions', *Supplement to IBM J. Res. Develop.*, **12**, 206–233

McLean, A. D. and Yoshimine, M. (1972) Unpublished results

Roothaan, C. C. J. (1951) 'New Developments in Molecular Orbital Theory', *Rev. mod. Phys.*, **23**, 69–89

Roothaan, C. C. J. and Bagus, P. S. (1963) 'Atomic Self-consistent Field calculations by the Expansion Method', *Methods in Computational Physics*, Vol. II, 47–94, Academic Press, New York

Schaefer III, H. F. (1971a) *The Electronic Structure of Atoms and Molecules: A Survey of Rigorous Quantum Mechanical Results*, Addison-Wesley, Menlo Park, Calif.

118   Wave Mechanics

Schaefer III, H. F. (1971b) 'Ab Initio Potential Curve for the $X^3\Sigma_g^-$ State of O_2', *J. chem. Phys.*, **54**, 2207–2211

Shavitt, I., Hsu, K. and Hosteny, R. P. (1972) Results reported in the *Summer Research Conference on Theoretical Chemistry*, Boulder, Colorado, 1972

Slater, J. C. (1963) *Quantum Theory of Molecules and Solids*, Vol. I and II, McGraw-Hill, New York

Yoshimine, M. (1972) 'Accurate Potential Curves and Properties for the $X^2\Pi$ and $A^2\Sigma^+$ States of LiO', *J.chem. Phys.*, **57**, 1108–1115

Spin and the Pauli Principle

A. T. AMOS
Department of Mathematics, The University, Nottingham

1. INTRODUCTION

There can be little doubt that the turning point in the formulation of the quantum theory was the suggestion by L. de Broglie (de Broglie, 1923, 1925) that electrons had wave as well as particle properties. Prior to this the first attempts to develop a theory of the microscopic behaviour of atoms and molecules produced a complicated and confusing picture but within a few years following the publication of de Broglie's postulate the theory had taken the shape in which it survives to this day.

This article is concerned with the electronic wave functions of atoms and molecules which can be derived from the Schroedinger non-relativistic equation, that is from the equation which represents the quantitative formulation of de Broglie's idea. As it happens the complete wave functions can be obtained from this equation provided two extra requirements are taken into account. The first of these is that electrons have spin co-ordinates as well as spatial ones so that the spatial wave functions for an atom or molecule derived from the Schroedinger equation must be supplemented by the inclusion of spin components. The second is that the Pauli principle (Pauli, 1925; see also Weyl, 1931 and Wigner, 1959) which states that electronic wave functions must be antisymmetric with respect to the interchange of electronic co-ordinates, has to be satisfied. This seemingly innocuous condition is in fact crucial since it determines the manner in which the spatial and spin components of the wave function have to be combined and, in the non-relativistic case, provides the only connection between the space and spin variables.

While the significance of the Pauli principle has been fully recognised from the time of its inception, there nevertheless seems to have been a great diminution in the amount of space devoted to it in many of the more modern textbooks. To some extent this is due to the widespread use of single Slater determinants as approximate wave functions since for these functions the

Pauli principle is automatically satisfied, but another difficulty is that there are relatively few simple yet realistic examples which can be given. A consequence of this neglect is a growing unawareness of the central and critical role played by the Pauli principle in the theory of atoms and molecules. A further contributory cause of this lack of understanding is that those few modern textbooks (for example, Hammermesh, 1962) which do treat the subject adequately almost invariably use the full group theoretical treatment based on the properties of that most complex object the symmetric group and thus are suitable only for experts.

Nevertheless some of the more recent developments in the use of relatively complicated approximate wave functions (for example, Matsen, 1964; Goddard, 1969, 1968; Musher, 1971) are derived from a consideration of the connection between the space and spin components of the wave function. In order that these developments can be followed the Pauli principle ought to be properly understood. For this reason, therefore, and also for pedagogical reasons, there seems to be a need for a fairly elementary treatment of the subject yet one which includes the essentials of the problem. It is the purpose of this article to try to provide such a treatment and to illustrate the major results using a not too complex yet quite realistic and important example. It seems appropriate for this to be a contribution to a book in honour of the anniversary of Professor de Broglie's discovery since it shows how proper solutions for the physically allowed states of atoms and molecules can be obtained from the quantitative formulation of that discovery.

2. COMMUTING OBSERVABLES: THE PERMUTATION SYMMETRY OF WAVE FUNCTIONS

The main concept used in this article concerns the eigenfunctions of two commuting observables. It is proved in all the quantum mechanics textbooks (Dirac, 1958, gives an excellent treatment of commuting observables) that if two observables commute they have a complete set of eigenfunctions in common. In the proof two cases have to be taken into account. The simplest one occurs if one of the observables has no degeneracies in its spectrum; in that case the non-degenerate eigenfunctions of that observable will automatically be eigenfunctions of any other observable which happens to commute with it. The more interesting case and the one most likely to occur in practice is when both of the commuting observables have a mixture of non-degenerate and degenerate eigenvalues in their spectra. If, for example, A and B commute and A has a d-fold degenerate eigenvalue a with eigenfunctions $\psi_1 \ldots, \psi_d$ so that

$$A\psi_i = a\psi_i \quad i = 1, \ldots, d \tag{1}$$

we can conclude that, if ψ is any function satisfying

$$A\psi = a\psi \tag{2}$$

then ψ can only be a linear combination of the $\{\psi_i\}$, i.e.

$$\psi = \sum_{i=1}^{d} c_i \psi_i \tag{3}$$

Using the commutation relation $[A, B] = 0$ and equation 1, it follows that each of $B\psi_i$, $i = 1, \ldots, d$ satisfies equation 2 and, therefore, by equation 3 is a linear combination of the original degenerate set. Thus the effect of the operator B on the functions $\{\psi_i\}$ is to transform them amongst themselves.

If one has a set of operators $C, D, E, \ldots$ as well as B which commute with A then, just as with B, the operators $C, D, \ldots$ will transform the degenerate functions amongst themselves. For any one of these, B say, it is always possible to transform the original $\{\psi_i\}$ into an equivalent set, which will be eigenfunctions of B as well as A. However, unless C, D, etc., commute with B as well as A this new set will not be eigenfunctions of $C, D, \ldots$ (at least not for every degenerate eigenvalue of A).

As an example of this consider the Schroedinger equation for an atom or molecule:

$$H\psi = E\psi \tag{4}$$

where H is the Hamiltonian. In the non-relativistic approximation, which is assumed here, H will be an operator on the spatial co-ordinates of the electrons and nuclei and so the solutions of equation 4 will be spatial functions only. Suppose there are n electrons which are labelled 1 to n, then we can write $H(r_1 \ldots r_n)$ and $\psi(r_1 \ldots r_n)$ to indicate explicitly the dependence on electron position co-ordinates $r_1 \ldots r_n$. However, because electrons are indistinguishable this labelling is a purely mathematical contrivance and can have no physical significance. Thus, if we relabel, for example by interchanging 1 and 2, H is left unaltered and, in some way, the wave functions must take into account this invariance. An alternative way of putting this is that H is symmetric in the co-ordinates of the electrons and the solutions of equation 4 which are derived solely from H must contain this symmetry in one form or another.

Any symmetries in a Hamiltonian imply that there exist other operators, which are the mathematical realisation of these symmetries, and which commute with H. In the case of the symmetry in the electron co-ordinates, these operators will be the permutation operators which permute the co-ordinates of the electrons. For example, P_{12} interchanges the co-ordinates of electrons 1 and 2 in any function of electron co-ordinates on which it operates. Similarly P_{123} interchanges the co-ordinates of electrons 1, 2 and 3 so that the electron co-ordinate labelled r_1 becomes r_2, that labelled r_2 becomes r_3 and r_3 becomes r_1. An n-electron system will have $n!$ such operators (including the identity, I); for example when $n = 3$ the six operators are $I, P_{12}, P_{13}, P_{23}, P_{123}, P_{132}$.

If we consider a degenerate eigenvalue of H, the situation will be that described in general terms in the early part of this section. The effect of the permutation operators on the degenerate set of eigenfunctions $\{\psi_i\}$ say, will be to transform the set amongst themselves. Corresponding to equations 2 and 3 we can write:

$$P\psi_i = \sum_{j=1}^{d} D_{ji}(P)\psi_j \qquad i = 1 \ldots d \qquad (5)$$

where P stands in turn for each of the $n!$ permutation operators. The coefficients $D_{ji}(P)$ in the expansion 5 satisfy

$$D_{ji}(P) = \langle \psi_j | P | \psi_i \rangle \qquad (6)$$

in the Dirac notation, where the degenerate set is assumed to be orthonormal. For each permutation operator there will be a different set of d^2 coefficients determining the transformations 5; usually these are written in matrix form, e.g. $\mathbf{D}(P)$ for the matrix with elements $D_{ji}(P)$.

It can sometimes happen that the set of functions can be divided into subsets with the property that the effect of the permutation operators on a member of a particular sub-set is to give a linear combination of members of that sub-set only. Thus the effect of the permutation operators would be to transform among the members of each sub-set but not between sub-sets. When this process of division has gone as far as possible the resulting sets of functions are said to be irreducible. Normally, in fact, the original degenerate set cannot be divided in this way and so is itself irreducible.

3. AN EXAMPLE: THE ALLYL RADICAL

In order to illustrate the results of the previous section we need an exactly soluble problem and the one we choose is in the context of pi-electron theory. This is a theory which is applied to conjugated hydrocarbons and similar molecules (for details see Parr, 1963 and Salem, 1966). It assumes, in effect, that the properties of such molecules are mainly determined by a limited number of electrons—the pi-electrons—one of which is provided by each carbon atom, the remaining electrons forming a kind of inert core localised around the nuclei or in CC and CH bonds. The wave function for the pi-electrons satisfies the Schroedinger equation

$$H\Psi = E\Psi \qquad (7)$$

where H is, in some sense, a model Hamiltonian incorporating the effects of interaction between the pi-electrons and the remaining localised ones. H contains both one- and two-electron operators:

$$H = \sum_i h(i) + \tfrac{1}{2} \sum_{i,j} g(i,j) \qquad (8)$$

where h and g are defined implicitly by specifying their matrix elements with respect to a finite number M, say, of basis orbitals $\{\omega_a\} a = 1 \ldots M$. The matrix elements of the operators with respect to any functions except the $\{\omega_a\}$ are zero. From this it follows that the $\{\omega_a\}$ form a complete set of functions for the problem and so the eigenfunctions of (8) can be expressed as sums of products of these basis functions (Amos and Woodward, 1969;

Amos and Burrows, 1970). The example we consider is the allyl radical C_3H_5. This has three pi-electrons and also three basis orbitals ω_1, ω_2, ω_3, one centred at each carbon atom. These functions are supposed to be highly localised, orthogonal to each other, normalised and very similar to Slater $2p_z$ atomic orbitals. Since there are 3 basis functions it follows that the eigenfunctions of H can be written as the linear sum of 27 product functions:

$$\Psi_A = \Sigma\, c_{abc}^A\, |abc\rangle \tag{9}$$

where

$$|abc\rangle = \omega_a(1)\omega_b(2)\omega_c(3) \tag{10}$$

and each of the indices a, b, c can take the values 1, 2 and 3. Writing the 27 coefficients $\{c_{abc}^A\}$ as a column vector c^A, it follows that

$$Hc^A = E_A c^A \tag{11}$$

where H is the 27×27 matrix whose elements are $\langle a'b'c' |H| abc\rangle$. The rules for forming these matrix elements follow from the orthogonality of the basis orbitals and the form of H in equation 8 and it can be shown that (Amos and Burrows, 1970)

$$\langle a'b'c' |H| abc\rangle = h_{aa'}\delta_{b'b}\delta_{c'c} + h_{bb'}\delta_{aa'}\delta_{cc'} + h_{cc'}\delta_{aa'}\delta_{bb'}$$
$$+ \delta_{aa'}\delta_{bb'}\delta_{cc'}(\gamma_{ab} + \gamma_{ac} + \gamma_{bc}) \tag{12}$$

with

$$h_{aa'} = \int \omega_a^*(1)h(1)\omega_{a'}(1)\,d\tau \tag{13}$$

and

$$\int \omega_a^*(1)\omega_b^*(2)g(1,2)\omega_{a'}(1)\omega_{b'}(2)\,d\tau_{12} = \gamma_{ab}\delta_{aa'}\delta_{bb'} \tag{14}$$

The $h_{aa'}$ and γ_{ab} have to be treated as semi-empirical parameters. They are chosen to have the values used by Snyder and Amos (1965).

Once the matrix H has been set up using equation 12 its eigenvalues and eigenvectors can be found so as to solve equation 11. This gives 27 distinct solutions and these will be the spatial solutions for the pi-electron wave functions of allyl. The results of this procedure are shown in Table 1 in which are listed the energy levels (i.e. the eigenvalues of H), the degeneracies of each level and the symmetries with respect to the action of the permutation operators (denoted by A, S and D). It will be noted that, although there are 27 distinct wave functions there are only 18 energy levels since 7 of these are doubly degenerate and one is triply degenerate.

Since the non-degenerate levels cannot be transformed into any other function by the permutation operators they must be eigenfunctions of these operators. Because of various relations between the operators† it follows that there can be only two possibilities. Either the non-degenerate eigenfunctions must be symmetric with respect to interchange of pairs of electron co-ordinates, i.e. their eigenvalues with respect to P_{12}, P_{13}, P_{23}, P_{123}, P_{132}

---

†For example $P_{12}^2 = 1$ so that the only eigenvalues of P_{12} are $+1$ and -1. $P_{23}\,P_{12}\,P_{23} = P_{13}$ so that the eigenvalues of P_{12} and P_{13} are the same, and so on.

are all $+1$, or antisymmetric, i.e. their eigenvalues with respect to P_{12}, P_{13}, P_{23} are all -1 and with respect to P_{123} and P_{132} $+1$ (for $P_{123} = P_{12}P_{23}$ and $P_{132} = P_{23}P_{12}$ correspond to the interchange of two pairs of electrons). These symmetries are indicated in Table 1 by S and A. It can be seen from the table that there is only one antisymmetric function. This is quite generally true. Given a set of N^N product functions for N electrons there is only one way to combine them to form a function antisymmetric with respect to the interchange of pairs of electron co-ordinates. The number of symmetric functions, however, increases rapidly with N.

The two functions belonging to a doubly degenerate level are transformed among themselves by the permutation operators as is implied by equation

Table 1. The Spatial Eigenvalues of the Allyl Radical

Eigenvalue (eV)	Degeneracy	Symmetry of the eigenfunctions	Spin property of the allowed states*
−26.254	1	S	
−24.257	2	D	doublet
−21.582	2	D	doublet
−20.777	1	S	
−19.630	1	A	quartet
−18.789	1	S	
−18.377	2	D	doublet
−15.730†	3	S, D†	doublet†
−15.656	1	S	
−15.484	2	D	doublet
−13.778	2	D	doublet
−11.413	2	D	doublet
−11.048	1	S	
−10.886	1	S	
−9.679	2	D	doublet
−5.233	1	S	
−2.717	1	S	
−2.359	1	S	

*The eigenvalues with no entry in this column correspond to physically unallowed solutions.
†This triply degenerate solution corresponds to one physically unallowed solution (S) plus a doublet state (D).

5 with $d = 2$. The two degenerate eigenfunctions of the lowest degenerate level at -24.257 eV as produced by the Nottingham computer do not have very good transformation properties due to the random fashion in which the computer chooses the members of the degenerate set. It turns out to be possible, and is clearly more convenient, to transform these to two equivalent functions ψ_1, ψ_2, say, with the property that $P_{12}\psi_1 = \psi_1$ and $P_{12}\psi_2 = -\psi_2$ so that the matrix $D(P_{12})$ defined by equation 5 is just

$$D(P_{12}) = \begin{bmatrix} 1 & 0 \\ 0 & -1 \end{bmatrix} \qquad (15)$$

With these functions ψ_1 and ψ_2 the matrices for P_{13} and P_{23} are

$$D(P_{13}) = \begin{bmatrix} -\dfrac{1}{2} & \dfrac{\sqrt{3}}{2} \\[2ex] \dfrac{\sqrt{3}}{2} & \dfrac{1}{2} \end{bmatrix}; \quad D(P_{23}) = \begin{bmatrix} -\dfrac{1}{2} & -\dfrac{\sqrt{3}}{2} \\[2ex] -\dfrac{\sqrt{3}}{2} & \dfrac{1}{2} \end{bmatrix} \tag{16}$$

and, since $P_{12}P_{23} = P_{123}$ and $P_{23}P_{12} = P_{132}$, the other matrices are found from $D(P_{123}) = D(P_{12})D(P_{23})$ and $D(P_{132}) = D(P_{23})D(P_{12})$.

The wave functions for the other degenerate levels are produced by the computer in an equally cumbersome form. However, they can all be transformed into equivalent pairs of functions with exactly the same transformation properties under the action of the permutation operators as the first set just discussed. Thus all the pairs of doubly degenerate functions have the same symmetry under the action of the permutation operators and so they are all labelled D in Table 1.

There remains the triply degenerate eigenvalue. This is rather unusual because it turns out that the three eigenfunctions are not irreducible. Instead they can be transformed into two sets containing one function and two functions, respectively, such that the permutation operators do not transform between the two sets. The one function is symmetric while the pair of functions have the same symmetry properties as the doubly degenerate functions. Therefore the triply degenerate level is labelled S and D to show the symmetry of its irreducible components.

4. ELECTRON SPIN

In addition to the three spatial co-ordinates of position an electron has a spin co-ordinate. There are three observables $S_x(j)$, $S_y(j)$, $S_z(j)$ for the spin of the j-th electron but, since these satisfy the same commutation relations as angular momentum components and so do not commute, it is not possible to find simultaneous spin eigenfunctions of all three. Conventionally the eigenfunctions of $S_z(j)$ are chosen to form the basis of a representation, and, as is implied by the Stern–Gerlach experiment and as is found theoretically in Dirac's relativistic theory (see Dirac, 1958), there are just two of them denoted by $\alpha(j)$ and $\beta(j)$ and corresponding to the eigenvalues $\frac{1}{2}$ and $-\frac{1}{2}$ (in units for which $\hbar = 1$). Rather than use $S_x(j)$ and $S_y(j)$ directly it is more convenient to use the operators $S^+(j) = S_x(j) + iS_y(j)$ and $S^-(j) = S_x(j) - iS_y(j)$ and it is easy to show that

$$\begin{aligned} S^+(j)\alpha(j) = 0 \quad &; \quad S^-(j)\alpha(j) = \beta(j) \\ S^+(j)\beta(j) = \alpha(j); \quad &S^-(j)\beta(j) = 0 \end{aligned} \tag{17}$$

The three operators $S_x(j)$, $S_y(j)$, $S_z(j)$ by themselves do not form a complete set for there is no operator corresponding to the unit operator. This turns out to be the compound operator

$$S^2(j) = S_x^2(j) + S_y^2(j) + S_z^2(j)$$

which satisfies

$$S^2(j)\alpha(j) = \tfrac{3}{4}\alpha(j); \quad S^2(j)\beta(j) = \tfrac{3}{4}\beta(j) \tag{18}$$

S^2 together with the previous three operators forms a complete set in that any spin operator can be written in terms of them.

For an n-electron system a complete set of spin functions can be obtained by taking products of one-electron spin functions:

$$\eta(1, 2, \ldots, n) = \theta_1(1)\, \theta_2(2) \ldots \theta_n(n) \tag{19}$$

where each of $\theta_1, \ldots, \theta_n$ can be either α or β so that there will be 2^n linearly independent functions. It is convenient to divide these into sets with the same numbers of α-functions and β-functions. If there are a α-functions and b β-functions we denote these by $\{\eta_k^m\} k = 1, \ldots, p$ where $m = \tfrac{1}{2}(a - b)$ and k is an index to identify the $p = {}^nC_a(= {}^nC_b)$ functions in each set. For example, if $a = n, b = 0$, we have the single function $\eta_1^{\frac{1}{2}n} = \alpha(1)\alpha(2) \ldots \alpha(n)$, while, if $a = n - 1, b = 1$, we have n functions $\beta(1)\alpha(2) \ldots \alpha(n), \alpha(1)\beta(2)\alpha(3) \ldots \alpha(n), \ldots, \alpha(1) \ldots \alpha(n - 1)\beta(n)$ denoted by $\eta_1^{\frac{1}{2}n-1}, \ldots, \eta_n^{\frac{1}{2}n-1}$.

Let us now introduce permutation operators which act on the spin variables and write them as Q rather than P to distinguish them from the permutation operators on spatial variables. It is trivial to see that the effect of any of the Q operators on a member of the set $\{\eta_k^m\}$ will only give other members of the same set. Thus the division we have made is into sets which are transformed amongst themselves by the action of the permutation operators. Unfortunately this division is not complete and the sets $\{\eta_k^m\}$ are not irreducible.

To see this, the n-electron operators S^+ and S^- defined by

$$S^+ = \sum_{j=1}^{n} S^+(j); \quad S^- = \sum_{j=1}^{n} S^-(j) \tag{20}$$

can be introduced. Since S^- decreases the number of α-spin functions by one and increases the β-spin functions by one its effect on the functions $\{\eta_k^m\}$ is to give linear combinations of the functions $\{\eta_k^{m-1}\}$. In the same way S^+ acts on the set $\{\eta_k^{m-1}\}$ to give functions in $\{\eta_k^m\}$.

As an example consider the three-electron spin functions listed in Table 2. The effect of S^- on the only function $\alpha(1)\alpha(2)\alpha(3)$ in the set $\{\eta_k^{\frac{3}{2}}\}$ ($k = 1$) is to give

$$\beta(1)\, \alpha(2)\, \alpha(3) + \alpha(1)\, \beta(2)\, \alpha(3) + \alpha(1)\, \alpha(2)\, \beta(3) \tag{21}$$

which is just the sum of the three functions comprising the set $\{\eta_k^{\frac{1}{2}}\}\ k = 1, 2, 3$. Similarly applying S^- to the three functions $\{\eta_k^{\frac{1}{2}}\}$ gives $\beta(1)\, \beta(2)\, \alpha(3) + \beta(1)\, \alpha(2)\, \beta(3), \beta(1)\, \beta(2)\, \alpha(3) + \alpha(1)\, \beta(2)\, \beta(3), \beta(1)\, \alpha(2)\, \beta(3) + \alpha(1)\, \beta(2)\, \beta(3)$ which are three linearly independent functions consisting of combinations of the functions $\{\eta_k^{-\frac{1}{2}}\}$. This type of result carries over to the general case. If m is positive, the set of functions $\{S^-\eta_k^m\}$ will be linearly independent and each $S^-\eta_k^m$ will consist of sums of functions in $\{\eta_k^{m-1}\}$.

The problem with our original division of the spin functions into the sets $\{\eta_k^m\}$ is that no account is taken of the possibility of transforming from functions in one set to those in another using the S^- operator. We can remedy this, however, by recombining functions in a particular set to form an equivalent set which displays this fact explicitly. In the three-electron case, for example, three linearly independent functions which involve two α-spin functions and one β-spin function are

$$\frac{1}{\sqrt{3}} \{\beta(1)\,\alpha(2)\,\alpha(3) + \alpha(1)\,\beta(2)\,\alpha(3) + \alpha(1)\,\alpha(2)\,\beta(3)\}$$

$$\frac{1}{\sqrt{2}} \{\beta(1)\,\alpha(2)\,\alpha(3) - \alpha(1)\,\beta(2)\,\alpha(3)\} \tag{22}$$

$$\frac{1}{\sqrt{6}} \{\beta(1)\,\alpha(2)\,\alpha(3) + \alpha(1)\,\beta(2)\,\alpha(3) - 2\alpha(1)\,\alpha(2)\,\beta(3)\}$$

Clearly these new functions are equivalent to the original set $\{\eta_k^{\frac{1}{2}}\}$ since any one of the three functions $\{\eta_k^{\frac{1}{2}}\}$ can be written as a linear sum of them. But

Table 2. Spin Functions for a Three-Electron Problem

$\eta_k^{\frac{3}{2}}$ $k = 1$: $\alpha(1)\alpha(2)\alpha(3)$

$\eta_k^{\frac{1}{2}}$ $k = 1, 2, 3$: $\beta(1)\alpha(2)\alpha(3),\ \alpha(1)\beta(2)\alpha(3),\ \alpha(1)\alpha(2)\beta(3)$

$\eta_k^{-\frac{1}{2}}$ $k = 1, 2, 3$: $\beta(1)\beta(2)\alpha(3),\ \beta(1)\alpha(2)\beta(3),\ \alpha(1)\beta(2)\beta(3)$

$\eta_k^{-\frac{3}{2}}$ $k = 1$: $\beta(1)\beta(2)\beta(3)$

These functions can be recombined into the following equivalent sets:

$\bar{\eta}_k^{\frac{3}{2},\frac{3}{2}}$ $k = 1$: $\alpha(1)\alpha(2)\alpha(3)$

$\bar{\eta}_k^{\frac{3}{2},\frac{1}{2}}$ $k = 1$: $\dfrac{1}{\sqrt{3}} \{\beta(1)\alpha(2)\alpha(3) + \alpha(1)\beta(2)\alpha(3) + \alpha(1)\alpha(2)\beta(3)\}$

$\bar{\eta}_k^{\frac{3}{2},-\frac{1}{2}}$ $k = 1$: $\dfrac{1}{\sqrt{3}} \{\alpha(1)\beta(2)\beta(3) + \beta(1)\alpha(2)\beta(3) + \beta(1)\beta(2)\alpha(3)\}$

$\bar{\eta}_k^{\frac{3}{2},-\frac{3}{2}}$ $k = 1$: $\beta(1)\beta(2)\beta(3)$

$\bar{\eta}_k^{\frac{1}{2},\frac{1}{2}}$ $k = 1, 2$: $\dfrac{1}{\sqrt{2}} \{\beta(1)\alpha(2)\alpha(3) - \alpha(1)\beta(2)\alpha(3)\},\ \dfrac{1}{\sqrt{6}} \{\beta(1)\alpha(2)\alpha(3) + \alpha(1)\beta(2)\alpha(3) - 2\alpha(1)\alpha(2)\beta(3)\}$

$\bar{\eta}_k^{\frac{1}{2},-\frac{1}{2}}$ $k = 1, 2$: $\dfrac{1}{\sqrt{2}} \{\alpha(1)\beta(2)\beta(3) - \beta(1)\alpha(2)\beta(3)\},\ \dfrac{1}{\sqrt{6}} \{\alpha(1)\beta(2)\beta(3) + \beta(1)\alpha(2)\beta(3) - 2\beta(1)\beta(2)\alpha(3)\}$

The set $\{\eta_k^{\frac{1}{2}}\}$ is equivalent to $\{\bar{\eta}_k^{\frac{3}{2},\frac{1}{2}}\}$ plus $\{\bar{\eta}_k^{\frac{1}{2},\frac{1}{2}}\}$ and $\{\eta_k^{-\frac{1}{2}}\}$ to $\{\bar{\eta}_k^{\frac{3}{2},-\frac{1}{2}}\}$ plus $\{\bar{\eta}_k^{\frac{1}{2},-\frac{1}{2}}\}$.

All of the functions $\{\bar{\eta}_k^{\frac{3}{2},m}\}$ $m = \frac{3}{2}, \frac{1}{2}, -\frac{1}{2}, -\frac{3}{2}$ are symmetric with respect to exchange of electron co-ordinates.

The functions $\{\bar{\eta}_k^{\frac{1}{2},\frac{1}{2}}\}$ are transformed among themselves by the action of the Q operators and so satisfy equation 24. The particular form of the functions (which is not unique) is chosen so that the $D(Q)$ matrices are $(-1)^p$ times the transposed inverses of those in equations 15 and 16. Similar remarks apply to $\{\bar{\eta}_k^{\frac{1}{2},-\frac{1}{2}}\}$

the first of these functions is, apart from a trivial normalising constant, just $S^-\eta_1^{\frac{3}{2}}$ while the other two are functions orthogonal to it and, therefore, cannot be obtained from the previous set by using S^-. In the general case, if m is positive, we can similarly recombine and rearrange the set $\{\eta_k^m\}$ to give an equivalent set which divides into two sub-sets, the first consisting of the ${}^nC_{\frac{1}{2}n+m+1}$ linearly independent functions obtained by applying S^- to each of the ${}^nC_{\frac{1}{2}n+m+1}$ functions in $\{\eta_k^{m+1}\}$ and the second consisting of ${}^nC_{\frac{1}{2}n+m} - {}^nC_{\frac{1}{2}n+m+1}$ functions orthogonal to the $\{S^-\eta_k^{m+1}\}$. This second group we write as $\{\bar{\eta}_l^{m,m}\}$.

If m is negative there will be more functions (or the same number of functions if $m = -\frac{1}{2}$) in the set $\{\eta_k^{m+1}\}$ than in $\{\eta_k^m\}$. It turns out for this reason, that a set equivalent to $\{\eta_k^m\}$ can be obtained by applying S^- to the set $\{\eta_k^{m+1}\}$. Hence, there will be no functions $\{\bar{\eta}_l^{m,\,m}\}$ which cannot be obtained from a previous set if m is negative.

To summarise, therefore, we can rearrange the sets $\{\eta_k^m\}$, $m>0$, in such a way as to extract from them functions $\{\bar{\eta}_l^{m,\,m}\}$ which have the property that they cannot be obtained from a previous set by using the S^- operator. The remaining functions which together with $\{\bar{\eta}_l^{m,\,m}\}$ make up a set equivalent to $\{\eta_k^m\}$ can be obtained by applying S^- to a previous set. Only if m is non-negative do the sets $\{\bar{\eta}_l^{m,\,m}\}$ exist; for negative m all the spin functions can be obtained by applying S^- to previously obtained functions. This implies, therefore, that we have in fact reduced the 2^n linearly independent functions into a smaller number consisting of the sets $\{\bar{\eta}_l^{m,\,m}\}m\geqslant0$ with the property that the remaining spin functions can be obtained by applying the S^- operator to them.

Certain properties of these sets $\{\bar{\eta}_k^{s,\,s}\}s\geq0$ can be proved without too much difficulty. We just quote the results here.

(1) From $\{\bar{\eta}_k^{s,\,s}\}$ can be obtained $2s$ sets of linearly independent functions of the form $\{(S^-)^p\bar{\eta}_i^{s,\,s}\}p = 1, 2, \ldots 2s$. We label these new functions $\{\bar{\eta}_i^{s,\,s-p}\}$ and each of them can be written as a linear sum of the original functions $\{\eta_k^{s-p}\}$.

(2) Conversely it follows that each of our original sets of functions $\{\eta_k^m\}$ can be recombined into the sub-sets $\{\bar{\eta}_k^{s,\,m}\}s = |m|, |m| +1, \ldots \frac{1}{2}n$, which can be obtained from $\{\bar{\eta}_i^{s,\,s}\}(s\geq |m|)$ by applying $(S^-)^{s-m}$. (The case $n = 3$ is worked through in detail in Table 2.)

(3) Each of the functions $\{\bar{\eta}_i^{s,\,m}\}s\geq0, m = -s, \ldots,+s$ is an eigenfunction of the n-electron operators S_z (defined analogously to 20) and S^2 (defined as $S^2 = S_x^2 + S_y^2 + S_z^2$) satisfying

$$S^2\bar{\eta}_l^{s,\,m} = s(s+1)\bar{\eta}_l^{s,\,m}; \quad S_z\bar{\eta}_l^{s,\,m} = m\bar{\eta}_l^{s,\,m} \tag{23}$$

In view of (23) we could have selected the functions by the requirement that the original set of functions be divided up into eigenfunctions of S^2 and S_z as is often done in the textbooks. However, this is misleading. The key properties of the $\{\bar{\eta}_l^{s,\,m}\}$ are the way they are transformed by the S^- operator. The fact that they are eigenfunctions of S^2 follows from this rather than vice versa.

The importance of the transformation properties of the $\{\bar{\eta}_l^{s,\,m}\}$ under the action of S^- is due to the fact that S^- is symmetric and its action does not destroy any symmetry properties with respect to the permutation operators Q. To see the significance of this we first prove a lemma. Suppose $\phi_1 \ldots \phi_k$ are functions which are transformed amongst themselves by the action of the Q operators and satisfy

$$Q\phi_j = \sum_{i=1}^{k} D_{ij}(Q)\phi_i \tag{24}$$

for every operator Q. Then the functions $\{S^-\phi_i\}$ are also transformed amongst themselves by the action of the Q operators and satisfy

$$Q(S^-\phi_j) = \sum_{i=1}^{k} D_{ij}(Q)(S^-\phi_i) \tag{25}$$

i.e. they have exactly the same transformation properties as the $\{\phi_i\}$ since the coefficients $D_{ij}(Q)$ are the same. The proof is simple indeed. We operate on (24) with S^- and use the fact that since S^- is symmetric in the electron spin co-ordinates it commutes with all the Q operators, i.e. $S^-Q = QS^-$, all Q.

We know that for m positive the set $\{\eta_k^m\}$ can be transformed to an equivalent set consisting of the two sub-sets $\{S^-\eta_k^{m+1}\}$ and $\{\bar\eta_l^{m,\,m}\}$. Since the original set is transformed amongst itself by the action of the Q operators so must the equivalent set. However, from the lemma, the sub-set $\{S^-\eta_k^{m+1}\}$ is transformed only within itself since the functions $\{\eta_k^{m+1}\}$ are. Hence it follows that the functions $\{\bar\eta_l^{m,\,m}\}$ are transformed only amongst themselves by the action of the Q's. Again, from the lemma, it must follow that each of the sets $\{\bar\eta_l^{s,\,m}\} m = -s, -s+1, \ldots, s, s = \frac{1}{2}n, \frac{1}{2}n-1, \ldots (s \geq 0)$, is transformed only within itself by the Q operators, i.e. the Q's do not transform between sets. Thus we have reduced the original sets $\{\eta_k^m\}$ which are transformed among themselves into components with the same property. In fact they satisfy

$$Q\bar\eta_l^{s,\,m} = \sum_{k} D_{kl}^{s}(Q)\bar\eta_k^{s,\,m} \tag{26}$$

where the sum is over the functions in $\{\bar\eta_l^{s,\,m}\}$ only, and the $D_{kl}^{s}(Q)$ are constants. Moreover, as follows from the lemma, the constants $D_{kl}^{s}(Q)$ are the same for all $m, m = s, s-1, \ldots -s$. For the case $n = 3$, see Table 2.

We have already seen that the $\{\bar\eta_k^{s,\,m}\}$ are eigenfunctions of S^2 and S_z, so that the division of the spin functions into sets of functions transformed amongst themselves by the permutation operators also gives rise to functions with the same properties under the action of S^2 and S_z. The important fact, however, is the transformation properties of the sets under the action of the Q operators.

Nevertheless, the eigenfunction property is useful if we ask whether the sets $\{\bar\eta_k^{s,\,m}\}$ could be further sub-divided into sub-sets which were transformed amongst themselves by the action of the Q operators. For, if this were possible then we could implicitly define a spin operator with these new sets as eigenfunctions belonging to different eigenvalues. Thus this new spin operator could not be written in terms of S^2, S_x, S_y and S_z. However, the new operator would have to be symmetrical in the spin co-ordinates since its eigenfunctions are transformed amongst themselves by the action of the Q operators. But $S^2(i), S_x(i), S_y(i), S_z(i)$ form a complete set of spin operators in one variable and so the only spin operators symmetrical in n spin co-ordinates are S^2, S_x, S_y, S_z. Hence there is a contradiction and the original assumption is wrong. Therefore, the functions $\{\bar\eta_k^{s,\,m}\}$ are irreducible with respect to the Q operators.

5. THE PAULI PRINCIPLE

To form complete wave functions we must multiply the spatial eigenfunctions of the spin-independent Hamiltonian by spin functions. If H has a d-fold degenerate eigenvalue E with degenerate eigenfunctions $\psi_1 \ldots \psi_d$ the most general wave function corresponding to that energy will have the form:

$$\psi = \sum_{i=1}^{d} \psi_i \phi_i \tag{27}$$

where the ϕ_i are spin functions which have yet to be determined. Of course, there may be more than one solution of this form and, indeed, if the only requirement were that ψ satisfied the Schroedinger equation there would be $d \cdot 2^n$ linearly independent solutions since there are, respectively, d spatial and 2^n spin functions which are linearly independent.

However, this is not the only requirement; for in addition, the wave function must satisfy the Pauli principle (Pauli, 1925). This states that, in the case of electronic wave functions, ψ must be antisymmetric with respect to the interchange of pairs of co-ordinates (space and spin). Suppose, therefore, P is a permutation operator on the spatial co-ordinates and Q is the same permutation operator on the spin co-ordinates and let their parity, i.e. the number of pairs of co-ordinates they interchange, be p. Then by the Pauli principle

$$PQ\psi = (-1)^p \psi \tag{28}$$

So that by substituting into (28) and using (5) gives

$$(-1)^p \sum_{i=1}^{d} \psi_i \phi_i = \sum_{i=1}^{d} (P\psi_i)(Q\phi_i)$$

$$= \sum_{j=1}^{d} \sum_{i=1}^{d} D_{ji}(P)(Q\phi_i)\psi_j \tag{29}$$

Since the $\{\psi_i\}$ are linearly independent it follows that ϕ_i must satisfy

$$(-1)^p \phi_j = \sum_{i=1}^{d} D_{ji}(P)Q\phi_i, \quad j = 1, \ldots, d \tag{30}$$

From this it follows that, since the $\boldsymbol{D}$ matrices are non-singular,

$$Q\phi_i = \sum_{j=1}^{d} C_{ji}(P)\phi_j, \quad i = 1, \ldots, d \tag{31}$$

where the coefficients $C_{ji}(P)$ are related to the coefficients $D_{ji}(P)$ by the matrix relationship $\boldsymbol{C}^T(P) = (-1)^p \boldsymbol{D}^{-1}(P)^*$ and this must hold for all the permutation operators.

Thus we see that the $\{\phi_i\}$ must be transformed amongst themselves by the permutation operators and their transformation properties are deter-

*$\boldsymbol{C}^T$ means the transposed matrix, i.e. $C_{ij}^T = C_{ji}$.

mined by the transformation properties of the spatial functions. Normally the degenerate spatial functions will be irreducible and, in that case, the $\{\phi_i\}$ will also be irreducible. This means that the set $\{\phi_i\}$ must be equivalent to one of the sets $\{\bar{\eta}_k^{s,\,m}\}$ since we have seen that these are the only sets of irreducible spin functions. Which particular set it must be is determined by the relation (31) which means, in effect, by the transformation properties of the spatial functions. There will usually be more than one choice since any one of the sets $\{\bar{\eta}_k^{s,\,m}\}$ $m = s, s-1, \ldots, -s$ has the same properties under the action of the Q operators so each in turn may be used giving $(2s+1)$ degenerate wave functions. Any degeneracy, however, is due to the spin functions, the spatial degeneracy being entirely removed because all of the functions $\psi_1, \ldots, \psi_d$ are used in the complete wave function. Since the spin functions which are used in ψ will all have the same s value, ψ will be an eigenfunction of S^2 with eigenvalue $s(s+1)$. Thus we have this extremely elegant and beautiful result that the Pauli principle serves to remove the degeneracy of the spatial functions and to force the total wave functions to be eigenfunctions of S^2 with eigenvalue $s(s+1)$.

Nor are these conclusions much altered in the rare event that the degenerate spatial functions are not irreducible. In that case they can be subdivided into their irreducible sub-sets and the previous remarks apply to each of the sub-sets separately.

However, this is not quite the end of the matter as can best be seen by example. Suppose we see how the process works out in the case of allyl. If we begin with the antisymmetric spatial function (A) it follows that $D(P) = (-1)^P$ and hence $C(P) = C(Q) = 1$ for all operators P and Q. Hence the required $\{\bar{\eta}_k^{s,\,m}\}$ are the symmetric spin functions which for $n = 3$ are the functions with $s = \frac{3}{2}$. Thus the single spatial functions can be multiplied by any of the four spin functions $\bar{\eta}_1^{\frac{3}{2},\,m}$, $m = \frac{3}{2}, \frac{1}{2}, -\frac{1}{2}, \frac{3}{2}$ to give a wave function and so we obtain the four functions corresponding to quartet states. Continuing the process with one of the pairs of doubly degenerate spatial functions (D), ψ_1, ψ_2 (say), we find that the required spin functions are $\{\bar{\eta}_k^{\frac{1}{2},\,m}\}$ each set containing two functions. Thus the wave functions are

$$\Psi = \psi_1 \bar{\eta}_1^{\frac{1}{2},\,m} + \psi_2 \bar{\eta}_2^{\frac{1}{2},\,m} \tag{32}$$

both of the spatial and both of the spin functions being used to form Ψ. However, since m can take either of the two values $\frac{1}{2}$ or $-\frac{1}{2}$ there will be two distinct functions corresponding to doublet states.

Finally if we turn to the symmetric spatial functions (S) it follows that $D(P) = 1$ all P and hence $C(P) = C(Q) = (-1)^P$ so that the required spin function should be antisymmetric. But there is no such spin function. Thus there is no way of using the symmetric spatial functions to form wave functions which satisfy the Pauli principle and hence these energy levels correspond to physically non-allowed states. This, therefore, explains the classification of the eigenvalues of allyl in Table 1 into energy levels of quartet and doublet states and the energy levels of non-allowed states.

The results just described are not special to a three-electron problem. Indeed as the number of electrons increase the number of non-allowed states will increase too for there will be more different types of symmetry under the action of the P's that spatial functions can possess and yet for which there will be no spin functions with the correct type of behaviour to form wave functions satisfying the antisymmetry requirement. The principles underlying this are exactly the same as those described in the case of allyl; it only becomes a more difficult computational and combinatorial problem to classify the spin states and work out their properties under the action of the Q's. For some indication of how this may be done see Kotani *et al.* (1955), Löwdin (1955, 1956).

However, our simple example has been enough to show the extraordinary power and importance of the Pauli principle. For it is this which selects from the eigenfunctions of H those lesser number of solutions of physical significance and which forces the classification of these allowed solutions as the multiplets of eigenstates of S^2 with which we are so familiar.

6. SUMMARY

A fairly elementary account of the Pauli principle is given to show how it enables physically allowed states to be selected from the spatial solutions of the Schroedinger equation and how, for it to be satisfied, these allowed spatial solutions must be combined with spin functions so as to form eigenfunctions of S^2. As a simple example, the full set of spatial eigenfunctions and eigenvalues of the pi-electron Hamiltonian for the allyl radical are obtained and classified into allowed and non-allowed solutions.

REFERENCES

Amos, A. T. and Burrows, B. L. (1970) *J. chem. Phys.*, **53**, 939
Amos, A. T. and Woodward, M. R. (1969) *J. chem. Phys.*, **50**, 119
de Broglie, L. (1923) *Nature*, **112**, 540
de Broglie, L. (1925) *Ann. Phys.*, **3**, 22
Dirac, P. A. M. (1958) *Quantum Mechanics*, Oxford University Press
Goddard, W. A. (1967) *Phys. Rev.*, **157**, 73, 81, 93
Goddard, W. A. (1968) *J. chem. Phys.*, **48**, 450, 1008
Goddard, W. A. (1969) *Phys. Rev.*, **169**, 120
Hammermesh, M. (1962) *Group Theory and Its Applications to Physical Problems*, Pergamon Press, Oxford
Kotani, M., Amemiya, A., Ishiguro, E. and Kimura, T. (1955) *Tables of Molecular Integrals*, Maruzen, Tokyo
Löwdin, P. O. (1955) *Phys. Rev.*, **97**, 1474
Löwdin, P. O. (1956) *Adv. Phys.*, **5**, 1
Matsen, F. A. (1964) *Adv. quantum Chem.*, **1**, 60
Musher, J. I. (1971) *J. Phys.*, **31**, C4, 51
Parr, R. G. (1963) *Quantum Theory of Molecular Electronic Structure*, Benjamin, New York
Pauli, W. (1925) *Z. Phys.*, **31**, 765
Salem, L. (1966) *The Molecular Orbital Theory of Conjugated Systems*, Benjamin, New York
Snyder, L. C. and Amos, A. T. (1965) *J. chem. Phys.*, **42**, 3670
Weyl, H. (1931) *Theory of Groups and Quantum Mechanics*, Dover, New York
Wigner, E. P. (1959) *Group Theory and its Applications to the Quantum Mechanics of Atomic Spectra*, Academic Press, New York

10

Time-dependent Hartree–Fock Theory and its Application to Calculations of Atomic Properties

MICHAEL J. JAMIESON

Department of Physics, The University of Exeter, Exeter

---

1. INTRODUCTION

Time-dependent Hartree–Fock (TDHF) theory has long been in the literature; the equations were derived by Dirac in 1930. Their application to the calculation of atomic response functions has been discussed recently by Bersuker (1960), McLachlan and Ball (1964), Dalgarno and Victor (1966a), and Heinrichs (1968). TDHF theory has been applied to many problems of nuclear physics as the Random Phase Approximation (RPA) (Nozières and Pines, 1958; Goldstone and Gottfried, 1959; Ehrenreich and Cohen, 1959; Thouless, 1960, 1961a, 1961b; Rowe, 1966, 1968; Brown, 1967 and Lemmer and Vénéroni, 1968). The infinite set of equations of the RPA is equivalent to the finite set of differential equations of Dalgarno and Victor. It would not be possible to summarise the many aspects of TDHF theory in one article. An outline of the theory is presented together with a description of its application to the calculation of some atomic properties. This is illustrated by calculations on two electron systems. Mathematical detail is kept to a minimum, especially where details have been given elsewhere. For these the reader should consult the appropriate original papers. I hope that this article will also provide a descriptive introduction to the subject for the uninitiated reader.

TDHF theory exists in coupled and uncoupled versions. The coupled version is usually referred to as TDHF theory and is equivalent to the RPA. The uncoupled version neglects certain self-consistency terms and contains no correlation (that part of the electron–electron interaction that is not included in static unrestricted Hartree–Fock theory). The coupled version contains first order correlation. The RPA may be derived by linearising the equations of motion of certain creation operators (Rowe, 1966, 1968;

Brown, 1967) or from a certain perturbation expansion of the two particle Green's function (Thouless, 1961a) knowledge of which is sufficient to describe excited states (Thouless, 1961b). The expansion is chosen with the aid of Bruckner–Goldstone diagrammatic perturbation theory. The reader should refer to Thouless's book (1961b) for an explanation of diagrammatic expansions. This expansion shows that the RPA and hence TDHF theory contains first order correlation. In the diagrammatic and linearisation procedures excited states are represented by linear combinations of single particle excitations and de-excitations of a Hartree–Fock ground state. The latter are necessary in order to include correlation; neglect of them leads to the Tamm–Dancoff equations (Dunning and McKoy, 1967). Ho, Segal and Taylor (1972) included the RPA in a comparison of several methods of computing excitation properties of some molecules. They concluded that for static rather than dynamic properties the RPA is not significantly better than a configuration interaction method where the configurations are single excitations of a Hartree–Fock ground state. Sections 5 to 8 below deal with dynamic properties. Caves and Karplus (1969) made diagrammatic studies of time-independent coupled Hartree–Fock theory.

In the linearisation procedure the operators create the excited states and are linear in single particle excitations. Inclusion of one single particle excitation provides the single particle approximation (Dunning and McKoy, 1967). One can include more correlation by retaining higher powers of single particle excitations or by including many particle excitation operators (Rowe, 1968). Such schemes lack the simplicity of the RPA.

The RPA has been used in scattering problems by Lemmer and Vénéroni (1968). They extended the equations to include scattering states of a single nucleon. They showed that correlations are incorporated and demonstrated schematically that these influence the decay widths of resonant states. Schneider, Taylor and Yaris (1970) and Csanak, Taylor and Yaris (1971) presented a method for elastic and inelastic scattering of electrons by atoms and some specific inelastic scattering by ions, based on an approximation to the equation of motion of Martin and Schwinger (1959). It involves the self-consistent solution of equations for the one-particle Green function and target response. For elastic scattering the procedure in a first approximation leads to a TDHF target response. Schneider (1970) studied inelastic scattering of electrons by He by a similar method. It reduces to TDHF description of the target. Schneider and Krugler (1971) studied low energy scattering of electrons by a method similar to that of Csanak, Taylor and Yaris (1971). To a first approximation it is the method of Lemmer and Vénéroni since it involves an RPA description of target *and* scattered electron.

A version of TDHF theory was given by Phythian (1970). He discussed the evaluation of the amplitude for the transition from one Hartree–Fock state to another over a time interval. One could use ordinary TDHF theory to examine the time evolution of the states and calculate the overlap at some time. This leads to a spurious time-dependence, avoided in Phythian's theory which is based on a functional whose stationary value with respect to variations of the wave functions is the transition amplitude. The theory involves the simultaneous calculation of the time evolution of the two states and requires two simultaneous solutions of the time-dependent Schroedinger equation corresponding to different boundary conditions. The TDHF theory discussed here concerns the calculation of the linear response for one state, from which details of the spectrum are calculated. Phythian's and ordinary TDHF theories lead to identical equations in the special case in Phythian's theory of examining the time evolution of just one state. The difference between the two theories is similar to the difference between the time-dependent and stationary approaches to collision theory. Ordinary TDHF theory is a stationary approach; one obtains time-independent equations. The study of the role of correlation is simpler for ordinary TDHF theory.

Time-independent coupled and uncoupled Hartree–Fock theories, developed by Dalgarno (1959, 1962), have been used in calculations of such atomic properties as dipole polarisabilities and shielding factors.

Uncoupled time-dependent Hartree–Fock calculations have been made by Karplus and Kolker (1963a, 1963b and 1964) for refractive indices, Verdet constants and long range forces. Kaveeshwar, Chung and Hurst (1968) made coupled polarisability calculations. Coupled and uncoupled calculations of polarisabilities and other properties have been made by Dalgarno and Victor (1966a, 1966b, 1967a), Victor (1967), Kaveeshwar, Chung and Hurst (1968) and Kaveeshwar, Dalgarno and Hurst (1968). These are based on variational methods. The time-dependent linear response provides the Green function and hence excited and scattering states of the atom. Variational solutions of the TDHF equations are unsatisfactory for this; they include a finite number of poles (Sengupta and Mukherji, 1967). Recently Alexander and Gordon (1971, 1972) made numerical solutions of coupled and uncoupled TDHF equations appropriate to the dipole perturbation of He. Their results are in good agreement with those of Jamieson (1971), summarised later.

The TDHF equations of Dalgarno and Victor (1966a) are presented in Section 2.

2. TIME-DEPENDENT HARTREE–FOCK EQUATIONS

Unrestricted Hartree–Fock theory (Nesbet, 1965) is used. The atomic wave function is a Slater determinant. For closed shell atoms unrestricted and traditional Hartree–Fock theories differ only in the separation of variables in the final equations.

In the time-independent case a one-electron perturbation may be included with the kinetic energy; the standard derivation of Hartree–Fock equations is applicable and Brillouin's theorem holds. The perturbed scheme yields one electron properties correct to first order in correlation. The orbitals are modified by the external field and by an internal self-consistency field; the latter is the coupling to the responses of the other orbitals. In uncoupled theory the internal field is omitted. The time-dependent case is similar.

In the notation of Dalgarno and Victor (1966a) the coupled perturbed equations for an N-electron system are

$$(H_i - \mathscr{E}_i^{(0)} \pm \omega)u_{i\pm}^{(1)}(\mathbf{r}_i) + [V_{i\pm}^{(1)}(\mathbf{r}_i) + v_i(\mathbf{r}_i) - \langle u_i^{(0)} | V_{i\pm}^{(1)} | u_i^{(0)} \rangle$$

$$\mp \omega \langle u_i^{(0)} | u_{i\pm}^{(1)} \rangle] u_i^{(0)}(\mathbf{r}_i) = 0 \qquad (1)*$$

where $V_{i\pm}^{(1)}(\mathbf{r}_i)$ are self-consistency terms coupling the positive and negative frequency components $u_{i\pm}^{(1)}(\mathbf{r}_i)$ of the response, $u_i^{(0)}(\mathbf{r}_i)$ is the static Hartree–Fock (HF) orbital satisfying

$$H_i u_i^{(0)}(\mathbf{r}_i) = \mathscr{E}_i^{(0)} u_i^{(0)}(\mathbf{r}_i) \qquad (2)$$

and $v_i(\mathbf{r}_i)$ is the space part of the external one-electron perturbation. The uncoupled equations are equations 1 with $V_{i\pm}^{(1)}(\mathbf{r}_i)$ omitted. The unperturbed function is

$$\Psi^{(0)} = \mathscr{A} \prod_{i=1}^{N} u_i^{(0)}(\mathbf{r}_i) \exp(-iE^{(0)}t) \qquad (3)$$

where $\mathscr{A}$ is the normalised antisymmetriser (Condon and Shortley, 1935) and $E^{(0)}$ is the ground state energy. The coupled perturbed function is

$$\Psi = \mathscr{A} \prod_{i=1}^{N} w_i(\mathbf{r}_i, t) \exp(-iE^{(0)}t) \qquad (4)$$

where

$$w_i(\mathbf{r}_i, t) = u_i(\mathbf{r}_i) + \lambda[u_{i+}^{(1)}(\mathbf{r}_i) \exp(i\omega t) + u_{i-}^{(1)}(\mathbf{r}_i) \exp(-i\omega t)] + \mathscr{O}(\lambda^2) \qquad (5)$$

The uncoupled function is

$$\Psi = \Psi^{(0)} + \lambda \sum_{j=1}^{N} \mathscr{A} \prod_{i \neq j}^{N} u_i^{(0)}(\mathbf{r}_i)[u_{j+}^{(1)}(\omega_j) \exp(i\omega t) + u_{j-}^{(1)}(r_j) \exp(-i\omega t)]$$

$$\times \exp[-iE^{(0)}t] \qquad (6)$$

The parameter λ is the size of the perturbation. The unperturbed Hamiltonian contains a sum over occupied orbitals which may or may not include a self-interaction term. If it is included there are at most a finite number of excited bound states of a neutral atom rather than the infinite number expected from the Coulomb interaction. If it is excluded the Hamiltonian is orbital dependent and the orbitals are not automatically orthogonal (Dalgarno, 1966; Langhoff, Karplus and Hurst, 1966).

3. THE RANDOM PHASE APPROXIMATION AND CORRELATION

Substitution of the expansion

$$u_{i\pm}^{(1)}(\mathbf{r}_i) = \sum_{m=N+1}^{\infty} K_{mi\pm} u_m^{(0)}(\mathbf{r}_i) \qquad (7)$$

* Atomic units (A.U.) are used throughout unless it is explicitly stated otherwise.

into equation 1 yields the infinite set of equations

$$(\mathscr{E}_m^{(0)} - \mathscr{E}_i^{(0)} \pm \omega)K_{mi\pm} + \sum_{j=1}^{N} \sum_{n=N+1}^{\infty} [K_{nj\pm}(V_{jm,ni} - V_{jm,in})$$

$$+ K_{nj\mp}^*(V_{nm,ji} - V_{nm,ij})] + v_{mi} = 0 \qquad (8)$$

where

$$V_{ij,kl} = \langle u_i(1)u_j(2) | r_{12}^{-1} | u_k(1)u_l(2) \rangle$$

The equations with the inhomogeneous term v_{mi} omitted are eigenvalue equations, with eigenvalue ω, identical to those of the RPA. The inhomogeneous equations are the response equations.

The RPA has been used in nuclear physics (Thouless, 1961b; Brown, 1967; Rowe, 1968) given in terms of second quantisation. Second quantisation is described for example in the book by Thouless. To derive the RPA by the linearisation procedure one seeks time-dependent creation operators Q_m^+ linear in single-particle excitations, which satisfy the equation of motion,

$$i\frac{\partial Q_n^+}{\partial t} = \omega_n Q_n^+ = [H, Q_n^+] \qquad (9)$$

with suitable choice of energy scale ($E_o = 0$), so that

$$HQ_n^+ | O_c \rangle = 0 \qquad (10)$$

The ground state $| O_c \rangle$ contains some correlation. The RPA equations follow on linearising equation 9 in single particle excitations with

$$Q_n^+ = \exp(-i\omega_n t) \sum_{m,i} [K_{mi-}a_m^+a_i - K_{mi+}^* a_i^+ a_m] \qquad (11)$$

In equation 11 a_m^+ and a_i are single particle creation and destruction operators. The assumption is that the true ground state differs little from the HF one. One seeks to include the single-particle part of the correlation. The terms neglected in the linearisation are multiple particle and those of higher order in correlation which are assumed to add incoherently. Neglect of the latter gives the approximation its name. The RPA generates single particle excitations of a partly correlated ground state. Correlation is introduced by including excitations and de-excitations of the ground state. If no de-excitations were included the Tamm–Dancoff equations would be obtained. The equations satisfied by the K's of equation 11 are identical to the TDHF eigenvalue equations, ω_n being the eigenvalue.

Equations 8 follow directly from TDHF theory in second quantisation. The time-dependent wave function (Thouless, 1960, 1961b; Rowe, 1968)

$$| \Psi(t) \rangle = \exp\left\{ \sum_{m,i} [K_{mi+}\exp(i\omega_n t) + K_{mi-}\exp(-i\omega_n t)]a_m^+a_i \right\} | HF \rangle \qquad (12)$$

$| HF \rangle$ being the static Hartree–Fock ground state, is used in Frenkel's principle. To first order in the K's expressions 7 and 12 are identical. The TDHF wave function cannot represent a correlated state yet it leads to the RPA equations. Rowe (1968) has shown that use of equation 12 in Frenkel's principle leads to equations identical to the equations of motion in which the linearisation is implied. Thus the approximations involved in deriving the RPA and TDHF theory are identical. In both only a zero order ground state, the static HF, is required. Identical terms are omitted.

Rowe interpreted the TDHF wave function as a hybrid of the Tamm–Dancoff and RPA wave functions. Physically it describes an oscillation of the HF ground state in a normal mode. He interpreted the TDHF model as a variation on the transformation $\exp\{\varepsilon[Q^+ \exp(-i\omega t) + Q \exp(i\omega t)]\}$ which generates a general time dependent wave function to at least first order in ε. The Q^+ are RPA single-particle operators. The TDHF function is thus an uncorrelated HF function operated on by the generator of an RPA function. The RPA may also be derived by summing certain diagrams in a perturbation expansion of the two-particle Green function (Thouless, 1961a; Rowe, 1968). This expansion shows that the RPA contains all single-particle excitations correct to first order in correlation.

4. TRANSITION MATRIX ELEMENTS AND FREQUENCIES

In the RPA transition matrix elements of one electron operators reduce to expressions involving the K's of equation 11 and expectation values, with respect to the ground state, of one-electron operators. The latter include first order correlation when the HF ground state is substituted (Brillouin's theorem) and so matrix elements are given to first order in correlation solely in terms of the K's.

$$\langle 0|v|n\rangle = \sum_{m,\,i} [K_{mi-}v_{im} + K^*_{mi+}v_{mi}] \quad \text{(Rowe, 1968)} \tag{13}$$

It is necessary to identify the excitation energies and matrix elements of TDHF theory with those of the RPA.

The TDHF response is (cf. equation 7)

$$\Psi_+ \exp(i\omega t) + \Psi_- \exp(-i\omega t) \tag{14}$$

where

$$\Psi_\pm = \sum_{m,\,i} K'_{mi\pm}\, a^+_m\, a_i\,|\,\mathrm{HF}\!> \tag{15}$$

The ground state energy is chosen as zero. The K''s satisfy the inhomogeneous equations 8.

$$\Psi_\pm = G(\mp\omega)v\Psi_0 \tag{16}$$

where Ψ_0 is the ground state

$$G(z) = 1/(z-H) \tag{17}$$

and H is the unperturbed Hamiltonian. The response is contained in the polarisability,

$$\alpha(\omega) = \langle \Psi_0 | v | \Psi_+ + \Psi_- \rangle = S_n \frac{| \langle 0 | v | n \rangle |^2 \omega_n}{\omega^2 - \omega_n^2} = S_n \frac{f_n}{\omega^2 - \omega_n^2} \qquad (18)$$

the ω_n being the excitation frequencies and f_n the oscillator strengths. The inhomogeneous equations may be written as (Thouless, 1961b),

$$\begin{bmatrix} A & B \\ B^*A^* \end{bmatrix} \begin{bmatrix} K'_+ \\ K'_-{}^* \end{bmatrix} = \omega \begin{bmatrix} K'_+ \\ -K'_-{}^* \end{bmatrix} - \begin{bmatrix} v \\ v \end{bmatrix} \qquad (19)$$

The homogeneous ones may be written

$$\begin{bmatrix} A & B \\ B^*A^* \end{bmatrix} \begin{bmatrix} K^{(n)}_- \\ K^{(n)*}_+ \end{bmatrix} = \omega_n \begin{bmatrix} K^{(n)}_- \\ -K^{(n)*}_+ \end{bmatrix} \qquad (20)$$

in an obvious notation. Thouless (1961a) and Rowe (1968) discussed ortho-normality and completeness of the solutions of equation 20. The K''s of TDHF theory satisfying equation 19 may be written in terms of the eigenvalues ω_n and K's of the RPA satisfying equation 20.

After some manipulation on equations 13–20 one finds that in TDHF theory

$$\alpha(\omega) = S_n \frac{| \langle 0 | v | n \rangle |^2 \omega_n}{\omega^2 - \omega_n^2} \qquad (21)$$

where the ω_n are the eigenvalues of equation 20 and $\langle 0 | v | n \rangle$ is given by equation 13. Thus the excitation energies and matrix elements of TDHF theory are identical with those of the RPA and hence contain first order correlation.

At frequencies greater than that associated with the ionisation energy of the atom, it is convenient to omit the external perturbation. The TDHF equations are eigenvalue equations equivalent to equations 20. In a full treatment there is no solution Ψ_+; Ψ_- is the eigenfunction to be used, for example, in a matrix element. In the TDHF approximation Ψ_+ and Ψ_- both exist. How are they to be used in a matrix element? The discussion above shows that the matrix element is given by equations 13 and 20, which implies that the combination $(\Psi_+ + \Psi_-)$ must be substituted for the true state.

The correlation in TDHF theory may be studied directly using the diagram technique as is done for the time independent HF theory by Caves and Karplus (1969). One may use diagrammatic perturbation theory (discussed, e.g. by Goldstone, 1957; Thouless, 1961b) to write down the time dependent response correct to first order in correlation. The single-particle part of the response is represented by the diagrams in Figure 1 which correspond to equation 15 to first order in correlation (cf. Jamieson, 1971). Hence the response contains all single-particle excitations correct to first order in correlation. This is equivalent to Brillouin's theorem. The one-electron external perturbation could be absorbed in kinetic energy terms in the HF

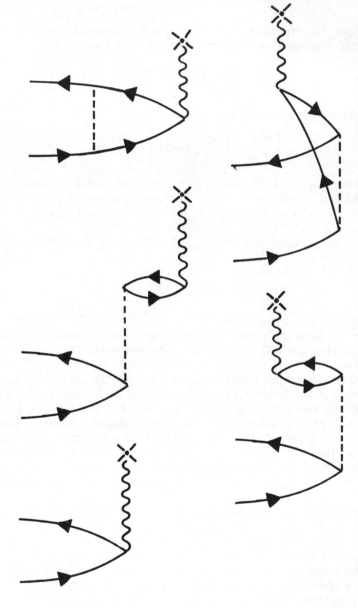

Figure 1 Diagrams included by TDHF theory. The external perturbation and correlation are denoted by ($\sim\!\sim\!\sim\!\!\times$) and (----)

equations. Consider the polarisability $\langle \Psi_+ + \Psi_- | v | \Psi_0 \rangle$. The HF Ψ_0 contains all single-particle excitation. The next order in correlation terms involve double excitations. Such terms can only combine with double excitations in the response. The result would be of second order in correlation. Thus the polarisability is correct to first order in correlation.

The poles of the response are at the transition frequencies. Comparison of the diagram expansion of these frequencies to first order in correlation with the perturbation expansion to first order in correlation for the locations of these poles shows that the TDHF frequencies are correct to first order in correlation. Hence the transition matrix elements between bound states are also correct to first order in correlation.

The matrix element $\langle \Psi_+ + \Psi_- | v | \Psi_0 \rangle$ contains all first order correlation when Ψ_+ are solutions appropriate to the inhomogeneous equations (cf. equations 8 and 15). Because the homogeneous and inhomogeneous equations contain the same terms in correlation the matrix element also contains all first order correlation when Ψ_+ are solutions to the homogeneous equations, i.e. the transition matrix elements to the continuum contain first order correlation.

In summary the coupled TDHF response provides transition frequencies, energies and transition matrix elements correct to first order in correlation. The procedures are equivalent to the RPA though a different interpretation is put on the wave function. The set of integrodifferential equations of TDHF theory is more convenient than the infinite set of equations of the RPA.

5. CORE POLARISATION IN TIME-DEPENDENT HARTREE–FOCK THEORY

The matrix element $\langle B | L | A \rangle$ of a one-electron operator, L, if calculated in HF approximation contains an error of first order in correlation (Cohen and Dalgarno, 1966). The correction to the dipole matrix element for the transition of an outer electron can be attributed to polarisation of the core by the electron. Several authors have studied core polarisation in various approximations containing correlation to at least first order. Cohen and Dalgarno (1966) and Dalgarno and Parkinson (1967) evaluated matrix elements to first order in inverse nuclear charge and showed that the corrections to the HF approximation correspond to dynamic polarisation of the core at the transition frequency. Hameed, Herzenberg and James (1968) calculated corrections to dipole oscillator strengths in alkali atoms. Their method is equivalent to the adiabatic approximation in which the state of the core is calculated in the presence of a stationary outer electron. Dalgarno, Drake and Victor (1968) discussed the adiabatic approximation. They found that the leading correction to the static interaction of a slowly moving electron with a spherically symmetric atom is as if the atom were polarised.

Bersuker (1960) used TDHF theory to predict core polarisation. His derivation is similar to that below but he excluded the transition electron's orbital from a sum in his time-dependent equations describing the influence of the time dependent perturbation (equivalent to the self-interaction terms in the Fock Hamiltonian). Use of the Fock Hamiltonian with self-interaction to calculate the unperturbed orbitals yields a non-physical spectrum (a finite number of bound states and continuum) for the excited states (Dalgarno, 1966; Langhoff, Karplus and Hurst, 1966). The Fock Hamiltonian without self-interaction is unsatisfactory because its eigenfunctions are not orthogonal. Dipole matrix elements for an outer electron are often calculated with orbitals of the Fock Hamiltonian without self-interaction. In the following derivation self-interaction is retained. When allowance is made for the non-physical nature of the excited Hartree–Fock orbital the usual dynamic polarisation of the core is predicted.

The matrix elements of an operator can be extracted from the response to that operator. Consider an initial time dependent state Φ_A of Hamiltonian H whose energy is chosen as zero for convenience. The linear response to perturbation

$$v[\exp(i\omega t) + \exp(-i\omega t)]$$

is

$$\Psi = \Psi_+ \exp(i\omega t) + \Psi_- \exp(-i\omega t) \tag{22}$$

where $\Psi_\pm$ satisfy equation 16.

The projector on to the subspace of eigenvalue E_i is (Messiah, 1964)

$$P_i = \frac{1}{2\pi i} \oint_{\gamma_i} G(z) dz \tag{23}$$

where γ_i is a contour containing $z = E_i$. Hence

$$\langle \Phi_B | v | \Phi_A \rangle = \frac{1}{2\pi i} \langle \Phi_B | \oint_{\gamma_B} \Psi_-(z) dz \rangle \tag{24}$$

Suppose v is a one electron dipole operator. The TDHF response is given by equations 15 and 19. Let Φ_B differ from Φ_A by having an electron excited from level i to level m (the eigenfunctions of the Fock Hamiltonian are called levels). The HF representation of Φ_B and TDHF response contain all single particle excitations and give the right-hand side of equation 24 to first order in correlation. Double excitations in Φ_B could only combine with double excitations in Ψ_- giving contributions of higher order in correlation.

$$\langle \Phi_B | R | \Phi_A \rangle = \frac{1}{2\pi i} \oint_{\gamma} K_{mi-} d\omega \tag{25}$$

where the contour γ encloses $\omega = E_B - E_A = \mathscr{E}_m^{(0)} - \mathscr{E}_i^{(0)}$ and R denotes the position vectors collectively. Solving equations 19 for K_{mi-} as a function of ω to first order in V one finds

$$\langle \Phi_B | R | \Phi_A \rangle = r_{mi} - \sum_{\substack{j=1 \\ j \neq i}}^{N} \sum_{\substack{n=N+1 \\ n=i}}^{\infty} \left[\frac{(V_{jm,ni} - V_{jm,in})}{(\mathscr{E}_n^{(0)} - \mathscr{E}_j^{(0)}) - (\mathscr{E}_m^{(0)} - \mathscr{E}_i^{(0)})} r_{nj} \right.$$

$$\left. + \frac{V_{nm,ji} - V_{nm,ij}}{(\mathscr{E}_n^{(0)} - \mathscr{E}_j^{(0)}) + (\mathscr{E}_m^{(0)} - \mathscr{E}_i^{(0)})} r_{jn} \right] - \sum_{s \neq m,i} \frac{(V_{im,si} - V_{im,is})}{\mathscr{E}_s^{(0)} - \mathscr{E}_m^{(0)}} r_{si} \tag{26}$$

The first term is that given by the frozen core approximation. Other treatments of this problem involve orbitals i and m calculated in the frozen core approximation with no self-interaction. The self-interaction affects orbital m but not i. Let $|m'\rangle$ denote orbital m calculated excluding the self-interaction. Treating the self-interaction as a perturbation we find

$$|m'\rangle = |m\rangle - \underset{n}{S} \frac{(V_{si,im} - V_{is,im})}{\mathscr{E}_s^{(0)} - \mathscr{E}_m^{(0)}} |s\rangle \tag{27}$$

Therefore the first term and first part of the last sum in equation 26 may be combined to yield $\langle m'|r|i\rangle$, the dipole matrix element used by others as the zeroth (in correlation) approximation for the transition matrix element.

Neglect of exchange terms and the dipole approximation for $1/r_{12}$ in equation 26 yields

$$\langle \Phi_B|R|\Phi_A\rangle = \langle m'|r|i\rangle - \sum_{\substack{j=1 \\ j \neq i}}^{N} \sum_{\substack{n=N+1 \\ n=i}}^{\infty} \left[\frac{2(\mathscr{E}_n^{(0)} - \mathscr{E}_j^{(0)})\langle m|r/r^3|i\rangle}{(\mathscr{E}_n^{(0)} - \mathscr{E}_j^{(0)})^2 - (\mathscr{E}_m^{(0)} - \mathscr{E}_i^{(0)})^2} r_{jn}r_{nj} \right]$$

$$= \langle m'|r - \underline{\alpha}_c(\omega_{mi})r/r^3|i\rangle \tag{28}$$

since level i is well outside the core and in the second term $\langle m|$ may be replaced by $\langle m'|$ to first order in correlation. This is the usual result implying that the core is polarised at the frequency of the transition $\omega_{mi} = \mathscr{E}_m^{(0)} - \mathscr{E}_i^{(0)}$ and gives rise to an extra dipole moment. In equation 28 $\underline{\alpha}_c$ is the uncoupled HF approximation to the dynamic polarisability tensor of the core.

6. CALCULATION OF ATOMIC PROPERTIES

Atomic properties such as dynamic multipole polarisabilities and shielding factors are readily calculated from the time-dependent response. The locations of the poles of the polarisability and their residues yield the bound part of the multipole spectrum (the transition matrix elements and excitation energies); cf. equation 18. Continuum oscillator strengths may be calculated from matrix elements of the multipole operator between the ground and continuum states of appropriate normalisation (Stewart, 1967). The continuum state is represented by the combination $\Psi_+ + \Psi_-$ of solutions to the homogeneous equations. In uncoupled theory Ψ_+ does not exist. The matrix element involves only Ψ_-. The continuum solutions of the $N+1$ electron problem contain phase shifts appropriate to the elastic scattering of electrons by the N electron core.

The bulk measurable properties of refractive index and the Verdet constant (Faraday rotation per unit path length per unit magnetic flux) are given directly by the dynamic polarisability (Born, 1960; Van Vleck, 1965). Many properties may be evaluated from the multipole spectrum. Some properties involve infinite sums of matrix elements which may be reduced to single matrix elements involving the response (Dalgarno, 1966).

The effect on a property of the correlation included in TDHF theory may be explored by comparing its coupled TDHF value, its uncoupled TDHF value (i.e. the uncorrelated value) and its fully correlated value (where available). The task of solving the TDHF equations is easiest for two-electron systems. Fully correlated values (based on variational calculations) are available for many properties of two-electron systems. Discussions, based on numerical

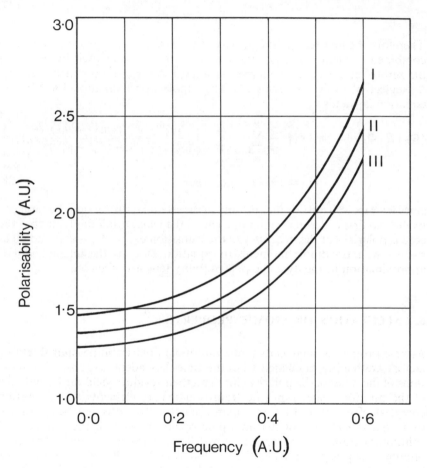

Figure 2   Dynamic dipole polarisabilities for He. *I: uncoupled Hartree–Fock ; II: corre-lated (Dalgarno and Victor, 1966a) ; III: coupled Hartree–Fock*

solution of the TDHF equations, are given elsewhere of the effect of the TDHF correlation on dipole properties of He and Li$^+$, elastic scattering phase shifts for e–He$^+$ and e–H and the complete dipole spectrum of He (Jamieson, 1971, 1972). These are summarised below together with other calculations to illustrate the discussion of the opening paragraph of this section. The HF ground state orbitals used were those of Bagus (1967) for He, of Roothaan, Sachs and Weiss (1960) for Li^+, and of Green *et al.* (1954) for H^-.

6.1 POLARISABILITY

The dynamic dipole polarisability for a two-electron atom is given by equation 18 with $v = z_1 + z_2$. Results for He and Li^+ are shown in Figures 2 and 3. There is good agreement with variational calculations (Dalgarno

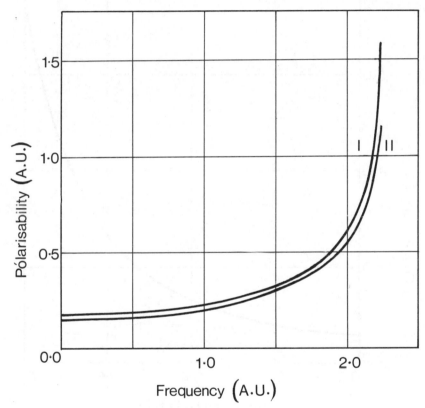

Figure 3   Dynamic dipole polarisabilities for Li^+. I: uncoupled Hartree–Fock; II: coupled Hartree–Fock

and Victor, 1966a; Kaveeshwar, Chung and Hurst, 1968) below and not near the first excitation frequency. This is as expected in a frequency range where $\alpha(\omega)$ does not depend much on the higher excited states which are not given well by the variational method. At frequencies below the $(1s)^2 1^1S \rightarrow (1s2p)2^1P$ transition frequency the results are identical to those of Alexander and Gordon (1972) for He. In the solution of the inhomogeneous coupled equations for $\omega > \omega(1^1S \rightarrow 2^1P)$ the convenient (though unnecessary) approximation of truncating the external perturbation was made. This approximation cannot shift the poles of $\alpha(\omega)$ and does not affect the values of the first few oscillator strengths. The significant contributions to the corresponding matrix elements are not affected by the truncation, because the ranges of the lower excited state functions are smaller than the truncated range. The uncoupled results contain no approximation and are identical with those of Alexander and Gordon (1971). Accurate coupled and uncoupled results

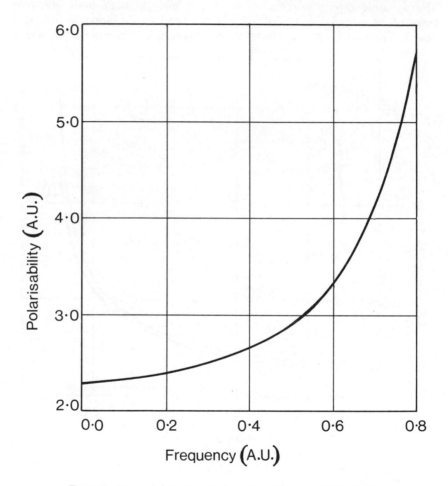

Figure 4   Uncoupled Hartree–Fock quadrupole polarisabilities for He

for $\omega > \omega(1^1S \rightarrow 1^1P)$ are given by Alexander and Gordon (1971, 1972) or by the details of the dipole spectrum. Static polarisabilities agree with established values (Dalgarno, 1959; Dalgarno and McNamee, 1961; Yoshimine and Hurst, 1964; Langhoff, Karplus and Hurst, 1966).

Figure 2 also shows the effect of correlation on $\alpha(\omega)$ for He. The accurate results are from refined variational calculations (Chan and Dalgarno,

1965a). Uncoupled quadrupole polarisabilities of He are shown in Figure 4. The static value agrees with that of Dalgarno and McNamee (1961).

6.2. TRANSITION FREQUENCIES AND OSCILLATOR STRENGTHS IN He AND Li^+

Transition frequencies and oscillator strengths for $(1s^2)1^1S \rightarrow (1snp)n^1P$ transitions were evaluated by fitting α to a function of the form of the right-hand side of equation 18. Of course one may calculate uncoupled eigenfunctions and energy levels as in a straightforward one-electron eigenvalue problem rather than by studying the linear response. There is no problem in interpreting the wave function as there is in the coupled case. This procedure provided a numerical check on the linear response method. Energy levels and transition frequencies are summarised by the quantum defect, σ_n. Coupled and uncoupled values of σ_n for the $(1snp)^1P$ series and oscillator strengths for $1^1S \rightarrow n^1P$ transitions are compared in Table 1 with the best

Table 1. Oscillator strengths and quantum defects in He and Li^+

	State $(1snp)^1P, n$	Coupled Hartree–Fock	Uncoupled Hartree–Fock	Accurate
		Oscillator strength		
He	2	0.2526	0.3318	0.27616[1]
	3	0.06973	0.0861	0.0734[1]
	4	0.02898	0.0345	0.0302[2]
	5	0.01478	0.0172	0.0151[3]
Li^+	2	0.4467	0.5181	0.4565[2]
		Quantum defect		
He	2	−0.0330	0.0191	−0.0094[4]
	3	−0.0330	0.0215	−0.0111[4]
	4	−0.0330	0.0222	−0.0116[4]
	5	−0.0328	0.0225	−0.0118[4]
Li^+	2	−0.0251	0.0160	−0.0135[5]

[1]Schiff and Pekeris (1964). [2]Weiss (1967). [3]Dalgarno and Parkinson (1967). [4]Martin (1960). [5]Moore (1949).

available theoretical and experimental results. The exchange part of σ_n (Bethe and Salpeter, 1957) is overestimated in coupled theory. Figures 5 and 6 show coupled and uncoupled phase shifts and continuum oscillator strengths $df/d\mathcal{E}$ for He. Figure 6 shows a comparison with $df/d\mathcal{E}$ derived from the photoionisation measurements of Lowry, Tomboulian and Ederer (1965). The coupled values are superior.

Low energy phase shifts were calculated for elastic scattering of electrons by H and He^+. As expected Levinson's theorem was found to hold in uncoupled theory. It also holds in coupled theory. For e–He$^+$ the spectral head relation

$$\delta(0) = \pi\sigma_\infty \tag{29}$$

between the zero energy phase shifts $\delta(0)$ and the spectral head value of the quantum defect σ_∞ was found to hold in coupled and uncoupled theories.

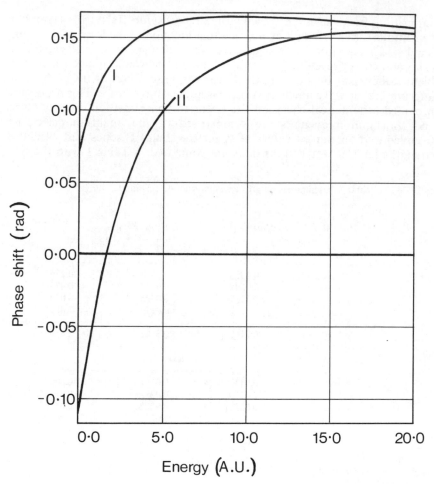

Figure 5   Elastic phase shifts for elastic scattering of electrons by He$^+$. *I: uncoupled Hartree–Fock; II: coupled Hartree–Fock*

It is the extension of Levinson's theorem to Coulomb potentials (Seaton, 1958). These low energy phase shifts have been compared with the results of correlated calculations (Jamieson, 1972). The coupled results are superior, but overestimate exchange.

Continuity of the dipole oscillator strength (Hargreaves, 1928) at the spectral head is expected in uncoupled theory. For He this was confirmed and found to hold in coupled theory. Asymptotic bound oscillator strengths are

$$f_n \sim \frac{1.82}{(n+0.033)^3} \qquad \text{(coupled)} \tag{30}$$

$$f_n \sim \frac{2.06}{(n-0.023)^3} \qquad \text{(uncoupled)} \tag{31}$$

The correlated calculations of Dalgarno and Parkinson (1967) yield

$$f_n \sim \frac{1.85}{n^3} \tag{32}$$

Table 2 shows values of the He oscillator strength sums (Dalgarno and Lynn,

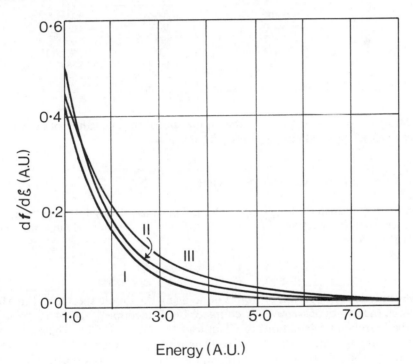

Figure 6   *Continuum dipole oscillator strengths in* He. *I: uncoupled Hartree–Fock; II: coupled Hartree–Fock; III: experimental (Lowry, Tomboulian and Ederer, 1965)*

1957) $S_k = \underset{n}{S} f_n \omega_n^k$. The continuum calculations were made for continuum energies up to 20.0 A.U. above which the asymptotic form

$$df/d\mathscr{E} \sim A\mathscr{E}^{-\frac{7}{2}}(1+B\mathscr{E}^{-1}) \tag{33}$$

was employed. The results are in good agreement with variational results (Victor, 1966) as expected for $k = -1, -2$. The $k = -2$ rule is the definition of $\alpha(0)$. Variational results are not so reliable for $k = 1, 2$ where the continuum is important. The Thomas–Kuhn sum rule ($k = 0$) holds in both coupled and uncoupled theories. Rowe (1968) has shown that it holds in the RPA. The $k = -1, 1$ rules of Dalgarno and Lynn are modified in TDHF theory.

The preceding discussion shows that in the uncoupled theory dipole oscillator strength is displaced to the bound and low continuum states. The coupled approximation overestimates correlation by displacing oscillator strength to the continuum. This is discussed by Jamieson (1971).

Table 2. Oscillator strength sums S_k in He (A.U.)

k	Coupled Hartree–Fock	Uncoupled Hartree–Fock	Accurate
2	28.104	26.890	30.33[1]
1	4.090	3.861	4.083725[1]
0	1.987	1.999	2.0[2]
−1	1.473	1.580	1.504994[1]
−2	1.320	1.488	1.3841[3]

[1]From Pekeris (1959). [2]Thomas–Kuhn sum rule. [3]From Chung and Hurst (1966).

Many properties are given by the dipole spectrum. The values of examples given reflect the oscillator strength distributions.

6.3. VAN DER WAALS INTERACTION

The leading term in the long range potential between atoms A, B at separation R such that retardation may be ignored is $\varepsilon_2 R^{-6}$ where (Dalgarno and Davison, 1966)

$$\varepsilon_2 = \frac{3}{\pi} \int_0^\infty d\omega \alpha_A(i\omega)\alpha_B(i\omega) \tag{34}$$

The analytic continuation of $\alpha(\omega)$ is done using the dipole spectrum. For He in A.U. the values of ε_2 are 1.365 coupled, 1.661 uncoupled, 1.4714 according to the correlated calculations of Chan and Dalgarno (1965b). These results are in good agreement with those of Dalgarno and Victor (1967a) based on variational solutions of the TDHF equations. This is as expected. Other calculations have been made for ε_2 but that of Chan and Dalgarno is the best.

6.4. RAYLEIGH SCATTERING

The Rayleigh scattering cross-section for unpolarised light of wavelength λ

$$Q = \frac{128\pi^5}{3\lambda^4} \left[\alpha\left(\frac{2\pi c}{\lambda}\right) \right]^2 a_0^6 \tag{35}$$

c being the speed of light and a_0 the Bohr radius, is readily calculated from values of $\alpha(\omega)$, calculated directly or from the dipole spectrum. Table 3

Table 3. Rayleigh scattering cross-sections, Q, for He

Wavelength $\lambda(\text{Å})$	$Q(\text{cm}^2)$ Coupled Hartree–Fock	$Q(\text{cm}^2)$ Uncoupled Hartree–Fock	$Q(\text{cm}^2)$ Correlated[1]
700	8.650E-25	1.256E-24	1.07E-24
800	3.096E-25	4.257E-24	3.59E-25
1000	8.407E-26	1.113E-25	9.51E-26
1500	1.215E-26	1.569E-26	1.33E-26
2000	3.498E-27	4.483E-27	3.83E-27
2500	1.374E-27	1.756E-27	1.50E-27
1216 (Lyman α)	3.179E-26	4.143E-26	3.53E-26

[1]Chan and Dalgarno (1965a). $En = 10^n$.

compares values of Q calculated from the coupled and uncoupled He dipole spectra with the correlated results of Chan and Dalgarno (1965a).

6.5. RANGE AND STRAGGLING AND LAMB SHIFT

The range and straggling of fast charged particles in a gas and the Lamb shift are characterised by mean powers of excitation energies weighted with oscillator strengths (Nicholls and Stewart, 1962). They are dS_k/dk. These energies I, I' and K_0 respectively correspond to $k = 0, 1, 2$ (Bethe and Salpeter, 1957; Dalgarno, 1966). For He the values of I in eV are 42.52 coupled, 39.64 uncoupled compared with the correlated value 42.1 of Chan and Dalgarno (1965c). Those of I' are 83.12 coupled, 80.21 uncoupled and 69.4 from Chan and Dalgarno. Contributions to I, I' change sign because of log terms. Cancellation of error makes the coupled result for I better than the individual oscillator strengths. The uncoupled value of I' is the more accurate because of fortuitous cancellation of error where the uncoupled oscillator strengths cross the correlated ones. Other authors have given values for I, I'; the most accurate are those of Chan and Dalgarno. Values, in rydberg, of K_0 are 76.24 coupled and 80.06 uncoupled. The most accurate of several other works is that of Schwartz (1961) giving $78.73 < K_0 < 79.36$.

6.6 SHIELDING FACTOR

Kaveeshwar, Dalgarno and Hurst (1968) used variational coupled solutions to calculate dynamic shielding factors $\beta_\infty(\omega)$ of He, Be, Ne and some ions. The shielding factor involves the matrix element of $\sum_{\text{electrons}} r^{-2} \cos \theta$ between the ground state and the linear response. This matrix element weighs the response close to the origin while that for the polarisability weighs it far from the origin. Kaveeshwar, Dalgarno and Hurst showed that for a real N-electron atom of nuclear charge z at frequency ω

$$z\beta_\infty(\omega) = N + \omega^2\alpha(\omega) \tag{36}$$

and that this holds in the static case for a Hartree–Fock atom. They concluded that their variational method gives a better representation far from the nucleus than near it, the calculations of β_∞ providing a check on the method used to solve the coupled equations. Coupled and uncoupled values of β_∞ calculated from the linear response of this work are in good agreement with variational results of Dalgarno (1962). The uncoupled theory overestimates β_∞ compared to the N/z value.

6.7. TWO-PHOTON DECAY

The predominant decay mechanism of the $(1s2s)2^1S$ state is the two-photon process

$$2^1S \rightarrow \hbar\omega_1 + \hbar\omega_2 + 1^1S$$

The probability of emission of a photon in the energy range $\Delta E[y, y+dy]$ (ΔE is the energy between 2^1S and 1^1S) is per unit time (α being the fine structure constant)

$$A(y)dy = \frac{\Delta E^7}{48\pi} \alpha^6 y^3 (1-y)^3 \, | \, M \, |^2 dy \tag{37}$$

(Breit and Teller, 1940). The Einstein A coefficient and lifetime τ are given by

$$\mathscr{A} = \frac{1}{\tau} = \frac{1}{2} \int_0^1 A(y)dy \tag{38}$$

$$M = \langle 2^1S \, | \, D_z[G(\omega_1 + E(1^1S)) + G(\omega_2 + E(1^1S))]D_z \, | \, 1^1S \rangle \tag{39}$$

where

$$D_z = z_1 + z_2 \tag{40}$$

$E(1^1S)$ being the energy of the 1^1S state. M is usually written as an infinite sum over intermediate states but may be written in terms of the linear response (Dalgarno, 1966).

$$\langle 2^1S \, | \, D_z \, | \, \Psi_-(\omega_1) + \Psi_-(\omega_2) \rangle \tag{41}$$

Correlated calculations for the He isoelectronic series were made variationally by Dalgarno and Drake (1968) and Drake, Victor and Dalgarno (1969). Uncoupled and coupled calculations of this work using the 2^1S functions of Cohen and Kelly (1965, 1966) and ΔE of Moore (1949) yield lifetimes in s of 0.02458, 0.02862 compared with the correlated value (Drake, Victor and Dalgarno) of 0.01953 for He and 0.0005199, 0.0005913 compared with 0.0005138 for Li$^+$. Figures 7 and 8 show the relative probabilities $A(y)$. Coupled and uncoupled theories underestimate $A(y)$, the coupled more than the uncoupled. Assuming the signs of the matrix elements in the infinite sum to be those for hydrogen, the term involving the 2^1P intermediate state differs in sign from the others, leading to a fortuitous cancellation of error in the uncoupled theory. The coupled and uncoupled $A(y)$ provide qualitative estimates of the distribution of oscillator strength for $2^1S \rightarrow (2snp)n^1P$. It is similar to that for $1^1S \rightarrow n^1P$. The complete $1^1S \rightarrow$

n^1P spectrum was not calculated for Li^+, but bound state calculations suggest similarities with that for He, suggesting also similarities of $2^1S \to n^1P$ spectra in He and Li^+.

A variational uncoupled TDHF result of Dalgarno and Victor (1966b) is

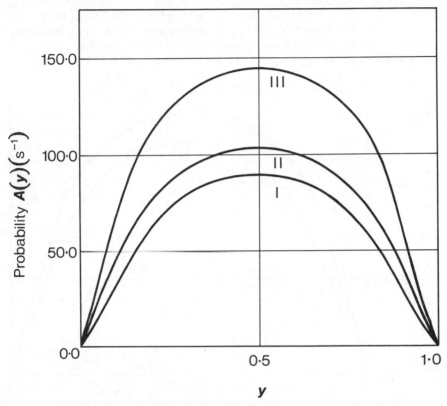

Figure 7   Relative probability of emission of a photon of energy $y\Delta E$ in the two-photon decay of He(1s2s)2^1S. I: coupled Hartree–Fock; II: uncoupled Hartree–Fock; III: refined variational (Drake, Victor and Dalgarno, 1969)

0.012 s for the lifetime of He, a factor of 2 less than that given above mainly because they adopted a different value for ΔE to the present calculation.

Two-photon ionisation may be explored in a similar way (Victor, 1967).

7. COLLECTIVE EFFECTS IN ATOMS

Brandt and Lundquist (1963, 1965, 1967) and Brown (1967) discussed collective effects in a factored RPA or TDHF approximation (FRPA) in which the correlation interaction is replaced by the dipole term in its multipole expansion. Then the infinite sum in the RPA (bubble diagrams) has closed form. Direct application of TDHF theory needs no such approximation.

Wendin (1971) analysed the role of collective effects in photoabsorption within the RPA. He obtained the spectral function $Im\alpha(\omega)$ in terms of the solutions of integral equations rather than the infinite sums of the RPA. He omitted double excitations. He has given an excellent discussion of the application of many body perturbation theory to atomic problems (Wendin, 1970a) and he showed that the coupling constant in the FRPA (the coefficient of the dipole–dipole term in the multipole expansion) is strongly frequency dependent so that the FRPA is inadequate. The FRPA suggests a collective plasma resonance at an energy of order Z rydberg, Z A.U. being the nuclear

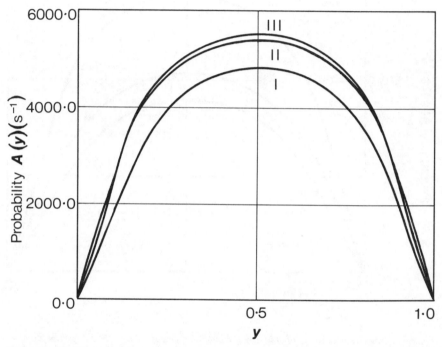

Figure 8   *Relative probability of emission of a photon of energy $y\Delta E$ in the two-photon decay of* Li$^+$ (1s2s)2^{1}S. *I: coupled Hartree–Fock; II: uncoupled Hartree–Fock; III: refined variational* (*Drake, Victor and Dalgarno, 1969*)

charge (Brandt and Lundquist, 1963, 1965). Wendin (1971) argued that there is no collective resonance in the conventional sense but collective effects redistribute the oscillator strength. He obtained results in agreement with the TDHF results of this work for He. He discussed corrections to the RPA. Brandt and Lundquist (1965, 1967) and Wendin (1970a, b), discussed correlation and collective effects on autoionisation where there is a discrete configuration interacting with other channels (a channel is a set of configurations differing only in the principal quantum number of one electron (Wendin, 1970a)) and derived the Fano formula describing the resonance in the photoabsorption cross-section. Wendin (1970a) gives a clear account of the situation. A full perturbation treatment includes coupling between all configurations. The couplings with the discrete resonance configuration, d, can be separated from the other couplings. In autoionisation the most important configurations coupled to d are the continuum ones degenerate with d. The collective influence arises from the coupling among other configurations. A shell of electrons consists of almost degenerate configurations. The coupling ('intrachannel' of Wendin) makes it possible for the electrons in the shell to respond coherently. Wendin (1970b) used the FRPA, with the coupling treated as an adjustable parameter, to study such effects. He concluded that they affect the parameters in the Fano formula, and got reasonable agreement with experiment for the line shapes for resonances in Ne, Ar, Kr, and Xe. The RPA is used for couplings not involving d. The coupling between d and the other states is treated separately. Indeed because the RPA includes no double excitations it could not account for coupling between a multiply excited resonance configuration and the continuum.

Weider and Borowitz (1966) and Ishmukhametov and Babanov (1967) made approximations to the solution of the TDHF equations to estimate possible collective frequencies. The former expanded in powers of inverse frequency, assuming a large value for the collective frequency. Their expression involves an infinite sum which may be reduced to closed form (Dalgarno and Victor, 1967b). Ishmukhametov and Babanov made a long wavelength approximation. Estimates of the collective frequency are not consistent. It would be interesting to solve the TDHF equations accurately for heavier systems and search for possible collective resonances.

8. AUTOIONISATION

TDHF theory cannot account for autoionisation of multiply excited states. No resonance was apparent in the variation of the elastic phase shift with energy for the elastic scattering of electrons by He^+. In heavier atoms shell–shell effects occur. TDHF theory, because it includes all single particle excitations, includes the coupling between a singly excited discrete state (inner shell excitation) and a singly excited continuum state (outer shell excitation) degenerate with it, and should predict autoionisation as the inner shell vacancy is filled, e.g. in Li

$$(1s^2 2s)^2 S \rightarrow (1s2s2p)^2 P$$
$$\rightarrow (1s^2 \mathscr{E} p)^2 P$$

Coupling through other states (Wendin, 1970a) influences the autoionisation and should be included in TDHF theory. The RPA with exchange has been used to study this by Amusia *et al.* (1972).

9. NUMERICAL PROCEDURE

The coupled and uncoupled equations were solved numerically by standard methods used for the ordinary Hartree–Fock equations (Hartree, 1957). The radial parts $\frac{1}{r}P_{\pm}^{(1)}(r)$ of the $u_{1\pm}^{(1)}(r) \equiv u_{2\pm}^{(1)}(r)$ were found numerically.

The boundary conditions at infinity are related to the asymptotic forms of the regular and irregular solutions of the Coulomb and Bessel equations (cf. Abramowitz and Stegun, 1964). At the origin $P_{\pm}^{(1)}(r)$ behaves as r^{l+1} for 2^l-pole external perturbation.

The uncoupled radial equations were integrated forwards and backwards (for bound solutions) to avoid exponential build-up of unwanted solutions, and forwards only (for unbound solutions) using Numerov's formula. The two coupled integro-differential equations were replaced by four coupled linear differential equations by the introduction of subsidiary functions for the integral terms (Hartree, 1957). They were solved forwards and backwards by a generalised Numerov method with matching at an intermediate point. Solution of such equations is discussed in detail by Burke and Smith (1962). At frequencies above the $1^1S \rightarrow 2^1P$ transition frequency the effective ranges and optimum matching points of the sought functions differ. The following procedure was satisfactory. The coupling and perturbation were neglected after the optimum matching point P, chosen just before the classical turning point, of $P_{+}^{(1)}(r)$, the function of shorter range. The neglect of the perturbation was convenient but unnecessary (cf. Section 6.1). At P the asymptotic form of the subsidiary functions sufficed. It is difficult to obtain linearly independent trials in the region $r > r_P$. The homogeneous equations were solved at frequencies sufficiently large that $P_{-}^{(1)}(r)$ is unbound. Outward trials were matched to the asymptotic subsidiary functions, one inward trial for $P_{+}^{(1)}(r)$ and the combination $F(r) + \tan \delta G(r)$ for $P_{-}^{(1)}(r)$, F and G being the regular and irregular continuum functions and δ the phase shift.

These methods are impractical for heavier systems where there are several coupled integro-differential equations. Iterative methods and matrix procedure, in which the derivatives and integrals are represented by difference formulae, did not prove satisfactory. The piecewise continuous method of Gordon (Gordon, 1969; Alexander and Gordon, 1971, 1972) in which the potentials rather than the solutions are approximated provides accurate solutions quickly. Alexander and Gordon (1972) used an iterative version successfully and obtained good convergence even near resonance. Much interest for heavier systems is in continuum solutions, where the resonances are. The variational-stabilisation method of Hazi and Taylor (1970), in which a variational solution corresponding to continuous energy is taken as the solution in a bounded region of space and matched to the asymptotic conditions, should be applicable.

10. CONCLUSION AND SUMMARY

Time dependent Hartree–Fock theory in its coupled form provides a way of determining the linear response and hence the Green function of a many body system such as an atom to an external perturbation such that first order correlation is included. It is equivalent to the RPA although the TDHF wave function cannot be interpreted in exactly the same way as the RPA function. Excitation energies and oscillator strengths may be determined correct to first order in correlation from the TDHF response. Many properties are deriveable in a straightforward way from the response or Green function, e.g. from the dipole spectrum. Elastic scattering phase shifts are given by the continuum part of the Green function. TDHF theory has advantages over more refined variational methods for the calculation of atomic properties. Variational methods are frequently specialised to specific properties, are complicated and unwieldly for heavy systems and do not easily provide information concerning excited states. TDHF theory predicts core polarisation in a transition of an outer electron, autoionisation and collective effects in an atom or molecule.

For the two electron systems investigated numerically, the coupled theory attributes excess oscillator strength to the continuum, while maintaining the Thomas–Kuhn sum rule, and overestimates exchange in quantum defects for (1snp) ^1P and low energy (1s$\mathscr{E}$p)^1P elastic phase shifts. This is reflected in the values of the calculated properties.

ACKNOWLEDGMENTS

I am very pleased to have had the privilege of contributing to this dedication to Professor Louis de Broglie.

I thank Professor A. Dalgarno, FRS for many useful, stimulating discussions. This research was supported mainly by the Smithsonian Institution, and NASA grant NGR-22-007-136 and partly by the Science Research Council (UK).

REFERENCES

Abramowitz, M. and Stegun, I. A. (1964) *Handbook of Mathematical Functions*, 55 Nat. Bur. Stand. App. Math Series, N.B.S. Washington, D.C.
Alexander, M. H. and Gordon, R. G. (1971) *J. chem. Phys.*, **55**, 4889
Alexander, M. H. and Gordon, R. G. (1972) *J. chem. Phys.*, **56**, 3823
Amusia, M. Ya., Kazachkov, M. P., Cherepkov, N. A. and Chernysheva, L. V. (1972) *Phys. Lett.*, **39A**, 93
Bagus, P. S. (1967) Private communication to A. Dalgarno
Bersuker, I. B. (1960) *Optics, Spectrosc., N.Y.*, **9**, 363
Bethe, H. A. and Salpeter, E. E. (1957) *Quantum Mechanics of One and Two Electron Atoms*, Springer-Verlag, Berlin
Born, M. (1960) *Atomic Physics*, Blackie, Glasgow
Brandt, W. and Lundquist, S. (1963) *Phys. Rev.*, **132**, 2135
Brandt, W. and Lundquist, S. (1965) *Ark. Fys.*, **28**, 399
Brandt, W. and Lundquist, S. (1967) *J. Quant. Spectrosc. Rad. Transfer*, **7**, 411
Breit, G. and Teller, E. (1940) *Astrophys. J.*, **91**, 215
Brown, G. E. (1967) *Unified Theory of Nuclear Models and Forces*, Wiley, New York
Burke, P. G. and Smith, K. (1962) *Rev. mod. Phys.*, **34**, 458
Caves, T. C. and Karplus, M. (1969) *J. chem. Phys.*, **50**, 3649
Chan, Y. M. and Dalgarno, A. (1965a) *Proc. phys. Soc.*, **85**, 227
Chan, Y. M. and Dalgarno, A. (1965b) *Proc. phys. Soc.*, **86**, 777
Chan, Y. M. and Dalgarno, A. (1965c) *Proc. R. Soc.*, **A285**, 457
Chung, K. T. and Hurst, R. P. (1967) *Phys. Rev.*, **152**, 35
Cohen, M. and Dalgarno, A. (1966) *Proc. R. Soc.*, **A293**, 359
Cohen, M. and Kelly, P. S. (1965) *Can. J. Phys.*, **43**, 1867

Cohen, M. and Kelly, P. S. (1966) *Can. J. Phys.,* **44,** 3227
Condon, E. U. and Shortley, G. H. (1935) *The Theory of Atomic Spectra,* Cambridge University Press
Csanak, G., Taylor, H. S. and Yaris, R. (1971) *Phys. Rev.,* **A3,** 1322
Dalgarno, A. (1959) *Proc. R. Soc.,* **A251,** 282
Dalgarno, A. (1962) *Adv. Phys.,* **11,** 281
Dalgarno, A. (1966) *Perturbation Theory and its Applications in Quantum Mechanics* (edited by Wilcox, C. H.), Wiley, New York
Dalgarno, A. and Davison, W. D. (1966) *Adv. at. mol. Phys.,* **2,** 1
Dalgarno, A. and Drake, G. W. F. (1968) *Mèmoires de la Soc. Roy. des Sciences de Liège,* ser. V, t. XVII, 69
Dalgarno, A., Drake, G. W. F. and Victor, G. A. (1968) *Phys. Rev.,* **176,** 194
Dalgarno, A. and Lynn, N. (1957) *Proc. phys. Soc.,* **A70,** 802
Dalgarno, A. and McNamee, J. M. (1961) *Proc. phys. Soc.,* **77,** 673
Dalgarno, A. and Parkinson, E. M. (1967) *Proc. R. Soc.,* **A301,** 253
Dalgarno, A. and Victor, G. A. (1966a) *Proc. R. Soc.,* **A291,** 291
Dalgarno, A. and Victor, G. A. (1966b) *Proc. phys. Soc.,* **87,** 371
Dalgarno, A. and Victor, G. A. (1967a) *Proc. phys. Soc.,* **90,** 605
Dalgarno, A. and Victor, G. A. (1967b) Private communication
Dirac, P. A. M. (1930) *Proc. Camb. phil. Soc.,* **26,** 376
Drake, G. W. F., Victor, G. A. and Dalgarno, A. (1969) *Phys. Rev.,* **180,** 25
Dunning, T. H. and McKoy, V. (1967) *J. chem. Phys.,* **47,** 1735
Ehrenreich, H. and Cohen, M. (1959) *Phys. Rev.,* **115,** 786
Goldstone, J. (1957) *Proc. R. Soc.,* **A239,** 267
Goldstone, J. and Gottfried, K. (1959) *Nuovo Cim.,* **13,** 849
Gordon, R. G. (1969) *J. chem. Phys.,* **51,** 14
Green, L. C., Mulder, M. M., Lewis, M. N. and Woll, J. W. (1954) *Phys. Rev.,* **93,** 757
Hameed, S., Herzenberg, A. and James, M. G. (1968) *J. Phys. B.,* **1,** 822
Hargreaves, J. (1928) *Proc. Camb. phil. Soc.,* **25,** 75
Hartree, D. R. (1957) *The Calculation of Atomic Structures,* Wiley, New York
Hazi, A. U. and Taylor, H. S. (1970) *Phys. Rev.,* **A2,** 1109
Heinrichs, J. (1968) *Chem. Phys. Lett.,* **2,** 315
Ho, J. C., Segal, G. A. and Taylor, H. S. (1972) *J. chem. Phys.,* **56,** 1520
Ishmukhametov, B. Kh. and Babanov, Yu. A. (1967) *Phys. Lett.,* **25A,** 98
Jamieson, M. J. (1971) *Int. J. quantum Chem. Symp. Issue.,* **4,** 103
Jamieson, M. J. (1972) *J. Phys. B.,* **5,** L26
Karplus, M. and Kolker, H. J. (1963a) *J. chem. Phys.,* **39,** 1493
Karplus, M. and Kolker, H. J. (1963b) *J. chem. Phys.,* **39,** 2997
Karplus, M. and Kolker, H. J. (1964) *J. chem. Phys.,* **41,** 3955
Kaveeshwar, V. G., Chung, K. T. and Hurst, R. P. (1968) *Phys. Rev.,* **172,** 35
Kaveeshwar, V. G., Dalgarno, A. and Hurst, R. P. (1968) *J. Phys. B.,* **2,** 984
Langhoff, P. W., Karplus, M. and Hurst, R. P. (1966) *J. chem. Phys.,* **44,** 505
Lemmer, R. H. and Vénéroni, M. (1968) *Phys. Rev.,* **170,** 883
Lowry, J. F., Tomboulian, D. H. and Ederer, D. L. (1965) *Phys. Rev.,* **137,** A1054
Martin, P. C. and Schwinger, J. (1959) *Phys. Rev.,* **115,** 1342
Martin, W. C. (1960) *J. Res. natn. Bur. Stand.,* **A64,** 19
Messiah, A. (1964) *Quantum Mechanics,* Vol II, North–Holland, Amsterdam
Moore, C. E. (1949) *Atomic Energy Levels,* Vol. I, Nat. Bur. Stand Circular 467, N.B.S., Washington, D.C.
McLachlan, A. D. and Ball, M. A. (1964) *Rev. mod. Phys.,* **36,** 844
Nesbet, R. K. (1965) *Adv. chem. Phys.,* **9,** 321
Nicholls, R. W. and Stewart, A. L. (1962) *Atomic and Molecular Processes* (edited by Bates, D. R.), Academic Press, New York
Nozières, P. and Pines, D. (1958) *Nuovo Cim.,* **9,** 470
Pekeris, C. L. (1959) *Phys. Rev.,* **115,** 1216–1221
Phythian, R. (1970) *J. Phys. B.,* **3,** L107
Roothaan, C. C. J., Sachs, L. M. and Weiss, A. W. (1960) *Rev. mod. Phys.,* **32,** 186
Rowe, D. J. (1966) *Nucl. Phys.,* **80,** 209

Rowe, D. J. (1968) *Rev. mod. Phys.,* **40,** 153
Schiff, B. and Pekeris, C. L. (1964) *Phys. Rev.,* **134,** A638
Schneider, B. (1970) *Phys. Rev.,* **A2,** 1873
Schneider, B. and Krugler, J. I. (1971) *Phys. Rev.,* **A4,** 1008
Schneider, B., Taylor, H. S. and Yaris, R. (1970) *Phys. Rev.,* **A1,** 855
Schwartz, C. (1961) *Phys. Rev.,* **123,** 1700
Seaton, M. J. (1958) *Mon. Not. R. Astr. Soc.,* **118,** 504
Sengupta, S. and Mukherji, A. (1967) *J. chem. Phys.,* **47,** 260
Stewart, A. L. (1967) *Adv. at. mol. Phys.,* **3,** 1
Thouless, D. J. (1960) *Nucl. Phys.,* **21,** 225
Thouless, D. J. (1961a) *Nucl. Phys.,* **22,** 78
Thouless, D. J. (1961b) *Quantum Mechanics of Many Body Systems,* Academic Press, New
    York
van Vleck, J. H. (1965) *The Theory of Electric and Magnetic Susceptibilities,* Oxford University
    Press
Victor, G. A. (1966) *Ph.D. Thesis,* The Queen's University of Belfast
Victor, G. A. (1967) *Proc. phys. Soc.,* **91,** 825
Weider, S. and Borowitz, S. (1966) *Phys. Rev. Lett.,* **16,** 724
Weiss, A. W. (1967) *J. Res. natn. Bur. Stand.,* **71A,** 163
Wendin, G. (1970a) *J. Phys. B.,* **3,** 455
Wendin, G. (1970b) *J. Phys. B.,* **3,** 466
Wendin, G. (1971) *J. Phys. B.,* **4,** 1080
Yoshimine, M. and Hurst, R. P. (1964) *Phys. Rev.,* **135,** A612

11

Influence of Wave Mechanics on Inorganic Chemistry

DAVID S. URCH
Chemistry Department, Queen Mary College, London

Mankind has been practising inorganic chemistry for many thousands of years. Gold and silver objects are known from the ruins of Ur. The invention of bronze enabled the great civilisations of Crete and Mycenean Greece to arise in the second millennium before Christ. Tribes that had discovered the use of iron destroyed the world of Homer's sagas but later laid the foundations for the civilisation of Classical Greece. By these times the inorganic technology of the ancients must have been considerable. Outcrops of gold, silver, copper, tin and iron were discovered and worked, not only in the Eastern Mediterranean and the near East but as far away as the British Isles. The ability to work iron enabled Celtic tribes to conquer Britain a few hundred years before the Roman invasions, invasions which brought Pax Romana to Britain. This peace was bought with and defended by the use of iron weapons. Iron by itself is, however, no guarantor of victory and the Roman civilisation perished at the hands of the barbarians. For the next one thousand years little or no useful progress was made in the field of inorganic chemistry. Alchemy, with its false hopes and false promises did little to advance human knowledge or to help mankind with practical inventions. However, during this period useful experimental techniques were developed (distillation, sublimation, etc.). It is only towards the end of the seventeenth century that inorganic chemistry begins to take on a familiar shape, whilst in the eighteenth century the now recognisable pattern of experimental scientific thought can truly be said to have been born. Experiments were carried out, events observed, rationalisations attempted and conclusions drawn. Of course sometimes the conclusions were wrong (e.g. the great phlogiston fiasco) but this can happen even in our time. By the nineteenth century enough experimental knowledge had been amassed to enable the concepts of element and compound to be clarified and then later, in the hands of Newlands (1863) and Mendeleeff (1869) for the properties of the elements themselves to be codified. In retrospect the Periodic Table really was a great and remarkable act of faith since the similarities between the elements of the first and second rows are much less obvious than their differences. The chemistries of carbon and silicon, of nitrogen and phosphorus really are quite different yet Mendeleeff boldly grouped these elements together. Subsequent discoveries have of course completely vindicated this arrangement.

Despite its ability to unite all the various elements in one table and its great power to predict properties (by inference) the Periodic Table remains only a system of classification, of itself it has no rationalisation and it gives us no understanding as to why it exists. Such understanding has only come in this century with the discovery of the structure of the atom. It is now known that chemistry is almost exclusively associated with the electrons that surround the nucleus of each atom and which occupy so very much more space than does the nucleus. Bohr's (1913) rationalisation of atomic spectra in terms of particular electronic orbits with discrete energies was swiftly followed (Sommerfeld, 1916) by more sophisticated treatments of the atom which also ascribed particular energies to electrons which occupied orbitals with specific spatial properties. These developments made possible the Electronic Theory of Valency of Lewis (1916) which was really an electronic theory of chemistry. The quantum mechanical picture of the atom immediately enabled the Periodic arrangement of the elements to be understood: elements in the same group of the Periodic Table had the same number of electrons in their outer orbitals, and these outer orbitals were of the same type (differing only in the principal quantum number). Chemistry was explained by the concept of an ionic or a covalent bond.

In an ionic bond an electron (or two, or three) was lost from one atom and gained by another; these charged species, ions, were then thought of as held together by purely electrostatic forces. The covalent bond on the other hand involved the overlapping, the interpenetration of the valence shell orbitals of one atom with those of another and the sharing of the electrons, originally in each separate orbital, between the two atoms. Such covalent bonds could be formed if each original orbital had one electron each or if one orbital had two electrons and the other was empty (although this latter situation is often treated as a 'special case' and given a 'special name', dative bond or donor bond).

These ideas work remarkably well throughout the whole of inorganic chemistry although in some cases it is necessary to invoke a composite covalent and ionic bond, as in sulphate for example,

$$
\begin{array}{ccc}
O^- & & O^- \\
& \diagdown\!\!\!\diagup & \\
& {}^{++}S & \\
& \diagup\!\!\!\diagdown & \\
{}^-O & & O^-
\end{array}
$$

The simple electronic theory of valency can be refined by the realisation that the designation into covalent and ionic bond types is arbitrary and that most bonds will in fact be best described as having a mixture of covalent and ionic character. A polar bond is one in which electrons spend rather more time near one atom than the other: apart from bonds in homonuclear diatomic molecules it can be truly said that 'all bonds are polar but some bonds are more polar than others' (cf. Orwell, 1945).

Whilst the electronic theory of valency provides an excellent framework for describing the general distribution of electron pairs in most molecules and ions, it does not provide understanding as to why bonds should be formed, or as to their nature: 'polar bond' is a rather vague concept and it really is quite important to know, when discussing the chemical properties of a bond, just how polar is polar!

The development of the Schroedinger (1926) equation and its subsequent application to chemical problems in the 1920s showed that the energies of bond formation and the subtleties of electronic distributions in molecules and ions could, at least in principle, be answered. The introduction of the ideas of wave mechanics by De Broglie (1924) made it easier for chemists to enter the rather esoteric mathematical domain of quantum mechanics. It is true one has to equate wave and particle description of an electron but this of course is our fault not the electron's. Mathematical expressions have been developed for macroscopic waves breaking on immense sea-shores and for the motion of particles the size of buses – is it any wonder it is rather difficult to adapt such equations to describe the properties of something as tiny as an electron? It is unfortunate, but not really surprising that we are not able to describe the unique nature of the electron with only one equation. The fact that we can and do use two types of equation reflects a deficiency in our existing mathematical apparatus not in any 'dual' character of the electron. (In a way this picture of the electron is somewhat like the concept of the Blessed Trinity which is three persons in one nature. The electron shows us but two persons, or two sets of characteristics, yet it has a single nature.)

The years around 1930 saw the application of quantum mechanical and wave mechanical ideas to a variety of chemical problems. Heitler and London (1927) showed how complicated a problem was, even for the hydrogen molecule (James and Coolidge, 1933). The history of theoretical chemistry is therefore the history of approximations. A series of approximations enabled Huckel (1931) to describe molecular orbitals for the π-electrons of benzene and to determine their relative energies. The foundations of molecular orbital theory had been laid. Another line of investigation sought to understand the electron pair bond, which lay at the heart of the original electronic theory of valency.

This led to the development of valence bond theory, especially by Slater (1931) and Pauling (1931, 1933). The applications of this approach to inorganic chemistry were extensively and vividly described in *The Nature of The Chemical Bond* (Pauling, 1935). The power of theory in rationalising the many newly determined crystal structures was most persuasive in convincing chemists for the following decades of the supremacy of the valence bond approach to problems of chemical bonding. This was rather unfortunate historically, because the MO theory (Lennard-Jones, 1929), whilst less sophisticated, is also capable of important applications in inorganic chemistry. That this is so was very powerfully and clearly demonstrated by Mulliken (1932, 1933) in a long series of papers in *The Physical Review* and *The Journal of Chemical Physics* (1933a, 1935) and also by Van Vleck (1933,

1935) who described all the essential features of crystal field theory in the mid-1930s.

Little real progress was made in the application of molecular orbital theory until after the end of World War II and then the progress was made in organic rather than inorganic chemistry. Coulson (1947, 1952), and Dewar (1949) demonstrated that molecular orbital theory, even in its most simple form, was widely applicable to organic problems, mostly in aromatic systems, but also heterocyclic and aliphatic systems as well. The success of MO theory in the organic field eventually spurred its reapplication in inorganic chemistry. Orgel (1952, 1955, 1960) and Griffith and Orgel (1957) developed the basic ideas of Van Vleck in their treatment of the chemical and spectroscopic properties of transition metal complexes. Finally in the 1960s MO ideas have been applied to main group systems, rationalising (Cruickshank, 1961) bond lengths and other structural problems in much the same way as was done a decade or more earlier in organic chemistry.

The effect of theoretical ideas upon the progress of inorganic chemistry has at times been tantalisingly slow, indeed it would seem that the cross-fertilisation of ideas and experiment, which has so marked the influence of wave mechanics upon organic chemistry, has only taken place in inorganic chemistry during the past fifteen years in transition metal chemistry and for an even shorter period in main group chemistry (if it has in fact yet started). The reasons for this delay are hard to understand. Maybe it was because the first theoretical ideas, the electronic theory of valency, were essentially descriptive and not predictive. Later with the advent of molecular orbital theory it was by no means obvious that the approximations made in such a simple one-electron model would not completely invalidate the value of any results. The coulomb integrals of the atoms involved in most inorganic molecules or ions will vary widely, in contrast to aromatic π-systems where to a good approximation all these integrals will be the same. Furthermore in π-electron systems the potential energy term in the Schroedinger equation is fairly constant, much more so than for σ-organic bonds and for almost all bonds in inorganic molecules.

'Faint heart n'er won fair lady' and faint hearts would seem to have thought these objections insurmountable. Mulliken (1933) did, however, show that with all its limitations MO theory could be applied to inorganic systems and yield results of qualitative value, i.e. the relative ordering of molecular orbitals. He also showed how symmetry could be applied to many simple inorganic systems and that here *no* approximations were involved. Molecular orbitals belonging to different irreducible representations do not interact, and this is a fact of nature, not an approximation conceived *in extremis* in a desperate attempt to simplify the Schroedinger equation. Apparently the chemical world was unimpressed, as it was equally unimpressed with the work of Van Vleck on transition metal complexes. These ideas had to wait twenty years before they were applied. Once, however, it was realised that theories did exist which could not only successfully rationalise the known chemical and spectral properties of transition metal complexes and predict many other properties, interest in inorganic chemistry was once more aroused. From that time progress in transition metal and organometallic chemistry has been swift and dramatic. The developments in main group chemistry have been rather less brilliant. Even now there are large numbers of compounds whose existence cannot be easily rationalised (sulphides of phosphorus to pick just one example). In the remainder of this chapter the emphasis will therefore be on problems in main group chemistry. The discussion will be simple and qualitative, seeking generalisations that may be of value in rationalising the properties of atoms and molecules. Such ideas must always underlie even the most sophisticated and complicated calculations: they are are also of value in teaching since they enable a unified view of chemistry to be presented at an early stage.

1. ATOMIC ORBITALS

An exact solution of the Schroedinger equation is possible for the hydrogen atom. As a result a series of orbitals is generated whose shapes are now familiar to all chemists (White, 1931). (Well, the angular parts of the wave functions in polar co-ordinate forms certainly are because they appear in almost every textbook.) In a poly-atomic atom these wave functions can be used as a first approximation to the true wave functions of that atom. However, the hydrogen-like wave functions are one-electron functions and as such take no account of the effects of mutual electron repulsion upon spatial electronic distribution (Dickens and Linnett, 1959). A hydrogen-like wave function gives no idea at all of how one electron in a 1s orbital of helium for example behaves in space relative to both the nucleus *and* the other electron. It is true, however, that a time average of its various positions does correspond quite closely to that which would be predicted by a 1s wave function with a suitable nuclear charge. Electron correlation is of fundamental importance when considering the problem of molecular shape (Dickens and Linnett, 1957) but, for the moment the essential point is that the usefulness of hydrogen-like wave functions is not destroyed by the problems of electron correlation. It will therefore be possible to use such wave functions in a discussion of the potential chemical properties of atoms. Much emphasis in the past (Pauling, 1935) has been placed on the directional aspect of orbitals as being a decisive factor in the shapes of molecules but at this stage let us consider in more detail the radial functions.

The general form of some radial functions is shown in Figure 1. The important feature is that the number of radial nodes increases as the principal quantum number increases. This means that whilst a 1s function has no such node a 6s orbital has five radial nodes. The 6s orbital therefore has five separate maxima. Calculations (Bratsev, 1966) suggest that even for occupied valence shell orbitals all these nodes will be at a distance of less than 1 Å from the nucleus. This means that only one of the five maxima could be in a region of space where overlap with other orbitals to form covalent bonds would be possible. It seems reasonable to assume that orbital overlap is a prerequisite for a covalent bond and that if overlap is small then only weak bonds can result. An orbital which has suffered five radial nodes will not be able to present a very large amplitude at the optimum distance for bond formation and only a weak covalent bond could therefore result. The general conclusion is therefore that in going down a group in the Periodic Table the strength of covalent bonds will decrease and also that the s orbitals will be more affected than p orbitals of the same principal quantum number.

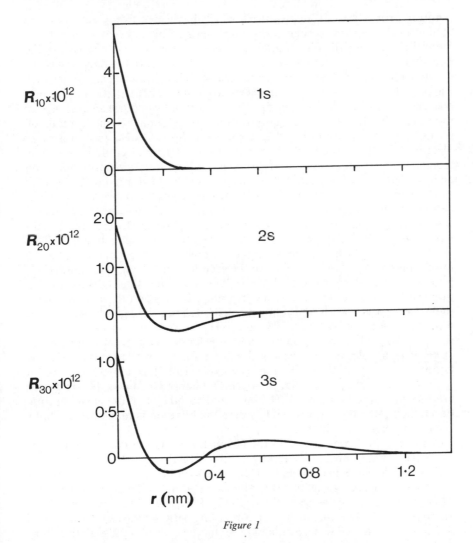

Figure 1

This effect is mostly clearly found in the lower rows of the Periodic Table where 'inert' lone-pairs are a distinctive feature of the chemistry. As a specific example consider the chlorides of Group IV. The tetrachloride of carbon has four strong covalent bonds, but on going down the table the strength of these bonds decreases; also the tendency to form stable ionic divalent chlorides in which the s orbital plays no part in bonding and the p orbitals merely lose electrons, increases.

2. ORBITAL IONISATION ENERGIES

Another property of atomic orbitals which has a decisive effect upon the chemistry of an atom is their ionisation energy. This is because one of the factors which controls the strength of a bond is the relative energies of the constituent atomic orbitals (Urch, 1970). The potential strength will be greatest when these energies are identical and diminishes as disparity between the orbital ionisation energies increases. To a first approximation ionisation energies determined for free atoms can be used but of course in each actual molecular situation the effective ionisation energy of an 'atomic' orbital will be modified by the number of electrons it holds and also the number of electrons in other valence shell orbitals. Even so general guide lines can be formulated since in going across a row in the Periodic Table, the ionisation energy of the valence shell s orbital increases quite dramatically (usually by a factor of $\times 3 \sim \times 5$) and this increase outweighs fluctuations that might be imposed by variations in electron repulsion terms. Thus one can conclude that if strong covalent bonds with a particular ligand are formed on the left-hand side of the table which involve s orbitals then these bonds will be weaker on the right-hand side. A simple example of the effect of the increasing ionisation energy of an s orbital can be seen in contrasting CH_4 and NH_4^+. Free atom ionisation energies (Jaffe, 1956) suggest that the 2s orbital ionisation energy increases by about 5 eV in going from carbon to nitrogen whilst the value for the 2p orbital is almost constant; since the ligands are also the same the differences between CH_4 and NH_4^+ must be due to the changes in the 2s orbital. That component of the C—H or N—H bond which is due to 2s orbital participation in the bond will be greater in the former than the latter since carbon 2s ionisation energy (~ 16 eV) is closer to that for hydrogen (13.2 eV than is the nitrogen 2s energy (~ 21 eV). N—H bonds are therefore weaker than C—H bonds and so NH_4^+ is more easily decomposed thermally. By contrast NH_3 where the bonding is much more wholly 2p in character is stable to 2000°C.

A similar effect can be seen in the chemistry of the tetrahedral oxyanions of the second and third rows. If only s and p orbitals are involved in bonding then it would be expected that MO_3 anions would become progressively more stable by contrast with MO_4 anions on going to the right in the table. That this is not so in the second row (SiO_4, PO_4, SO_4, ClO_4) can be rationalised (Urch, 1963) as due to 3d—2p π-bonding between the central atom and the oxygen lone pairs perpendicular to the M—O σ bonds. If this is so the importance of such bonding should be least for silicon and greatest for chlorine. This is in complete agreement with the chemistry of these anions since the readiness with which covalent bonds are formed by oxygen outside the basic tetrahedron (thus breaking down the π-bonds) decreases in going from silicon to chlorine. The corresponding anions of the third row present an interesting contrast. The stability of the MO_4 unit decreases on going to the right; selenate and perbromate are very much stronger oxidising agents than sulphate and perchlorate. It is as though the 3d—2p π-bonding postulated for the second row were very much weaker when 4d orbitals are involved. This is quite reasonable because of the radial node present in 4d orbitals that is absent in 3d orbitals. In the fourth row 5d or 4f orbitals can possibly be used in bonding with electronegative ligands. It seems reasonable to suggest that it is the 4f orbitals that will be eventually shown to participate in bonding and not the 5d orbitals.

The inability of valence shell s orbitals in the bottom right-hand corner of the Periodic Table to play an effective part in covalent bond formation (due to the presence of many radial nodes and because of its being relatively tightly bound) has interesting stereochemical effects (Urch, 1964). Complex anions of the type MX_6^{n-} where X is a halogen are quite well characterised and are in most cases regular octahedra. This is because the interaction between the ligand symmetry orbital ($a_{1g}\sigma$) and the valence shell s orbital of M is small; the bonding MO is almost wholly Ms in character and the anti-bonding orbital is almost wholly concentrated on the electronegative ligands and is not very antibonding either. It is this anti-bonding a_{1g} orbital that can house the extra pair of electrons without detriment to the stability or the shape of the MX_6 unit. Crystal forces may coerce such anions to adopt the regular configuration but clearly there can be very little energetic difference between the octahedral and distorted forms. This is most clearly seen in the case of gaseous XeF_6 where of course crystal forces are absent. The regular octahedron is not the most stable configuration but a study (Bartell and Gavin, 1968) of the infra-red spectrum shows that one of the Xe−F bending modes has a very broad peak. If such a vibrational mode, in which three fluorines nearest the lone pair move towards the lone pair and the other three move away from the three-fold axis that passes through the lone pair, then a regular octahedron is easily achieved, and if such a vibration proceeds further the lone pair will be found at the other side of the molecule. Such inversion coupled with interactions with the corresponding vibrational modes of the regular octahedron might well give rise to a complex array of vibrationally excited levels. Such complexity would be a direct result of near degeneracy of the energy of the distorted and regular octahedral forms for XeF_6.

3. MOLECULAR SHAPE

Apart from the interesting example discussed above the shape of most molecules and ions can be rationalised using the simple principle enunciated by Gillespie and Nyholm (1957): the strength of electron pair repulsions decreases in the series lone pair, lone pair > lone pair, bonding pair > bonding pair, bonding pair. These rules of course apply to those type of molecules or ions where lone pairs are most often found, i.e. to the right of the middle of the Periodic Table and this in turn prompts speculation about the shapes of molecules formed by other elements, shapes which are sometimes rather bizarre.

Single molecular orbital theory shows that if two orbitals interact with each other one bonding molecular orbital and one anti-bonding molecular orbital are formed. If three orbitals interact then geometry is important, a linear arrangement gives a bonding, a non-bonding and an anti-bonding orbital but a triangular juxtaposition in which each orbital interacts with the other two gives only one bonding orbital and two anti-bonding orbitals. This situation can be easily generalised and if n orbitals can be so arranged in space that each orbital interacts with *every* other orbital then only *one* bonding orbital will result. There will be $(n-1)$ anti-bonding orbitals. Spatial arrangements of this type will be very attractive to elements on the left hand side of the Periodic Table, endowed as they are with the normal complement of valence orbitals but with very few electrons in them. By adopting such structures as will permit many orbitals to mutually overlap only a small number of bonding orbitals will be generated and these can then be completely filled with the small number of electrons that are available. This simple argument provides the basic explanation for the formation of cluster compounds by the elements of Groups I, II and III when they attempt to make covalent bonds with each other (bunchy bonds are best for boron!). The so-called three-centre bond (Eberhardt, Crawford and Lipscomb, 1954) is, of course, just the simplest example of this general idea:

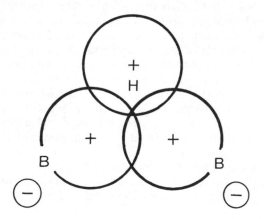

Each boron orbital overlaps with the other boron orbital and the hydrogen orbital, and the hydrogen orbital overlaps with both boron orbitals, each orbital overlaps with all the others and only one bonding orbital is formed. Thus two electrons can hold the $\overset{\displaystyle H}{\underset{\displaystyle B \qquad B}{\diagup \diagdown}}$ structure together (Longuet-Higgins, 1946). The tendency to form cluster compounds should not, however, be confined to Groups I, II and III, any element which has fewer electrons than effective valence orbitals will tend to form compounds with some degree of clustering — because only in this way can structures be formed which do not have empty bonding orbitals. (Whilst compounds with empty bonding orbitals will not be necessarily unstable they will be very reactive chemically towards electron donating species, e.g. water, oxygen, etc.)

Transition metals fall into this category since s, p and d orbitals constitute the valence shell for these elements, i.e. nine orbitals. For the most part the cations formed by transition elements overcome their chronic electron-deficiency, in the same way as the ions of Groups I and II, by accepting donated electron pairs from electron rich ligands (e.g. water, ammonia, carbon monoxide, etc.). However, cluster compounds are also known (Johnston, 1970) in which the spatial arrangement of atoms generates an excess of anti-bonding orbitals. Even so the number of electrons required can be quite considerable and will require ligands not only to be joined directly to the transition metal atoms but also to be present at the faces of cluster structure. Furthermore it will usually be found that each transition metal atom will contribute many electrons (i.e. formally d^8 or d^9 or d^{10}). Compounds based on clusters of transition metal ions will therefore be found most commonly on the right side of the Periodic Table. It is also to be anticipated that such structures might be found in main Groups IV, V and to the right in the Periodic Table where the presence of an electronegative ligand confers bonding potential upon d-orbitals, thus generating a situation formally analogous to that of a transition metal. Sometimes the ligand will also confer electrons (e.g. $-O^-$), which would destroy the potential electron-deficiency and if new cluster compounds of this type are to be found, fluorine would seem to be the most suitable ligand (e.g. might $:SiF_2$ make a series of compounds analogous to the boron hydrides?).

It is an interesting reflection that there would seem to be no special spatial arrangement of orbitals which will generate an excess of bonding over anti-bonding orbitals. Thus atoms on the right hand side have to sublimate their excess electrons either by forming bonds with electropositive elements, or by going to higher formal valence states, or by reducing the number of covalent bonds formed leaving the other electrons in non-bonding lone pairs. The only exception is to be found in $(4n+1)$ aromatic systems where $2n+1$ bonding but only $2n$ anti-bonding π-orbitals are formed (e.g. pyrrole, cyclopentadienyl anion).

This then is the basic reason why most molecules formed by elements in group four and to the right in the Periodic Table are based on electron pair bonds; this in fact corresponds to the best spatial arrangement of atoms that will generate a suitable number of bonding and, at worst, non-bonding orbitals. The rules of Gillespie and Nyholm, which are based on simple electrostatics then apply. Exceptions are rare and usually present interesting bonding features, e.g. the $:MX_6$ anions considered above; the dithionite anion $S_2O_4^{--}$ in which the S—O bonds from the two sulphurs are opposite each other (i.e. the ion is *cis*, not *trans*) (Dunitz, 1956); $Re_2Cl_8^{--}$, which adopts the eclipsed and not staggered conformation (Cotton, 1965), etc. In both the latter examples the observed geometry may be understood by postulating d–d bonding.

4. HYBRIDISATION

The above discussion on the shape of molecules and ions has been conducted without any reference to hybridisation, which may seem a strange omission since molecules and ions are often said to adopt a particular shape simply because of the hybridisation of the atomic orbitals of the central atom. The omission was deliberate because the oft made assertion is unfortunately based on a logical fallacy. It is quite true that it is possible to hybridise the orbitals of an atom to generate hybrid orbitals which have particular spatial characteristics and point in specific directions (Kimball, 1940). There is, however, no basis for the converse assertion, i.e. that if a particular orientation of groups about a central atom is observed that it necessarily follows that the orbitals of that atom *must* have the corresponding type of hybridisation.

When atomic orbitals are hybridised, and the corresponding valence state of the atom is constructed it is then possible to consider the interaction of the hybrid orbitals with ligand orbitals to form electron pair bonds. A particular valence-bond structure will result which will have a particular energy. Many other possible structures based on other possible hybridisation schemes, also ionic structures and partially non-bonded structures can and should be considered. If it can be shown that all other reasonable structures are much less stable than the first then it can be concluded that the first structure contributes overwhelmingly to the final wave function for the molecule and as such it is a good approximation to the 'real' molecule. Under these circumstances one could truly speak of the central atom having a particular hybridisation. If, however, many of the other structures have energies comparable with the first then there will be strong interaction (resonance) between these structures and the final wave function will contain contributions from all those structures of similar energy. For example it can be shown that for SF_6 structures such as,

Sulphur, d^2sp^3 hybrid	Sulphur$^+$, ionic bond,	Sulphur$^+$, ionic bond and non-bonded fluorine atoms

have quite similar energies. When the statistical weighting of some structures is taken into account it becomes apparent that the structure based on the octahedral sp^3d^2 hybrid contributes only about 1–2% to the final wave function. SF_6 is definitely octahedral but this is certainly *not* due to the hybridisation of s, p and d orbitals!

5. EXPERIMENTAL METHODS

The application of wave mechanics to inorganic chemistry has led to various roles being attributed to particular atomic orbitals; in some cases it is proposed that electrons are fairly well localised on particular atoms, or not at all — as in ionic bonds; in other cases electrons are said to be shared between two or more atoms. Whilst the ideas that follow from these concepts are in very good agreement with known chemical properties it is unfortunately true that until very recently no means existed to experimentally verify the assertions that were made as to the role of particular orbitals in bond formation. Thus the shapes of the sixth group hydrides can be easily rationalised by assuming that the valence shell s orbital plays a decreasing bonding role in going down the group. Very reasonable but is it true? A whole edifice of explanation has been built up in inorganic and also organic chemistry in which the *primary* assertions (how electrons from particular orbitals are distributed in molecular orbitals) have never been experimentally tested but only the secondary conclusions that can reasonably be held to follow from those primary assertions.

Various experimental probes are fortunately now available. The simplest, which merely gives an indication of overall asymmetry of charge distribution, is the dipole moment. Much more detailed knowledge of the over-all charge distribution can be obtained from X-ray crystallographic data. A knowledge of the effective charge of each atom can now be obtained using X-ray photo-electron spectroscopy (Baker, Brundle and Thompson, 1972) (more precisely changes in effective charge can be correlated with changes in inner orbital ionisation energy).

Mössbauer spectroscopy permits a quite detailed discussion of the bonding role of s-electrons to be made (IAEA, 1966), and so by inference of the role of p- and d-electrons. The p-electron distribution about certain nuclei can be investigated with nuclear quadrupole spectroscopy (Orville–Thomas, 1957; Lucken, 1969). Unfortunately although these latter two techniques give quite detailed and intimate details of the bonding involvements of certain atomic orbitals they are limited in their application to a rather small range of nuclei.

For a knowledge of the energies of molecular orbitals photoelectron spectroscopy (using helium resonance or some similar radiation) has, in recent years, provided dramatic confirmation of the predictions of wave mechanics (Baker, 1970). The other parameter besides energy, that can be determined by wave mechanics, is the coefficient, a_{ri}, for a particular atomic orbital (ϕ) in the LCAO equation for a molecular orbital (ψ_r),

$$\psi_r = \Sigma a_{ri}\phi_i$$

It seems reasonable to hope that these coefficients will now be susceptible to a direct experimental determination using high resolution X-ray emission spectroscopy (Urch, 1971). The atomic selection rules specify that $\Delta l = \pm 1$, so that a vacancy in an inner s orbital for example, will attract electronic transitions from outer p orbitals. If the valence shell p orbitals are engaged in bond formation the corresponding p $\rightarrow$ s transition will show a structure which reflects the various molecular orbitals which have p character. Furthermore the intensities of these peaks will be directly related to the values of $(a_{ri})^2$. A detailed study of such spectra should therefore enable these co-efficients to be found experimentally.

6. CONCLUSION

By the application of wave mechanics inorganic chemistry has been transformed over the past fifty years from being a subject in which a vast mass of empirical knowledge was rather roughly collated by means of the Periodic Table into one where it is possible to understand the general features of how and why the elements behave as they do. Much work still remains to be done. The challenge presented by inorganic chemistry to the theoretician is indeed great; the elements are so varied that it is very difficult to find common factors that will simplify the calculations. It seems the progress will probably take place at two quite distinct levels, that of sophisticated *ab init.*, and that of the quick and unashamedly approximate calculation. The latter approach being simpler is of more use to the experimentalist but in its use lies the danger of drawing wrong conclusions by not fully appreciating the nature of the approximations that have been made. Even so it does seem reasonable to conclude that wave mechanical ideas can be used to provide a real understanding of the chemical behaviour of inorganic compounds. It also seems reasonable to expect that wave mechanics will be at the heart of future calculations destined to clarify inorganic chemistry.

REFERENCES

Baker, A. D. (1970) *Accounts Chem. Res.,* **3,** 17
Baker, A. D., Brundle, C. R. and Thompson, M. (1972) *Chem. Soc. Revs.,* **1,** 355
Bartell, L. S. and Gavin, R. T. (1968) *J. chem. Phys.,* **48,** 2460, 2466
Blair, E. (pseudo. G. Orwell) (1945) *Animal Farm,* 87, Secker and Warburg, London
Bohr, N. (1913) *Phil. Mag.,* **26,** 476
Bratsev, B. F. (1966) *Tables of Atomic Orbital Functions,* Acad. Nauk, Moscow
Cotton, F. A. (1965) *Inorg. Chem.,* **4,** 337
Coulson, C. A. (1947) *Quart. Rev.,* **1,** 144
Coulson, C. A. (1952) *Valence,* Oxford University Press
Cruickshank, D. W. J. (1961) *J. chem. Soc.,* **1961,** 5486
De Broglie, L. (1924) *Thesis,* Paris
Dewar, M. J. S. (1949) *The Electronic Theory of Valency,* Oxford University Press
Dickens, P. G. and Linnett, J. W. (1957) *Quart. Rev.,* **11,** 291
Dunitz, J. D. (1956) *Acta Cryst.,* **9,** 579
Eberhardt, W. H., Crawford, B. L. and Lipscomb, W. N. (1954) *J. chem. Phys.,* **22,** 989
Gillespie, R. J. and Nyholm, R. S. (1957) *Quart. Rev.,* **9,** 339
Griffiths, J. S. and Orgel, L. E. (1957) *Quart. Rev.,* **11,** 381
Heitler, W. and London, F. (1927) *Z. Phys.,* **44,** 455
Hückel, E. (1931) *Z. Phys.,* **70,** 204
IAEA (1966) *Applications of the Mossbauer Effect in Chemistry and Solid State Physics,* Tech. Report No. 50, LAEA, Vienna
Jaffe, H. H. (1956) *J. chem. Ed.,* **33,** 25 calculated from spectroscopic data
James, H. M. and Coolidge, A. S. (1933) *J. chem. Phys.,* **1,** 825
Johnston, R. D. (1970) *Adv. inorg. Radiochem.,* **13,** 471
Kimball, G. E. (1940) *J. chem. Phys.,* **8,** 198
Lennard-Jones, J. E. (1929) *Trans. Faraday Soc.,* **25,** 668
Lewis, G. N. (1916) *J. Am. chem. Soc.,* **38,** 762
Longuet-Higgins, H. C. (1946) *J. chem. Soc.,* **1946,** 139
Lucken, E. A. C. (1969) *Nuclear Quadrupole Coupling Constants,* Academic Press, New York
Mendeléeff, D. I. (1869) *J. Russ. phys. Chem. Soc.,* **1869,** 60
Moore, C. E. (1949–52) *Tables of Atomic Energy Levels,* N. B. S. Circular No. 467, U.S. Gov. Print. Office, Washington D.C., U.S.A.
Mulliken, R. S. (1932) *Phys. Rev.,* **40,** 55
Mulliken, R. S. (1932a) *Phys. Rev.,* **41,** 49, 751
Mulliken, R. S. (1933) *Phys. Rev.,* **43,** 279

Mulliken, R. S. (1933a) *J. chem. Phys.*, **1,** 492; (1935) **3,** 375, 506, 514, 517, 564, 573, 586, 635, 720

Newlands, J. A. R. (1863) *Chemical News,* 7, 70

Orgel, L. E. (1952) *J. chem. Soc.,* **1952,** 4756; (1955) *J. chem. Phys.,* **23,** 1004

Orgel, L. E. (1960) *An Introduction to Transition Metal Chemistry—Ligand Field Theory,* Methuen, London

Orville-Thomas, W. J. (1957) *Quart. Rev.,* **11,** 162

Pauling, L. (1931) *J. Am. chem. Soc.,* **53,** 1367, 3225

Pauling, L. (1933) *J. chem. Phys.,* **1,** 280

Pauling, L. (1935) *The Nature of the Chemical Bond,* Cornell University Press, Ithaca, N.Y.

Schroedinger, E. (1926) *Ann. Phys.,* **79,** 361, 489

Slater, J. C. (1931) *Phys. Rev.,* **38,** 1109

Sommerfeld, A. (1916) *Ann. Phys.,* **51,** 1

Urch, D. S. (1963) *J. inorg. nucl. Chem.,* **25,** 771

Urch, D. S. (1964) *J. chem. Soc.,* **1964,** 5775

Urch, D. S. (1970) *Orbitals and Symmetry,* 54, Penguin Education, Harmondsworth, Mddx.

Urch, D. S. (1971) *Quart. Rev.,* **25,** 343

Van Vleck, J. H. (1932) *Phys. Rev.,* **41,** 208

Van Vleck, J. H. (1933) *J. chem. Phys.,* **1,** 177, 219

Van Vleck, J. H. (1935) *J. chem. Phys.,* **3,** 807

Van Vleck, J. H. and Sherman, A. (1935) *Rev. mod. Phys.,* **7,** 167

White, H. E. (1931) *Phys. Rev.,* **37,** 1416

12

Ligand Fields—Some Recent Symmetry Developments

CLAUS E. SCHÄFFER

Chemistry Department I (Inorganic Chemistry),
H. C. Ørsted Institute, University of Copenhagen

1. WAVE MECHANICS AND TRANSITION METAL CHEMISTRY. THE BETHE–VAN VLECK MODEL

The language of wave mechanics has made an impact upon inorganic chemistry. One of its most fruitful subjects has been the complexes and compounds containing transition metal ions, i.e. ions with a partially filled d or f shell.

The early names are the physicists Bethe (1929) and van Vleck (1932) and the chemist Pauling (1940). It is essential historically to mention the educational background of these pioneers, because the way they spoke gave them audiences accordingly. In 1930 few chemists listened to the physicists (Schäffer, 1970).

Pauling (1940) formed the opinion that, even though much of the progress in chemistry had been due to quantum mechanics, it should be possible to describe the new developments in a thoroughgoing and satisfactory manner without the use of advanced mathematics. And Pauling proved his case aided by his scientific brilliance, artistic journalism and personal radiation. For more than 20 years his views dominated inorganic chemistry and he was the main contributor to a development which brought this branch of chemistry from a natural history state to a well classified branch of science.

Unfortunately, Bethe and van Vleck were at that time not able to speak to the chemists who had no mathematical background. Even when Finkelstein and van Vleck (1940), for the first time tackled the problem of the colour of a complex, a property that had interested chemists for a century, this impact took place in a too spectroscopic way. The narrow spin-forbidden transitions of chromium alum, which have little influence on its colour, were correctly assigned, while the broad spin-allowed transitions, which are responsible for the colour, were ignored as absorption edges.

Nevertheless it is the Bethe–van Vleck model which has been analysed and applied in the last 20 years by physicists and chemists, still with a barrier between them, but much more hand in hand than ever before. It is their model that has inspired most of the research on transition metal complexes and made possible a whole new classification of this branch of chemistry. It is a model, mathematically equivalent to theirs, which this paper is concerned with.

The Bethe–van Vleck model of transition metal complexes is based upon the assumption that the l electrons (d or f electrons) of the partially filled shell can be recognised in the complexes. The quantum number l remains a reasonably good quantum number during complex formation. Quantitatively this is expressed by assuming that the ground states of these complexes and their lower excited states can be described by an l^n configuration, perturbed somehow by the ligands, i.e. by the ions or molecules surrounding the central ion. This is the same as assuming that the effect of the ligands can be accounted for by a field, a ligand field, using a first order perturbation treatment with a basis of central ion l orbitals. Physically the assumption is that this field of the ligands is essentially spherical. However, it is important to emphasise that it is only the non-spherical part of the perturbation which is of interest for the applications.

It is not known how one can represent this ligand field by a charge distribution which is physically adequate. It has been realised that any simple kind of point charge distribution is fallacious (Freeman and Watson, 1960). The model is therefore taken as a semi-empirical model. This means that the observable quantities are expressed in terms of empirical parameters whose coefficients are determined quantitatively by the above-mentioned assumptions.

The perturbing potential is thought of as expanded into terms belonging to irreducible representations of the three-dimensional rotation-inversion group R_{3i}. This means that it has the form

$$\mathbf{V} = \sum_{k, q, i} \mathfrak{B}_q^k(r_i)\mathfrak{C}_q^k(\theta_i, \phi_i) \tag{1}$$

where i labels the electrons of the partially filled shell and $\mathfrak{C}_q^k$ is a real surface harmonic [normalised to $4\pi/(2k+1)$], k being their degree and q specifying their components. Since the matrix elements are to be evaluated either within a d basis or an f basis, all terms of odd degree (k odd) must vanish, and since further only energy differences are of interest for the model, also the totally symmetric term, the s term, can be left out. Further all terms with $k > 2l$ vanish because the direct product of two l spaces spans no space higher than $2l$. The l set of functions, $|l\rangle$, [$(2l+1)$ in all, the individual functions characterised by t] has the form of a product of a radial function and a real angular one $\mathfrak{Z}_t^l$ (both normalised to unity).

$$Mlt = M(r)\mathfrak{Z}_t^l(\theta, \phi) \tag{2}$$

The coefficient $\mathfrak{B}_q^k$ to $\mathfrak{C}_q^k$ (equation 1) and the radial factor of the l functions only play a formal role in the semi-empirical model since they may be used to express the empirical parameters. However, the angular factors of the potential (equation 1) and of the l functions (equation 2) give rise to integrals which are evaluated and which make the model quantitative. These integrals occur in the theory as coefficients to the empirical parameters.

2. THE EXPANDED RADIAL FUNCTION MODEL

(a) GENERAL REMARKS

In the introduction the one-electron operator part $\mathbf{V}$ of the theory, the model field representing the surroundings of the central ion, was mentioned. It is the beauty of ligand field theory that the two-electron operator part of the theory, the interelectronic repulsion, $\sum_{i<j} 1/r_{ij}$, may be accounted for within the same framework of assumptions.

Here the Slater–Condon–Shortley (Condon and Shortley, 1935) first order perturbation parametrisation based upon spherical harmonic expansions is used to obtain the empirical parameters F^2 and F^4, the spherical term F^0 again vanishing in the applications where only energy differences are considered. Sometimes the Racah parameters B and C, which are linearly related to F^2 and F^4, are used instead of these.

It is characteristic, quite analogously to the one-electron operator part of the theory, that the empirical parameters are formally expressed as radial integrals, which are not evaluated, whereas their coefficients are integrals over the angular co-ordinates and are evaluated exactly.

The values of the interelectronic repulsion parameters are invariably found from experiments on metal ion complexes to be smaller than those of the naked gaseous ions (Jørgensen, 1962a). This phenomenon which represents an apparent expansion of the radial function of Ml (equation 2) has been named (Schäffer and Jørgensen, 1958) nephelauxetism (cloud expansion) and the whole ligand field theory in the form of two superimposed first order perturbations, $\mathbf{V}$ and $\sum_{i<j} 1/r_{ij}$, has accordingly been named (Jørgensen, 1958) the expanded radial function model. Still it is essentially the Bethe–van Vleck model.

Most of the chemical systems of the d periods that have been studied from a ligand field point of view are central ions co-ordinated by six ligands whose co-ordinating atoms, ligators, form an approximate octahedron with the central ion at the point which remains invariant under the operations of the octahedral group. When the co-ordination is octahedral, it is a symmetry property that a d orbital will split into two symmetry classes, an $e_g(O_h)$ and a $t_{2g}(O_h)$ class, where the symbols refer to irreducible representations of the octahedral group O_h. The energy difference between these two classes is called, for historical reasons, the spectrochemical parameter and is defined by

$$\Delta \equiv h(e_g) - h(t_{2g}) \tag{3}$$

where h stands for orbital energy.

The statement that the partially filled, approximately d shell splits into two sub-shells, e_g and t_{2g}, has more general validity than the first order perturbation model, which assumes a pure d $(l = 2)$ shell. When the molecule has a centre of inversion, only gerade l orbitals can mix with the d orbitals, and since in the octahedron s $(l = 0)$ orbitals span $a_{1g}(O_h)$, the lowest l value that can mix with d $(l = 2)$ is g $(l = 4)$ spanning a_{1g}, e_g, t_{1g}, and t_{2g}. Therefore for octahedral systems it is understandable from a symmetry point of view that the first order perturbation assumption works so well.

(b) WEAK FIELD, STRONG FIELD, AND ALMOST DIAGONAL INTERMEDIATE FIELD SCHEMES

When considering a d^n system in a ligand field, one has to set up a matrix of $\mathbf{V}$ and a matrix of $\sum\limits_{i<j} 1/r_{ij}$, using the same n electron function basis (with signs), add these matrices and diagonalise in such a way as to make differences between certain of the characteristic values of the sum matrix equal to observed energy differences (Maegaard, Mønsted and Schäffer, 1973). This happens for particular values of the empirical parameters, F^2, F^4, and Δ, say, which thereby become determined from experiments.

For f^n systems (and for $4d^n$ and $5d^n$ systems) the spin-orbit coupling is also important. We then have three independent first order perturbations and three matrices to add in order to obtain the energy matrix which is relevant to a comparison with experiment. Also here the order in which the perturbations are taken into account is immaterial provided the full energy matrix is evaluated and used. However, particularly for f^n systems for which the number of states is considerable, it may for reasons of computation of the matrix elements themselves as well as for reasons of diagonalisation of the matrices be quite important which order is chosen.

In setting up the above matrices two classes of approach are natural and have both been used extensively in the literature. However, it is historically interesting that the fact that the two classes of approach are equivalent was apparently not realised (Hartmann, 1958) and their connection by a unitary transformation not formulated until quite late (Schäffer, 1958).

In both weak field and strong field schemes, when spin-orbit coupling is not included (cf. the above paragraph in small print) the d^n function basis is first classified according to the total spin S and its projection quantum number M_S, so that both matrices of $\mathbf{V}$ and $\sum\limits_{i<j} 1/r_{ij}$ obtain block forms with no interaction between different S values and between different M_S values.

In the weak field scheme the d^n function basis is chosen to belong to irreducible representations of the full three-dimensional rotation-inversion group R_{3i}, i.e. chosen to have a well defined L and parity, g. This has the consequence that the matrix of $\sum\limits_{i<j} 1/r_{ij}$ is diagonal except for the cases where more terms of the type ^{2S+1}L occur within d^n, e.g. the two ^2D terms in d^3. The weak field scheme is not uniquely defined, but depends on the choice of components for the irreducible representations L. Two possibilities are of special importance. The component L functions are chosen so that they belong to irreducible representations of the sub-group $D_{\infty h}$ of R_{3i} (Harnung and Schäffer, 1972a) or to those of the sub-group O_h (Harnung and Schäffer, 1973). In each case one may use a real or a complex basis. Sometimes it is useful to choose the L component functions to belong to irreducible representations of the true molecular point group, even though this group must

be a sub-group of either $D_{\infty h}$ or O_h. It has been shown how it is possible to define the empirical ligand field parameters uniquely also with respect to sign through the application of the Wigner–Eckart theorem, using phase-fixed 3-l symbols of R_{3i} (Harnung and Schäffer, 1972b). The 3-l symbols (and 3-Γ symbols for the point groups) correspond to the Wigner 3-j symbols and are chosen in such a way that the reduced matrix elements are the same in the real and the complex basis. Using Racah's so-called irreducible tensor method and his coefficients of fractional parentage, it is possible to arrive at the weak field energy matrices themselves, without evaluating the d^n functions. These matrices need not be made by hand. They can be constructed, including spin-orbit coupling, by use of a computer. A useful feature of the method is that the signs of all the matrices are well defined, so that when a calculation of an energy matrix has been done in one laboratory it is possible in another laboratory, to supplement this matrix with that of some extra perturbation, by simple addition. This has never been possible before except with so much effort that it was easier to do the work already done in the literature all over again.

To conclude the remarks about the weak field matrices it is noted that once the choice of type of component L functions has been made, all the functions are determined with phase so that the energy matrices are uniquely defined, also with respect to sign.

In the strong field scheme the d^n function basis is chosen to belong to pure sub-configurations $\gamma_1^p \gamma_2^q \gamma_3^r \gamma_4^s \gamma_5^t$ $(p+q+r+s+t=n)$, where each γ_i represents the same or different irreducible representations of some suitably chosen point group G, say, not necessarily the true molecular point group. This choice of function basis has the consequence that the matrix of $\hat{\mathbf{V}}$ by symmetry may be made diagonal with respect to all terms which are totally symmetric in G, and if G is the true molecular point group, the matrix may be made diagonal with respect to the whole of $\hat{\mathbf{V}}$.

For the strong field scheme no general principles of phase-fixation have till now been proposed, but for cubic symmetry a proposal for the signs of the coefficients of fractional parentage required for a phase-fixation of the energy matrices of d^n systems, has been made (Harnung, 1973). The strong field scheme will not be discussed further here, but from Figure 1 it will appear that sometimes a cubic strong field scheme is more nearly diagonal for a D_{4h} system than is a tetragonal scheme.

The concepts of the weak field and the strong field approximations have been widely used. These concepts mean that the expanded radial function model is approximated by the diagonals of the weak field and strong field energy matrices, respectively. Particularly the strong field approximation has a historical interest because it has often been identified with the model itself and used to test the model against experiments. In the computer age this is no longer necessary and not to be recommended. On the other hand it may be physically important to find intermediate field schemes which are almost diagonal, i.e. matrix representations of $\left[\hat{\mathbf{V}} + \sum\limits_{i<j} 1/r_{ij} \right]$ which are as close as possible to being diagonal and at the same time have·functions, referring to this diagonal, which can be classified properly. The column,

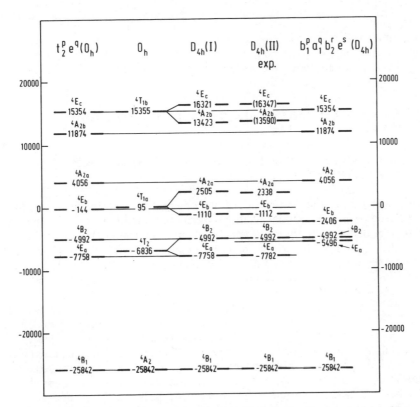

Figure 1   Barycentred analysis of the ligand field absorption spectrum arising from quartet energy levels of the trans-difluorotetramminechromium (III) ion

The figure is based upon barycentred parameters. For the cubic ligand field one has $\Delta(d)$, or $E_{a_1}^g$, and for the tetragonal fields $\Delta(e)$ and $\Delta(t_2)$, or $E_{e\theta}^d$ and $E_{e\theta}^g$ (equations 12–15). For the interelectronic repulsion the barycentre parameter $b = (3/2)B$(Racah) has been introduced. Expressed in b units, $\hbar[^4F] = -3$ and $\hbar[^4P] = +7$. The parameters corresponding to the one-electron energies (equation 12), $\hbar(b_1) = 10926$ K; $\hbar(a_1) = 11882$ K; $\hbar(b_2) = -9924$ K; $\hbar(e) = -6442$ K and the repulsion parameter $b = 1011$ K, were determined from the four experimental transitions of lowest energy (in K units), which are the energies of the excited states in column $D_{4h}(II)$, measured relative to the ground state. The two uppermost energies are calculated values. A broad band is observed here, corresponding to an excited state, associated with 15000 K when referred to the position of the ground state on the figure.

The wing columns, $t_2^p e^q(O_h)$ and $b_1^p a_1^q b_2^r e^s(D_{4h})$, whose energy levels are characterised by D_{4h} reps, represent the cubic and tetragonal strong field approximations, respectively. The column O_h represents the perturbation arising from the repulsion plus the cubic part of $\mathbf{\hat{V}}(D_{4h})$. When the O_h column is considered the unperturbed situation, the column $D_{4h}(I)$, which are the diagonal elements of the almost diagonal intermediate field scheme, contains energies which are barycentred relative to those of the O_h column, and represents the first order tetragonal perturbation, while the column $D_{4h}(II)$ represents the first plus the second order tetragonal perturbation. All the perturbations are first order when referred to the three d electrons as the unperturbed situation.

$D_{4h}(I)$, of Figure 1 may serve as an illustration. When this column is taken to represent the diagonal elements of a $[\,\bar{\mathbf{V}}(D_{4h}) + \sum_{i<j}(1/r_{ij})\,]$-energy matrix, the functions corresponding to the diagonal are at the same time basis functions for reps of O_h and D_{4h}, but the mixing of cubic sub-configurations by $\sum_{i<j} 1/r_{ij}$ has already been accounted for.

3. THE GENERAL LIGAND FIELD MODEL OR NON-ADDITIVITY MODEL

THE SPHERICAL HARMONIC AND THE ORBITAL ENERGY PARAMETRISATIONS

The most general field model does not assume that the ligand field can be expressed as a sum of contributions from the individual ligands. In many cases it is useful to analyse an experiment first in terms of this general model and then perhaps afterwards in terms of an additivity model (Section 4), sometimes misleadingly called a superposition model (Bradbury and Newman, 1967). There are thousands of papers using empirical field parameters of the general field model, and a parameter expressing the same physical quantity often appears in the literature in different ways. Further, it is sometimes not possible from a given paper to see how the parameters are defined, and consequently results from different papers are difficult to compare.

It is therefore desirable to make a rational choice which can be universally adopted. A symmetry choice, based upon an expansion on general spherical harmonics using the Wigner–Eckart theorem and independent of whether the basis is real or complex, has been proposed (Harnung and Schäffer, 1972b). In the present section this choice is mentioned and an alternative choice, directly referring to orbital energies and simply linearly related to the general spherical harmonic choice, is proposed.

Consider a one-electron matrix element of $\mathbf{V}$ (equation 1) connecting the central ion orbitals Mlu and Mlv, where M represents the radial factor of the l orbitals (equation 2) whose components are specified by u and v. One may then write the general matrix element as (Schäffer, 1973a)

$$\langle Mlu \,|\, \mathbf{V} \,|\, Mlv \rangle = \langle Mlu \,|\, \sum_{k,\,q} \mathcal{B}_q^k \mathbb{C}_q^k \,|\, Mlv \rangle$$

$$= \sum_{k,\,q} \langle M \,|\, \mathcal{B}_q^k \,|\, M \rangle \langle lu \,|\, \mathbb{C}_q^k \,|\, lv \rangle \qquad (4)$$

$$= \sum_{k,\,q} I_q^k \langle lu \,|\, \mathbb{C}_q^k \,|\, lv \rangle$$

or, alternatively (Harnung and Schäffer, 1972b)

$$\langle Mlu \,|\, \mathbf{V} \,|\, Mlv \rangle = \sum_{k,\,q} E_q^k \big(\begin{smallmatrix} l & k & l \\ u & q & v \end{smallmatrix}\big) \qquad (5)$$

where I_q^k and E_q^k make up the proposed symmetry choice of parameters. $\langle lu \,|\, \mathfrak{C}_q^k \,|\, lv \rangle$ of equation 4 is an angular integral which is symmetry determined because its three factors are symmetry determined. It is, by the Wigner–Eckart theorem, equal to

$$\langle lu \,|\, \mathfrak{C}_q^k \,|\, lv \rangle = \langle l \,||\, \mathfrak{C}^k \,||\, l \rangle \left(\begin{smallmatrix} l & k & l \\ u & q & v \end{smallmatrix} \right) \tag{6}$$

where according to equations 4 and 5

$$E_q^k/I_q^k = \langle l \,||\, \mathfrak{C}^k \,||\, l \rangle \tag{7}$$

which always is a positive quantity. For d electrons the two relevant reduced matrix elements are incidentally equal and

$$\langle d \,||\, \mathfrak{C}^2 \,||\, d \rangle = \langle d \,||\, \mathfrak{C}^4 \,||\, d \rangle = \sqrt{70/7} \tag{8}$$

is a special value of the reduced matrix element $\langle l \,||\, \mathfrak{C}^k \,||\, l \rangle$ for which a closed expression exists. $\left(\begin{smallmatrix} l & k & l \\ u & q & v \end{smallmatrix} \right)$ of equations 5 and 6 is a 3-l symbol as mentioned in Section 2b. The set of these symmetry quantities, which may be determined once and for all time, also with respect to sign, makes up group invariants, in much the same way as group characters, except for the fact that they depend on a choice of (standard) components for the reps. In this particular case the group is the three-dimensional rotation group R_3, but 3-Γ symbols* have been defined for other groups.

The parameter E_q^k of equation 5 has been called the reduced ligand field parameter even though it depends on the component q of $\mathfrak{B}^k$. The reason for this is that 3-l symbols rather than 3-Γ symbols (with Γ referring to the symmetry group of the physical system) are used in its definition. By doing this, full advantage is taken of the fact that the ligand field model is a first order perturbation model acting upon a basis set adapted to spherical symmetry. The number of linearly independent E_q^k parameters depends on the number of times the totally symmetric representation of the point group occurs in $\mathbf{D}^{(k)}, 0 < k \leq 2l$ (k even), where $\mathbf{D}^{(k)}$ is an irreducible representation of the three-dimensional rotation-inversion group R_{3i}. Usually the standard choice of l components u and v for spherical symmetry is not symmetry-adapted to the sub-group in question, but standard ways of making this symmetry adaptation have been advocated, and since $\langle l \,||\, \mathfrak{C}^k \,||\, l \rangle$ of equation 6 is independent of the choice of components, this equation states that the 3-l symbol is proportional to a simple integral and thereby specifies its transformation properties.

The general spherical harmonic parametrisation of the ligand field $\mathbf{V}$ may be illustrated by an example, worked in some detail also in the rest of the paper. Consider the hexaco-ordinated chromium(III) complex trans-$[\text{Cr(NH}_3)_4\text{F}_2]^+$ which is based upon the octahedron and has the symmetry D_{4h} (apart from the hydrogen atoms) and a d shell with three electrons in it, a d^3 configuration. In order to determine the number of independent spherical harmonic parameters one has to reduce the representation spanned by

---

*These were first defined by Griffith (1962) who did not consider the relationships between the 3-l symbols of R_{3i} and the 3-Γ symbols of its sub-groups. Later studies of these relationships gave rise to a redefinition and a fixation of the phases of all these symbols (Harnung and Schäffer, 1972a).

$D^{(2)} + D^{(4)}$ in the group D_{4h} and count the number of $a_1(D_{4h})$ terms. The result is three. The terms may be classified according to the usual real spherical harmonics which makes up one real standard basis set for the group R_3, transforming standard in the group R_{3i}.

In a previous paper (Harnung and Schäffer 1972a), a distinction was made for pure rotation groups, between a restricted concept of a set of standard basis functions for the reps of a group, giving rise to a phase fixation of the 3-Γ symbols of the group, and a less restricted concept of a set of basis functions transforming standard. Both sets will generate identically the same irreducible matrix representations, but will sometimes define the 3-Γ symbols with different signs as a consequence of a lemma by Racah.

These functions quite generally also transform standard in the group $C_{\infty v}$, or with certain phase changes also in the groups $D_{\infty h}$ and D_{4h}, all three groups with their main axis taken as the Z axis. Omitting the label g (for gerade) in all reps the terms kq then are

$$d\sigma = d[\sigma\,(R_{3i}) \quad a_1(D_{\infty h}) \quad\quad\quad e\theta(O_h) \quad\quad\quad a_1(D_{4h})]$$
$$g\sigma = g[\sigma\,(R_{3i}) \quad a_1(D_{\infty h}) \quad (\sqrt{21}/6)a_1(O_h)+(\sqrt{15}/6)e\theta(O_h) \quad a_1(D_{4h})] \quad\quad (9)$$
$$g\gamma c = g[\gamma c(R_{3i}) \quad e_4 e_1(D_{\infty h}) \quad (\sqrt{15}/6)a_1(O_h)-(\sqrt{21}/6)e\theta(O_h) \quad a_1(D_{4h})]$$

giving rise (equation 5) to the parameters E_σ^d, E_σ^g, and $E_{\gamma c}^g$ and the energy expressions

$$\bar{h}[d\delta c(R_{3i})] = \bar{h}[e_2 e_1(D_{\infty h})] = \bar{h}[de\varepsilon(O_h)] = \bar{h}[db_1(D_{4h})]$$
$$= E_\sigma^d(-2/\sqrt{70}) + E_\sigma^g(1/3\sqrt{70}) + E_{\gamma c}^g(1/3\sqrt{2})$$

$$\bar{h}[d\sigma(R_{3i})] = \bar{h}[da_1(D_{\infty h})] = \bar{h}[de\theta(O_h)] = \bar{h}[da_1(D_{4h})]$$
$$= E_\sigma^d(2/\sqrt{70}) + E_\sigma^g(2/\sqrt{70}) \quad\quad (10)$$

$$\bar{h}[d\delta s(R_{3i})] = \bar{h}[de_2 e_2(D_{\infty h})] = \bar{h}[dt_2\zeta(O_h)] = \bar{h}[db_2(D_{4h})]$$
$$= E_\sigma^d(-2/\sqrt{70}) + E_\sigma^g(1/3\sqrt{70}) + E_{\gamma c}^g(-1/3\sqrt{2})$$

$$\bar{h}[d\pi c(R_{3i})] = \bar{h}[de_1 e_1(D_{\infty h})] = \bar{h}[dt_2\eta(O_h)] = \bar{h}[dee_1(D_{4h})]$$
$$= E_\sigma^d(1/\sqrt{70}) + E_\sigma^g(-4/3\sqrt{70})$$
$$= \bar{h}[d\pi s(R_{3i})] = \bar{h}[de_1 e_2(D_{\infty h})] = \bar{h}[dt_2\xi(O_h)] = \bar{h}[dee_2(D_{4h})]$$

In equation 10 h stands for energy as in equation 3, or more specifically the diagonal elements of $\bar{\mathbf{V}}(D_{4h})$. The bar over h means that the energy is measured relative to the average energy of the d shell. It means for a more general, non-diagonal case, that the diagonal elements, summed over the $(2l+1)$ orbitals of $|l\rangle$, always are vanishing. The potential terms in equation 9 and the functions in equation 10 have been characterised by their transformation properties under R_{3i} and under the three relevant sub-groups. The nomenclature for the e reps of $D_{\infty h}$ and D_{4h} is such that the first e refers to the rep and the second to the component e_1 or e_2, where the sub-indices mean symmetrical or antisymmetrical, respectively, with respect to C_2^Y. The numerical coefficients to the E parameters, the 3-l symbols in equation 10, were taken from Schäffer (1973a).

As it will appear from the example, when further worked below, it is useful to choose an alternative spherical harmonic parametrisation based upon harmonics which are chosen so as to transform standard under the octahedral group O_h. These are often called cubic harmonics and a phase convention for them (Harnung and Schäffer, 1973) is necessary in order to define 3-l symbols with cubic component bases.

The corresponding spherical harmonic parametrisation, which is physically equivalent to that of equation 10 but which is based upon the cubic harmonics, give rise to the parameters

$$E_{e\theta}^d = E_\sigma^d$$
$$E_{e\theta}^g = +(\sqrt{15/6})E_\sigma^g - (\sqrt{21/6})E_{\gamma c}^g \tag{11}$$
$$E_{a_1}^g = (\sqrt{21/6})E_\sigma^g + (\sqrt{15/6})E_{\gamma c}^g$$

and the energy expressions

$$
\begin{aligned}
\hbar[de\varepsilon(O_h)] = \hbar[db_1(D_{4h})] &= E_{e\theta}^d(-2/\sqrt{70}) + E_{e\theta}^g(-6\sqrt{15/18}\sqrt{70}) + \\
&\quad E_{a_1}^g(6\sqrt{21/18}\sqrt{70}) \\
&= I_{e\theta}^d(-2/7) + I_{e\theta}^g(-\sqrt{15/21}) + I_{a_1}^g(\sqrt{21/21}) \\[6pt]
\hbar[de\theta(O_h)] = \hbar[da_1(D_{4h})] &= E_{e\theta}^d(2/\sqrt{70}) + E_{e\theta}^g(6\sqrt{15/18}\sqrt{70}) + \\
&\quad E_{a_1}^g(6\sqrt{21/18}\sqrt{70}) \\
&= I_{e\theta}^d(2/7) + I_{e\theta}^g(\sqrt{15/21}) + I_{a_1}^g(\sqrt{21/21}) \tag{12} \\[6pt]
\hbar[dt_2\zeta(O_h)] = \hbar[db_2(D_{4h})] &= E_{e\theta}^d(-2/\sqrt{70}) + E_{e\theta}^g(8\sqrt{15/18}\sqrt{70}) + \\
&\quad E_{a_1}^g(-4\sqrt{21/18}\sqrt{70}) \\
&= I_{e\theta}^d(-2/7) + I_{e\theta}^g(4\sqrt{15/63}) + I_{a_1}^g(-2\sqrt{21/63}) \\[6pt]
\hbar[dt_2\eta(O_h)] = \hbar[dee_1(D_{4h})] &= E_{e\theta}^d(1/\sqrt{70}) + E_{e\theta}^g(-4\sqrt{15/18}\sqrt{70}) + \\
&\quad E_{a_1}^g(-4\sqrt{21/18}\sqrt{70}) \\
&= I_{e\theta}^d(1/7) + I_{e\theta}^g(-2\sqrt{15/63}) + I_{a_1}^g(-2\sqrt{21/63})
\end{aligned}
$$

$$= \hbar[dt_2\xi(O_h)] = \hbar[dee_2(D_{4h})]$$

where in equation 12 cubically based 3-l symbols have been used together with equation 5 and the I parameters introduced through equations 7 and 8.

The important thing here is that when the symmetry is increased to O_h only the term with the parameter $E_{a_1}^g$ remains non-vanishing. The Wigner–Eckhart-3-l-symbol formalism, the so-called irreducible tensor method, is not necessary (Schäffer, 1967), but particularly apt to express a perturbation as a sum of individual terms of decreasing symmetry so that the centre of gravity rule (the barycentre rule) applies to the energy levels at any stage when the perturbations are taken in order of decreasing symmetry (Figure 1). This is one reason why the parametrisation of equations 11 and 12 is more illustrative than that of equation 10. The other reason is physical and also illustrated in Figure 1, where the first column corresponds to equation 12 and the last one to equation 10.

The fact that the parametrisation of $\bar{\mathbf{V}}(D_{4h})$ comprises three parameters has till now been attributed to the symmetry result that in the expansion of $\bar{\mathbf{V}}(D_{4h})$, only three terms among those which may combine two d functions, will be totally symmetrical in D_{4h}. It is possible to arrive at the same number of parameters using an apparently different, but yet equivalent symmetry result (Griffith, 1961): In D_{4h} the five d functions span only four different reps (b_1, a_1, b_2, and e) of D_{4h}, resulting in three independent energy differences.

This leads to the alternative symmetry parametrisation of the general field model which one might call the orbital energy parametrisation. Corresponding to equation 12 one defines

$$\Delta(d) \equiv h[e(O_h)] - h[t_2(O_h)] = \tfrac{1}{2}\{h[db_1(D_{4h})] + h[da_1(D_{4h})]\}$$
$$- \tfrac{1}{3}\{h[db_2(D_{4h})] + 2h[de(D_{4h})]\}$$

$$\Delta(e) \equiv h[db_1(D_{4h})] - h[da_1(D_{4h})] \tag{13}$$

$$\Delta(t_2) \equiv h[db_2(D_{4h})] - h[de(D_{4h})]$$

where $\Delta(d)$ accounts for the whole cubic field and $\Delta(e)$ and $\Delta(t_2)$ for the tetragonal fields. The sign of $\Delta(d)$ is the conventional one which leads to positive parameter values for all known 6-co-ordinated complexes, based upon the octahedron. $\Delta(t_2)$ is conventionally chosen positive when the non-degenerate orbital, the $d\delta s$ orbital (see equation 10) with angular maxima in the XY-plane, has the higher energy, and $\Delta(e)$ analogously when the $d\delta c$ orbital has the higher energy.

The relationship between the spherical harmonic parametrisation of equation 12 and the orbital energy parametrisation of equation 13 is obtained by a comparison of the two equations giving

$$\Delta(d) = \qquad 5\sqrt{3}/9\sqrt{10}\,\mathrm{E}^g_{a_1} \qquad = (5\sqrt{21}/63)\mathrm{I}^g_{a_1}$$
$$\Delta(e) = -(4\sqrt{70})\mathrm{E}^d_{e\theta} - (2\sqrt{3}/3\sqrt{14})\mathrm{E}^g_{e\theta} = -(4/7)\mathrm{I}^d_{e\theta} - (2\sqrt{15/21})\mathrm{I}^g_{e\theta} \tag{14}$$
$$\Delta(t_2) = -(3/\sqrt{70})\mathrm{E}^d_{e\theta} + (2\sqrt{3}/3\sqrt{14})\mathrm{E}^g_{e\theta} = -(3/7)\mathrm{I}^d_{e\theta} + (2\sqrt{15/21})\mathrm{I}^g_{e\theta}$$

or the reverse relationships

$$\mathrm{E}^g_{a_1} = (9\sqrt{10}/5\sqrt{3})\Delta(d); \qquad\qquad \mathrm{I}^g_{a_1} = (63/5\sqrt{21})\Delta(d)$$
$$\mathrm{E}^g_{e\theta} = -(3\sqrt{14}/14\sqrt{3})[3\Delta(e) - 4\Delta(t_2)]; \quad \mathrm{I}^g_{e\theta} = -(3/2\sqrt{15})[3\Delta(e) -$$
$$4\Delta(t_2)] \tag{15}$$
$$\mathrm{E}^d_{e\theta} = -(\sqrt{70}/7)[\Delta(e) + \Delta(t_2)]; \qquad \mathrm{I}^d_{e\theta} = -[\Delta(e) + \Delta(t_2)]$$

The fact that there are two linearly independent parameters expressing the tetragonal part of the ligand field means that in the non-additivity model the statement that one complex is more tetragonal than another one is hardly meaningful and probably never useful. Such a statement does not even have a meaning when both the tetragonal parameters are larger for one of the complexes than for the other, unless the parameter choice is specified. It is more appropriate to speak about two tetragonal fields, in which case one may use either the spherical harmonic parameters $\mathrm{E}^d_{e\theta}$, $\mathrm{E}^g_{e\theta}$ or the orbital energy parameters $\Delta(e)$, $\Delta(t_2)$. The two choices are of course linearly dependent through equations 14 or 15.

Some tentative results (Glerup and Schäffer, 1968) for chromium(III) complexes of the type trans-$[\mathrm{Cr(NH_3)_4X_2}]^{m+}$ ($X = \mathrm{Br^-, Cl^-, F^-, OH^-}$) are given in Table 1 for both types of parameter sets. It is noteworthy that all four parameters vary monotonically in the order of ligands given above, but if X is taken as NH_3, so that the complex has cubic symmetry with all four parameters vanishing, NH_3 falls in the $\Delta(e)$ and $E^d_{e\theta}$ series between Cl^- and F^-, in the $\Delta(t_2)$ series above Br^-, and the $E^g_{e\theta}$ series below OH^-. These parametrisation results have very little to say chemically, but when re-analysed in the light of the additivity model they become interesting (Schäffer, 1970).

The non-additivity ligand field model has now been characterised by two types of symmetry parametrisation, a general spherical harmonic para-metrisation and an orbital energy parametrisation, where one is linearly

Table 1. Tentative empirical spherical harmonic and orbital energy parameters (in kK units) of the tetragonal fields in chromium(III) D_{4h}-complexes of the type trans-$[Cr(NH_3)_4X_2]^{m+}$

E^k_q	X Br^-	Cl^-	F^-	OH^-
$E^d_{e\theta}$	−4.2	−1.2	5.3	9.6
$E^g_{e\theta}$	−8.1	−7.7	−5.1	−3.5
$\Delta(e)$	4.5	3.0	−1.0	−3.5
$\Delta(t_2)$	−1.0	−2.0	−3.5	−4.5

dependent on the other. An analogous double symmetry characterisation is possible for the additivity model and this is the subject of Section 4.

4. THE ADDITIVITY LIGAND FIELD MODEL

(a) GENERAL CONSIDERATIONS

In the additivity model it is assumed that the ligand field operator $\bar{V}$ can be written as a sum of contributions from the individual ligands, in which case the operator is called $\bar{A}$ (for angular overlap) (Schäffer, 1973). It is known from calculations as well as from experiments that the ligands do interact significantly. It is therefore important to realise that the additivity assumption only concerns itself with the barred operator which does not contain the major term of the ligand field, the spherically symmetrical term. This means that the assumption of additivity only assumes that the energetic consequences of the interaction between any pair of ligands do not depend on their individualities.

In order to describe the additivity model a Cartesian co-ordinate system XYZ is chosen to define the standard l functions, $|l|$, of the central ion M. A ligand J is situated on the positive Z^j axis of an $X^j Y^j Z^j$ system which is rotated with respect to the XYZ system, and in such a way that the direction cosines of Z^j in XYZ are $(\alpha_j, \beta_j, \gamma_j)$. In the next sub-section and throughout the paper it is assumed that the ligand J is linearly ligating, i.e. that the M–J system has the linear symmetry $C_{\infty v}$ with the consequence that only two parameters are required to specify the necessary knowledge about the orientations of $X^j Y^j Z^j$ relative to XYZ, the parameter ψ_j to be mentioned below being unnecessary. These parameters may be chosen as the polar co-ordinates (θ_j, ϕ_j) of Z^j in the XYZ system given by

$$(\alpha_j, \beta_j, \gamma_j) = (\sin\theta_j\cos\phi_j, \sin\theta_j\sin\phi_j, \cos\theta_j) \tag{16}$$

with the condition $\alpha_j^2 + \beta_j^2 + \gamma_j^2 = 1$, where $(\alpha_j, \beta_j, \gamma_j)$ are the direction cosines of Z^j in the XYZ system or the Cartesian XYZ co-ordinates of the crossing point between Z^j and the unit sphere.

It is noted that the ligand's individuality is characterised by J and its position by j. Different J may refer to the same type of ligand, but at different radial distances from the central ion.*

In identically the same way (Schäffer, 1968) in which a standard central ion l basis $|l\}$ is defined with respect to XYZ, another rotated standard central ion l basis $|(l)^j\}$ is defined with respect to $X^jY^jZ^j$. The linear relationship between $|l\}$ and $|(l)^j\}$ is expressed by

$$\mathbf{R}(\phi_j, \theta_j, \psi_j)\,|\,lu\rangle = |(lu)^j\rangle = \sum_t |\,lt\rangle\, \mathbf{D}_{tu}^{(l)}[\mathbf{R}(\phi_j, \theta_j, \psi_j)] \tag{17}$$

with the reverse relationship, using the orthogonality of $\mathbf{D}$,

$$|\,lu\rangle = \sum_t |(lt)^j\rangle\, \mathbf{D}_{ut}^{(l)}[\mathbf{R}(\phi_j, \theta_j, \psi_j)] \tag{18}$$

$\mathbf{R}(\phi_j, \theta_j, \psi_j)$ is the operator which rotates the contour of any standard function $|\,lu\rangle$ over into its associated function $|(lu)^j\rangle$ which has the same orientation in space relative to $X^jY^jZ^j$ as $|\,lu\rangle$ relative to XYZ. $\mathbf{R}(\phi_j, \theta_j, \psi_j)$ means a rotation of ψ_j about Z, followed by one of θ_j about Y and finally followed by a rotation of ϕ_j about Z. This composite rotation is also equal to the two last rotations performed first and finally followed by a rotation of ψ_j about Z^j (Schäffer, 1968). The latter rotation is of no importance for all the results which will be derived for the additivity model for linearly ligating ligands, because the assumed linear symmetry about the M–J bonds makes the perturbation energies independent of the angle ψ_j. $\mathbf{D}^{(l)}$ of equations 19 and 20 is the rotation matrix for l orbitals, $|l\}$, or the standard $(2l+1)$-dimensional real irreducible matrix representation of the three-dimensional rotation group R_3 and $\mathbf{D}^{(l)}$ is a function of $\mathbf{R}(\phi_j, \theta_j, \psi_j)$ as indicated by the square bracket. Putting $\psi_j = 0$, the $\mathbf{D}$ matrix goes over into the so-called angular overlap matrix $\mathbf{F}$ (Schäffer, 1970).

$$\mathbf{D}^{(l)}[\mathbf{R}(\phi_j, \theta_j, 0)] = \mathbf{F}^{(l)}(\phi_j, \theta_j) \tag{19}$$

The necessary prerequisites for handling the additivity model is now available. First it is noted that the matrix of the operator $\bar{\mathbf{A}}^{(ZJ)}$ [the index Z meaning that the ligand J is placed upon the positive Z axis, $(\alpha_j, \beta_j, \gamma_j) = (0, 0, 1)$], calculated with respect to $|l\}$ of XYZ, is equal to the matrix of $\bar{\mathbf{A}}^{(jJ)}$ calculated with respect to $|(l)^j\}$ of $X^jY^jZ^j$. It is useful to consider this matrix first in the latter way, so that it is expressed as $\langle(lu)^j|\,\bar{\mathbf{A}}^{(jJ)}\,|(lv)^j\rangle$, and then translate it into the matrix of $\bar{\mathbf{A}}^{(jJ)}$ with respect to $|l\}$, $\langle lu|\,\bar{\mathbf{A}}^{(jJ)}\,|lv\rangle$.

---

*Often in the following, summations over j and J will take place simultaneously, J being the index for the empirical parameter, j that of the angular — geometrically determined — coefficient.

After that a summation may be performed over all the single ligand perturbation contributions, whereby in the finish a matrix of

$$\bar{\mathbf{A}} = \sum_{(jJ)} \bar{\mathbf{A}}^{(jJ)} \tag{20}$$

with respect to $|l\}$ is obtained.

(b) THE ADDITIVITY MODEL FOR LINEARLY LIGATING LIGANDS. AXIAL SPHERICAL HARMONIC AND ANGULAR OVERLAP SYMMETRY PARAMETRISATIONS

When the symmetry about the central ion-to-ligand bonds is $C_{\infty v}$ (linearly ligating ligands), this symmetry causes

$$\langle Mlu \mid \bar{\mathbf{A}}^{(ZJ)} \mid Mlv \rangle = \langle (Mlu)^j \mid \bar{\mathbf{A}}^{(jJ)} \mid (Mlv)^j \rangle \tag{21}$$

to be diagonal. Further, when $\bar{\mathbf{A}}^{(ZJ)}$ is expanded into spherical harmonics, only axial harmonics, σ terms, are non-vanishing and one may speak about the axial spherical harmonic single ligand parametrisation.

Using equation 5 for $\bar{\mathbf{V}} = \bar{\mathbf{A}}^{(ZJ)}$, one obtains for d electrons

$$\bar{h}_J[d\sigma] = \langle (M\,d\sigma)^j \mid \bar{\mathbf{A}}^{(jJ)} \mid (M\,d\sigma)^j \rangle$$

$$= \langle M\,d\sigma \mid \bar{\mathbf{A}}^{(ZJ)} \mid M\,d\sigma \rangle = \sum_k E_J^k \begin{pmatrix} d\,k\,d \\ \sigma\,\sigma\,\sigma \end{pmatrix}$$

$$= E_J^2(2/\sqrt{70}) + E_J^4(2/\sqrt{70}) = I_J^2(2/7) + I_J^4(2/7)$$

$$\bar{h}_J[d\pi] = \{(M\,d\pi)^j \mid \bar{\mathbf{A}}^{(jJ)} \mid (M\,d\pi)^j\} = \langle (M\,d\pi s)^j \mid \bar{\mathbf{A}}^{(jJ)} \mid (M\,d\pi s)^j \rangle \tag{22}$$

$$= \langle (M\,d\pi c)^j \mid \bar{\mathbf{A}}^{(jJ)} \mid (M\,d\pi c)^j \rangle = \{M\,d\pi \mid \bar{\mathbf{A}}^{ZJ} \mid M\,d\pi\}$$

$$= \langle M\,d\pi s \mid \bar{\mathbf{A}}^{(ZJ)} \mid M\,d\pi s \rangle$$

$$= \langle M\,d\pi c \mid \bar{\mathbf{A}}^{(ZJ)} \mid M\,d\pi c \rangle = \sum_k E_J^k \begin{pmatrix} d\,k\,d \\ \pi s\,\sigma\,\pi s \end{pmatrix} = \sum_k E_J^k \begin{pmatrix} d\,k\,d \\ \pi c\,\sigma\,\pi c \end{pmatrix}$$

$$= E_J^2(1/\sqrt{70}) + E_J^4(-4/3\sqrt{70}) = I_J^2(1/7) + I_J^4(-4/21)$$

$$\bar{h}_J[d\delta] = E_J^2(-2/\sqrt{70}) + E_J^4(1/3\sqrt{70}) = I_J^2(-2/7) + I_J^4(1/21)$$

where $\bar{h}_J[d\delta]$ refers to the degenerate set $|d\delta\}$, consisting of $|d\delta s\rangle$ and $|d\delta c\rangle$, in the analogous way in which $\bar{h}_J[d\pi]$ refers to $|d\pi\}$. In the last steps of equation 22 the same 3-l symbols have been used as in the respective equations 10, but the sub-indices σ, which apply everywhere on the E parameters, have been dropped in the additivity model for linearly ligating ligands, and the degree super index replaced by its number symbol. Instead a sub-index J, referring to the type of ligand, has been introduced.

The orbital energy parametrisation of $\mathbf{A}^{(ZJ)}$ is identical to that of the angular overlap model, applied to linearly ligating ligands. The energy expressions for the unbarred operator are

$$h_J[d\sigma] = e_J(r) + e_{\sigma J}$$

$$h_J[d\pi] = e_J(r) + e_{\pi J} \tag{23}$$

$$h_J[d\delta] = e_J(r) + e_{\delta J}$$

where $e_J(r)$ is a formal representation of most of the spherically symmetrical term of the unbarred operator $A^{(ZJ)}$. This term vanishes completely for the barred operator $\bar{A}^{(ZJ)}$ (Schäffer, 1973a). Equation 23 may be rewritten for $\bar{A}^{(ZJ)}$ by subtracting the weighted average of the five orbital energies.

$$
\begin{aligned}
\bar{h}_J[d\sigma] &= (4/5)e'_{\sigma J} - (2/5)e'_{\pi J} \\
\bar{h}_J[d\pi] &= -(1/5)e'_{\sigma J} + (3/5)e'_{\pi J} \\
\bar{h}_J[d\delta] &= -(1/5)e'_{\sigma J} - (2/5)e'_{\pi J}
\end{aligned}
\tag{24}
$$

where the two angular overlap model parameters e'_σ and e'_π, which measure the energies of the σ and the π orbitals, respectively, relative to the δ orbitals, are defined by

$$
\begin{aligned}
e'_\sigma &\equiv e_\sigma - e_\delta \\
e'_\pi &\equiv e_\pi - e_\delta
\end{aligned}
\tag{25}
$$

In order to obtain the relationship between the axial spherical harmonic parameters and the angular overlap symmetry parameters, one may now compare equations 22 and 24. Alternatively one may, using equation 25, express the two independent energy differences $h_J[d\sigma] - h_J[d\delta]$ and $h_J[d\pi] - h_J[d\delta]$, respectively, in terms of equations 22 and 23 to obtain

$$
\begin{aligned}
e'_{\sigma J} &= E_J^2(4/\sqrt{70}) + E_J^4(5/3\sqrt{70}) = I_J^2(4/7) + I_J^4(5/21) \\
e'_{\pi J} &= E_J^2(3/\sqrt{70}) - E_J^4(5/3\sqrt{70}) = I_J^2(3/7) - I_J^4(5/21)
\end{aligned}
\tag{26}
$$

Equation 26 corresponds to equation 14 for the non-additivity model and the reverse relationship

$$
\begin{aligned}
E_J^d = E_J^2 = (\sqrt{70}/7)(e'_{\sigma J} + e'_{\pi J}); &\qquad I_J^2 = e'_{\sigma J} + e'_{\pi J}; \\
E_J^g = E_J^4 = (3\sqrt{70}/35)(3e'_{\sigma J} - 4e'_{\pi J}): &\qquad I_J^4 = (3/5)(3e'_{\sigma J} - 4e'_{\pi J})
\end{aligned}
\tag{27}
$$

corresponds to equation 15.

In equation 22 the matrix of $\bar{A}^{(jJ)}$ representing the single ligand J with angular position (θ_j, ϕ_j) has been given with respect to the basis $|(l)^j\}$. This matrix is now, using equation 18, transformed so as to apply to the basis $|l\}$. The general matrix element connecting lu and lv is then

$$
\begin{aligned}
\langle Mlu | \bar{A}^{(jJ)} | Mlv \rangle &= \sum_t \langle (Mlt)^j | \bar{A}^{(jJ)} | (Mlt)^j \rangle D_{ut}(j) D_{vt}(j) \\
&= \sum_t \bar{h}_J[lt] D_{ut}(j) D_{vt}(j)
\end{aligned}
\tag{28}
$$

where l has been omitted in D and $D[R(\phi_j, \theta_j, \psi_j)]$ abbreviated $D(j)$. The double summation, which arises from equation 18, reduces to a single summation because of the diagonality of $\bar{A}^{(jJ)}$ with respect to $|(l)^j\}$. So equation 28 contains $(2l+1)$ summation terms.

The only stage left now, in order to obtain the final expression for the additivity model, is the summation over the ligands (equation 20), giving

$$
\langle Mlu | \bar{A} | Mlv \rangle = \sum_{(jJ)} \sum_t \bar{h}_J[lt] D_{ut}(j) D_{vt}(j)
\tag{29}
$$

The axial spherical harmonic parametrisation and the angular overlap parametrisation for d electrons may now be obtained by substituting equations
22 and 24, respectively, into equation 29. It may be noted that the energy
expression for $\bar{\mathbf{A}}$ in terms of $\bar{h}_{\mathrm{J}}[lt]$ parameters is identical to that of $\mathbf{A}$ in
terms of $h_{\mathrm{J}}[lt]$ parameters (equation 23). Therefore the results of Schäffer
and Jørgensen (1965, Table 4) may directly be used to obtain the two new
barycentre parametrisations.

The relationship between the spherical harmonic parametrisations of the
non-additivity model (equations 5–7) and the additivity model (equation 22)
may (Schäffer, 1973a) be expressed by

$$E_q^k = \sum_{(jJ)} E_J^k \mathfrak{H}_q^k(\alpha_j, \beta_j, \gamma_j); \quad I_q^k = \sum_{(jJ)} I_J^k \mathfrak{H}_q^k(\alpha_j, \beta_j, \gamma_j) \tag{30}$$

where $\mathfrak{H}_q^k$ is a solid harmonic of the degree k, normalised to $4\pi/(2k+1)$
over the unit sphere. The ligator's positions, projected upon the unit sphere,
are given by the Cartesian co-ordinates $(\alpha_j, \beta_j, \gamma_j)$. It should be remembered
that the sub-index q refers to the component of the set $\mathfrak{H}^k$ (equation 1)
while the sub-index J is a labelling of the ligator J.

An O_h chromium(III) complex, CrJ_6^{m+}, may serve as an introductory
example on the relationship. Either by using equations 14 and 30, or by
working out the appropriate energy difference (equation 13) through
equation 29 and inserting into equations 24 and 26 one obtains

$$\Delta_J = (5\sqrt{21/63})I_{a_1}^g = (5/3)I_J^4 = 3e_{\sigma J}' - 4e_{\pi J}' \tag{31}$$

where $\Delta(d)$ of equation 13, has been called Δ_J for the cubic complex. For a
D_{4h} complex, trans-$\mathrm{Cr(NH_3)_4 X_2^{m+}}$, one obtains

$$E_{a_1}^g = (\sqrt{30}/3)(I_X^4 + 2I_N^4) = (\sqrt{30}/5)(\Delta_X + 2\Delta_N)$$
$$E_{e\theta}^g = (5\sqrt{42}/21)(I_X^4 - I_N^4) = (\sqrt{42}/7)(\Delta_X - \Delta_N) \tag{32}$$
$$E_{e\theta}^d = (2\sqrt{70}/7)(I_X^2 - I_N^2) = (2\sqrt{70}/7)[(e_{\sigma X}' - e_{\sigma N}') + (e_{\pi X}' - e_{\pi N}')]$$

where $\Delta(d)$ of equation 13, has been called Δ_J for the cubic complex. For a
and equation 27 in that for the E^d parameter. Equation 32 may be taken to
illustrate the most important aspect of the additivity model, the possibility
of transferability. If it is assumed that the single ligand parameters of the
additivity model are transferable from one complex to another the two first
equations of equation 32 express the parameters, corresponding to the axial
spherical harmonics of the degree 4, in terms of the cubic field parameters
Δ of the two octahedral chromium(III) systems $[\mathrm{CrX}_6]^{m+}$ and $[\mathrm{Cr(NH_3)_6}]^{3+}$,
while only the parameter, corresponding to the spherical harmonic of the
degree 2, contains new terms of the single ligand parametrisation.

So under this transferability assumption of the additivity model one of
the tetragonal parameters is observable from cubic complexes and is proportional to a difference between Δ values, and the other one is a lower symmetry
parameter which only can be observed in a non-cubic complex. So in this
special sense one might speak about one tetragonal field, associated with
$E_{e\theta}^d$, and in the additivity model proportional to the difference between I_X^2
and I_N^2 (equation 32).

In this light Table 1 may be looked upon again. The series of increasing $E_{e\theta}^g$ parameters is now that of increasing cubic field for the complex $[CrX_6]^{m+}$. Similarly, the series of increasing $E_{e\theta}^d$ parameters is that of increasing lower symmetry field, or I_X^2 parameters. In the angular overlap model increasing $\Delta(e)$ and $\Delta(t_2)$ parameters are interpreted as decreasing participation of the d-electrons of the chromium(III) ion in σ-bond and π-bond formation, respectively, with the ligand X, and the position of ammonia in the upper end of the $\Delta(t_2)$ series is associated with the fact that this ligand has no π-electrons available.

5. THE BETHE–VAN VLECK MODEL. ITS LIMITATIONS, ITS CLASSIFICATORY ACCOMPLISHMENTS, ITS PEDAGOGICAL ASPECTS AND CHEMICAL APPEAL

When the Bethe–van Vleck model of transition metal systems, is taken, not as a physical, but as a semi-empirical model, and defined as a general first order perturbation model embracing the ligand field and the inter-electronic repulsion, it is identical to the expanded radial function model.

This model has its obvious limitations from a theoretical point of view. One only needs to mention such a model's deficiencies when the ligand field is vanishing and the system becomes the gaseous ion. Further there are various types of experiments to indicate that the eigenfunctions of the partially filled shell cannot be characterised properly as radially modified central ion functions. This is, however, energetically not so serious in view of the perturbation principle which leaves the functions one step behind the energies calculated from them (see Figure 1).

It is hard theoretically as well as experimentally to specify the limits within which the Bethe–van Vleck model can be applied with reason, but the question about the model's accomplishments within inorganic chemistry as a classificational and educational tool, is without dispute.

The classificatory value of the expanded radial function model is associated with regularities in the variations of its empirical parameters. For systems of cubic symmetry one has the orbital energy ligand field parameter Δ, the Racah repulsion parameter B and the parameter $\Sigma = \Delta/B$, which determines the extent of the mixing of cubic sub-configurations by the repulsion (Jørgensen, 1962 and 1971; Schäffer, 1967). When arranged in order of increasing magnitude these parameters give rise to the spectrochemical, the nephelauxetic, and the field strength series of ligands, respectively, rather independently of the central ion. Similarly, series of central ions, which do not depend much upon the ligands, can be established. For tetragonal and trigonal systems the single ligand parameters, Δ_σ and Δ_π based upon the angular overlap model, give rise to the two-dimensional spectrochemical series of ligands (McClure, 1961; Glerup and Schäffer, 1968). Even exceptions to the mentioned regularities have been of classificatory value and have been explained along general, i.e. non-*ad-hoc*, lines (Orgel, 1960).

The pedagogical aspect of the model is worth emphasising further. It has provided a framework within which various wave mechanical principles can be illustrated in a motivating way. Under the headings perturbation theory and group theory all kinds of wave mechanical interactions between stationary states caused by intra- or extramolecular influences, can be illustrated.

In the present paper the symmetry aspect of the expanded radial function model has been discussed and it has been shown that independent of whether one assumes additivity of single ligand contributions to the ligand field or not, it is always possible to parametrise in two ways which are equivalent and linearly related to one-another. One of these ways is associated with a spherical harmonic expansion of the ligand field and the other with the orbital energy splittings, which the ligand field imposes upon the partially filled shell.

It is the additivity model that has most chemical appeal and classificatory capacity. It leads to an orbital energy parametrisation identical to that of the angular overlap model which relates the single ligand parameters to the central ion-to-ligand bonding. This additivity model stands comparisons with experiments well. This may be associated with the circumstances that one only needs to assume that the non-spherical terms of the ligand field are additive.

6. SUMMARY

The language of wave mechanics has made an impact on inorganic chemistry. For co-ordination compounds consisting of a metal ion with a partially filled d or f shell surrounded by ligands, the model concept of a ligand field represents the effects of the ligands without taking their electrons into account explicitly. This concept leads to a semi-empirical model which is a first order perturbation model using a basis set of d or f orbitals. It comprises two independent perturbations, the ligand field and the interelectronic repulsion, both of which give rise to parameters to be obtained from experiments (empirical parameters). The coefficients to these parameters are calculated exactly and are based essentially upon the symmetry of three-dimensional space. Two symmetry-based mathematically equivalent parametrisations of the ligand field are discussed, one based upon an expansion of the field into spherical surface harmonics, the other upon the energy differences within the partially filled l shell. In either case additivity of single ligand contributions may or may not be assumed. The empirical parameters of the additivity model (the Angular Overlap Model) are the chemically more interesting ones and because of their regularity in behaviour they serve classification purposes in transition metal chemistry. The fact that the energy observables correspond to energy differences rather than absolute energies has the consequence that the additivity assumption is only required for the non-spherical part of the ligand field. When no assumptions are made regarding the physical meaning of the empirical parameters, the old Bethe–van Vleck model is equivalent to Jørgensen's expanded radial function model, and linear relationships exist between their ligand field parameters. The weak field and strong field schemes are discussed and almost diagonal intermediate field schemes are proposed.

REFERENCES

Bethe, H. (1929) *Annln Phys.*, (5) **3**, 133
Bradbury, M. I. and Newman, D. J. (1967) *Chem. Phys. Letters*, **1**, 44
Condon, E. U. and Shortley, G. H. (1935) *The Theory of Atomic Spectra*, Cambridge University Press
Finkelstein, R. and Van Vleck, J. H. (1940) *J. chem. Phys.*, **8**, 790
Freeman, A. J. and Watson, R. E. (1960) *Phys. Rev.*, **120**, 1254
Glerup, J. and Schäffer, C. E. (1968) *Proc. XI Int. Conf. Coord. Chem.*, 500; *Progress in Coordination Chemistry* (edited by Cais. M.), Elsevier, Amsterdam
Griffith, J. S. (1961) *The Theory of Transition Metal Ions*, Cambridge University Press
Griffith, J. S. (1962) *The Irreducible Tensor Method of Molecular Symmetry Groups*, Prentice Hall, New Jersey
Harnung, S. E. (1973) *Mol. Phys.* (in press)
Harnung, S. E. and Schäffer, C. E. (1972a) *Struct. & Bond.*, **12**, 201
Harnung, S. E. and Schäffer, C. E. (1972b) *Struct. & Bond.*, **12**, 257
Harnung, S. E. and Schäffer, C. E. (1973). To be published
Hartmann, H. (1958) *Inorg. nucl. Chem.*, **8**, 64
Jørgensen, C. K. (1958) *Disc. Faraday Soc.*, **26**, 110
Jørgensen, C. K. (1962a) *Progr. inorg. Chem.*, **4**, 73
Jørgensen, C. K. (1962b) *Absorption Spectra and Chemical Bonding in Complexes*, Pergamon Press, Oxford
Jørgensen, C. K. (1971) *Modern Aspects of Ligand Field Theory*, North-Holland, Amsterdam
McClure, D. S. (1961) *Proc. VI Int. Conf. Coordination Chem., Advances in the Chemistry of Coordination Compounds*, 498 (edited by Kirschner, S.), Macmillan, New York
Maegaard, H., Mønsted, O. and Schäffer, C. E. To be published
Orgel, L. E. (1960) *An Introduction to Transition Metal Chemistry*, Methuen, London
Pauling, L. (1940) *The Nature of the Chemical Bond*, Oxford University Press
Schäffer, C. E. (1958) *Acta Vålådalensia*, 68, 'Proceedings of a Panel Meeting Held in Vålådalen, Sweden', Quantum Chemistry Group, Uppsala University
Schäffer, C. E. (1967) *Proc. R. Soc.*, **A297**, 96
Schäffer, C. E. (1968) *Struct. & Bond.*, **5**, 68
Schäffer C. E. (1970) *Pure appl. Chem.*, **24**, 361
Schäffer, C. E. (1973) *Struct. & Bond.*, **14**, 69
Schäffer, C. E. and Jørgensen, C. K. (1958) *Inorg. nucl. Chem.*, **8**, 143
Schäffer, C. E. and Jørgensen, C. K. (1965) *Mol. Phys.*, **9**, 401
Van Vleck, J. H. (1932) *The Theory of Electric and Magnetic Susceptibilities*, Chapter XI, Oxford University Press

13

A Group-Theoretic View of Weakly Interacting Sites

DOUGLAS J. KLEIN
Department of Physics, The University of Texas, Austin, Texas

1. INTRODUCTION

We consider a collection of sites, each site being an atom or molecule. Starting in the limit of completely isolated sites, we may view the evolution of the total system, as the intersite interaction is turned on, to occur in a number of steps:

(0) the *intrasite* interactions, with no intersite interactions in the isolated site limit, yield wave functions as simple products of isolated site kets;

(1) the *crystal field* (or classical electrostatic and induction) interactions give rise to energy level shifts and/or splittings with the system wave function still expressed as a simple product;

(2) the *resonance* interactions give rise to further energy terms and the wave function is no longer necessarily a simple product, though specific electrons are still assigned to given sites;

(3) the *dispersion* interactions give rise to the Van der Waals terms and the wave function which is no longer a simple product also lacks local point group symmetry;

(4) the lowest order (resonant) *charge transfer* interactions may give rise to energy terms and a wave function in which some of the electrons are no longer identified with specific sites;

(5) the *direct exchange* and higher order *charge transfer* interactions may lift exchange degeneracies, distort the wave function and destroy the local permutation or spin symmetries.

In fact these different types of effects all take place simultaneously and the distinction of different cases is not always justified. Frequently one of these types of interactions alone may account for the lifting of a physical degeneracy associated with the isolated site limit, in which case a treatment which primarily focuses on this type of interaction might be employed; if a parametric theory is used, the parameters might in fact reflect the contributions of the other effects while still formally ascribing all the level splittings to the one effect. Thus we identify different theories associated with each one of these individual interactions:

(1) crystal field theory (see Bethe, 1929, or Griffith, 1961);
(2) Frenkel exciton theory (see Frenkel, 1931, or Davydov, 1970);
(3) theory of Van der Waals or London forces (see London, 1937, or Dalgarno, 1967);
(4) Wannier exciton theory (see Wannier, 1937) and Mulliken charge transfer theory (see Mulliken, 1952, or Mulliken and Pearson, 1969);
(5) exchange Hamiltonian theory (see Heisenberg, 1928, Dirac, 1929, Anderson, 1959, or Herring, 1963).

All of these theories have been known since the early days of quantum mechanics, though in all cases there have been important modern developments in the theory.

Here we wish to consider all these theories from a uniform viewpoint. We describe different groups associated with each of these cases, and in so doing formally distinguish among these different types of interactions. Having accomplished this distinction, we may then see how to treat several types of interactions simultaneously thus leading to hybrid versions of the various theories. We consider the Hamiltonian at each step to commute with a different group and in so doing find that the group of the perturbed Hamiltonian is sometimes a subgroup of the group of the zero-order Hamiltonian, whereas in other cases the perturbed group contains the zero-order group. These situations are referred to as either *descent in symmetry* or *ascent in symmetry*. The descent in symmetry situation involves subduction of group representations while the ascent in symmetry situation involves induction of group representations. In both situations effective Hamiltonians may be obtained as linear combinations of group elements of the larger group, and decompositions in terms of irreducible tensorial operators or in terms of double cosets may be found convenient.

Here we let A, B, C, ..., label the sites and assign electrons to these sites by grouping them into disjoint sets, also denoted by A, B, C, ... The Hamiltonians for individual sites A, B, C, ... determine the isolated site Hamiltonian

$$H^0 \equiv H_A(A) \oplus H_B(B) \oplus H_C(C) \oplus \ ...$$

Both H^0 and the perturbing intersite interaction V are assumed to be spin-free operators, and the motion of the nuclei is neglected.

2. GROUPS OF INTEREST

Each isolated site Hamiltonian commutes with the corresponding spin-free symmetric group permuting electrons assigned to that site. We denote these symmetric groups as $\mathscr{S}_A, \mathscr{S}_B, \mathscr{S}_C, \ldots$ for sites A, B, C, ... so that elements of the group

$$\mathscr{S}^0 \equiv \mathscr{S}_A \otimes \mathscr{S}_B \otimes \mathscr{S}_C \otimes \ldots$$

commute with the isolated site Hamiltonian H^0. We see $\mathscr{S}^0$ involves only those permutations which do not interchange any electrons among sites. The fully perturbed N-electron Hamiltonian $H \equiv H^0 + V$ commutes with the symmetric group $\mathscr{S}_N$ of all permutations on electronic indices. In proceeding from H^0 to H we clearly do not have a case of descent in symmetry, and if point group considerations are totally neglected, it is a case of ascent in symmetry. Denoting the isolated site point groups as $\hat{\mathscr{G}}_A, \hat{\mathscr{G}}_B, \hat{\mathscr{G}}_C, \ldots$ for sites A, B, C, ... we see that the elements of the group

$$\hat{\mathscr{G}}^0 \equiv \hat{\mathscr{G}}_A(A) \otimes \hat{\mathscr{G}}_B(B) \otimes \hat{\mathscr{G}}_C(C) \otimes \ldots$$

commute with H^0. The fully perturbed Hamiltonian H commutes with the full molecular point group $\hat{\mathscr{G}}$.

Having identified groups for the two extreme Hamiltonians H^0 and H, we next identify some groups for intermediate Hamiltonians. The point groups $\hat{\mathscr{G}}$ and $\hat{\mathscr{G}}_A$ for H and H_A have, in general, some elements in common, where we disregard the difference between the elements of $\hat{\mathscr{G}}$ and $\hat{\mathscr{G}}_A$ due to the fact that they act on N and N_A-electron spaces. The subgroup of $\hat{\mathscr{G}}_A$ which has elements in common with $\hat{\mathscr{G}}$ is termed the *crystal field* (or site) group for site A and is denoted $\hat{\mathscr{G}}_A^{CF}$. Similarly defining $\hat{\mathscr{G}}_B^{CF}, \hat{\mathscr{G}}_C^{CF}, \ldots$, the crystal field group for the full system is taken as

$$\hat{\mathscr{G}}^{CF} = \hat{\mathscr{G}}_A^{CF}(A) \otimes \hat{\mathscr{G}}_B^{CF}(B) \otimes \hat{\mathscr{G}}_C^{CF}(C) \otimes \ldots$$

Now introducing the intersection

$$\hat{\mathscr{G}} \equiv \hat{\mathscr{G}}^0 \wedge \hat{\mathscr{G}}$$

it may be shown that $\hat{\mathscr{G}}$ is a direct product of $\hat{\mathscr{G}}$ and a factor group $\hat{\mathscr{F}} \subseteq \hat{\mathscr{G}}$,

$$\hat{\mathscr{G}} = \hat{\mathscr{G}}\hat{\mathscr{F}}$$

To prove this we first consider linear, planar non-linear, and non-planar arrangements of atomic sites. In the linear case $\hat{\mathscr{G}} = \mathscr{C}_{\infty v}$ and $\hat{\mathscr{G}} = \mathscr{C}_{\infty v}$ or $\hat{\mathscr{D}}_{\infty h}$; hence $\hat{\mathscr{F}} = \{1\}$ or $\{1, \hat{\imath}\}$ where the inversion $\hat{\imath}$ commutes with all elements of $\hat{\mathscr{G}}$. For planar molecules $\hat{\mathscr{G}} = \{1, \hat{\sigma}_h\} \equiv \mathscr{C}_{\sigma h}$ and the horizontal reflection $\hat{\sigma}_h$ commutes with rotations in the horizontal molecular plane, with rotations by π about an axis in the molecular plane, with reflections perpendicular to the molecular plane, and with the inversion operator; hence $\hat{\mathscr{F}}$ may be chosen as the subgroup of all proper rotations. For non-planar molecules $\hat{\mathscr{G}} = \{1\}$ and $\hat{\mathscr{F}} = \hat{\mathscr{G}}$. Having established that $\hat{\mathscr{G}} = \hat{\mathscr{G}}\hat{\mathscr{F}}$ is a direct product for the instance the sites are atoms, we note that when the sites are molecules $\hat{\mathscr{G}}$ and $\hat{\mathscr{G}}$ are subgroups of the corresponding groups for the atomic case. Consequently one may easily show that $\hat{\mathscr{G}}$ and $\hat{\mathscr{F}}$ may be chosen as subgroups of the corresponding groups for the atomic case. Hence the result is established for the case of molecular sites also.

We note that for any non-identity element, say $\hat{G}$, of $\hat{\mathscr{F}}$ there is a coordinate system centred on some site, say J, such that $\hat{G}$ transforms it to a coordinate system on another identical site, say K. If the identical sites whose co-ordinates are so transformed all have the same number of electrons, the corresponding permutation of $\mathscr{S}_N$ which transforms all the indices of site J to site K is denoted P_G. We see that the set of all permutations P_G, $\hat{G} \in \hat{\mathscr{F}}$, forms a group isomorphic to $\hat{\mathscr{F}}$ as does

$$\hat{\mathscr{F}} \star \equiv \{\hat{G} P_G; \hat{G} \in \hat{\mathscr{F}}\}$$

Clearly the elements of $\hat{\mathscr{F}}\star$ commute with H. They also commute with H^0, and indeed there may even be more elements not yet identified which commute with H^0. For instance, there may be simultaneous co-ordinate transformations and permutations which change a site Hamiltonian $H_J(J)$ into a second as $H_K(K)$ even though the identical sites J and K may not be in point group equivalent positions. We, however, for most purposes consider only $\hat{\mathscr{G}}^0 \mathscr{S}^0$ as the group of H^0.

In the case when different identical sites have different numbers of electrons, a special type of group may be constructed. We consider the case when all sites identical to site J are in singlet states except one with an additional mobile electron. We assign a set J of electron indices to fill out this singlet core for each such site J. The remaining unassigned electron index may be associated with any one of these identical sites. We then define

$$\bar{F} \equiv P_G \hat{G}, \hat{G} \in \hat{\mathscr{F}}$$

where P_G is chosen similarly to the previous manner and in such a way that it does not affect the mobile electron index. The group generated by these $\bar{F}$ is denoted as $\hat{\mathscr{F}}\star$. The $\mathscr{S}^0$ group is taken to permute only the assigned electrons and then, as usual, only among those assigned to the same site.

3. GENERAL DEVELOPMENT

We make the following identifications of symmetry group and wave function with the various theories mentioned in the introduction:

    (0) $\psi^0 = \psi_A^0(A) \otimes \psi_B^0(B) \otimes \ldots$ symmetry adapted to $\hat{\mathscr{G}}^0 \mathscr{S}^0$
    (1) $\psi^{CF} = \psi_A^{CF}(A) \otimes \psi_B^{CF}(B) \otimes \ldots$ symmetry adapted to $\hat{\mathscr{G}}^{CF} \mathscr{S}^0$
    (2) ψ^{FE} symmetry adapted to $\hat{\mathscr{G}}^{CF} \hat{\mathscr{F}} \star \mathscr{S}^0$
    (3) $\psi^{dispersion}$ symmetry adapted to $\hat{\mathscr{G}}\,\hat{\mathscr{F}} \star \mathscr{S}^0$
    (4) ψ^{WE} symmetry adapted to $\hat{\mathscr{G}}\,\hat{\mathscr{F}} \star \mathscr{S}^0$
    (5) $\psi^{exchange}$ symmetry adapted to $\hat{\mathscr{G}} \mathscr{S}_N = \hat{\mathscr{G}}\,\hat{\mathscr{F}} \star \mathscr{S}_N$

It is seen that a number of these theories involve descent in symmetry and a number involve ascent in symmetry.

The treatment of the descent in symmetry cases is well-known. First-order corrections to the energies are obtained merely by diagonalisation of the perturbation in the zero-order eigenspaces. In general, one may derive (see des Cloizeaux, 1963, Primas, 1961, or Buleavski, 1966) a higher-order effective Hamiltonian

$$\mathscr{H}_{eff} \equiv E^0 \mathscr{P}^0 + \mathscr{P}^0 V \mathscr{P}^0 + \mathscr{P}^0 V \mathscr{R}^0 V \mathscr{P}^0 + \ldots$$

defined on the zero-order eigenspace with projection operator $\mathscr{P}^0$. Here E^0 and $\mathscr{R}^0$ are the zero-order energy and resolvent. Decomposition of $\mathscr{H}_{\text{eff}}$ into irreducible tensorial components of the zero-order group leads to useful results, as physical arguments usually yield the approximate but reasonable truncation of such expansions.

In the case of ascent in symmetry we have the zero-order group $\mathscr{G}^0$ as a subgroup of the perturbed group $\mathscr{G}$ corresponding to $H = H^0 + \lambda V$ at some particular value of λ, say $\lambda = 1$. (If $\mathscr{G}$ were to be the group of H for arbitrary λ, we see that $\mathscr{G}$ would necessarily be a subgroup of $\mathscr{G}^0$.) A zero-order eigenspace $\mathscr{V}(\alpha^0)$ generates an irreducible representation α^0 of $\mathscr{G}^0$ and thus has a basis

$$\{\,|\,\alpha^0 r^0\,\rangle\,;\, r^0 = 1 \text{ to } f^{\alpha^0}\}$$

of kets each transforming as some one of the rows, labelled by r^0, of the f^{α^0}-dimensional (α^0)th representation of $\mathscr{G}^0$. Now as the perturbation is slowly turned on, the states of H^0 evolve eventually into states of H at $\lambda = 1$, where the larger group $\mathscr{G}$ is obtained. But since larger groups imply degeneracies equal to or, as is frequently the case, greater than a subgroup, it follows that several zero-order eigenspaces may necessarily come together to form a single $\lambda = 1$ perturbed eigenspace.

For H^0 to provide a good, but approximate, description of H one might infer that the spectrum of H^0 is similar to that of H and in addition the perturbed eigenkets of H are composed primarily of zero-order eigenkets from the corresponding region of the spectrum. Should several zero-order eigenspaces merge to a single perturbed eigenspace, satisfaction of this criterion would imply that these zero-order eigenspaces are all similar in energy to the perturbed eigenspace. Although such a criterion is usually that implied in descent in symmetry cases, it is often blatantly violated in ascent in symmetry cases (see Claverie, 1971).

In the case of ascent in symmetry we consider rather than this first criterion, a second in which, instead of the individual zero-order eigenspaces $\mathscr{V}(\alpha^0)$, the corresponding *induced* spaces, as $\mathscr{V}(\alpha^0\uparrow)$ with a basis

$$\{G_c\,|\,\alpha^0 r^0\,\rangle\,;\, c, r^0 \text{ ranging}\}$$

are employed. Here c labels the left coset multipliers G_c for the left coset decomposition of $\mathscr{G}^0$ in $\mathscr{G}$,

$$\mathscr{G} = \sum_c \oplus\, G_c \mathscr{G}^0$$

In this alternative criterion the perturbed eigenkets of H are to be composed primarily of kets from the induced spaces obtained from zero-order eigenspaces close in energy to the perturbed eigenket. Since some of the kets in the induced spaces are not eigenkets to H^0, this criterion does not imply that the zero-order eigenspaces of H^0 which merge to a single eigenspace of H are close in energy. The induced basis kets $G_c\,|\,\alpha^0 r^0\rangle$ are evidently eigenkets to the different zero-order Hamiltonians $G_c H^0 G_c^{-1}$ so that, in a certain sense, these induced basis kets are degenerate.

For the ascent in symmetry case there are very many suggestions for the explicit form of the perturbation expansion. Part of the problem arises in that the union of all induced spaces becomes overcomplete if every state of

H^0 is considered. One solution would be to diagonalise H^0 on an incomplete subspace such that the corresponding induced space is exactly complete. Such an incomplete subspace is implicit in the perturbation formalism of Jansen (1967) and Byers Brown (1968). Further, the use of such an incomplete subspace would rectify some of the criticism (see Musher and Amos, 1967) of the methods described by Anderson (1959) or Murrell, Randic and Williams (1965). However, in such procedures the detailed form of the perturbation expansion depends on the initial choice of subspace and the manner in which the perturbation order is defined in the induced space. Other alternative forms which may in principle apply to the use of all the states of H^0 are reviewed by Amos (1970) and Klein (1971). We do not deal with such explicit perturbation formalisms here.

Here we shall merely consider a first-order treatment, which may be described as the diagonalisation of the perturbed Hamiltonian in the induced space. Many of the formal perturbation formalisms yield this result at least in the case of two weakly interacting hydrogen atoms, which often is the initial problem to which these formalisms are applied (see, for instance, Certain *et al.*, 1968). Indeed a possible form of an ascent in symmetry perturbation formalism might diagonalise an effective Hamiltonian in the induced space in all orders. This would be desirable particularly in cases where more than one state of a given perturbed symmetry arises, as these states, although not necessarily exactly degenerate after first-order, are still nearly degenerate; hence it would be advisable to take into account the interactions between such states through a high order.

4. EFFECTIVE HAMILTONIANS FOR ASCENT IN SYMMETRY

In the case of ascent in symmetry from the zero-order group $\mathscr{G}^0$ up to the larger perturbed group we may construct an effective Hamiltonian accurate through first order by restricting H to the induced subspace. We assume that the zero-order ket $|0\rangle \equiv |\alpha^0 r^0\rangle$ transforms as the r^0th row of the α^0th irreducible representation (IR) of $\mathscr{G}^0$ and that all kets degenerate to $|0\rangle$ may be generated from $|0\rangle$ by the action of an element of the larger perturbed group algebra $\mathscr{A}(\mathscr{G})$ of $\mathscr{G}$ such that this subalgebra has symmetry $\alpha^0 r^0$ on both $\langle 0|HX|0\rangle$ or $\langle 0|X|0\rangle$ where X is an element of the subalgebra of the group algebra $\mathscr{A}(\mathscr{G})$ of $\mathscr{G}$ such that this subalgebra has symmetry $\alpha^0 r^0$ on both the left and right. Letting $e^{\alpha^0}_{r^0 r^0}$ be the projector onto the space of symmetry $\alpha^0 r^0$, we see that this subalgebra is $e^{\alpha^0}_{r^0 r^0} \mathscr{A}(\mathscr{G}) e^{\alpha^0}_{r^0 r^0}$ and is semisimple, so that it has a matric basis (see Klein, Carlisle, and Matsen, 1970)

$$\{e^{\alpha}_{\rho\sigma}; \alpha\rho\sigma \text{ ranging}\}$$

$$e^{\alpha}_{\rho\sigma} e^{\beta}_{r\omega} = \delta_{\alpha\beta}\delta_{\sigma r} e^{\alpha}_{\rho\omega}$$

These matric basis elements are merely an abbreviation for some of the matric basis elements of the form $e^{\alpha}_{(\rho\beta^0 s^0)(\sigma\gamma^0 t^0)}$ of the group algebra $\mathscr{A}(\mathscr{G})$

$$e^{\alpha}_{\rho\sigma} \equiv e^{\alpha}_{(\rho\alpha^0 r^0)(\sigma\alpha^0 r^0)}$$

These matric basis elements may be expanded in terms of group elements

$$e^{\alpha}_{(\rho\beta^0 s^0)(\sigma\gamma^0 t^0)} = \frac{f^{\alpha}}{g} \sum_{G\in\mathscr{G}} [G^{-1}]^{\alpha}_{(\sigma\gamma^0 t^0)(\rho\beta^0 s^0)}$$

Here the $[G^0]^{\alpha}_{(\rho\beta^0 s^0)(\sigma\gamma^0 t^0)}$ are IR matrix elements for $\mathscr{G}$ such that

$$[G]^{\alpha}_{(\rho\beta^0 s^0)(\sigma\gamma^0 t^0)} = \delta_{\rho\sigma}\delta_{\beta^0\gamma^0}[G^0]^{\beta^0}_{s^0 t^0}, \ G^0 \in \mathscr{G}^0$$

and $[G^0]^{\beta^0}_{s^0 t^0}$ is an IR matrix element for $\mathscr{G}^0$.

We now introduce the operator

$$\mathscr{H}^{(1)} \equiv \sum_{\alpha\rho\sigma} \frac{g f^{\alpha^0}}{f^{\alpha} g^0} \langle 0 | e^{\alpha}_{\rho\sigma} H | 0 \rangle e^{\alpha}_{\sigma\rho}$$

defined on the space with a basis

$$\{e^{\alpha}_{\sigma\rho} | 0 \rangle; \sigma \text{ ranging}\}$$

An alternative spanning set for this space is

$$\{e^{\alpha^0}_{r^0 r^0} G_c | \alpha^0 s^0 \rangle; c, s^0 \text{ ranging}\}$$

where the G_c are left coset multipliers of $\mathscr{G}^0$ in $\mathscr{G}$. Matrix elements of this effective operator are

$$\langle 0 | e^{\alpha}_{\rho\sigma} \mathscr{H}^{(1)} e^{\beta}_{\tau\rho} | 0 \rangle = \langle 0 | e^{\alpha}_{\tau\sigma} H | 0 \rangle \frac{g f^{\alpha^0}}{f^{\alpha} g^0} \langle 0 | e^{\alpha}_{\rho\sigma} e^{\alpha}_{\sigma\tau} e^{\beta}_{\tau\rho} | 0 \rangle$$

$$= \delta_{\alpha\beta} \langle 0 | e^{\alpha}_{\tau\sigma} H | 0 \rangle \frac{g f^{\alpha^0}}{f^{\alpha} g^0} \langle 0 | e^{\alpha}_{\rho\rho} | 0 \rangle$$

Assuming $\langle 0 | e^{\alpha}_{\rho\rho} | 0 \rangle$ is normalised to $f^{\alpha} g^0 / g f^{\alpha^0}$ we see that $\mathscr{H}^{(1)}$ is the transpose of the restriction of H to the space with a basis $\{e^{\alpha}_{\rho\sigma} | 0 \rangle; \sigma \text{ ranging}\}$. This restriction of H yields, however, our first-order estimate of the perturbed energies. Thus the solution of $\mathscr{H}^{(1)}$ yields the first-order energies also. If $\langle 0 | e^{\alpha}_{\rho\rho} | 0 \rangle$ were not normalised to $f^{\alpha} g^0 / g f^{\alpha^0}$, then one would merely need consider an effective overlap matrix not equal to the identity; transformation via the inverse square root of the overlap matrix as in Löwdin's (1950) symmetric orthogonalisation method, would then yield a new effective Hamiltonian of the same general form.

Instead of writing $\mathscr{H}^{(1)}$ in terms of matric basis elements we may write it in terms of group elements

$$\mathscr{H}^{(1)} = \sum_{\alpha\rho\sigma} \sum_{\beta^0 s^0} \sum_{\gamma^0 t^0} \frac{g f^{\alpha^0}}{f^{\alpha} g^0} \langle 0 | e^{\alpha}_{(\rho\beta^0 s^0)(\sigma\gamma^0 t^0)} H | 0 \rangle e^{\alpha}_{(\sigma\gamma^0 t^0)(\rho\beta^0 s^0)}$$

$$= \sum_{\alpha} \sum_{\rho\beta^0 s^0} \sum_{\sigma\gamma^0 t^0} \frac{f^{\alpha} f^{\alpha^0}}{g g^0} \sum_{G, G' \in \mathscr{G}} [G^{-1}]^{\alpha}_{(\sigma\gamma^0 t^0)(\rho\beta^0 s^0)} [(G')^{-1}]^{\alpha}_{(\rho\beta^0 s^0)(\sigma\gamma^0 t^0)} \langle 0 | GH | 0 \rangle G'$$

$$= \sum_{\alpha} \frac{f^{\alpha} f^{\alpha^0}}{g g^0} \sum_{G, G' \in \mathscr{G}} \chi^{\alpha}(G^{-1}(G')^{-1}) \langle 0 | GH | 0 \rangle G'$$

Here we have used the fact that, for instance,

$$e^{\alpha}_{(\rho\beta^0 s^0)(\sigma\gamma^0 t^0)} | 0 \rangle \sim \delta_{\gamma^0 \alpha^0} \delta_{t^0 r^0}$$

Now using the orthogonality relation

$$\frac{1}{g}\sum_\alpha f^\alpha \chi^\alpha(G) = \delta(G, 1)$$

we obtain

$$\mathcal{H}^{(1)} = \sum_G \frac{f^{\alpha 0}}{g^0} \langle 0|GH|0\rangle G^{-1}$$

This form of the effective Hamiltonian is found in the work of Serber (1934) for the case $\mathcal{G} = \mathcal{S}_N$ and $\mathcal{G}^0 = \{1\}$.

Next using the *double coset* (DC) decomposition of $\mathcal{G}$ with respect to the subgroup $\mathcal{G}^0$,

$$\mathcal{G} = \sum_q \oplus \mathcal{G}^0 G_q \mathcal{G}^0$$

we may cast $\mathcal{H}^{(1)}$ into another useful form. Each element of $\mathcal{G}$ occurs in one and only one DC. In carrying out the $(g^0)^2$ multiplications indicated in the DC $\mathcal{G}^0 G_q \mathcal{G}^0$ one finds each element repeated d_q times, where d_q is the order of the intersection $\mathcal{G}^0 \wedge G_q \mathcal{G}^0 G_q^{-1}$. Hence $\mathcal{G}^0 G_q \mathcal{G}^0$ contains $(g^0)^2/d_q$ distinct elements. Using this DC decomposition, we have

$$\mathcal{H}^{(1)} = \sum_q \frac{1}{d_q} \frac{f^{\alpha 0}}{g^0} \sum_{G^0, G^{0\prime} \varepsilon \mathcal{G}^0} \langle 0|G^0 G_q G^{0\prime} H|0\rangle (G^0 G_q G^{0\prime})^{-1}$$

$$= \sum_q \frac{f^{\alpha 0}}{g^0 d_q} \sum_{G^0, G^{0\prime} \varepsilon \mathcal{G}^0} \sum_{s^0, t^0} [G^0]^{\alpha 0}_{r^0 s^0} [G^{0\prime}]^{\alpha 0}_{t^0 r^0} \langle \alpha^0 s^0|G_q H|\alpha^0 t^0\rangle (G^{0\prime})^{-1} G_q^{-1} (G^0)^{-1}$$

$$= \sum_q \frac{f^{\alpha 0}}{g^0 d_q} \left(\frac{g^0}{f^{\alpha 0}}\right)^2 \sum_{s^0 t^0} \langle \alpha^0 s^0|G_q H|\alpha^0 t^0\rangle e^{\alpha 0}_{r^0 t^0} G_q^{-1} e^{\alpha 0}_{s^0 r^0}$$

This development is closely related to that by Matsen, Klein and Foyt (1971) and is particularly convenient since the different DC's can usually be associated with the approximate magnitude of the matrix element $\langle \alpha^0 s^0|G_q H|\alpha^0 t^0\rangle$, and hence used in deleting smaller contributions. Such Hamiltonians are applicable in the cases of a single Frenkel or Wannier excitation, although we shall not consider these cases here.

5. EXCHANGE HAMILTONIAN

Here we consider in some greater detail the ascent in symmetry from $\mathcal{S}^0$ up to $\mathcal{S}_N$. The intersite exchange integrals $\langle \alpha^0 s^0|G_q H|\alpha^0 t^0\rangle$, which appear in the effective Hamiltonian, decrease as the differential overlap between $\langle \alpha^0 s^0|$ and $G_q|\alpha^0 t^0\rangle$ decreases. Further this differential overlap should be smaller the more electrons G_q interchange between sites and the farther apart these sites are. Noting that the DC's of $\mathcal{S}^0$ in $\mathcal{S}_N$ are in one-to-one correspondence with the pattern of electron interchanges among the sites (see Kramer and Seligman, 1969 and 1972, or Junker and Klein, 1971), we thus may truncate the DC expansion in a physically meaningful manner. Restricting our attention to only the identity DC, with $G_q = 1$, and the nearest neighbour pair interchange DC's, with G_q a simple transposition, we obtain the truncated form

$$\mathcal{H}^{(1)} \cong \frac{g^0}{f^{\alpha^0}} \frac{1}{g^0} \langle 0 | H | 0 \rangle \sum_{s^0} e^{\alpha^0}_{r^0 s^0} e^{\alpha^0}_{s^0 r^0}$$

$$+ \sum_{j<k} \frac{1}{d_{\mathrm{JK}}} \frac{g^0}{f^{\alpha^0}} \sum_{s^0 t^0} \langle \alpha^0 s^0 | (\ )H | \alpha^0 t^0 \rangle \, e^{\alpha^0}_{r^0 t^0} (jk) e^{\alpha^0}_{s^0 r^0}$$

where $j \in J$ and $k \in K$. The DC for the pair interchanges between sites J and K has $d_q = d_{\mathrm{(JK)}} = g^0 / N_\mathrm{J} N_\mathrm{K}$ with N_J and N_K the number of electrons on sites J and K.

We next consider the simplification of this truncated Hamiltonian as it involves the $s^0 t^0$-summation. To accomplish this we use a number of special properties of the symmetric group and the Young diagram representation of its IR's, as discussed, for instance, by Rutherford (1968) or Klein, Carlisle and Matsen (1970). For many purposes the present derivation is less important than the final result, which is very similar to the spin space result of Herring (1962 and 1963) or Matsen, Klein and Foyt (1971). However, the present derivation in effect evaluates some spin space reduced matrix elements that appear in the earlier derivations and aids in the derivation of the less thoroughly considered exciton-exchange Hamiltonian of the next section. Without further apology we thus pursue these simplifications.

We choose the matric basis for the symmetric group $\mathcal{S}_\mathrm{J}$ for site J to be sequence adapted to the subgroup $\mathcal{S}_\mathrm{J}^- \subseteq \mathcal{S}_\mathrm{J}$ which does not involve the index $j \in J$ (see Klein, Carlisle and Matsen, 1970). Then the elements of $\mathcal{S}_\mathrm{J}^-$ commute with (jk). This choice is accomplished if j is taken as the last index in J and the Young–Yamanouchi orthogonal representation is used. Now

$$e^{\alpha_\mathrm{J}}_{r_\mathrm{J} t_\mathrm{J}} (jk) e^{\alpha_\mathrm{J}}_{s_\mathrm{J} r_\mathrm{J}} \sim \delta_{s_\mathrm{J} t_\mathrm{J}}$$

since the Young tableaux $T^{\alpha_\mathrm{J}}_{s_\mathrm{J}}$ and $T^{\alpha_\mathrm{J}}_{t_\mathrm{J}}$ corresponding to s_J and t_J must be identical on deleting $j \in J$. Hence our effective Hamiltonian now appears as

$$\mathcal{H}^{(1)} \cong \langle 0 | H | 0 \rangle e^{\alpha^0}_{r^0 r^0} + \sum_{\mathrm{J}<\mathrm{K}} \frac{N_\mathrm{J} N_\mathrm{K}}{f^{\alpha^0}} \sum_{s^0} \langle \alpha^0 s^0 | (jk) H | \alpha^0 s^0 \rangle \, e^{\alpha^0}_{r^0 s^0} (jk) e^{\alpha^0}_{s^0 r^0}$$

Further, a number of the matrix elements $\langle \alpha^0 s^0 | (jk) H | \alpha^0 s^0 \rangle$ may be equal.

Even more simplification of $\mathcal{H}^{(1)}$ will result if we restrict our attention to only the physically significant spin-free permutational symmetries (see Matsen, 1970) and their corresponding (1- or) 2-columned Young diagrams. Letting s_J range, we find only two different $e^{\alpha_\mathrm{J}}_{r_\mathrm{J} s_\mathrm{J}} (jk) e^{\alpha_\mathrm{J}}_{s_\mathrm{J} r_\mathrm{J}}$, since j may be deleted in only two ways from (either the first or second columns of) $T^{\alpha_\mathrm{J}}_{s_\mathrm{J}}$. If j is in the second column of $T^{\alpha_\mathrm{J}}_{s_\mathrm{J}}$ then the operator $e^{\alpha_\mathrm{J}}_{r_\mathrm{J} s_\mathrm{J}} (jk) e^{\alpha_\mathrm{J}}_{s_\mathrm{J} r_\mathrm{J}}$ is effectively scalar in $e^{\alpha_\mathrm{J}}_{r_\mathrm{J} r_\mathrm{J}}$ so long as the only overall permutational symmetries considered are 2-columned also. Hence

$$\mathcal{H}^{(1)} \cong (\text{constant}) e^{\alpha^0}_{r^0 r^0} + \sum_{\mathrm{J}<\mathrm{K}} \frac{N_\mathrm{J} N_\mathrm{K}}{f^{\alpha^0}} \sum_{s^0}^{\mathrm{JK}} \langle \alpha^0 s^0 | (jk) H | \alpha^0 s^0 \rangle \, e^{\alpha^0}_{r^0 s^0} (jk) e^{\alpha^0}_{s^0 r^0}$$

where the s^0 sum is over only those s^0 involving s_J and s_K such that j and k

are deleted from the first columns of $T_{SJ}^{\alpha J}$ and $T_{SK}^{\alpha K}$. The matrix element $\langle \alpha^0 s^0 | (jk)H | \alpha^0 s^0 \rangle \equiv H_{JK}$ is then independent of these values of s^0, so that

$$\mathcal{H}^{(1)} \cong (\text{constant})e_{r^0 r^0}^{\alpha^0} + \sum_{J<K} \frac{N_J N_K}{f^{\alpha^0}} H_{JK} \sum_{s^0}^{JK} e_{r^0 s^0}^{\alpha^0}(jk)e_{s^0 r^0}^{\alpha^0}$$

Although the s^0-summation is still restricted as indicated above, we may lift this restriction merely by changing the scalar constant multiplying $e_{r^0 r^0}^{\alpha^0}$ so as to cancel the additional terms. Next we note that

$$\sum_{s^0} e_{r^0 s^0}^{\alpha^0}(jk)e_{s^0 r^0}^{\alpha^0} = \sum_{\beta^0 t^0 s^0} \frac{f^{\alpha^0}}{f^{\beta^0}} e_{r^0 r^0}^{\alpha^0} e_{t^0 s^0}^{\beta^0}(jk)e_{s^0 t^0}^{\beta^0}$$

$$= \sum_{\beta^0 t^0 s^0} \frac{f^{\alpha^0} f^{\beta^0}}{(g^0)^2} \sum_{P^0, P^{0'} \in \mathcal{S}^0} [(P^0)^{-1}]_{s^0 t^0}^{\beta^0} [(P^{0'})^{-1}]_{t^0 s^0}^{\beta^0} \, e_{r^0 r^0}^{\alpha^0} P^0 (jk) P^{0'}$$

$$= \sum_{P^0, P^{0'} \in \mathcal{S}^0} \frac{f^{\alpha^0}}{(g^0)^2} \sum_{\beta^0} f^{\beta^0} \chi^{\beta^0}((P^0)^{-1}(P^{0'})^{-1}) e_{r^0 r^0}^{\alpha^0} P^0 (jk) P^{0'}$$

and then use a character orthogonality theorem, to obtain

$$\sum_{s^0} e_{r^0 s^0}^{\alpha^0}(jk)e_{s^0 r^0}^{\alpha^0} = \frac{f^{\alpha^0}}{g^0} \sum_{P^0, P^{0'}} \delta(P^0 P^{0'}, 1) e_{r^0 r^0}^{\alpha^0} P^0 (jk) P^{0'}$$

$$= \frac{f^{\alpha^0}}{N_J N_K} e_{r^0 r^0}^{\alpha^0} \vec{J} \cdot \vec{K}$$

Here we have defined

$$\vec{J} \cdot \vec{K} \equiv \sum_{m\epsilon J} \sum_{n\epsilon K} (mn)$$

The truncated effective Hamiltonian

$$\mathcal{H}^{(1)} \Rightarrow (\text{constant})e_{r^0 r^0}^{\alpha^0} + \sum_{J<K} H_{JK} e_{r^0 r^0}^{\alpha^0} \vec{J} \cdot \vec{K}$$

thus yields the physical eigenvalues correct through 'lowest order' terms (where account of an effective overlap may also be necessary).

The result is equivalent to the spin space result of Herring (1962 and 1963) and Matsen, Klein and Foyt (1971) if we replace spin-free transpositions (mn) by spin transpositions $-(mn)^\sigma$. The Dirac identity may also be used to express it in terms of spin operators $\vec{S}_J$. Higher order DC's may also be treated in the manner employed here.

We note that $\mathcal{H}^{(1)}$ may lack some important terms if $|0\rangle \equiv |\alpha^0 r^0\rangle$ is not of the appropriate form. In particular, charge transfer matrix elements, $\langle \alpha^0 r^0 | H | \alpha^{0'} r^{0'} \rangle$ with $|\alpha^{0'} r^{0'}\rangle$ having one more and one fewer electrons on sites J and K, may be important. These off-diagonal matrix elements are linear in intersite differential overlap so that 'second order' terms

$$\langle \alpha^0 r^0 | H | \alpha^{0'} r^{0'} \rangle \frac{1}{E(\alpha^0) - E(\alpha^{0'})} \langle\!\langle \alpha^{0'} r^{0'} | H | \alpha^0 r^0 \rangle$$

will be quadratic in differential overlap, as the intersite exchange integrals are. These contributions are implicit in Herring's treatment (1962) and appear explicitly in the treatments of Anderson (1959), Buleavski (1966), and others. The general form of the effective Hamiltonian is, however, unchanged in lowest order, except that H_{JK} is replaced by a modified value.

6. EXCITON-EXCHANGE HAMILTONIAN

We consider the hybrid case of a single Frenkel exciton in a lattice of para-magnetic sites when both exciton and exchange interactions are important. We assume that the local point group symmetry of the individual sites does not influence the problem, so that the induction process of interest is from $\mathscr{S}^0$ up to $\hat{\mathscr{F}} \star \mathscr{S}_N = \hat{\mathscr{F}} \mathscr{S}_N$. We assume all the (valence) electrons to be unpaired in the ground state for an isolated site while the excited state has two fewer unpaired electrons. Further we consider all the single exciton kets to be generated by the action of elements of $\hat{\mathscr{F}} \star \mathscr{S}_N$ on the zero-order ket $|0\rangle \equiv |\alpha^0 r^0\rangle$ with the exciton on site A.

The most important matrix elements which do not involve exciton transfer and which appear in our first-order effective Hamiltonian are

$$\langle 0 | H | 0 \rangle$$

$$\langle \alpha^0 s^0 | (jk)H | \alpha^0 t^0 \rangle \equiv \delta(s^0, t^0) \mathscr{J}_{\text{ex}}, \quad j \in J \neq A, k \in K \neq A, J \sim K$$

$$\langle \alpha^0 s^0 | (aj)H | \alpha^0 t^0 \rangle \sim \delta(s^0, t^0), \quad a \in A, j \in J \sim A$$

Here $J \sim K$ indicates that sites J and K are nearest neighbours. We have also noted that all local permutational symmetries are one-dimensional except α_A so that $\langle \alpha^0 s^0 | (jk)H | \alpha^0 s^0 \rangle$ is independent of $s^0 \backsimeq s_A$ if $j, k \notin A$. Further, $e_{r^0 s^0}^{\alpha^0}(aj)e_{s^0 r^0}^{\alpha^0}$ is effectively a scalar multiple of $e_{r^0 r^0}^{\alpha^0}$ if s_A involves $a \in A$ in the second column of the Young tableau $T_{s_A}^{\alpha_A}$. Hence we define

$$\Delta \equiv \langle \alpha^0 s^0 | (aj)H | \alpha^0 s^0 \rangle - \mathscr{J}_{\text{ex}}$$

where $s^0 \backsimeq s_A$ is such that $a \in A$ appears in the first column of $T_{s_A}^{\alpha_A}$. The effective Hamiltonian then becomes

$$\mathscr{H}^{(1)} \backsimeq (\text{constant})e_{r^0 r^0}^{\alpha^0} + \mathscr{J}_{\text{ex}} \tfrac{1}{2} \sum_{J \sim K} \vec{J} \cdot \vec{K} e_{r^0 r^0}^{\alpha^0}$$

$$+ \Delta \sum_{J \sim A} \vec{A} \cdot \vec{J} e_{r^0 r^0}^{\alpha^0}$$

$$+ \text{exciton transfer terms}$$

The elements of $\hat{\mathscr{F}} \star$ will accomplish the transfer of the exciton while retaining the assigned electrons on their corresponding sites. We note that integrals $\langle 0 | P_G \hat{G} H | 0 \rangle$, $P_G \hat{G} \in \hat{\mathscr{F}} \star$ and $P_G \hat{G} \neq 1$ are zero since the exciton is of different local permutational symmetry than the ground state for a site. Letting $\hat{G}_J \in \hat{\mathscr{F}}$ transfer co-ordinates on site A to site J, we anticipate that the most important exciton transfer matrix elements are

$$\langle \alpha^0 s^0 | (aj)P_{G_J} \hat{G}_J H | \alpha^0 t^0 \rangle, s^0 t^0 \text{ ranging}$$

Temporarily suppressing reference to all sites except A and J, we write $\alpha^0 = \alpha_A \otimes \alpha_J$, $t^0 = t_A \otimes 1 = t_A$, and

$$|\alpha^0 s^0\rangle = |\alpha_A s_A \otimes \alpha_J\rangle$$
$$P_{G_J}\hat{G}_J |\alpha^0 t^0\rangle = |\alpha_J \otimes \alpha_A t_A\rangle$$

Hence, if we let j and a be the last indices in J and A, our exciton transfer matrix element

$$\langle \alpha_A s_A \otimes \alpha_J | (a_j)H | \alpha_J \otimes \alpha_A t_A\rangle$$

is seen to be zero unless $T^{\alpha_A}_{s_A}(A)$ is identical to $T^{\alpha_J}(A)$ on deleting their last indices and also $T^{\alpha_J}(J)$ is identical to $T^{\alpha_A}_{t_A}(J)$ on deleting their last indices. This matrix element will thus be non-zero only when s_A and t_A are such that a and j occupy the lone position in the second column of $T^{\alpha_A}_{s_A}(A)$ and $T^{\alpha_A}_{t_A}(J)$. This particular value of s_A or t_A is denoted as 1_A. Hence defining

$$\mathscr{I}_{ex-ex} \equiv \langle \alpha_A 1_A \otimes \alpha_J | (aj)H | \alpha_J \otimes \alpha_A 1_A\rangle = \langle \alpha^0 1^0 | (aj)P_{G_J}\hat{G}_J H | \alpha^0 1^0\rangle$$

we obtain

$$\mathscr{H}^{(1)} \simeq (\text{constant})e^{\alpha^0}_{r^0 r^0} + \mathscr{I}_{ex} \; \tfrac{1}{2}\sum_{J \sim K} \vec{J} \cdot \vec{K} e^{\alpha^0}_{r^0 r^0} + \Delta \sum_{J \sim A} \vec{J} \cdot \vec{A} e^{\alpha^0}_{r^0 r^0}$$

$$+ \mathscr{I}_{ex-ex} \sum_{J \sim A} \frac{g^0}{f^{\alpha^0} d_q} e^{\alpha^0}_{r^0 1 0} P^{-1}_{G_J}(aj)e^{\alpha^0}_{1 0 r^0}\hat{G}_J^{-1}$$

Just such exciton-exchange Hamiltonians are of interest to describe spectroscopic observations in MnF_2 (see, for instance, Meltzer et al., 1968) and $CsMnCl_3 \cdot 2H_2O$ (see, for instance, Marzacco and McClure, 1969). Indeed theoretical treatments of MnF_2 have already been described by Elliot et al. (1968) and Freeman and Hopfield (1968). Another method for treating exciton-exchange interactions is described by Van Vleck (1962).

As a further example of the present construction we consider the binuclear complex $[(NH_3)_5CrOCr(NH_3)_5]^{+4}$. Each Cr^{+3} site is in an octahedral crystal field which is distorted to the lower symmetry $\mathscr{C}_{4v}$. For the ground multiplet both Cr^{+3} ions are in $^4B_1$ states (with respect to $\mathscr{C}_{4v}$ or $^4A_{2g}$ with respect to $\mathscr{O}_h$). The usual exchange interaction applies, as is verified (see Earnshaw and Lewis, 1961, and Kobayashi, Haseda and Kanda, 1960) from magnetic susceptibility data. For the exciton multiplet with one Cr^{+3} ion in the lowest excited $^2B_1$ state (or ^2E_g with respect to $\mathscr{O}_h$), we expect exciton-exchange interations as indicated above to apply. The effective Hamiltonian becomes

$$\mathscr{H}^{(1)} \simeq Je^{D}_{r^0 r}{}^{x2} \; \vec{A} \cdot \vec{B} + J_{ex-ex} \frac{36}{2.8}e^{D\alpha}_{r^0 r^0}{}^{\mathcal{Q}}(14)(25)e^{D\alpha}_{j^0 r^0}{}^{\mathcal{Q}}$$

Here we have neglected the scalar term and considered electrons 1, 2, 3 on site A and 4, 5, 6 on site B. We have denoted the doublet and quartet permutational symmetries, $[2, 1]$ and $[1, 1, 1]$, for sites A and B by D and Q. The factor group is simply the inversion group $\hat{\mathscr{F}} = \{1, \hat{1}\}$. These D and Q states couple to triplet and quintet permutational symmetries. There are several ways $\mathscr{H}^{(1)}$ may be solved; here we note that the pertinent irreducible representation matrix elements are given in Section XII.C of Klein, Carlisle and Matsen (1970). The eigenvalues are

$$E(^2B_1 \otimes ^4B_1; ^5A_{1g}) = -6J + J_{ex-ex}$$
$$E(^2B_1 \otimes ^4B_1; ^5A_{1u}) = -6J - J_{ex-ex}$$
$$E(^2B_1 \otimes ^4B_1; ^3A_{1g}) = -2J + 3J_{ex-ex}$$
$$E(^2B_1 \otimes ^4B_1; ^3A_{1u}) = -2J - 3J_{ex-ex}$$

where we have used the $\mathscr{D}_{4h}$ symmetry to label the states. From the ground multiplet states $^4B_1 \otimes ^4B_1; ^3A_{1u}$ and $^4B_1 \otimes ^4B_1; ^5A_{1g}$ we would expect to see transitions to the $^3A_{1g}$ and $^5A_{1u}$ exciton states found above. Indeed the spectral data of Dubicki and Martin (1970) has been interpreted in nearly this manner by König (1971), although he did not discuss the exciton exchange interaction J_{ex-ex}.

7. CONCLUSION

We have described the different groups associated with various theories of weakly interacting sites. The identification of these groups aids in the construction of effective Hamiltonians on zero-order eigenspaces. These effective Hamiltonians are conveniently expanded in terms of irreducible tensorial operators in the case of descent in symmetry and in terms of double cosets in the case of ascent in symmetry. As the case of ascent in symmetry is usually less familiar, it was considered in greater detail. Cases of mixed ascent and descent in symmetry where neither the zero-order or perturbed group is a subgroup of the other were solved by first descending in symmetry from the zero-order group to the intersection of the zero-order and perturbed groups; then having solved this, ascending in symmetry from the intersection up to the perturbed group.

There is some lack of generality in our present approach. In the Frenkel exciton case our formal description of the effective Hamiltonian fails in part if there are two or more excitons present; this difficulty arises since not all the zero-order degenerate kets may be generated by the action of group elements on a single primitive ket. While the case of a single Frenkel exciton is often of primary interest, even if more than one exciton is present, one may generalise the present technique. In this generalisation a minimal set of primitive kets is introduced and different effective Hamiltonians between the different spaces induced from these primitive kets are constructed in much the same manner as indicated in our earlier discussion here. One case of particular interest in which there are many Frenkel excitations even at low temperatures is that of Wurster's blue perchlorate and related organic crystals in which aromatic free radicals alternate in separation down a linear stack; in this case close pairs prefer to couple to a singlet although the excited triplet is only $\sim 100 \, \text{cm}^{-1}$ higher in energy (see Nordio, Soos and McConnell, 1966, or Soos, 1968).

In the Wannier exciton case our present identification of a symmetry group becomes ambiguous if there are two or more sites with a mobile electron or if the 'core' above which the mobile electron moves is not a singlet. If the number of electrons assigned to a particular type of site is not an integral multiple of the number of these sites, as in mixed-valence compounds, we see that the case of many mobile electrons may occur even in the ground multiplet case. We thus leave the non-dilute Wannier exciton and Mulliken charge transfer cases in an incomplete state of affairs here.

There are of course numerous other methods of constructing effective Hamiltonians. However, whatever method is applied it would seem that the present sequence of groups should be identifiable within the development. For example in perturbation formalisms for weakly interacting sites one should expect to be able to identify an approximation equivalent to the effective Heisenberg exchange Hamiltonian.

REFERENCES

Amos, A. T. (1970) *Chem. Phys. Letters*, **5**, 587
Anderson, P. W. (1959) *Phys. Rev.*, **115**, 2
Bethe, H. (1929) *Ann. Phys.*, **3**, 133
Buleavski, L. N. (1966) *Zh. eksp. teor. Fiz.*, **51**, 230 [English trans.: (1967) *JETP*, **24**, 154]
Byers Brown, W. (1968) *Chem. Phys. Letters*, **2**, 105
Certain, P. R., Hirschfelder, J. O., Kolos, W. and Wolniewicz, L. (1968) *J. chem. Phys.*, **49**, 24
Claverie, P. (1971) *Int. J. quantum Chem.*, **5**, 273
Dalgarno, A. (1967) *Adv. chem. Phys.*, **12**, 143
Davydov, A. S. (1970) *Theory of Excitons*, Plenum Press, New York
des Cloizeaux, J. (1963) *Phys. Rev.*, **129**, 554
Dirac, P. A. M. (1929) *Proc. R. Soc.*, **A123**, 714
Dubicki, L. and Martin, R. L. (1970) *Aust. J. Chem.*, **23**, 215
Earnshaw, A. and Lewis, J. (1961) *J. chem. Soc.*, **1961**, 397
Elliot, R. J., Thorpe, M. F., Imbusch, G. F., Loudon, R. and Parkinson, J. B. (1968) *Phys. Rev. Letters*, **21**, 147
Freeman, S. and Hopfield, J. (1968) *Phys. Rev. Letters*, **21**, 910
Frenkel, Y. I. (1931) *Phys. Rev.*, **37**, 17
Griffith, J. S. (1961) *The Theory of Transition-Metal Ions*, Cambridge University Press
Heisenberg, W. (1928) *Z. Physik*, **49**, 619
Herring, C. (1962) *Rev. mod. Phys.*, **34**, 631
Herring, C. (1963) *Magnetism* **2B**, (G. T. Rado and H. Suhl, Eds.), p. 1, Academic Press, New York
Jansen, L. (1967) *Phys. Rev.*, **162**, 63
Junker, B. R. and Klein, D. J. (1971) *J. chem. Phys.* **55**, 5533
Klein, D. J. (1971) *Int. J. quantum Chem.*, **4S**, 271
Klein, D. J., Carlisle, C. H. and Matsen, F. A. (1970) *Adv. quantum Chem.*, **5**, (edited by Löwdin, P. O.). p. 218, Academic Press, New York
Kobayashi, H., Haseda, T. and Kanda, E. (1960) *Phys. Soc. Japan*, **15**, 1646
König, E. (1971) *Chem. Phys. Letters*, **9**, 31
Kramer, P. and Seligman, T. H. (1969) *Nucl. Phys.*, **A136**, 545
Kramer, P. and Seligman, T. H. (1972) *Nucl. Phys.*, **A186**, 49
London, F. (1937) *Trans. Faraday Soc.*, **56**, 753
Löwdin, P. O. (1950) *J. chem. Phys.*, **18**, 365
Marzacco, C. J. and McClure, D. S. (1969) *Sym. of the Faraday Soc.*, 106
Matsen, F. A. (1970) *J. Am. chem. Soc.*, **92**, 3525
Matsen, F. A. and Klein, D. J. (1971) *J. phys. Chem.*, **75**, 1860
Matsen, F. A., Klein, D. J. and Foyt, D. C. (1971) *J. phys. Chem.*, **75**, 1867
Meltzer, R. S., Chen, M. Y., McClure, D. S. and Lowe-Pariseau, M. (1968) *Phys. Rev. Letters*, **21**, 913

Mulliken, R. S. (1952) *J. Am. chem. Soc.*, **64,** 811
Mulliken, R. S. and Pearson, W. B. (1969) *Molecular Complexes,* Wiley, New York
Murrell, J. N., Randic, M. and Williams, D. R. (1965) *Proc. R. Soc.*, **A284,** 566
Musher, J. I. and Amos, A. T. (1967) *Phys. Rev.*, **164,** 31
Nordio, P. L., Soos, Z. G. and McConnell, H. M. (1966) *Ann. Rev. phys. Chem.*, **17,** 237
Primas, H. (1961) *Helv. Phys. Acta,* **34,** 331
Rutherford, D. E. (1968) *Substitutional Analysis,* Hafner, New York
Serber, R. (1934) *Phys. Rev.*, **45,** 461 ; *J. chem. Phys.*, **2,** 697
Soos, Z. G. (1968) *J. chem. Phys.*, **46,** 4284
Van Vleck, J. H. (1962) *Revista Mat. y Fis. Teor.*, **14,** 189
Wannier, G. H. (1937) *Phys. Rev.*, **52,** 191

14

The Role of Symmetry and Spin Restrictions in Reactive and Non-Reactive Collisions and Molecular Decompositions

JOYCE J. KAUFMAN
Department of Chemistry The John Hopkins University
Baltimore, Maryland

1. INTRODUCTION

In principle, it is possible at present to carry out configuration interaction calculations for the potential energy surfaces of triatomic and simple polyatomic systems. In practice, there are only a handful of such results available since these computations require extensive amounts of computer time, computer capacity and research effort. There are many more interesting observed phenomena and experiments which are being carried out which need theoretical interpretation and guidance than will be possible in the foreseeable future to compute the corresponding potential energy surfaces. It is just in this area that extremely valuable insight can be gleaned from the application of the symmetry and spin restrictions on what electronic states of an intermediate quasimolecule are permissible from the symmetries and spins of the separated reactants or products or on what electronic states of fragments are permissible from a particular electronic state of a dissociating molecule. These restrictions are the three-dimensional extensions of the original Wigner–Witmer (1928) quantum mechanical derivations for diatomic molecules and are discussed in great and comprehensible detail in the recent book by Herzberg (1966).

In the case of collisional phenomena, reactive or non-reactive, consideration of these surfaces is vital whether the intermediate complex is a stable bound entity or even a completely repulsive state since the reactants must approach and the products must recede along the potential energy curves of the quasimolecule. The symmetries and stabilities of these intermediates have a profound effect on the paths along which and the rates at which reactions proceed. Over the past years, in collaboration with Professor W. S.

Koski (with whose experimental group I am associated), we have been able to demonstrate the usefulness of this concept to ion–neutral reactions (Kaufman and Koski, 1969; Pipano and Kaufman, 1970, 1971, 1972; Catlow et al., 1970; Kaufman, in press) neutral–neutral reactions and hot atom reactions (Kaufman, in press; Kaufman, Harkins and Koski, 1969).

The course of the reaction is determined by the shapes of these curves and by the kinds of curve crossings which do or do not take place. In this regard there are apparently two types of behaviour followed: (1) adiabatic, where the reactants approach each other slowly and, at least for states of the same symmetry, the curves do not cross; (2) diabatic, where the reactants approach each other rapidly and diabatic transitions occur even between states of the same symmetry.

For the case of molecular dissociative phenomena we have noted an additional restriction which we have called a 'uniqueness' criterion. For any specified electronic state of a molecule along a particular dissociation co-ordinate there is one and only one set of dissociation products permitted to it by the symmetry and spin restrictions. When one is trying to decide the electronic states of the fragments into which a specified molecular state can dissociate, it is not sufficient to say that such an electronic state could dissociate into fragments with various possible combinations of symmetries and spins. Rather it is important to realise that only one particular set of fragments is permissible and it is often possible from the energetics to determine uniquely this set of fragments. The only way in which this specified molecular state can dissociate into products in different electronic states is by a curve crossing. In collaboration with Professor Koski we applied these symmetry and spin restrictions combined with only semi-rigorous molecular orbital calculations for the interpretation of the non-appearance of a parent peak in the mass spectrum of CF_4 where the photoelectron spectroscopy indicates at least five peaks for a parent CF_4^+ (Kaufman, Kerman and Koski, 1971). Upon further consideration it appears that the 'uniqueness' restriction is of wide applicability and should be especially useful for determining the products in the case of dissociation from excited states versus dissociation from the ground state.

For states of different symmetries of diatomics the probability of crossing was derived by Landau and Zener. Even for diatomics the Landau–Zener formula has been shown to be inapplicable in certain instances and has been revised. For polyatomic molecules the Landau–Zener formula is not applicable as such and modifications have been made to permit its use for specified cases.

2. POTENTIAL ENERGY SURFACES

It is not the purpose of this present chapter to go into exhaustive detail either on the correlation rules or on the various quantum chemical techniques available for calculating potential energy surfaces. However, the important points will be highlighted here to enable the reader to follow the crucial arguments and with sufficient references to enable the reader to go back to the original sources for the necessary background material to carry through such analyses himself.

A. CORRELATION RULES–GENERAL

1. DIATOMICS

There are three different schemes for building up the electronic states of diatomic molecules: (a) from separated atoms; (b) from the united-atom and (c) from the molecular orbitals of the diatomic molecule itself. It is the correlation between the electronic states of the diatomic molecule as built up from the separated atoms and as determined from the molecular orbitals of the diatomic which is most valuable for any general consideration of reactions and excited states. The correlation of molecular states obtained by these two methods is not limited solely to diatomic molecules but also forms a valid approach for polyatomic molecular systems. The correlation of separated atoms with the hypothetical united atom has value for diatomics and has been applied to simple polyatomic molecules, especially those with a heavy atom or two and a number of hydrogen atoms. However, it is conceptually less appealing even for simple polyatomic molecules and completely inapplicable for complex polyatomic molecules.

The correlation rules for determining what types of molecular electronic states result from given electronic states of the separated atoms were derived quantum mechanically by Wigner and Witmer (1928) and are discussed in great detail in Herzberg (1950). These correlation rules hold for the adiabatic potential curves of the electronic states* when Russell–Saunders coupling is valid for the separated atoms as well as the molecule.

2. POLYATOMICS

There are four different schemes for building up the electronic states of polyatomic molecules: (a) from the united atom or molecule; (b) by starting from a geometrical conformation of different symmetry (either higher or lower); (c) from separated atoms or molecular fragments; (d) from the molecular orbitals of the polyatomic molecule. In general, the most useful correlations for elucidation of the mechanism of collisions involving polyatomic intermediates are those between the states of the molecule formed from separated atoms or molecular fragments and those built up from the molecular orbitals of the polyatomic intermediate. For the elucidation of molecular decomposition as well as formation, in addition to these two correlations, the correlations by starting from a geometrical conformation of different symmetry (either higher or lower) is also extremely valuable. In all of the correlations, as long as the spin–orbit coupling is small, there must be spin

---

*In general, electronic motions are very fast compared to nuclear motions. For slowly varying changes in the internuclear distance the electronic motion adjusts itself to each change in nuclear configuration without undergoing electronic transitions. This behaviour is called adiabatic.

correlation; singlets correlate with singlets, doublets with doublets, triplets with triplets, etc. In a later section examples of both of these applications will be presented from our own research.

(a) From a geometrical conformation of different symmetry
For the question of collisions this correlation may be of importance because during the collisional process the symmetry of the intermediate often changes. For example, even if a completely symmetrical intermediate complex results during the reaction of say an atom plus a diatomic, the complex is only completely symmetrical when it is in its equilibrium conformation. During the formation and breakup, the intermediate has had a conformation of lower symmetry. A table in Herzberg Polyatomics gives the correlation of species of different point groups corresponding to different conformations of a given molecule: $D_{2h} \rightarrow D_2 \rightarrow D_{2d}$; $D_{3d} \rightarrow D_3 \rightarrow D_{3h}$; $C_{2v} \rightarrow C_2 \rightarrow C_{2h}$ (Herzberg, 1966). Herzberg points out that while the correlations of species of a point group of higher symmetry into those of a point group of lower symmetry are always unambiguous, the reverse correlations are not always unambiguous. The actual correlation depends on the path chosen. It is straightforward to decipher which states of a distorted molecule arise from certain states of a more symmetrical molecule; the reverse correlation is not always unambiguous.

In the case of molecular decompositions this correlation takes on even more importance. Oftentimes, the original molecule does possess a certain high or moderately high degree of symmetry. As the molecule decomposes, this symmetry distorts to that of a point group of lower symmetry but it is always possible to trace the correlation uniquely.

(b) From separated atoms or molecular fragments
The correlation between the electronic states of a polyatomic molecule and those of the separated atoms or molecular fragments can be derived from a generalisation of the Wigner–Witmer correlation rules for diatomic molecules.

(1) Linear Molecules
(a) Unsymmetrical molecules $(C_{\infty v})$
The possible values for the quantum number Λ of the molecule is obtained by algebraic addition of the M_{L_i} values of all atoms (or fragments)

$$\Lambda = |\Sigma M_{L_i}|$$

where $M_{L_i} = L_i, L_i-1, L_i-2, ..., -L_i$. When an atom of angular momentum L approaches a linear diatomic of angular momentum Λ_d with the formation of a linear complex, the resultant Λ is given by

$$\Lambda = |M_{L_i}+M_{L_2}|$$

where $M_{L_i} = L, L-1, ..., -L$ and $M_{L_2} = \pm\Lambda_d$. When two diatomics or two linear polyatomics approach to form a linear intermediate, the resultant Λ is given by the sum of the $M_{L_i} (= \pm\Lambda_i)$. A most useful table of both of these correlations is given in Herzberg (1966). The resultant permitted spins $\overline{S}$ are obtained by vector addition of the individual spins $\overline{S}_i$

$$\overline{S} = \Sigma \overline{S}_i$$

If the states are built up from separated atoms then partial resultants have to be formed according to

$$S_{ik} = S_i + S_k, S_i + S_k - 1, \ldots, |S_i - S_k|$$

which are then added, also using partial resultants. The permitted multiplicities of the states are $|2S+1|$. Each of the resultant states occurs with each of the possible multiplicities.

A table of molecular electronic states of linear molecules resulting from certain states of the separated atoms is given in Herzberg (1966).

(b) Symmetrical molecules ($D_{\infty h}$)

If two like linear fragments (but in different electronic states) are brought together, the correlations given in the previous section and listed in Herzberg (1966) are the ones used. Each state of that table now occurs twice since resolution of the resonance degeneracy leads to splitting into a g and a u state. If two identical linear fragments (in identical electronic states) are brought together, there is no resonance degeneracy and only the same total of states arise as for unequal groups. The symmetry character of the states alternates for different possible spin values and, therefore, some of the states are g and some are u. A table in Herzberg (1966) lists the electronic states of symmetrical linear molecules ($D_{\infty h}$) resulting from identical states of the separated equal groups.

(2) Non-linear molecules

The building up of non-linear molecules from individual atoms leads to a very great number of molecular states. The procedure is outlined in Herzberg (1966) but will not be discussed here since it is not easily usable for the purposes of studying possible collisional intermediate complexes. Much more useful for this purpose is the building up of the intermediate molecules from separated fragments.

(a) Unlike groups

If neither of the two parts which are to be combined has a symmetry lower than that of the final molecule and if the full symmetry is retained during the approach of the two parts, then the correlation is the same as if the molecule were built up from separated atoms.

If the symmetry of at least one of the parts is lower than that of the complete molecule and if during the approach of the parts the molecule has the geometrical conformation of the part with the lower symmetry, the resulting states are obtained in terms of the lower symmetry. If in the final molecule the symmetry is now higher, the correlations rules between deformed conformations are used to find the relation of the states of the intermediate to the states of the final molecule.

A third case is where the symmetry of both parts is higher than that of the final molecule being built up but where the symmetry of the intermediate during the formation process is lower than that of the final molecule. The states of both parts must be resolved into the species of the final molecule.

The resulting molecular states of the final molecule are obtained by multiplication of the states of the two parts. For symmetrical final molecules this resolution is not completely unambiguous.

(b) Like groups
If two like fragments are brought together (even if the symmetry of the parts is lower than that of the final molecule) it is usually possible to determine unambiguously the resulting molecular states. This is possible because even at large separation of the two fragments the full symmetry of the molecule may exist.

If the two like fragments are in different electronic states there is a resonance degeneracy which leads to a splitting into a symmetrical and an anti-symmetrical state. If the two fragments are identical (in the identical electronic states) there is no resonance degeneracy. The symmetries of the resulting states are either g or u or else $'$ and $''$. These symmetries are determined by using the correlation with the corresponding diatomic or linear polyatomic molecules. [A paper by Shuler (1953) in which adiabatic orbital and spin correlation rules applicable to study of elementary chemical reactions involving non-linear polyatomic intermediate complexes were formulated is often quoted. The same paper also presented some pertinent correlation tables. Two other useful references discussing the same general topic are a paper by Laidler and Shuler (1951) and Laidler's book (1955)].

B. CORRELATIONS WITH MOLECULAR ORBITALS

It is not the intent of this chapter to present a detailed review of quantum chemical theory. The few following portions on the theoretical background serve merely to put the discussion of the quantum chemical computations into the proper perspective.

1. THEORETICAL BACKGROUND

(a) Separation of electronic and nuclear motion
Because, in general, electrons move with much greater velocities than nuclei to a first approximation electron and nuclear motions can be separated (Born–Oppenheimer, 1927). The validity of this separation of electronic and nuclear motions provides the only real justification for the idea of a potential energy curve of a molecule. The eigenfunction Ψ for the entire system of nuclei and electrons can be expressed as a product of two functions Ψ_e and Ψ_n where Ψ_e is an eigenfunction of the electronic co-ordinates found by solving Schroedinger's equation with the assumption that the nuclei are held fixed in space and Ψ_n involves only the co-ordinates of the nuclei (Eyring, Walter and Kimball, 1944).

The exact Hamiltonian operator may be written as

$$\mathcal{H} = -\sum_A \frac{\hbar^2}{2M_A} \nabla_A^2 - \sum_i \frac{\hbar^2}{2m} \nabla_i^2 + V_{nn} + V_{ne} + V_{ee}$$

where the first term represents the kinetic energy of the nuclei, the second represents the kinetic energy of the electrons and V_{nn}, V_{ne} and V_{ee} are the contributions to the potential energy arising from nuclear, nuclear–electronic and electronic interactions respectively. If the nuclei were assumed to be fixed in space, the Hamiltonian for the electrons would be

$$\mathcal{H}_e = -\sum_i \frac{\hbar^2}{2m} \nabla_i^2 + V_{ne} + V_{ee}$$

The remaining terms are represented by $\mathcal{H}_n$

$$\mathcal{H}_n = -\sum_A \frac{\hbar^2}{2M_A} \nabla_A^2 + V_{nn}$$

and

$$\mathcal{H} = \mathcal{H}_e + \mathcal{H}_n$$

Ψ_e is defined as the function which satisfies the equation

$$\mathcal{H}_e \Psi_e = E_e \Psi_e$$

where E_e is the electronic energy. For the total system

$$\mathcal{H} \Psi_e \Psi_n = E \Psi_e \Psi_n$$

(This holds only in a 'loose' perturbation sense for $\mathcal{H}\Psi = E\Psi$. Ψ should be expanded in a complete set.) In the Born–Oppenheimer approximation

$$(\mathcal{H}_n + E_e)\Psi_n = E\Psi_n$$

since small terms coupling the electronic and nuclear motion have been neglected. (Often by convention V_{nn} is included in $\mathcal{H}_e$ rather than in $\mathcal{H}_n$, so that with $\mathcal{H}_e$ defined in this second manner $\mathcal{H}_e\Psi$ gives rise to the molecular energy.) Thus the molecular energy as a first approximation can be considered as the sum of two terms: the energy that the electrons would have if the nuclei were at rest plus the repulsive mutual energies of these fixed nuclei. The neglected terms may be treated as a perturbation and will give rise to energy terms representing the interaction of electronic and nuclear motions. These perturbations prove of importance later when discussing probabilities for crossing from one potential curve to another.

(b) *Group theory and quantum mechanics* (Eyring, Walter and Kimball, 1944)
If the Schroedinger equation

$$\mathcal{H}\Psi_i = E_i \Psi_i$$

for an atomic or molecular system is subjected to some transformation of co-ordinates R, which interchanges like particles in the system, then

$$R\mathcal{H}\Psi_i = RE_i\Psi_i$$

Since R interchanges only like particles it can have no effect on the Hamiltonian, so that $R\mathscr{H} = \mathscr{H}R$. R commutes with the constant E_i, therefore

$$\mathscr{H}R\Psi_i = E_i R\Psi_i$$

that is, the function $R\Psi_i$ is a solution of the Schroedinger equation with the eigenvalue E_i. If E_i is a non-degenerate eigenvalue, then Ψ_i or constant multiples of Ψ_i are the only eigenfunctions satisfying the above equation. If E_i is k-fold degenerate then any linear combination of the functions $\Psi_{i1}, \Psi_{i2}, \ldots, \Psi_{ik}$ will be a solution of the above equation, in this case

$$R\Psi_{il} = \sum_{j=1}^{k} \Psi_{ij} a_{jl}$$

where the a_{jl}'s must satisfy the relation

$$\sum_{j=1}^{k} a_{jl}^2 = 1$$

The matrices obtained from the coefficients in the expansion of $R\Psi_{il}$ are unitary and form a representation of the group of operations which leave the Hamiltonian unchanged. The set of eigenfunctions $\Psi_{i1}, \ldots, \Psi_{ik}$ forms a basis for the representation of the group. The representations generated by the eigenfunctions corresponding to a single eigenvalue are irreducible representations.

If Γ_j is an irreducible representation of dimension k, and if $\Psi_1^j, \Psi_2^j, \ldots, \Psi_k^j$ is a set of degenerate eigenfunctions which form the basis for the jth irreducible representation of the group of symmetry operations, these eigenfunctions transform according to the relation

$$R\Psi_i^j = \sum_{l=1}^{k} \Gamma_j(k)_{li} \Psi_l^j$$

For a symmetrical atomic or molecular system, these considerations place a severe restriction on the possible eigenfunctions of the system. All possible eigenfunctions must form bases for some irreducible representation of the group of symmetry operations. The form of the possible eigenfunctions is also determined to a large extent since they must transform in a quite definite way under the operations of the group.

(c) Various quantum chemical computational methods
In order to render tractable the problem of determining the molecular electronic eigenfunction, Ψ_e, it is customary to assume the individual molecular orbitals to be functions of the atomic electron eigenfunctions, χ_r, centred on each atom. The molecular orbitals (MO's), ϕ_i, are taken to be linear combinations of the atomic orbitals (LCAO's), χ_r

$$\phi_i = \sum_r \chi_r c_{ri}$$

A configurational wave function Φ (where here Φ denotes the molecular electronic eigenfunction Ψ_e) is represented by antisymmetrised product wave function (Slater, 1963)

$$\Phi = (N!)\phi_1{}^{[1}\phi_2 \ldots \phi_N{}^{N]}$$

since only states can occur whose eigenfunctions are antisymmetric with respect to exchange of any two electrons.

The total wave function Φ is normalised

$$(\Phi \mid \Phi) = \int \Phi^* \Phi \, d\tau = 1$$

and

$$E_{e1} = \frac{\int \Phi^* \mathscr{H}_{e1} \Phi \, d\tau}{\int \Phi^* \Phi \, d\tau}$$

The Rayleigh–Ritz variational method is used to determine the coefficients, c_{ri}, corresponding to the best approximation to the minimum energy, E_{e1}, for the system. It is convenient to make linear combinations of the atomic orbitals which form symmetry orbitals of the molecule in question. The allowed combinations of atomic eigenfunctions to form molecular eigenfunctions of the various symmetry types are determined (for a molecule of any given symmetry) directly from the character table of the irreducible representations of this group. (The various symmetry operations, the classification of molecular symmetry types according to the behaviour of molecular electronic eigenfunctions with respect to the symmetry operations and character tables for all the groups of diatomic and polyatomic molecules are listed in Herzberg, 1950, 1966).

Since one of the most useful correlations is with the molecular orbitals for the intermediate species involved in collisions or in molecular decompositions, it seems appropriate now to indicate the various calculational methods in most common usage.

(1) Non-empirical

These methods under the category of non-empirical fall into two subclasses. The first is the well-known Hartree–Fock–Roothaan (1951, 1960) LCAO–MO–SCF (self-consistent field) method. The second is an even more rigorous technique which includes either the effect of configuration interaction on the eigenfunction obtained in a non-empirical calculation or the multiconfiguration SCF method. Configuration interaction or the multiconfiguration SCF method correct for the deficiency in the Hartree–Fock method which is necessary for calculating correlation energy corrections to the molecular energies.

A single determinantal wave function does not represent an exact solution of Schroedinger's equation; by determining the spin-orbitals according to the Hartree–Fock equations, one gets the best approximate solution in the form of a single determinant. To obtain a better solution, one must set up many determinantal functions, formed from different spin-orbitals, and must use an approximate wave function which is a linear combination of these determinantal functions, with coefficients to be determined by minimising the energy. This process is called configuration interaction (CI) or superposition of configurations.

The determinantal functions must be linearly independent, and be eigenfunctions of the spin operators S^2 and S_z, and preferably belong to a specified row of a specified irreducible representation of the symmetry group of the molecule (Gershgorn and Shavitt, 1967; Pipano and Shavitt, 1968). Definite spin states can be obtained by applying a spin-projection operator to the spin-orbital product defining a configuration (Harris, 1967). Suppose Φ_0 to be the solution of the Hartree–Fock equation. From functions of the same symmetry as Φ_0 one can build a wave function Φ,

$$\Phi = a_0\Phi_0 + a_1\Phi_1 + a_2\Phi_2 + \dots$$

By optimising the orbitals in each function and by variationally selecting the CI coefficients $a_0, a_1, a_2, \dots$, one gets an eigenfunction as good as or better than Φ_0. If the series is sufficiently long one can reach an exact solution. The trouble is that the necessary series is too long and usually converges slowly. In practice a truncated series is used. [A recent useful book by Schaefer (1972) includes examples of configuration interaction calculations.]

If when Φ_0 is constructed, Φ_1, Φ_2, etc., are constructed at the same time from a common orthonormal set of orbitals and one solves, not for the best possible Φ_0, but for the best Φ, then the variational principle, used simultaneously on both the a's (the CI coefficients) and the c's (the atomic orbital coefficients), will insure that the Φ_i will overlap as much as possible.

Complete multiconfiguration–self-consistent field (CMC–SCF) technique designates the method where a given occupied molecular orbital of the set is excited to all unoccupied molecular orbitals. If an occupied orbital is excited to one or more, but not all, of the unoccupied orbitals, the technique is described as incomplete MC–SCF (IMC–SCF). The reader is referred to Clementi (1968) and Hinze and Roothaan (1967) for details of the derivation. The CMC–SCF formalism differs from most many-body techniques presented to date insofar as the Hartree–Fock energy is not assumed to be the zero-order energy.

(2) Semi-rigorous

Semi-rigorous LCAO–MO–SCF methods start with the complete many-electron Hamiltonian and make certain approximations for the integrals and for the form of the matrices to be solved. Several years ago the author derived such a method starting with the correct many-electron Hamiltonian (in which interelectronic interactions are included explicitly) and the LCAO–MO–SCF equations of Roothaan and then making a consistent series of systematic approximations for the integrals involved (Kaufman, 1965). The molecular orbitals ϕ_i were still represented by a linear combination of atomic orbitals χ_r

$$\phi_i = \sum_r \chi_r c_{ri}$$

and the notation for the integrals is defined as

$$(a'c''' \mid G \mid b''d^{\mathrm{IV}}) = \iint a'^*(1)c'''^*(2)(r_{12})^{-1}b''(1)d^{\mathrm{IV}}(2)\,\mathrm{d}v_1\,\mathrm{d}v_2$$

There were several different levels of approximation which evolved, depending on how restrictive one made the conditions for neglecting $a'(1)a''(1)$ where a' and a'' are two orbitals on atom A; and $a'(1)b''(1)$ where a is an orbital on atom A and b is an orbital on atom B, $A \neq B$ and for neglecting

$(a'c''' \mid G \mid b''d^{IV})$ where a, b, c, d are orbitals on atoms A, B, C and D respectively. The following set of successively less restrictive approximations was outlined:

Method A $a'^*(2)a''(2) = 0 \; a' \neq a''$ for any atom A

Method B $a'^*(2)b''(2) = 0$ B $\neq$ A and $(a'c''' \mid G \mid b''d^{IV}) = 0$ unless B $=$ A and D $=$ C but in general $b'^*(2)b''(2)$ need not be zero
all one-centre integrals and two-centre Coulomb integrals allowed

Method C $(a'c''' \mid G \mid b''d^{IV} \neq 0$ also in the case A $=$ D and B $=$ C. In addition all two-centre exchange integrals allowed

Method D $(a'c''' \mid G \mid b''d^{IV}) \neq 0$ also in the case if A $=$ B $=$ C. In addition all two-centre hybrid integrals allowed.

In the article procedures were presented for estimating the various integrals involved: $H_{a'a'}^{core}$, $H_{a'a''}^{core}$ and $H_{a'b''}^{core}$. Preliminary calculations were made on a test system and the results compared both with experiment and with those of rigorous non-empirical computations. These results indicated that Method A seemed not to be a physically justifiable method and that one should go at least to Method B. The two most restrictive approximations A and B of this semi-rigorous scheme are similar conceptually to the CNDO method published by Pople at the same time and his later revision, the NDDO and INDO methods published subsequently (Pople and Beveridge, 1970). Method A resembles Pople's CNDO (complete neglect of differential overlap) method. Method B resembles Pople's NDDO (neglect of diatomic differential overlap) or INDO (intermediate neglect of diatomic differential overlap) methods. [There are several additional useful review articles on all-valence electron semi-rigorous self-consistent field calculations (Jaffe, 1969; Jug, 1969; Nicholson, 1970).]

(3) Semi-empirical

The semi-empirical extended Hückel method, which takes into account all valence electrons in a molecule, was introduced by Wolfsberg and Helmholz (1952), has been used over the years extensively by Lipscomb and co-workers, especially Hoffmann (Eberhardt, Crawford and Lipscomb, 1954; Hoffman and Lipscomb, 1962, 1962a; Hoffman, 1963, 1964, 1966), and has been applied by a large number of investigators.

From a molecular orbital ϕ_i built up as a linear combination of atomic orbitals χ_r

$$\phi_i = \sum_r \chi_r c_{ri}$$

and by application of the variation principle for the variation of energy, the following set of equations for the expansion coefficients is obtained:

$$(\alpha_r - ES_{rr})c_r + \sum_{r \neq s} (\beta_{rs} - ES_{rs})c_s = 0$$

$$S_{rs} = \int \chi_r^* \chi_s \, dv = \text{overlap integral}$$

$$H_{rr} = \alpha_r = \int \chi_r^* \mathcal{H} \chi_r \, dv = \text{Coulomb integral}$$

$$H_{rs} = \beta_{rs} = \int \chi_r^* \mathcal{H} \chi_s \, dv = \text{resonance integral} \; (r \neq s)$$

$\mathscr{H}$ is an effective one-electron Hamiltonian representing the kinetic energy, the field of the nuclei and the smoothed-out distribution of the other electrons. Electron–electron repulsion is neglected in this method.

The diagonal elements are set equal to the effective valence state ionisation potentials of the orbitals in question. The off-diagonal elements, H_{rs}, can be evaluated in several ways. The two expressions in most common usage are the original Wolfsberg–Helmholz expression

$$H_{rs} = 0.5\, k\, (H_{rr} + H_{ss}) S_{rs}$$

with $k = 1.75 - 2.00$ (Hoffman set $k = 1.75$) and the more recent Cusachs (1965, 1966) expression

$$H_{rs} = \frac{(H_{rr} + H_{ss})}{2} S_{rs}(2 - |S_{rs}|)$$

The total electronic energy of a particular system is taken to be the sum of the orbital energies, ε_i, times their occupation numbers. For a closed shell system $E_{elect} = 2\Sigma\varepsilon_i$. The total molecular energy can be written as

$$E = 2\Sigma\varepsilon_i + \sum_{n,\,n'} E_{nn'} - \sum_{e,\,e'} E_{ee'}$$

where $E_{nn'}$ and $E_{ee'}$ are nuclear–nuclear and electron–electron repulsion energies. The success of these extended Hückel calculations in predicting preferred geometrical conformations from calculated minimum energies lies in the fact that the method of selecting the H_{rs} values must simulate, within the calculated electronic energies, the contribution of nuclear repulsions to the total energy (Hoffman, 1963). The nuclear–nuclear and electron–electron repulsion energies cancel approximately (Slater, 1963) and thus the simple sum of one-electron energies behaves similarly to the true molecular energy.

Due to neglect of electron–electron repulsion, the calculated energy values are, unfortunately, equal for the identical molecular states with different multiplicities.

2. ELECTRONIC CONFIGURATION

The electrons are fed into the molecular orbitals according to the aufbau principle. First the allowed orbitals are determined and then the electrons are fed one at a time into these levels, beginning with the lowest, and satisfying the Pauli exclusion principle by allowing only two electrons to each of the orbitals. For the ground state of the molecule all of the lowest molecular orbitals are filled and in the case of non-totally filled degenerate orbitals, the electrons are fed in according to Hund's rule where the state of lowest energy is almost invariably the state of highest multiplicity. For excited states the electrons are fed into all of the low-lying molecular orbitals with now one or more of the electrons from higher-lying filled orbitals being promoted into upper levels.

To derive the term type (species) from the electronic configuration, Russell–Saunders coupling is assumed as defined earlier

$$\vec{\Lambda} = \Sigma \vec{\lambda}_i$$

$$\vec{S} = \Sigma \vec{S}_i$$

where, for non-equivalent electrons, the quantities $\vec{\lambda}_i$ and $\vec{S}_i$ are added vectorially in all possible combinations. If the electrons are equivalent, the Pauli principle must be taken into account when adding the $\vec{\lambda}_i$ and the $\vec{S}_i$; the electrons must differ in m_l or m_s. The energy difference between corresponding states of different multiplicity is due to the electrostatic interaction of the electrons. The state with the greatest multiplicity almost invariably lies lower in energy.

If equivalent as well as non-equivalent electrons are present the resulting states are found by first forming the resulting states of each group of equivalent electrons and then forming the direct product of the species so obtained. Since closed shells always give a single totally symmetric singlet state they can be entirely neglected in the determination of the resulting states.

Herzberg (1950, 1966) gives tables of the terms arising from non-equivalent electrons, equivalent electrons, and equivalent as well as non-equivalent electrons.

In addition, for diatomics consisting of like atoms (or certain symmetrical linear polyatomic molecules) the resulting states are also either even (g) or odd (u). They are even if the number of 'odd' electrons (σu, πu, ...) is even, whereas they are odd if the number of 'odd' electrons is odd.

These terms of electron configurations arising from different partially filled orbitals are very valuable in deciphering what excited electronic states of molecules and intermediates are possible and lie in an energy range accessible to the particular experiment.

The ordering of the molecular orbitals is dependent on the relative positions of the nuclei. In analysing a collisional problem care must be taken to feed the electrons into the appropriate molecular orbitals correlating the reactants or products with the proper intermediate state as permitted by the symmetry and spin restrictions outlined above.

3. CROSSING OR PSEUDO-CROSSING OF MOLECULAR POTENTIAL ENERGY SURFACES

The course of behaviour followed in a collisional process is greatly influenced by the crossings or pseudo-crossings of relevant molecular potential energy surfaces. For diatomic molecules these surfaces are simply curves, crossing at a point. But for a general molecule with N atoms, the surfaces are in a configuration space of dimension $3N-6$ (for a linear molecule $3N-5$) (Coulson and Zalewski, 1962). For the convenience of visualising these energy surfaces are often referred to as curves, and that practice will be followed in the present section.

The well known Landau–Zener formula (Landau, 1932, 1932a; Zener, 1932; Stueckelberg, 1932) relating to the probability of an electronic jump near the crossing point of two potential energy curves or surfaces has been seriously critiqued (Coulson and Zalewski, 1962; Eyring, Walter and Kimball, 1944). New treatments of greater validity have been formulated (Bates, 1960, 1964).

A fairly recent paper which discusses avoided crossings in bound potential-energy curves of diatomic molecules (Lewis and Hougen, 1968) presents a mathematically well defined procedure for going from potential curves exhibiting an avoided crossing (such as would be obtained from solving exactly the eigenvalue problem associated with the complete electronic Hamiltonian for fixed nuclei) to two crossing potential curves and an interaction function.

Theoretical justification for the non-crossing rule for diatomic molecules is a special case of a more general treatment which was presented by von Neumann and Wigner (1929). The rule states that vibrational potential energy curves for electronic states of like symmetry do not cross and, as a result, situations of avoided crossings occur where two potential energy curves approach each other closely and then move apart. In Figure 1a is presented an avoided crossing and in Figure 1b two normal potential curves constructed from two potential curves exhibiting an avoided crossing. A time-dependent treatment is particularly appropriate when considering scattering or predissociation problems, but less appropriate when considering the discrete rotational and vibrational levels of two bound electronic states exhibiting an avoided crossing. The derivation of a time-independent vibrational Hamiltonian for a molecule with a pair of potential energy curves exhibiting an avoided crossing is presented in Lewis and Hougen (1968). The functions $V_1(R)$, $V_2(R)$, $\Psi_1(R)$ and $\Psi_2(R)$ shown in Figure 1a are those obtained from an exact calculation of the electronic energies and wave functions for fixed nuclei, using the complete electronic Hamiltonian plus the internuclear repulsion (total Hamiltonian with the nuclear kinetic energy removed), and thus have an exact physical meaning. It is sometimes convenient to use an approximate Hamiltonian which leads to potential curves like those shown by the solid lines in Figure 1b for which the crossing is not avoided. At some stage the approximate Hamiltonian must be supplemented by a perturbation operator which reintroduces the required crossing. [Often, unavoided crossings and their associated wave functions arise naturally in zeroth order molecular orbital calculations (Davidson, 1961).]

A series of papers by Lichten (1963, 1965, 1967, 1968; Fano and Lichten, 1965) on the general topic of molecular wave functions and inelastic atomic collisions introduced a new descriptive terminology which has since been used extensively. In Figure 1b if the atoms approach each other slowly in state ϕ_1 an adiabatic transition from ϕ_1 to ϕ_2 will occur. Lichten coined the expression 'diabatic' to describe the behaviour if the two atoms approached each other rapidly and a transition from ϕ_1 to ϕ_1 occurred. This type of a transition is called a 'diabatic' transition. Recent results in low-energy ion–atom and ion–molecule collisions indicated that these processes can be highly effective in producing optical excitation in contradiction to the well-known 'adiabatic criterion' (Dworetsky et al., 1967). The concept of diabatic transitions has proved particularly useful in delineating the processes in resonant

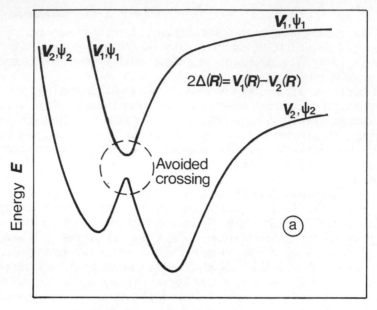

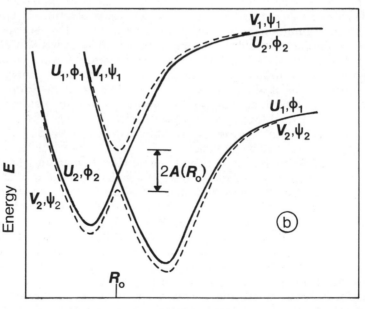

Figure 1(a) Potential energy curves (E v. internuclear distance). An avoided crossing. The curves are labelled by the potential functions they represent and by the corresponding electronic wave functions. (b) An avoided crossing—two normal potential curves constructed from two potential curves exhibiting an avoided crossing

and quasi-resonant charge exchange processes (Fano and Lichten, 1965; Dworetsky et al., 1967; Lichten, 1968). Lichten (1968) discusses some of the effects which cause transitions among diabatic MO's at crossings: electron penetration, electron correlation and electronic interaction with nuclear motion.

The role of potential curve crossing in sub-excitation molecular collisions has been investigated numerically by exact (two-state) computations v. decoupling approximations (Levine, Johnson and Bernstein, 1968). In this work, attention was restricted to elastic scattering, in low energy (sub-excitation) collisions, when the (approximate) potential energy curves cross. The results showed that even in the sub-excitation region a considerable range of behaviour is to be expected, from a fully adiabatic description (non-crossing) to the pure distortion one. (In that work are a number of references to related theoretical investigations by those authors.)

The role of curve crossing in atomic and molecular collisions is a subject of considerable interest at the present time. Fairly recently there was an entire international conference devoted to Transitions Non-Radiative Dans Les Molecules' (1969). A fine review article on radiationless molecular electronic transitions has also appeared rather recently (Kasha and Henry, 1968).

4. TWO EXAMPLES OF THE APPLICATION OF THESE CONCEPTS

A. THEORETICAL EXPLANATION FOR THE APPARENTLY ANOMALOUS BEHAVIOUR OF THE $O^+ + N_2 \rightarrow NO^+ + N$ REACTION

The utility of the concepts described earlier in this chapter as an aid in deciphering and predicting experimental results in molecular collisional processes may be illustrated by the following theoretical justification derived for the experimentally observed apparently anomalous low-energy behaviour of the $O^+ + N_2$ ion–molecule reaction.

It is generally observed that cross-sections for exothermic ion–molecule reactions decrease with increasing ion kinetic energies indicating a negligible activation energy. An important exception to this behaviour is exhibited by the exothermic reaction $O^+ + N_2 \rightarrow NO^+ + N$, the cross-section of which rises with energy, goes through a broad maximum at 10 eV and then decreases (Giese, 1966). This is contrary to the expected behaviour of an exothermic ion–molecule reaction since on the basis of the simple considerations of the Gioumousis and Stevenson (1958) picture the cross-section is expected to decrease with increasing energy. A recent review had mentioned that there was no satisfactory explanation for this apparent anomalous behaviour. More recently Schmeltkopf, Ferguson and Fehsenfeld (1968) studied this reaction in the thermal region and found it proceeded with no activation energy and furthermore the rate of reaction increased rapidly with vibrational excitation of the N_2. It was very desirable to have detailed knowledge of the potential energy surfaces of this system since it would permit a more definitive interpretation of the experimental results. When Professor Koski and the author initiated our theoretical investigation, no accurate quantum chemical calculations had yet been made on these surfaces. However, application merely of the concepts described earlier in this chapter combined with the available theoretical, optical spectroscopic and mass spectroscopic data which existed at that time for this system permitted Professor Koski and the author to outline schematically some of the pertinent potential energy curves (Kaufman and Koshi, 1969). From these curves it was possible to explain qualitatively the apparently anomalous features of this reaction. Our recent extensive configuration interaction calculations have validated the qualitative theoretical explanation based on the symmetry and spin analyses.

The intermediate of the $O^+ + N_2$ reaction, N_2O^+, has 15 valence electrons and, thus, is assumed to have a linear configuration (N-N-O^+) in accordance with Walsh's (1963) rules for a 15 valence electron triatomic. Experimentally the ground state of N_2O^+ is known to be $^2\Pi_i$ and its first excited state, lying at 3.49 eV, to be $^2\Sigma^+$ (Callomon, 1959). A second excited state with no symmetry assigned had been reported at 7.19 eV (Herzberg, 1966). McLean and Yoshimine (1967) at IBM had done accurate LCAO–MO–SCF calculations on the parent system N_2O itself. The bond lengths of ground state N_2O^+ and N_2O are quite similar (Callomon, 1959). Since N_2O is a closed shell system Koopmans' (1933) theorem is considered valid for the ionisation potential and thus also for the total energy of the remaining N_2O^+ system. The ordering of the topmost filled levels for N_2O^+ ($\tilde{X}\ ^2\Pi_i$) is $(1\pi)^4(7\sigma)^2(2\pi)^3$ (McLean and Yoshimine, 1967); and the first excited state $^2\Sigma^+$ must have the configuration $(1\pi)^4(7\sigma)^1(2\pi)^4$; and the second excited state must be a $^2\Pi_i$ state arising from $(1\pi)^3(7\sigma)^2(2\pi)^4$. From recent photoelectron spectroscopic measurements on N_2O (Brundle and Turner, 1969; Natalis and Collin, 1969) it is substantiated that the ground and first excited states of N_2O^+ are $\tilde{X}\ ^2\Pi_i$ and $^2\Sigma^+$ (3.49 eV above ground state N_2O^+) and that the second excited state, $^2\Pi_i$, lies only about 1.3 eV above the $^2\Sigma^+$ state. This $^2\Pi_i$ state is attractive since transitions to at least fourteen different vibrational levels were reported. The third excited state, $^2\Sigma^+$, corresponds to the earlier state reported at ~ 7.19 eV above the ground state of N_2O^+ and is also attractive since transitions are reported to five different vibrational levels. From these levels for N_2O^+, the experimental data for the reactants $O^+ + N_2$ (or $O + N_2^+$ which must also be taken into account when drawing a full potential curve) (Table 1), the products $NO^+ + N$ (or $NO + N^+$) (Table 2), the experimental dissociation energy of N_2O^+ to $NO^+ + N$ (Gilmore, 1967) and *most importantly* from the rules for the possible symmetry- and spin-permitted combinations of these reactants and products into the intermediate states of N_2O^+, the following schematic potential diagram was drawn for the system (Figure 2). The energies indicated are those relative to the separated neutral atoms. [The N–N distance in N_2O^+ ($\tilde{X}\ ^2\Pi_i$ 1.115 Å; $A\ ^2\Sigma^+$ 1.140 Å) is not very different from the N–N distance in N_2 (1.094 Å) or N_2^+ (1.1162 Å) (Herzberg, 1950) and the N–O distance in N_2O^+ ($\tilde{X}\ ^2\Pi_i$ 1.185 Å; $A\ ^2\Sigma^+$ 1.141 Å) is similar to those in NO (1.1508 Å) (Herzberg, 1966) or NO^+ (~ 1.08 Å) (extrapolated graphically from the data in Gilmore, 1967). Figure 2 depicts two potential energy profiles through a section of a cube (Laidler, 1955): the one for the reactants corresponding to the approximate N–N distance in N_2 and the one for the products corresponding to the approximate N–O distance in NO. Since there is very little change in the bond distances involved, these two profiles present an adequate representation for a qualitative understanding of the potential energy surfaces involved.]

The stippled curves designate the reactants $O^+ + N_2$ or $O + N_2^+$; in these cases, the abscissa represents the N–O co-ordinate. The unstippled curves

Table 1. NNO^+ reactant energy levels and symmetry- and spin-permitted intermediates

Reactants			Energy scale (relative to separated neutral atoms)
$N_2 \quad + O^+$ $\tilde{X}^1\Sigma_g^+ \quad {}^4S_u$	$\longrightarrow$	N–N–O$^+$ ${}^4\Sigma^-$	3.859 eV
$N_2^+ \quad + O$ $\tilde{X}^2\Sigma_g^+ \quad {}^3P_g$	$\longrightarrow$	N–N–O$^+$ ${}^{2,4}\Sigma^-, \, {}^{2,4}\Pi$	5.821 eV
$N_2^+ \quad + O$ $A^2\Pi_u \quad {}^3P_g$	$\longrightarrow$	N–N–O$^+$ ${}^{2,4}\Sigma^+, \, {}^{2,4}\Sigma^-, \, {}^{2,4}\Pi, \, {}^{2,4}\Delta$	6.941 eV
$N_2 \quad + O^+$ $\tilde{X}^1\Sigma_g^+ \quad {}^2D_u$	$\longrightarrow$	N–N–O$^+$ ${}^2\Sigma^-, \, {}^2\Pi, \, {}^2\Delta$	7.184 eV

Table 2. NNO^+ product energy levels and symmetry- and spin-permitted intermediates

Products			Energy scale (relative to separated neutral atoms)
$NO^+ + N$ $\tilde{X}^1\Sigma^+ \quad {}^4S_u$	$\longrightarrow$	N–N–O$^+$ ${}^4\Sigma^-$	2.760 eV
$NO^+ + N$ $\tilde{X}^1\Sigma^+ \quad {}^2D_u$	$\longrightarrow$	N–N–O$^+$ ${}^2\Sigma^-, \, {}^2\Pi, \, {}^2\Delta$	5.144 eV
$NO^+ + N$ $\tilde{X}^1\Sigma^+ \quad {}^2P_u$	$\longrightarrow$	N–N–O$^+$ ${}^2\Sigma^+, \, {}^2\Pi$	6.336 eV
$NO \quad + N$ $\tilde{X}^2\Pi_i \quad {}^3P_g$	$\longrightarrow$	N–N–O$^+$ ${}^{2,4}\Sigma^+, \, {}^{2,4}\Sigma^-, \, {}^{2,4}\Pi, \, {}^{2,4}\Delta$	8.025 eV

designate the products $NO^+ + N$ or $NO + N^+$; here the abscissa represents the N–N distance. It becomes very obvious from examination of this potential diagram (and Table 2) that ground state $O^+ + N_2$ can combine uniquely only along the repulsive ${}^4\Sigma^-$ N_2O^+ curve (represented by a _____ _____ curve)* which is then cut by the attractive ground state $\tilde{X}\,^2\Pi_i$ N_2O^+ curve [the solid curve arising from $N_2^+({}^2\Sigma_g^+) + O({}^3P_g)$]. The $\tilde{X}\,^2\Pi_i$

*The curves leading to each individual intermediate state either from reactants or products have different graphic representations.

N_2O^+ curve indicates that predissociation from the $\tilde{X}\,^2\Pi_i\,N_2O^+$ to ground state $NO^+ + N$ can take place since the repulsive $^4\Sigma^-$ lower curve (also a _____ curve), which separates to ground state $NO^+(^1\Sigma^+)$ and $N(^4S_u)$, cuts the attractive $\tilde{X}\,^2\Pi_i$ ground state solid curve which would separate to $NO^+(^1\Sigma^+)$ and $N(^2D_u)$. This behaviour is substantiated by the experimental observation that (although spin-forbidden) predissociation to ground state products $NO^+ + N$ is noted to occur in the mass spectroscopic decomposition of ground state N_2O^+ (Herzberg, 1966).

Three experimental facts were in need of explanation. (1) The absence of

N–N–O$^+$ POTENTIAL ENERGY DIAGRAM

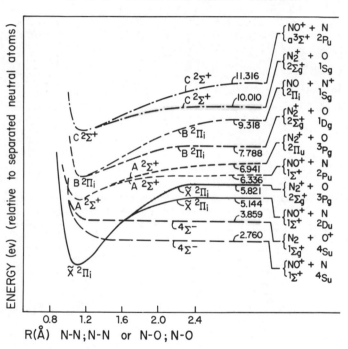

Figure 2   N–N–O$^+$ *potential energy diagram (schematic)*

an activation energy in the reaction $O^+ + N_2 \rightarrow NO^+ + N$ when studied at thermal energies and its very low reaction efficiency under these conditions, (2) the increase in rate constant with increasing vibrational energy of N_2, and (3) the higher energy data which indicate an activation energy for the reaction.

Referring to Figure 2, since the ground state of O^+ and N_2 can combine uniquely only along the indicated $^4\Sigma^-$ curve, the lack of an activation energy for the production of NO^+ at thermal energies implies that this $^4\Sigma^-$ curve, even though repulsive, must come in with an almost flat slope until the O–N distance is in the region of the NNO^+ equilibrium distance. The very low reaction efficiency under these conditions when N_2 is in its ground vibrational state indicates that the reactants in the $^4\Sigma^-$ state must make a forbidden curve crossing into another state (in this case into the $\tilde{X}\ ^2\Pi_i$ state of NNO^+). This is in harmony with experiment since the appearance potential for the process $N_2O \rightarrow N_2(\tilde{X}\ ^1\Sigma_g^+)+O^+(^4S_u)$ is reported as 15.3 eV (Tanaka, Jersa and Le Blanc, 1960; Curran and Fox, 1961) (which coincides with the energy level of $N_2 + O^+$ on the same relative energy scale). The appearance potential for the process $N_2O \rightarrow NO^+(\tilde{X}^1\Sigma^+)+N(^4S_u)$ is 15.01 eV (Dibeler, Walker and Liston, 1967). This scheme permits the curve of the ground state reactants to cross into the attractive ground state of N_2O^+ with no activation energy and then to cross over into the $^4\Sigma^-$ state of the ground state products. The spin forbiddenness of the cross overs then accounts for the observed low probability of reaction as suggested by Schmeltkopf, Ferguson and Fehsenfeld (1968).

The increase in the rate constant at low energies with increasing vibrational energy of the N_2 indicates that there is probably a barrier in the exit channel (Polanyi, 1971). Since the ground state products $NO^+ + N$ also must arise uniquely from a $^4\Sigma^-$ state of NNO^+, the schematic curve Figure 2 would have indicated that the thermal energy reaction when N_2 is vibrationally excited either went through the $\tilde{X}\ ^2\Pi_i\ NNO^+$ state with another curve crossing back into the dissociative $^4\Sigma^-$ state or possibly through the $^4\Sigma^-$ state for the entire reaction. The accessibility of the portion of the $^4\Sigma^-$ state leading to the products would be enhanced with increasing vibrational excitation of the N_2. The appearance potential for $N_2O \rightarrow NO^+ + N$ of 15.01 eV indicates the presence of some type of barrier in this exit channel since the energy of the separated $NO^+ + N$ is lower than 15.01 eV on the same relative energy scale. The detailed quantum chemical calculations for $O^+ + N_2 \rightarrow NO^+ + N$ (Pipano and Kaufman, 1971, 1972), which will be described below, indicated an unsuspected channel, a low lying $^4\Pi$ state which cuts through both the $\tilde{X}\ ^2\Pi_i$ and the $^4\Sigma^-$ states in the exit channel. To exit from the $^2\Pi_i$ state via the $^4\Pi$ state to the $^4\Sigma^-$ state would also be facilitated by an increase in the vibrational energy of the N_2.

The activation energy indicated by the data for higher kinetic energies of the O^+ is indicative of the fact that higher states of NNO^+ become accessible and thus more channels open up through which the reaction can proceed.

In addition to the curves shown on Figure 2 there are 27 more possible intermediate $N-N-O^+$ states resulting from reactants with energies from 3.859 to 10.01 eV (relative to separated neutral atoms) and 39 more possible intermediate states resulting from the products with energies from 2.760 eV to 11.316 eV (relative to separated neutral atoms). A great many of these will undoubtedly be repulsive states. It should be noted that the potential energy curves are compatible with the appearance potential of 16.4 eV (Weissler et al., 1959) for the process $NO^+(\tilde{X}\ ^1\Sigma^+)+N(^2D_u)$, of 17.4 eV (Weissler et al., 1959) for the process $N_2^+(\tilde{X}\ ^2\Sigma_g^+)+O(^3P_g)$ and 20 eV (Curran and Fox, 1961; Weissler et al., 1959) for the process $NO(\tilde{X}\ ^2\Pi)+N^+(^3P_g)$.

A more comprehensive interpretation of this reaction depended on the availability of more detailed and more accurate potential curves. Dr. Pipano and the author have performed *ab initio* large scale configuration interaction calculations of some of the pertinent states for the above reaction (Pipano and Kaufman, 1971, 1972). Within the limit of the computer time so far available we have used a minimal Slater orbital basis set and computed only the same two cuts along the reaction surface as our schematic curves (Figure 2). A configuration interaction calculation (including all single and double excitations) was carried out at each point and for each state. For each state the following number of configurations were included:

$$
\begin{array}{ll}
{}^2\Pi_i & 123 \\
{}^2\Sigma^+ & 177 \\
{}^4\Pi & 1379 \\
{}^4\Sigma^- & 1310
\end{array}
$$

The orbital composition of each configuration of each state was identified with a serial number. The configurations which have the largest weight in the resulting wave functions are listed below according to their serial numbers:

$$
\begin{array}{lll}
{}^1\Sigma^+ & (1) & (1\sigma)^2(2\sigma)^2(3\sigma)^2(4\sigma)^2(5\sigma)^2(6\sigma)^2(7\sigma)^2(1\pi)^4(2\pi)^4 \\
{}^2\Pi & (1) & (1\sigma)^2(2\sigma)^2(3\sigma)^2(4\sigma)^2(5\sigma)^2(6\sigma)^2(7\sigma)^2(1\pi)^4(2\pi)^3 \\
& (7) & (1\sigma)^2(2\sigma)^2(3\sigma)^2(4\sigma)^2(5\sigma)^2(6\sigma)^2(7\sigma)^1(8\sigma)^1(1\pi)^4(2\pi)^3 \\
{}^2\Sigma^+ & (1) & (1\sigma)^2(2\sigma)^2(3\sigma)^2(4\sigma)^2(5\sigma)^2(6\sigma)^2(7\sigma)^1(1\pi)^4(2\pi)^4 \\
& (8) & (1\sigma)^2(2\sigma)^2(3\sigma)^2(4\sigma)^2(5\sigma)^2(6\sigma)^2(7\sigma)^2(1\pi)^4(2\pi)^2(8\sigma)^1 \\
{}^4\Pi & (1) & (1\sigma)^2(2\sigma)^2(3\sigma)^2(4\sigma)^2(5\sigma)^2(6\sigma)^2(7\sigma)^2(1\pi)^4(2\pi)^2(3\pi)^1 \\
& (2) & (1\sigma)^2(2\sigma)^2(3\sigma)^2(4\sigma)^2(5\sigma)^2(6\sigma)^2(7\sigma)^1(1\pi)^4(2\pi)^3(8\sigma)^1 \\
& (4) & (1\sigma)^2(2\sigma)^2(3\sigma)^2(4\sigma)^2(5\sigma)^2(6\sigma)^2(7\sigma)^2(1\pi)^3(2\pi)^3(3\pi)^1 \\
{}^4\Sigma^- & (1) & (1\sigma)^2(2\sigma)^2(3\sigma)^2(4\sigma)^2(5\sigma)^2(6\sigma)^2(7\sigma)^2(1\pi)^4(2\pi)^2(8\sigma)^1
\end{array}
$$

The calculated curves are presented in Figures 3 and 4. An even greater wealth of structure than would have been anticipated prior to such a detailed calculation is evident from these figures. The energy distances between curves near the separation limit are in reasonable agreement with experiment. The two apparent discrepancies in case of the two $^2\Pi$ states of reactants $[^2\Pi(1)$ and $^2\Pi(7)]$ and the $^2\Sigma^+(8)$ and $^2\Pi(1)$ states of products can be accounted for by looking at the corresponding curves in Figures 3 and 4. In the case of reactants, the energy of the $^2\Pi(7)$ state decreases and the energy of the $^2\Pi(1)$ state increases while increasing the N–O distance, so that it is quite possible that the corresponding curves cross at larger N–O separations. This explains the experimental observation that metastable $O^+(^2D_u)$ ions react with N_2 in the low energy range to form principally $N_2^+(^2\Pi_u)$ while the ion–molecule reaction to form NO^+ has a very small probability (Rutherford and Vroom, 1971). The $^2\Pi(1)$ curve $[O^+(^2D_u)+N_2(^1\Sigma_g^+)]$ is near resonant with the $^2\Pi(7)$ curve $[O(^3P_g)+N_2^+(^2\Pi_u)]$ for N_2^+ in the $A\ ^2\Pi_u$, $v=1$ state. Thus charge exchange takes place at very large distances and the $O(^3P_g)+N_2^+(^2\Pi_u)$ are formed where they then dissociate along a slightly repulsive curve.

The same argument holds for the $^2\Sigma^+(8)$ and the $^2\Pi(1)$ states, which separate to the same products. As no attempt was made simultaneously to optimise the N–O distance, while changing the N–N distance or vice versa, the resulting lower ground state energy of reactants, compared to the ground

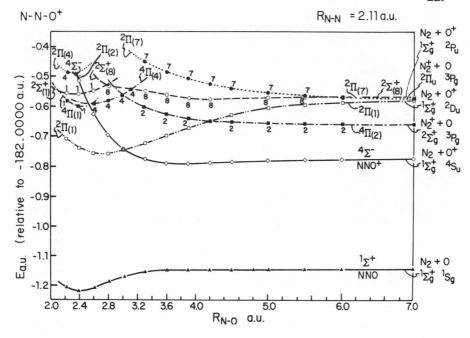

Figure 3   CI energies of NNO$^+$ *v.* R$_{N-O}$

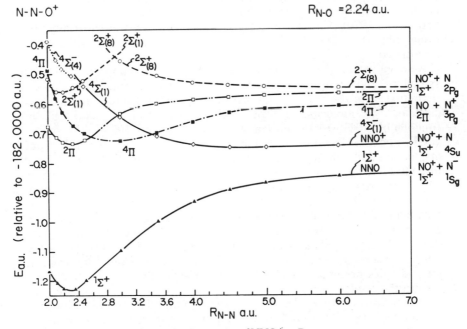

Figure 4   CI energies of NNO$^+$ *v.* R$_{N-N}$

state energy of products, near the dissociation limits has no significance. Moreover, due to a steeper energy drop along the $^4\Sigma^-$ curve, when the N–O distance is increased (compared to the effect of increasing the N–N distance) it is anticipated that such a simultaneous optimisation would yield the correct results. Although the energy separations between potential curves are in reasonable agreement with experimental data, the results inevitably reflect the deficiencies of a minimal basis set. However, the results are of more than sufficient accuracy to validate our original hypothesis that the ground state reactants initially proceed along the one potential surface available to them, the $^4\Sigma^-$ surface, until they either curve cross into another surface or proceed far enough along that surface to enable reaction to take place. Likewise, the ground state products can only arise from the $^4\Sigma^-$ surface.

A situation similar to that of $O^+ + N_2$ appears to exist in the reaction $C^+(^2P_u) + O_2(\tilde{X}\ ^3\Sigma_g^-) \rightarrow CO^+(\tilde{X}\ ^2\Sigma^+) + O(^3P_g)$ (Lao, Rozett and Koski, 1968) and it also can be explained in the same manner.

B. THEORETICAL EXPLANATION FOR THE NON-APPEARANCE OF THE PARENT CF_4^+ ION UPON ELECTRON OR PHOTON IMPACT ON CF_4

The CF_4 molecule gives no parent ion under electron or photon impact (Walter et al., 1969). The onset of ionisation of CF_3^+ from CF_4 has a long tail which the authors in Walter et al. (1969) ascribe to the fact that the transition goes to a steeply rising part of a repulsive CF_4^+ state. However, in the photo-electron spectrum there are a number of peaks observed (Brundle, Robin and Basch, 1970). The two peaks at lowest energy do not show any vibrational fine structure. This implies that these peaks are probably pre-dissociated.

As a guide for the interpretation of this phenomenon we derived the symmetry and spin restrictions governing what intermediate states are permitted from the various states of the separated product fragments and calculated the molecular energies and symmetries of CF_4 and CF_4^+ as a function of r (CF_3–F′) (Kaufman, Kerman and Koski, 1970).

CF_4 and CF_4^+ become molecules of C_{3v} symmetry when one F atom is displaced axially along its bond direction. Only when all four F atoms are in equivalent positions do these molecules assume their full T_d symmetry. Thus they are considered as being formed from a YZ_3 fragment combined with an X atom. [Here the notation follows that of Herzberg (1966) with YZ_3 being CF_3 and X being the odd F atom which we shall denote F′.] To determine the symmetry states of the resulting molecular states from the states of the separated parts, one must resolve the various states of YZ_3 and X into the species of point groups T_d or C_{3v}. This case is more involved than the examples presented in Herzberg so we shall indicate the analysis and present the results.

CF_3 is known experimentally to have a pyramidal structure (Fessenden and Shuler, 1965). Its ground state is $^2A_1$ in C_{3v} symmetry. [Our calculations indicated that there are also fairly low lying excited CF_3 states of symmetries

$^2E_1$, $^2A_2$ and ^2E. In a ^2E state, since this is doubly degenerate, there could be a static Jahn–Teller distortion to a lower spatial symmetry (Jahn and Teller, 1932). In all subsequent discussions of CF_4 and CF_4^+ potential curves any further Jahn–Teller distortions (dynamic ones) due to the splitting of degenerate ^2E states will be neglected since these are expected to be very small compared to the physical phenomena we are discussing.] The photoionisation experiments of Chupka indicate that the geometry of the ground state of CF_3^+ is expected to be planar (Walter et al., 1969). The ground state of CF_3^+, if pyramidal and of C_{3v} symmetry, would be $^1A_1$; if planar and of D_{3h} symmetry, would be $^1A_2''$.

CF_4 is T_d in its symmetrical equilibrium conformation and has a $^1A_1$ ground state. Since the highest occupied molecular orbital in our calculations turns out to be a triply degenerate f_2 orbital,* this splits into an a_1 and a pair of e orbitals as the molecule is distorted into C_{3v} symmetry. The first excited state at the equilibrium conformation would be expected to be an F_2 state [with an electron configuration $(f_2)^5$ (a_1) — this would be subject to a Jahn–Teller distortion]. As the CF_3–F' distance is increased our results indicate that the first excited state would be an A_1 state.

CF_4^+ in its tetrahedral conformation would have a $^2F_2$ ground state. Because of the triple degeneracy of the f_2 orbitals this $^2F_2$ state would be subject to a Jahn–Teller distortion. However, Coulson has estimated that CF_4^+ would have a small lengthening of the C–F' bond and an almost negligible Jahn–Teller distortion (Coulson and Strauss, 1962). Coulson assumed in his arguments that the highest occupied orbital of CF_4 would be a triply degenerate f_1 orbital. Our molecular orbital results indicate that the highest occupied molecular orbitals in CF_4 are the f_2 orbitals with the f_1 orbitals lying somewhat lower in energy. However, this reversal would have virtually no effect on Coulson's arguments and thus it is still expected that CF_4^+ would have an almost negligible static Jahn–Teller distortion. Coulson further concluded that any pseudo Jahn–Teller effect arising from the closeness of the e, f_1 and f_2 orbitals would probably also be negligible. As the CF_3–F' distance is increased, the $^2F_2$ state splits into a $^2A_1$ and a ^2E state, and the $^2F_1$ state splits into a $^2A_2$ and a ^2E state.

The F atom and its positive and negative ions have the states shown in Table 3, which resolve into the species of point group T_d or C_{3v}.

In Table 4 are listed some of the lower lying states of CF_4 which are permitted from the spins and symmetries of the separated $CF_3 + F$ or $CF_3^+ + F^-$ fragments. Only magnitudes of some of the lower lying energies are indicated. (For the purposes of these correlations it makes no difference if we assume the CF_3^+ to be planar or pyramidal since one can always relax the pyramidal form to the planar and resolve the species of the one form into those of the other.)

The states of CF_4^+ which are permitted from the combination of $CF_3^+ + F$ and $CF_3 + F^+$ are given in Table 5.

*A recent calculation (Brundle, Robin and Basch, 1970) indicated that *ab initio* Gaussian type calculations gave f_1 as the highest occupied molecular orbital with f_2 lower (but by only 0.2 eV). For reasons which we will discuss later we still feel that f_2 may be the highest occupied molecular orbital.

Table 3. Resolution of the species of F, F$^+$ and F$^-$ into those of T$_d$ and C$_{3v}$

Atom	Electron configuration	Resolution into species of	
		T$_d$	C$_{3v}$
F(^{2}P$_u$)	s^{2}p^{2}p^{2}p	F$_2$	A$_1$ + E
(^{2}S$_g$)	sp^{2}p^{2}p^2	A$_1$	A$_1$
F$^+$(2) valent			
(^{3}P$_g$)	s^{2}p^{2}pp	F$_1$	A$_2$ + E
(^{3}P$_u$)	sp^{2}p^{2}p	F$_2$	A$_1$ + E
(0) valent			
(^{1}S$_g$)	s^{2}p^{2}p^2	A$_1$	A$_1$
(^{1}S$_g$)	p^{2}p^{2}p^2	A$_1$	A$_1$
(^{1}D$_g$)	s^{2}p^{2}pp	E + F$_2$	A$_1$ + 2E
(^{1}P$_u$)	sp^{2}p^{2}p	F$_2$	A$_1$ + E
F$^-$(^{1}S$_g$)	s^{2}p^{2}p^{2}p^2	A$_1$	A$_1$

Table 4. CF$_4$ states permitted from the spin and symmetry combinations of the separated fragments

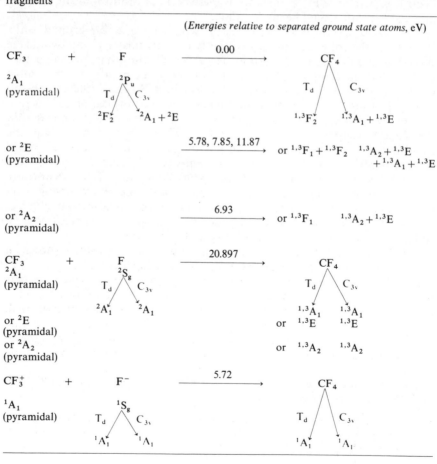

(Energies relative to separated ground state atoms, eV)

While it would have been gratifying to have been able to calculate the CF_4 and CF_4^+ potential curves with the same accuracy as our recent *ab init.* LCAO–MO–SCF calculation of CF_2 (Sachs, Geller and Kaufman, 1969) which used 72 uncontracted atom optimised Gaussian basis functions – it was simply unfeasible in light of the computer time available to us for the present problem. We turned thus to using semi-rigorous LCAO–MO–SCF calculations by the CNDO and INDO methods of Pople and Beveridge (1970).

Table 5. CF_4^+ states permitted from the spin and symmetry combinations of the separated fragments

(Energies relative to separated neutral fragments, eV)

Calculations were performed by the CNDO and INDO methods for CF_4 and CF_4^+ as a function of CF_3–F' distance (keeping the CF_3 fragment fixed as it was in the tetrahedral conformation).

Based on the calculated behaviour of the CF_4^+ states, the relative thermochemical energies of the dissociated products, the spin and symmetry restrictions of the states of the intermediate CF_4^+ states permissible from these products, the experimental appearance potential of CF_3^+ from CF_4 of

15.35 eV (Walter *et al.*, 1969) and the complete absence of a CF_4^+ ion under any situation, we were able to sketch the curves for CF_4 and CF_4^+ states relative to their dissociation products (Figure 5).

Certain features of the curve are determined unambiguously. The non-appearance of a CF_4^+ ion under any circumstances and the appearance of a CF_3^+ ion at 15.35 eV (Walter *et al.*, 1969) places the minimum of even the adiabatic portion of the lowest state above the energy of the lowest set of dissociated products $CF_3^+(\tilde{X}\,^1A_1) + F(^2P_u)$ which lies 14.317 above the ground state of CF_4 [or 9.17 eV above the separated neutral fragments $CF_3(\tilde{X}\,^2A_1) + F(^2P_u)$]. If the adiabatic minimum lay below 14.317 eV then in the photo-ionisation experiments where an adiabatic transition takes place to the 0, 0 state of the ion a CF_4^+ ion should have been observed. It is the fact that the lowest CF_4^+ state lies above the energy of the separated products that permits dissociation by the repulsive $^2A_1$ branch of the $^2F_2$ state. Examination of the behaviour of the excited CF_4^+ curves shows that the $^2A_1$ curve is by far more strongly repulsive. The ^2E curve always lies higher in energy than the $^2A_1$ curve. This argument would place the $^2F_1$ state above the $^2F_2$ state since the dissociative $^2A_1$ and ^2E curves never cross. The ground doublet $^2F_2$ state dissociates to the repulsive $^2A_1$ curve and to a strongly attractive ^2E curve. This ^2E curve must be cut very quickly by the ^2E curve leading to the $^2F_1$ state and an avoided crossing should take place.

In the schematic potential energy diagram in Figure 5 we had placed the minimum of the potential curve for the lowest energy state of the CF_4^+ ion (which we assign as $^2F_2$) at ~ 15.35 eV, since the value of ≥ 15.35 eV was reported by Brundle, Robin and Basch (1970) as the lowest photoelectron peak for CF_4. Very recent results with a special mass spectrometer, which has the capability of measuring positive and negative ions in coincidence, indicates that an ion-pair process to form CF_3^+ and F^- is also present with a threshold of about 13 eV (Alnnell and Koski, 1973). The most probable way in which this ion pair could arise would be from the excited $^1A_1$ ion-pair state of CF_4, dissociating to $CF_3^+(^1A_1)$ and $F^-(^1S_g)$. This state is quite probably a regular state. Further experiments are in progress in which it is hoped that the various mechanisms of CF_3^+ formation will be elucidated.

Any highly repulsive CF_4^+ curve would have to arise from molecular states in which one or two electrons are excited from filled CF_4^+ orbitals into the lowest lying unoccupied a_1 orbital. This is the only orbital whose energy decreases sharply enough with increasing $r(CF_3 - F')$ to produce a highly repulsive curve.

What is perhaps most important is that the theoretical symmetry and spin treatment outlined in this section is completely general and applicable to the entire field of molecular decompositions.

SUMMARY

The most significant conclusion of the present research is the unambiguous confirmation of the overriding role that spin and symmetry restrictions play

CF₄ and CF₄⁺   POTENTIAL ENERGY CURVES

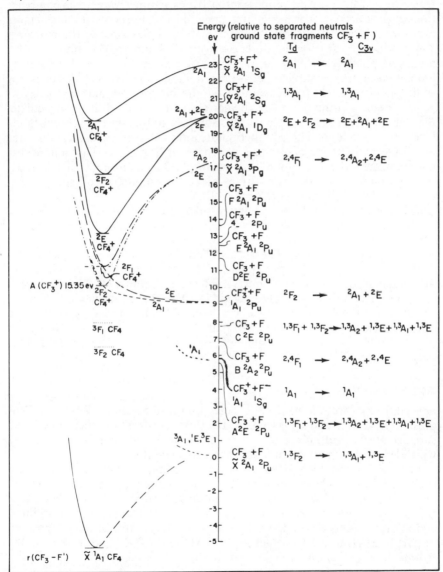

Figure 5 CF_4 and CF_4^+ potential energy curves (schematic)

in any type of collisional or decompositional process. Correlation rules based on the group theoretical approach are indicated in the chapter. These govern the potential energy surfaces for intermediate complexes or molecules which can be formed from any symmetry or spin combinations of reactants or which can decompose into various symmetry and spin combinations of products. It does not matter, in a collisional process, if the reactants ever form a transiently bound intermediate complex or not, the reactants and products must proceed along the symmetry and spin permitted paths, mitigated possibly by curve crossings, which are themselves also fixed by the potential energy surfaces. However, it should be stressed that it is not sufficient merely that the reactants have the proper spin and symmetry to permit them to combine to a particular electronic state of the intermediate, they must be the exact pair which does combine to this state. The same holds true for the products. There is an absolute uniqueness in the correspondence of an intermediate state to a pair of separated fragments. This uniqueness has permitted the derivation of a corresponding comprehensive general theory for molecular decompositions based on symmetry and spin restrictions (Kaufman, Kerman and Koski, 1971).

The relationship of these correlation rules to correlations with molecular orbitals is discussed. A brief description of various quantum chemical computational methods is given to facilitate the comparison.

Two examples are given. The apparently anomalous low energy behaviour of the $O^+ + N_2 \rightarrow NO^+ + N$ reaction is shown to arise from the symmetry and spin restrictions which permit ground state reactants $O^+(^4S_u) + N_2(^1\Sigma_g^+)$ to go uniquely only to a highly excited $^4\Sigma^-$ state of the intermediate NNO^+. The basis for the non-appearance of a parent CF_4^+ ion in mass spectroscopy and the lack of vibrational fine structure for the two lowest photoelectron peaks of CF_4 also lies in the symmetry and spin restrictions for molecular decompositions.

ACKNOWLEDGEMENT

Special thanks are due to the Air Force Office of Scientific Research, Propulsion Division, and to its Chief, Dr. Joseph Masi, for their long time support of the author's research in theoretical and quantum chemistry which led to the understanding necessary to write this chapter.

The author should also like to thank Professor Walter S. Koski who suggested the $O^+ + N_2$ problem to her and who collaborated in its theoretical treatment, and Dr. Aaron Pipano who collaborated in the configuration interaction computation of the $O^+ + N_2$ reaction. The configuration interaction calculations on the $O^+ + N_2$ potential energy surface were supported in part by BRL under Contract No. DAAD05–70–0027 and in part by the Atomic Energy Commission.

REFERENCES

Alnnell, J. E. and Koakin, W. S. (1973). Private communication
Bates, D. R. (1960) *Proc. R. Soc.,* **A257,** 22
Bates, D. R. (1964) *Proc. phys. Soc.,* **84,** 517

Born, M. and Oppenheimer, R. (1927) *Ann. Phys.*, **84**, 457

Brundle, C. and Turner, D. W. (1969) *J. Mass Spectrometry Ion Phys.*, **2**, 195

Brundle, C. R., Robin, M. B. and Basch, H. (1970) *J. chem. Phys.*, **53**, 2196

Callomon, J. H. (1959) *Proc. chem. Soc.*, **1959**, 313

Catlow, G. W., McDowell, M. R. C., Kaufman, Joyce J., Sachs, L. M. and Chang, E. S. (1970) *J. Phys. B: atom. molec. Phys.*, **3**, 833

Clementi, E. (1968) *Chem. Rev.*, **68**, 341

Coulson, C. A. and Strauss, H. L. (1962) *Proc. R. Soc.*, **A269**, 443

Coulson, C. A. and Zalewski, K. (1962) *Proc. R. Soc.*, **A268**, 437

Curran, R. K. and Fox, R. E. (1961) *J. chem. Phys.*, **34**, 1590

Cusachs, L. C. (1965) *J. chem. Phys.*, **43**, S157

Cusachs, L. C. (1966) *J. chem. Phys.*, **45**, 2717

Davidson, E. R. (1961) *J. chem. Phys.*, **35**, 1189

Dibeler, V. H., Walker, J. A. and Liston, S. K. (1967) *J. Res. Natn. Bur. Stds.*, **71A**, 371

Dworetsky, S., Novick, R., Smith, W. W. and Tolk, N. (1967) *Phys. Rev. Letters*, **18**, 939

Eberhardt, W. H., Crawford, Bryce Jr. and Lipscomb, W. N. (1954) *J. chem. Phys.*, **22**, 989

Eyring, H., Walter, J. and Kimball, G. E. (1944) *Quantum Chemistry*, Wiley, New York

Fano, U. and Lichten, W. (1965) *Phys. Rev. Letters*, **4**, 627

Fessenden, R. W. and Shuler, R. H. (1965) *J. chem. Phys.*, **43**, 2704

Gershgorn, Z. and Shavitt, I. (1967) *Int. J. quantum chem.*, **15**, 403

Giese, C. F. (1966) *Adv. Chem. Ser.*, **58**, 20

Gilmore, F. (1967) *DASA Reaction Rate Handbook*, Chap. 4

Gioumousis, G. and Stevenson, D. P. (1958) *J. chem. Phys.*, **29**, 294

Harris, F. E. (1967) *J. chem. Phys.*, **46**, 2769

Herzberg, G. (1950) *Molecular Spectra and Molecular Structure, I. Spectra of Diatomic Molecules*, 2nd Edn, p. 315, Van Nostrand, New York

Herzberg, G. (1966) *Molecular Spectra and Molecular Structure, III. Electronic Spectra and Electronic Structure of Polyatomic Molecules*, Van Nostrand, New York. See Subject Index for tables for individual symmetry groups and for tables for symmetry correlations

Hinze, J. and Roothaan, C. C. J. (1967) *Prog. theor. Phys. Osaka, Suppl.*, **40**, 37

Hoffman, R. (1963) *J. chem. Phys.*, **39**, 1397

Hoffman, R. (1964) *J. chem. Phys.*, **40**, 2472, 2480

Hoffman, R. (1966) *Tetrahedron*, **22**, 521

Hoffman, R. and Lipscomb, W. N. (1962) *J. chem. Phys.*, **36**, 2179

Hoffman, R. and Lipscomb, W. N. (1962a) *J. chem. Phys.*, **37**, 2872

Jaffe, H. H. (1969) *Acc. Chem. Res.*, **2**, 136

Jahn, H. A. and Teller, E. (1932) *Proc. R. Soc.*, **A161**, 220

Jug, K. (1969) *Theor. Chim. Acta*, **14**, 91

Kasha, M. and Henry, B. R. (1968) *Ann. Rev phys. Chem.*, **19**, 161

Kaufman, Joyce J. (1965) *J. chem. Phys.*, **43**, S152

Kaufman, Joyce J. (in press) 'Potential Energy Surface Considerations for Reactions Involving Excited States', *Excited States in Chemical Physics* (edited by McGowan, J. W.), Wiley-Interscience, New York

Kaufman, Joyce J., Harkins, J. J. and Koski, W. S. (1969) *J. chem. Phys.*, **50**, 771

Kaufman, Joyce J., Kerman, Ellen and Koski, W. S. (1970) *Implications of Photoelectron Spectroscopic Measurements for Compounds which Produce No Parent Ion*, Division of Physical Chemistry, 160th American Chemical Society National Meeting, September, Abstract 176

Kaufman, Joyce J., Kerman, Ellen and Koski, W. S. (1971) *Int. J. quantum Chem.*, **4S**, 391

Kaufman, Joyce J. and Koski, W. S. (1969) *J. chem. Phys.*, **50**, 1942

Koopmans, T. (1933) *Physica*, **1**, 104

Laidler, K. J. and Shuler, K. E. (1951) *Chem. Rev.*, **48**, 157

Laidler, K. J. (1955) *The Chemical Kinetics of Excited States*, Clarendon Press, Oxford

Landau, L. (1932) *Soviet Phys.*, **1**, 89

Landau, L. (1932a) *Z. Phys., Sowj.*, **2**, 46

Lao, R. C. C., Rozett, R. W. and Koski, W. S. (1968) *J. chem. Phys.*, **49**, 4202

Levine, R. D., Johnson, B. R. and Bernstein, R. (1968) WIS-TCI-305, Theoretical Chemistry Institute, University of Wisconsin, July 12

238　Wave Mechanics

Lewis, J. K. and Hougen, J. T. (1968) *J. chem. Phys.*, **48**, 5329

Lichten, W. (1963) *Phys. Rev.*, **131**, 229

Lichten, W. (1965) *Phys. Rev.*, **139A**, 27

Lichten, W. (1967) *Phys. Rev.*, **164**, 131

Lichten, W. (1968) *Advances in Chemical Physics*, Volume XIII (edited by Prigogine, I.), 41, Interscience, New York

McLean, A. D. and Yoshimine, M. (1967) Supplement to *IBM J. Res. Dev.*, November

Natalis, P. and Collin, J. E. (1969) *J. Mass Spectrometry Ion Phys.*, **2**, 221

Nicholson, B. J. (1970) *Adv. chem. Phys.*, **18**, 249

Pipano, A. and Kaufman, Joyce J. (1970) *Chem. Phys. Letters*, **7**, 99

Pipano, A. and Kaufman, Joyce J. (1971) *Int. J. quantum Chem.*, **5**, 233

Pipano, A. and Kaufman, Joyce J. (1972) *J. chem. Phys.*, **56**, 5258

Pipano, A. and Shavitt, I. (1968) *Int. J. quantum Chem.*, *II*, 741

Polanyi, J. C. (1971) *Proceedings of the Conference on Potential Energy Surfaces in Chemistry* (edited by Lester, W. A. Jr.), p. 10, IBM Research Lab., San Jose, Cal.

Pople, J. A. and Beveridge, D. A. (1970) *Approximate Molecular Orbital Theory*, McGraw-Hill, New York

Roothaan, C. C. J. (1951) *Rev. mod. Phys.*, **23**, 69

Roothaan, C. C. J. (1960) *Rev. mod. Phys.*, **32**, 179

Rutherford, J. A. and Vroom, D. A. (1971) *J. chem. Phys.*, **55**, 5622

Sachs, L. M., Geller, J. and Kaufman, Joyce J. (1969) *J. chem. Phys.*, **51**, 2771

Schaefer, H. F., III, (1972) *The Electronic Structure of Atoms and Molecules. A Survey of Rigorous Quantum Mechanical Results*, Addison-Wesley, Reading, Mass.

Schmeltkopf, A. L. Ferguson, E. E. and Fehsenfeld, F. C. (1968) *J. chem. Phys.*, **48**, 2966

Shuler, K. E. (1953) *J. chem. Phys.*, **21**, 624

Slater, J. C. (1963) *Quantum Theory of Molecules and Solids*, Vol. 1. McGraw-Hill, New York

Stueckelberg, E. G. C. (1932) *Helv. Phys. Acta.*, **5**, 369

Tanaka, Y., Jursa, A. S. and LeBlanc, F. J. (1960) *J. chem. Phys.*, **32**, 1205

Transitions non Radiative Dans Les Molecules (1969) Société de Chimie Physique, Paris, France, May 27–30

von Neumann, J. and Wigner, E. (1929) *Phys. Z.*, **30**, 467

Walsh, A. B. (1953) *J. chem. Soc.*, 2260

Walter, T. A., Lifshitz, C., Chupka, W. A. and Berkowitz, J. (1969) *J. chem., Phys.*, **51**, 3531 and earlier experimental references cited therein

Weissler, G. L., Samson, J. A. R., Ogawa, M. and Cook, G. R. (1959) *J. opt. Soc. Am.*, **49**, 338

Wigner, E. and Witmer, E. E. (1928) *Z. Phys.*, **51**, 859

Wolfsberg, M. and Helmholz, L. (1952) *J. chem. Phys.*, **29**, 837

Zener, C. (1932) *Proc. R. Soc.*, **137A**, 696

15

The Role of Semi-empirical SCF MO Methods

MICHAEL J. S. DEWAR, F.R.S.
Robert A. Welch Professor of Chemistry
The University of Texas at Austin,
Austin, Texas

The most fundamental problem in chemistry, and certainly one of the hardest, is to find out exactly what happens during the course of chemical reactions. We can study at leisure the reactants and products in a reaction but we have no way of telling the detailed way in which the reaction takes place. Even if we accept the transition state theory, which is in itself an approximation, we can do no more by experiment than estimate the thermodynamic properties of the transition state. We have no direct way of estimating its geometry. This therefore is an area where quantum mechanical calculations are likely to be of very real practical value to chemists, providing important information that cannot be obtained by experiment.

Organic chemists have indeed developed a very satisfactory qualitative theory of chemical reactivity on the basis of very simple quantum mechanical treatments. This, however, is inevitably limited by its qualitative nature. One can for example predict in this way that conrotatory opening of cyclobutene to butadiene should be favoured over disrotatory opening but one has no way of predicting how big the difference in rates should be. Likewise steric effects on rates of reactions depend critically on the detailed geometry of the transition state which again cannot be predicted. There is therefore an urgent need for a more quantitative approach.

In order to be useful in this connection such a procedure must satisfy three conditions.

First, it must provide results of 'chemical' accuracy. In particular, it must reproduce changes in energy of a collection of atoms with changes in geometry to an accuracy of the order of ± 1 kcal/mol (0.05 eV).

Secondly, the calculations must be possible for systems of real chemical interest. There is usually little point in carrying out quantitative calculations for simple 'model' systems in the hope of deducing qualitative principles because organic chemists already have a very successful and comprehensive qualitative theory of reactivity.

Thirdly, the calculations must not be too expensive. We are concerned here with quantum mechanical calculations as a chemical tool. We must be sure that the cost of a calculation is not out of proportion to the chemical value of the results it yields.

There is at present no question of obtaining *a priori* answers by this kind of approach because even the best available *ab initio* quantum mechanical procedures give energies that are in error by chemically speaking huge amounts. There are only two ways in which we can hope for success. Either we may be able to show empirically that the existing *ab initio* methods work sufficiently well for some specific range of structures, due to a fortuitous cancellation of errors, or we may be able to make some empirical correction that will enable them to give satisfactory results. Any such procedure will of course be purely empirical and so must be judged solely by the results it gives. From the point of view of quantum chemistry, there is no other criterion than this. All empirical methods are empirical.

The first possibility is exemplified by the procedure initially introduced by Boys in 1950 (see Boys *et al.*, 1956), i.e. the Roothan (1951) SCF method using a basis set of Gaussian orbitals to simplify the calculation of integrals. Thanks to recent developments in computers and programming, calculations of this type can now be carried out for quite large molecules. The calculated energies are inevitably in error by very large amounts, due to neglect of electron correlation. The hope is that the correlation energy for a given set of atoms may remain sensibly constant so that the calculated changes in total energy during the course of chemical reactions may be satisfactory.

This, however, is not the case. Not only are the heats of atomisation calculated by the best available SCF procedures in error but the error can be as much as $\pm 100\%$. We must certainly be careful in using energies calculated in this way as a guide to chemical behaviour. In other connections the SCF method seems to give good results. Thus predicted molecular geometries and dipole moments usually agree well with experiment and the calculated orbital energies correspond quite nicely to observed ionisation potentials, as Koopman's theorem predicts.

Recently Hehre *et al.* (1970) have shown that the calculated heats of reactions involving small molecules agree quite well with experiment if the reactants and products contain similar numbers of localised bonds. This suggests that the change in correlation energy when such bonds are formed may be a function of the number of bonds only. If this result holds generally, SCF calculations should be trustworthy for calculating heats of reaction and also for studying the course of reactions in which the atoms in the reactants remain bound throughout by a constant number of localised bonds. Typical problems of this kind are the barriers to rotation about single bonds and to pyramidal inversion in compounds such as ammonia. Here the SCF method has proved very successful. The accuracy attained, both here and in the calculation of geometries, dipole moments, etc., does, however, depend on the size of the basis set. Thus to get a good value for the inversion barrier in ammonia requires the use of an extended basis set including 3d orbitals (Rauk, *et al.*, 1970). It should also be pointed out that experimental tests have so far been limited to very simple molecules, mostly acyclic, because of the cost involved.

This work does not, however, justify the extension of *ab initio* SCF calculations of energies to systems of other kinds. Indeed, given that there is a very large change in correlation energy when atoms combine by two-centre bonds, one would certainly expect significant changes when two-centre bonds are replaced by three-centre ones, as in 'non-classical' carbonium ions and other π-complexes, or when bonds are weakened in forming transition states. The available evidence seems to suggest that this is the case. Thus two very recent and very detailed SCF calculations for molecular rearrangements (cyclobutene → butadiene) (Hsu, Buenker and Peyerimhoff, 1972) and methyl isocyanide → acetonitrile (Rauk, Allen and Clements, 1970) gave activation energies that were too large by more than 20 kcal/mol and similar errors have appeared in analogous calculations for other reactions (e.g. rotation about C=C bonds; see Hehre *et al.*, 1970). Problems of this kind also seem to have arisen in SCF calculations (Radom, *et al.*, 1972) of the species formed by protonation of cyclopropane, the relative energies of the classical and non-classical ions being wrongly (Bodor, Dewar and Lo, 1972) predicted. These errors seem significant since the calculations were carried out by procedures that have given good heats of reaction and have also been used successfully in the case of reactions where the bonding remains essentially unchanged throughout (rotation about single bonds, inversion of pyramidal nitrogen). It therefore seems very likely that single-configuration SCF calculations will prove generally unsatisfactory for the study of 'non-classical' species and transition states.

These uncertainties make even more cogent the major objection to *ab initio* SCF calculations of reaction rates, namely their cost. In order to calculate the rate of a chemical reaction, we must at the very least estimate the difference in energy between the reactants and the transition state. If moreover the results are to be of quantitative value, we must estimate this difference very accurately, to the order of ± 1 kcal/mol. To do this we must find the optimum geometry of both species by minimising their energies with respect to all geometrical variables, subject only to the condition that the gradient of the energy is zero. We cannot safely make any assumptions about bond lengths, angles, etc., even for parts of the molecule apparently not involved in the reaction, because the small errors so introduced, summed over a molecule of quite moderate size, can easily add up to 20 kcal/mol or more. Now calculations of molecular geometries are sufficiently difficult even for normal molecules, involving the optimisation of (3n-6) variables in the case of a molecule containing n atoms. The corresponding calculation for a transition state is even harder because an extensive survey of the complete (3n-6)-dimensional potential surface is necessary to locate the transition state. The situation is made still worse by the fact that the best minimisation procedures are inapplicable. These involve a knowledge of the derivatives of the energy with respect to the geometrical variables and to calculate such derivatives of the Roothaan SCF energy would involve so much computation that the whole advantage of using them would be lost.

The cost of such calculations is indicated by that of a recent very detailed study (Horsley et al., 1972) of a very simple chemical reaction, conformational isomerisation of the trimethylene biradical ($\cdot CH_2CH_2CH_2 \cdot$). Solution of this problem, which incidentally was ideally suited to the *ab initio* SCF approach since the number and type of bonds remain unchanged throughout, took 250 hours (Salem, 1972) on a CDC 6600 computer. At current commercial rates, this would cost about $200 000.

In this case the cost was made more acceptable by the fact that the results are likely to be accurate and reliable. The authors moreover carried out every possible check against experiment to test the applicability of their procedure to this problem. If it were possible in some way to increase the accuracy of the *ab initio* SCF method so that the results had a similar reliability attached to them, they would clearly be of great value since they would be definitive and indeed could serve as a basis for checking experiment rather than the reverse. Attempts have been made to achieve this by introducing configuration interaction (CI) into the SCF treatment but while this represents an improvement it seems insufficient (cf. Hehre et al., 1970; Hsu, Buenker and Peyerimhoff, 1972). Although one could in principle allow for correlation in this way, the convergence is too slow. One might remark in this connection that Boys was also the first to introduce CI into Roothaan SCF calculations for polyatomic molecules and that he abandoned this whole approach some time ago as too inaccurate for chemical purposes. If really accurate and reliable results are to be obtained, it will be necessary to find some entirely novel approach. Although several promising-looking possibilities have appeared (e.g. Conroy, 1964; Boys and Handy, 1969) in recent years, success still seems far off.

If therefore theoretical calculations are to be used as a practical and quantitative guide in the study of chemical reactivity, they must involve some procedure both simpler and more accurate than the *ab initio* SCF approach. These mutually contradictory requirements can only be met, if indeed they can be met at all, by some kind of empirical or semi-empirical procedure. The *ab initio* SCF treatment itself falls into this category since it can be used only in an empirical sense but it is too complicated and it also fails to predict heats of atomisation. If we are to have any real trust in a treatment of reactivity, it must give good estimates of the energy changes in *all* possible chemical processes, including the breaking of bonds. In our studies we will certainly need to extend our calculations to types of systems that have not yet been examined in detail experimentally. We cannot make such extrapolations if there is any area at all where our procedure is known to fail.

The simplest solution of our problems would be provided by the purely empirical approach introduced by Westheimer (1956) in which interatomic forces are represented by empirical potential functions. This has proved very successful (see e.g. Lifson and Warshel, 1970) in the case of 'normal' molecules containing localised bonds. However, it is difficult to see how it could be extended to the intermediate phases of reactions where unusual types of bonding occur. A very large number of parameters are needed even for normal molecules and many more would be required for systems containing the partial bonds characteristics of reaction intermediates. Since we have no experimental data for the intermediate phases of reactions, apart from the thermodynamic properties of transition states, it would be extremely difficult to determine the additional parameters in a convincing way. Besides, the more parameters there are in a given treatment, the less certain the results obtained by extrapolating it beyond the region where it has been tested. A further drawback of this kind of approach is that it gives no information concerning the distribution of electrons. Even if this is not strictly required, it is likely to be useful in relating the results of calculations of reaction paths to existing qualitative 'electronic' theories of reactivity.

The most hopeful approach therefore seems to lie in some simplification of the *ab initio* SCF approach, leading to a treatment which can be applied quickly even to quite large molecules. While less accurate than the full SCF treatment itself, the results should not be too far removed from reality. Introduction of quite a limited number of parameters may then enable it not merely to recover the accuracy of the full *ab initio* treatment but even to surpass it. The basic philosophy, that of upgrading a very approximate treatment by introducing parameters, has of course often been applied in chemistry, the Debye–Hückel theory of strong electrolytes being a classic example.

Semi-empirical treatments of this kind have been little regarded in the past for several reasons. First, many people still think of them in terms of primitive HMO calculations. Secondly, hardly any attempt has been made to parametrise them properly. Thirdly, as a result of this inadequate parametrisation, many people still believe that semi-empirical methods can be useful only in specific and limited areas, the parameters for each particular application having to be chosen specifically for it. Fourthly, as a result of this mistaken belief, many assume that the results of *ab initio* calculations are inherently 'better' and 'nearer to reality' because one has no way to tell which of the spectrum of parametric treatments to believe. And finally, semi-empirical treatments have less mathematical status value than ones based on rigorous mathematical treatments even if horrifying approximations are necessary for such a treatment to be possible.

The Hückel method, using the Hückel σ, π approximation, was of course very crude and its main value lay in qualitative applications, in particular those involving the use of perturbation theory (Dewar, 1952). Attempts to extend the HMO treatment to all-valence-electron calculations for inorganic compounds by Wolfsberg and Helmholz (1952, 1953, 1955) proved unsatisfactory in a quantitative sense. Recently Hoffmann (1963) has used this approach to study a number of organic chemical problems, renaming it the Extended Hückel (EH) method. However, it seems clear that it is not in fact reliable for quantitative purposes, giving poor estimates of the geometries and energies of molecules and also leading to unrealistic electron distributions. Although it has the attraction of requiring very little computation time, it is probably too crude to be rescued even by the introduction of additional parameters.

Some years ago we showed (Chung and Dewar, 1965; Dewar and Gleicher,

1965; Dewar and de Llano, 1969; Dewar and Harget, 1970) that the properties of conjugated molecules could be calculated with remarkable accuracy using the Hückel σ, π approximation together with the Pople (1953) approximation for the π electrons, provided that the parameters were correctly chosen. The calculated geometries and heats of atomisation for conjugated hydrocarbons of all types were in almost perfect agreement with experiment. This suggested that an analogously simplified version of the Roothaan method might be equally successful for calculations in which all the valence electrons were included. Several such approximate treatments have been suggested (Pople *et al.*, 1965, 1966, 1967; Dewar and Klopman, 1967; Dixon, 1967) in which calculations are carried out for all the valence electrons, both σ and π. These are assumed to move in a fixed core composed of the nuclei together with the inner shell electrons and the calculations are further simplified by neglecting some or all of the integrals involving overlap between AOs.

The purpose of earlier work in this area was to find a cheap and simple substitute for *ab initio* SCF calculations. The parameters were therefore chosen to reproduce as closely as possible *ab initio* calculations for the few simple molecules for which the latter were then available. Since there are reasons for doubting if the Hartree–Fock method itself is accurate enough for our purpose, it is not surprising that attempts to use these even more approximate procedures (CNDO/2, INDO) for calculations of chemical reactivity have proved unsatisfactory. Much of the present prejudice against semi-empirical methods is due to the failure of these mistaken attempts.

If a semi-empirical treatment is to be used to estimate the geometries and energies of molecules, the parameters in it must clearly be chosen to reproduce those quantities rather than the results of some rather crude quantum mechanical approximation. One proviso should be added. Since the SCF approach seems to give a reasonably good approximation to the first order density matrix, we also should expect our procedure to give overall charge distributions similar to those predicted by good *ab initio* SCF calculations. Otherwise the results of these calculations can be disregarded.

Most semi-empirical treatments in the past have been developed in a very perfunctory manner. Some arbitrary scheme for parametrisation has been set up and the parameters then determined by casual comparison with experiment. There are two likely reasons for this. First, most of those concerned were probably convinced that no procedure of this kind could possibly give really accurate results and so were unwilling to put much effort into it. Secondly, the development of satisfactory parametric procedures of this kind is a far more difficult matter than most people realise. The consequent failure of a few hurried attempts was taken as supporting evidence for the inadequacy of semi-empirical methods in general rather than an indication of the inadequacy of the effort put into their parametrisation.

In developing *ab initio* SCF methods, the procedure to be followed is clear cut and the problems encountered are mainly technical ones of programming. In a semi-empirical approach, where integrals are replaced by parametric functions, there is an unnerving degree of arbitrary choice. Ultimately the functions have to be selected by trial-and-error and conclusions drawn from past failures. This is a very time-consuming process and can be very frustrating. Even the choice of parameters in a given set of parametric functions is far from trivial. The various quantities in the SCF treatment are so closely interrelated that it is impossible to predict what the effect of a given change in a given parameter will be. We were only able to make progress after writing a computer program (Dewar and Haselbach, 1970) to optimise parameters automatically, by a least squares fit between the calculated and observed heats of atomisation and geometries of a carefully selected set of reference molecules.

Given these difficulties and given the current distrust of semi-empirical methods, it is not surprising that so little attempt has been made to develop them. Indeed our own efforts in this direction represent almost the only serious effort so far. It should be added that our earlier endeavours were very disappointing and frustrating. Had it not been for the extraordinary and unexpected success of our earlier work using the π approximation, we would certainly not have persevered. As it was, we were convinced that success must be attainable and this belief kept us going. While we are still far from satisfied with out present achievements, they are at least ample to justify our faith in the semi-empirical approach. Our calculations reproduce all the ground state properties of molecules with reasonable accuracy, including heats of atomisation, using a single set of parameters and at a minute fraction of the cost of *ab initio* calculations. Even in its present state this procedure has shown itself capable of providing useful information concerning the course of chemical reactions and of course it is capable of almost unlimited development. Better systems of parametrisation than our present one will undoubtedly be found and the way is also open to the parametrisation of more elaborate and accurate SCF approximations. There seems little doubt that this kind of approach must remain for a good many years to come by far the most promising for the practical study of chemical reactivity.

Since so many misconceptions still exist, we may perhaps usefully outline the strategy we have followed in our own work and indicate the kind of results that can be obtained by semi-empirical procedures in the present state of the art. The approach we have followed may very well not be the best but at present we have no competition.

The success of our earlier work with the π approximation provided an obvious starting point. It seemed likely that we would be able to allow for the effects of electron correlation in an all-valence-electron treatment in a similar way, i.e. by empirical adjustment of electron repulsion integrals as was first suggested by Pariser and Parr (1953). The idea here is first to choose the one-centre integrals so that our single-determinant SCF treatment gives good energies for various states of atoms and derived ions. Use of these integrals in a molecular calculation should then automatically allow for correlation between pairs of electrons in the vicinity of the same nucleus. Long range correlation is then taken into account by equating the two-centre repulsion integrals to suitable functions of internuclear distances (R), chosen so that for $R \rightarrow 0$ the integral approaches the corresponding empirical one-centre value while for $R \rightarrow \infty$ it has the correct asymptotic form [e.g. e^2/R for integrals of the type (ii,jj)]. In our work we have used expressions of the type first introduced by Dewar and Hojvat (1961) and Dewar and Sabelli (1962) and subsequently used by Ohno (1964), viz.:

$$(ii, jj) = e^2[R^2 + 1/4(\rho_i + \rho_j)^2]^{-1/2} \tag{1}$$

where:

$$\rho_i = e^2(ii, ii)^{-1}; \rho_j = e^2(jj, jj)^{-1} \tag{2}$$

This approximation was first used in all-valence electron calculations by Klopman (1964).

Three procedures have been used to estimate one-centre integrals; Klopman's (1964) barycentre method, a method (Pople, Santry and Segal, 1965; Pople and Segal, 1966) based on fitting the Slater–Condon parameters, and a generalised version of (Dewar and Lo, 1972) Oleari's method. The last seems clearly superior, as it should be since in it all the one-centre integrals are determined individually without any simplifying assumptions.

The only other quantities to be considered are the one-electron core resonance integrals (β^c_{ij}), the core-electron attractions, and the core–core repulsions.

For reasons that are well understood (see Dewar, 1969) it is almost essential in treatments of this kind to represent the β^c_{ij} by some parametric function. This in any case is a logical thing to do in the case of a semi-empirical treatment of ground state properties since terms involving the β^c_{ij} play a predominant role in chemical bonding (Ruedenberg, 1962). Of several dozen parametric functions we have tried, including all the various forms that have been suggested in the literature, the best seems to be the simple Mulliken or Wolfsberg–Helmholz approximation, i.e.

$$\beta^c_{ij} = B_{mn}S_{ij}(I_i + I_j) \tag{3}$$

where I_i and I_j are the valence state ionisation potentials of the AOs ϕ_i and ϕ_j, S_{ij} is the overlap integral, and B_{mn} is a parameter characteristic of the pair of atoms m and n of which ϕ_i and ϕ_j are AOs.

Since the electron/core attractions and repulsions are very large, and since we have used a parametric expression for the electron–electron repulsions, care has to be taken in estimating the core–electron attractions and core–core repulsions; for if these get out of balance, the calculations are disrupted by spurious long range coulombic interactions. In the π approximation (Chung and Dewar, 1969), this was achieved by setting all the coulombic attractions and repulsions equal for pairs of neutral atoms, equivalent in the case of the core–electron attraction to using the Goeppert–Mayer–Sklar potential with neglect of penetration integrals. The result was an overall attraction between neutral atoms at all distances, due to one-electron and exchange terms. This was satisfactory in the case of the π approximation since the strength of a π bond does increase continuously with decreasing length and since the σ bonds were treated by the localised bond model. In an all-valence-electron treatment, however, it leads to molecular collapse.

One solution would be to use theoretical values for the core–electron attractions and core–core repulsions. Here, however, the overall coulombic interactions get out of balance if we use our empirical scheme for the electron–electron repulsions. The same is true if we try to set the core–electron attractions equal to the electron–electron repulsions but use theoretical values for the core–core repulsion. Pople *et al.* (1965) and Pople and Segal (1966) used this scheme in CNDO/2 and INDO. A third alternative, which we have used successfully, is to keep the electron–electron repulsions and core–electron attractions equal but to use a parametric expression for the core–core repulsions. The core–core repulsion is required to approach the electron–electron repulsion as the internuclear distance (R) becomes large, thus avoiding spurious long range coulomb interactions, but to become larger at short distances, thus preventing the molecule from collapsing. A simple expression of this kind for the core repulsion (CR_{mn}) between the cores of atoms m and n is:

$$CR_{mn} = ER_{mn} + \left(\frac{Z_m Z_n e^2}{R} - ER_{mn} \right) f(R) \qquad (4)$$

when ER_{mn} is the interelectronic repulsion between the two atoms in their neutral valence states at internuclear distance R and $f(R)$ is a function of R which contains an adjustable parameter (α_{mn}) and which tends to unity as $R \to 0$ and to zero as $R \to \infty$. We thus have two adjustable parameters for each atom pair (α_{mn} and B_{mn}). By adjusting these we should be able to fit the corresponding bond lengths and bond energies.

This scheme was first applied (Dewar and Klopman, 1967) to an approximation (PNDO) intermediate between the INDO and NDDO approximations. At that time we were still trying to fit the parameters by trial and error. We were therefore unable to fit molecular geometries and heats of atomisation simultaneously though the method gave good energies if geometries were assumed. The same was true of an analogous version (Baird and Dewar, 1969) of INDO which we call MINDO (Modified INDO) and now termed MINDO/1. With the development of our parametrisation program we succeeded in developing a new version of MINDO which gave geometries as well as energies, MINDO/2 (Dewar and Haselbach, 1970; Bodor *et al.*, 1970) and since then we have reparametrised PNDO and parametrised CNDO/2* and NDDO. Our results suggest that there is little point in pursuing CNDO/2 or PNDO since INDO and NDDO are no more complicated to use and seem to give somewhat better results.

The original form of MINDO/2 was generally quite successful, heats of atomisation usually being reproduced to within a few kcal/mol, bond lengths to $\pm 0.02\,\text{Å}$, and bond angles to a few degrees. However, it had several quite bad shortcomings, viz.: (1) bond lengths involving hydrogen were systematically too long, by $0.1\,\text{Å}$ for CH and NH and $0.15\,\text{Å}$ for OH; (2) dipole moments were uniformly too large by *c.* 50%; (3) heats of formation of simple diatomic molecules, in particular H_2, N_2 and CO, were badly in error; (4) strain energies in small rings were underestimated; (5) bond lengths for bonds between atoms with lone pairs of electrons (NN, OO, etc.) were too short and the heats of atomisation correspondingly too large; (6) bond angles in compounds of the type R_2O or R_3N were too large.

The first objection was overcome (Bodor, Dewar and Lo, 1972) by a

*Fischer and Kollmar (1969) have developed an alternative parametrisation of CNDO/2 which seems to give results very similar to those of our version and to those of the early versions of MINDO/2.

minor change in the parametrisation and the second by adopting an improved procedure (Dewar and Lo, 1972) for estimating one-electron integrals. This version was termed MINDO/2′. Both MINDO/2 and MINDO/2′ have been used extensively in the study of reaction paths. Some results are summarised below. Recently MINDO has been completely reparametrised. The new version (MINDO/3) seems to avoid all the systematic errors (1)–(6) of MINDO/2. The predicted geometries are now in good agreement with experiment in all cases. The average error in the calculated heats of atomisation (5 kcal/mol) is a little larger than it was for MINDO/2′ in cases where MINDO/2′ was applicable. Here, however, the errors seem to be comparable for compounds of all types. MINDO/3 has been parametrised for C, H, O, N, F, Si, P, S, and Cl, though applications to third row elements are at present limited since we have not yet included 3d AOs.

In the approach we have used, the SCF equations are solved using our parametric values for the various integrals. Concern has been expressed (see Dike, 1971) at this kind of procedure on the grounds that it may lead to unrealistic MOs. It is argued that since *ab initio* SCF methods apparently give good descriptions of electron distributions, a semi-empirical treatment can be realistic only if it reproduces the *ab initio* eigenvectors. It has therefore been suggested that it might be better to use a two-step procedure, first calculating the MOs with parameters chosen to give the best fit to the *ab initio* ones, and then using these MOs together with a second set of parameters to calculate the energy. We have not tried this approach because the MINDO/2′ and MINDO/3 electron distributions are very similar to those given by good *ab initio* calculations. However, it may well prove better. Indeed, one of the major attractions, and at the same time a major frustration, of semi-empirical methods is that they are always capable of further improvement without any necessary increase in complexity. One might add that the accuracy of MINDO could undoubtedly be improved considerably by introducing more adjustable parameters; at present we have two only per atom pair. Since our objective is purely pragmatic, i.e. to find a procedure which gives results of practical value to chemists, there is no reason why we should not increase the number of parameters. We ourselves have avoided this partly from an admittedly chauvinistic and irrational feeling that parameters should be kept to a minimum, partly because the computer time required to optimise parameters is surprisingly great and increases rapidly with their number, and partly because the lack of good thermochemical data for compounds containing elements other than C, H, and O would make the determination of additional parameters rather difficult.

While MINDO was parametrised to fit observed geometries and heats of atomisation, it has also been surprisingly successful in other connections. MINDO/2′ and MINDO/3 give good estimates of dipole moments and also of ionisation potentials, the latter being found either from Koopmans' theorem or by difference between the calculated heats of atomisation of a molecule and the derived ion radical. Earlier calculations (Bodor *et al.*, 1970) of force constants also gave results apparently in good agreement with experiment. We have now written a computer program for calculating complete molecular potential functions and hence vibration frequencies.

A detailed calculation (Dewar and Metiu, 1972) for ethane, using MINDO/2, gave vibration frequencies in surprisingly good agreement with experiment and it seems likely that the results for MINDO/3 will be better still. If so, this kind of approach could provide a solution to the unsolved problem of calculating molecular entropies.

The time required for MINDO/2 calculations is remarkably modest. Pople (1972) has stated that his simplest *ab initio* SCF treatment (STO-3G) takes about thirty times longer for a single calculation for a molecule the size of benzene than does CNDO/2, and CNDO/2 in turn is slightly slower than MINDO. The discrepancy increases as the square of the number of orbitals in the basis set and so is very much greater for the more elaborate *ab initio* treatment in which large basis sets are used. It also of course increases rapidly with increasing molecular size. Thus a recent calculation by Clementi, Mehl and von Niessen (1971) for the guanine–cytosine nucleotide base pair, using a small basis set and an assumed geometry, took eight days of central processor time on an IBM 360-195 computer. The same system could be calculated on our somewhat slower computer (CDC 6600) in about two minutes, using MINDO/3.

For chemical purposes, where geometries have to be optimised, the discrepancy becomes much greater because far more efficient optimisation procedures can be used in the case of semi-empirical methods. The best of them depend on a knowledge of the derivatives of the energy with respect to the geometrical variables. In the case of *ab initio* SCF procedures, calculation of the derivatives involves so much extra computation that the whole advantage of the procedure would be lost. In the case of semi-empirical methods the derivatives can be found very quickly and easily. This approach was first introduced by McIver and Korminicki (1971) for MINDO/2. A good comparison is possible because they calculated several systems which we had previously studied using a less efficient non-gradient method of optimisation. The gain in speed for a single structure, in the case of a molecule the size of cyclobutene, is about a factor of twenty. The advantage of the gradient method moreover increases as the size of the molecule increases because the number of function evaluations (i.e. SCF calculations) increases much more slowly with an increasing number of parameters than it does for non-gradient methods.

An even greater gain is possible in the case of transition states because, as McIver and Korminicki (1971) have also shown, these can be located directly by gradient methods. Previously transition states had been located only by laborious procedures involving a point-by-point plotting of reaction paths. Their ingenious approach seems at one stroke to have opened up the whole of organic chemistry to this kind of method and the possibility of calculating reaction paths for relatively large systems at reasonable cost also has obvious biochemical implications. We have developed an alternative gradient procedure for optimising geometries which we believe to be even faster than that of McIver and Korminicki and no doubt even better methods will appear.

In conclusion some recent results may be quoted to show the current status of the semi-empirical SCF MO treatments outlined above.

Table 1 compares calculated and observed heats of formation for a number of molecules. Note that the quantity calculated is in fact the heat of atomisation, this being converted for convenience to a heat of formation using experimental values for the heats of formation of gaseous atoms. The errors quoted in Table 1 are therefore errors in the total bonding energy of the molecule.

The compounds listed in Table 1 fall into two groups. The first twelve are 'ordinary' molecules included to show the general level of accuracy of

Table 1. Calculated (MINDO/3) and observed heats of atomisation of various molecules

Compound	Heat of atomisation (kcal/mol at 25°C)		
	Calcd	Obsd	Error
$H_3C—CH_3$	−19.8	−20.2	0.4
$H_2C=CH_2$	19.2	12.4	6.8
$HC\equiv CH$	57.8	54.3	3.5
$H_3C—CH_2—CH_3$	−26.5	−24.8	−1.7
$H_3CCH_2CH_2CH_3$	−30.4	−30.4	0
$(H_3C)_3CH$	−24.9	−32.4	7.5
$H_2C=CH—CH=CH_2$	31.9	26.1	5.8
Benzene	28.8	19.8	9.0
Cyclohexane	−36.6	−29.5	−7.1
NH_3	−13.3	−10.9	−2.4
CH_3NH_2	−4.0	−5.5	1.5
Pyridine	39.6	34.6	5.0
Aniline	19.8	20.8	−1.0
H_2O	−53.6	−57.8	4.2
CH_3OH	−52.2	−48.1	−4.1
CO_2	−96.4	−94.1	−2.3
Cyclopropane	8.6	12.7	−4.1
Cyclopropene	59.4	66.2	−6.8
Cyclobutene	33.1	37.5	−4.4
Bicyclobutane	49.8	51.9	−2.1
Cubane	139.7	148.7	−9.0
H_2	0.1	0	0.1
N_2	5.2	0	5.2
O_2	21.1	22.0	−0.9
$HOOH$	−39.7	−32.5	−7.2

MINDO/3. The rest are molecules for which MINDO/2 and MINDO/2' give poor energies.

Table 2 shows a similar comparison for the geometries of a dozen molecules of various kinds. The MINDO/3 results are satisfactory in a chemical sense.

Table 3 compares the observed vibration frequencies of ethane with those calculated using the original version of MINDO/2. The agreement is clearly quite good (Dewar and Metiu, 1972). Given that better semi-empirical treatments are already available, the calculation of partition functions of molecules, and also their infra-red and Raman spectra, is clearly within sight. Such a treatment could be parametrised to give equilibrium energies of molecules, these being used in conjunction with calculated partition functions to estimate absolute thermodynamic properties. It should again be emphasised that goals of this kind already seem feasible so far as semi-empirical procedures are concerned while quite out of reach of any known *ab initio* procedures in the foreseeable future.

Table 4 compares observed activation energies for several reactions with values calculated by various methods. It will be seen that the MINDO results are uniformly better than the *ab initio* SCF ones, except for barriers

Table 2. Calculated (MINDO/3) and observed geometries of molecules

Compound	Bond	Bond lengths (Å) Calcd	Obsd	Angle	Bond angles (degrees) Calcd	Obsd
$H_2C=CH_2-CH_3$	C=C	1.333	1.336	C—C—C	129.0	124.3
	C—C	1.4800	1.501			
	=C—C	1.100	1.086			
	H_2C—H	1.114	1.095			
	C—C	1.504	1.510	H—C—H	108.6	115.1
	C—H	1.104	1.089			
	C=C	1.326	1.300	C=C—H	152.7	149.5
	C—C	1.480	1.515	H—C—H	105.2	114.7
	=C—H	1.085	1.070			
	HC—H	1.110	1.087			
$H_2C=C=CH_2$	C=C	1.310	1.308	C=C—H	124.2	120.9
	C—H	1.105	1.087			
	C=C	1.330	1.340	C—C—C	130.5	123.3
	C—C	1.461	1.463			
	C_1—H	1.100	1.090			
	C_2H	1.113	1.090			
NH_3	N—N	1.040	1.012	H—N—H	103.1	106.7
H_2O	H—O	0.944	0.957	H—O—H	100.1	104.5
	C—N	1.351	1.370	C—N—H	124.2	125.1
	C_2—C_2	1.408	1.382	N—C—C	107.7	107.7
	C_2—C_3	1.427	1.417	C—C—C	106.6	107.4
	N—N	1.000	0.996			
	C—O	1.334	1.371	O—C—C	111.9	110.9
	C_1C_2	1.389	1.354	C—O—C	107.0	106.0
	C_2—C_3	1.443	1.440	C—C—C	104.6	106.1
H_2N—CH=O	C—N	1.327	1.343	N—C—O	128.3	123.6
	C=O	1.210	1.243	N—C—N	110.3	103.9
	C—H	1.143	1.094	C—N—H	124.2	120.5
	N—N	1.029	0.995			
$CH_3-C\overset{O}{\underset{OH}{\diagup}}$	C—C	1.503	1.520	C—C—O	110.3	110.6
	C—O	1.332	1.364	C—C=O	120.5	126.6
	C=O	1.215	1.214	C—O—H	106.4	107.0
	O—H	0.956	0.97			

to rotation about single bonds and to pyramidal inversion. As we have already noted, these reactions are particularly appropriate to study by *ab initio* SCF methods since they involve no change in bonding. In the case of the cyclobutene → butadiene rearrangement, both MINDO and *ab initio* SCF gave poor results. Whereas, however, the error in the *ab initio*

Table 3. Calculated (MINDO/2) and observed vibration frequencies (cm^{-1}) of ethane

Calcd	Obsd	Calcd	Obsd
3281	3175	1349, 1334[a]	1449
3143, 3096[a]	3139	1178	1438
3029, 3024[a]	3061	934	1246
2917	3042	863	1016
2043	1551	794	822
1852	1526	345	303

(a) The lack of degeneracy in the calculated frequencies is due to a failing of the eigenvalue subroutine in the program we used for calculating vibration frequencies from force constants.

Table 4. Calculated and observed activation energies for reactions

Reaction	Activation energy (kcal/mol)		
	Obsd	MINDO	Ab init. SCF
Rotation about CC bond in:			
$H_2C{=}CH_2$	65[a]	64[b]	139[c]83[d]
$H_2C{=}C{=}CH_2$	47[e]	47[b]	92[c]
$H_2C{=}C{=}C{=}CH_2$	30[a]	38[b]	—
$H_2C{=}C{=}C{=}C{=}CH_2$	—	30[b]	—
$H_2C{=}C{=}C{=}C{=}C{=}CH_2$	20[a]	27[b]	—
	36[f]	53[g]	65[h]49[i]
$CH_3NC \rightarrow CH_3CN$	38.4[j]	34.3[k]	59[l]
1,5H shift in 1,3-pentadiene	35.4[m]	28.3[n]	—
$CH_3{\cdot}+CH_4 \rightarrow CH_4+{\cdot}CH_3$	14.6[a]	11.3[b]	—
Cope rearrangement of:			
1,5-hexadiene via chair	33.5[o]	31.8[b]	
1,5-hexadiene via boat	44.7[o]	39.9[b]	
Bullvalene	11.8, 12.8[p]	11.3[g]	
Barbaralane	8.6[p]	5.9[q]	

(a) For references see Dewar and Haselbach (1970)
(b) Unpublished MINDO/3 calculations by Dr. R. C. Bingham
(c) Radom, L. and Pople, J. A. (1970) *J. Am. chem. Soc.*, **92**, 4786
(d) SCF with CI; Kaldor, U. and Shavitt, I. (1968) *J. chem. Phys.*, **48**, 191; Buenker, R. L. (1968) *J. chem. Phys.*, **48**, 1368
(e) Personal communication from Professor W. R. Roth
(f) Frey, H. M. and Skinner, R. F. (1965) *Trans. Faraday Soc.*, **61**, 1918
(g) MINDO/2′ calculations by S. Kirschner; see Dewar, M. J. S. and Kirschner, S. (1971) *J. Am. chem. Soc.*, **93**, 4290 for preliminary results
(h) Hsu *et al.* (1971)
(i) SCF with CI; Hsu *et al.* (1972)
(j) Schneider, F. W. and Rabinovitch, B. S. (1962) *J. Am. chem. Soc.*, **84**, 4215
(k) MINDO/2′, Dewar, M. J. S. and Koh, M. S. (1972) *J. Am. chem. Soc.*, **94**, 2704
(l) Liskow *et al.* (1972)
(m) Roth, W. R. and König, J. (1966) *Ann. chem.*, **699**, 24
(n) MINDO/CI, Bingham, R. C. and Dewar, M. J. S. *J. Am. chem. Soc.*, in press
(o) Goldstein, M. J. and Benzon, M. S. (1972) *J. Am. chem. Soc.*, **94**, 7147
(p) For references see ref. q below.
(q) MINDO/2, Dewar, M. J. S. and Schoeller, W. W. (1971) *J. Am. chem. Soc.*, **93**, 1481

results probably reflects (see above) an inherent failing of the Hartree–Fock approximation, that in the MINDO one was due to the use of an early version (MINDO/2) which was known to overestimate the stability of small rings. Indeed the error in the case of cyclobutene was exceptionally large (25 kcal/mol). It seems likely that MINDO/3 will give a much better result but we have not yet completed the calculation.

SUMMARY

(1) The available evidence suggests that current *ab initio* methods are too inaccurate to be useful in the quantitative study of chemical reactivity.

(2) These procedures are in any case so time-consuming that they are limited to very simple reactions and even then the cost of the calculations is out of proportion to the chemical value of the results obtained. No foreseeable technical developments could alter this situation significantly.

(3) Semi-empirical procedures are also of little value unless they are properly parametrised.

(4) Contrary to much current opinion, it is possible to reproduce all the ground state properties of molecules with reasonable accuracy by a simple semi-empirical treatment, using a single set of parameters, provided the parametrisation is carried out properly and the parameters properly chosen. The properties in question include heats of atomisation.

(5) Calculations for a number of reactions show that reaction paths can already be calculated with reasonable accuracy.

(6) It is also already possible to calculate molecular vibration frequencies and so to estimate absolute entropies of molecules.

(7) The semi-empirical approach is moreover capable of virtually unlimited development, either by better parametrisation of existing methods or by parametrisation of better ones. Our preliminary version of NDDO already seems superior to MINDO/3.

(8) Even with Simplex optimisation of geometries, we can carry out complete calculations for reactions involving eight atoms other than hydrogen at reasonable cost on a CDC 6600 computer. Use of existing gradient methods for optimisation should increase this by a factor of at least five and use of faster computers currently available by a further factor of between two and three. It should therefore be possible to carry out calculations at the present time for reactions involving 80 heavy atoms, with complete optimisation of the geometries of the reactants and the transition states.

(9) The methods described here have been mostly developed by one group working in virtual isolation and composed mainly of chemists rather than theoreticians at a time when everyone else was concerned with the development of the *ab initio* SCF approach. If one-tenth of the effort put into the latter had been diverted to the former, quantitative calculations of chemical reaction paths and thermodynamic properties might by now have become a matter of simple routine.

REFERENCES

Baird, N. C. and Dewar, M. J. S. (1969) *J. chem. Phys.*, **50**, 1262
Bodor, N., Dewar, M. J. S., Harget, A. and Haselbach, E. (1970) *J. Am. chem. Soc.*, **92**, 3854
Bodor, N., Dewar, M. J. S. and Lo, D. H. (1972) *J. Am. chem. Soc.*, **94**, 5303
Boys, S. F., Cook, G. B., Reeves, C. M. and Shavitt, I. (1956) *Nature*, **178**, 1207
Boys, S. F. and Handy, N. C. (1969) *Proc. R. Soc.*, **A310**, 43, 63
Brown, A., Dewar, M. J. S., Metiu, H., Student, P. J. and Wasson, J. (in press) *Proc. R. Soc., A*
Chung, A. L.-H. and Dewar, M. J. S. (1969) *J. chem. Phys.*, **42**, 756
Clementi, E., Mehl, J. and von Niessen, W. (1971) *J. chem. Phys.*, **54**, 508
Conroy, H. (1964) *J. chem. Phys.*, **41**, 1327, 1331, 1336, 1341
Dewar, M. J. S. (1952) *J. Am. chem. Soc.*, **74**, 3341, 3345, 3350, 3353, 3355, 3357
Dewar, M. J. S. (1969) *The Molecular Orbital Theory of Organic Chemistry*, Section 3.4,
    McGraw-Hill, New York
Dewar, M. J. S. and Gleicher, G. J. (1965) *J. Am. chem. Soc.*, **87**, 692
Dewar, M. J. S. and Harget, A. (1970) *Proc. R. Soc.*, **A315**, 443, 457
Dewar, M. J. S. and Haselbach, E. (1970) *J. Am. chem. Soc.*, **92**, 590
Dewar, M. J. S. and Hojvat, N. L. (1961) *J. chem. Phys.*, **34**, 1232
Dewar, M. J. S. and Hojvat, N. L. (1961) *Proc. R. Soc.*, **A264**, 431
Dewar, M. J. S. and Klopman, G. (1967) *J. Am. chem. Soc.*, **89**, 3089
Dewar, M. J. S. and Kohn, M. S. (1972) *J. Am. chem. Soc.*, **94**, 2704
Dewar, M. J. S. and de Llano, C. (1969) *J. Am. chem. Soc.*, **91**, 352
Dewar, M. J. S. and Lo, D. H. (1972) *J. Am. chem. Soc.*, **94**, 5296
Dewar, M. J. S. and Metiu, H. (1972) *Proc. R. Soc.*, **A329**, 173
Dewar, M. J. S. and Sabelli, N. L. (1962) *J. phys. Chem.*, **66**, 2310
Dike, B. J. (1971) *Ann. Rep. Chem. Soc.*, **A68**, 3
Dixon, R. N. (1967) *Molec. Phys.*, **12**, 83
Fischer, H. and Kollmar, H. (1969) *Theoret. Chim. Acta.*, **13**, 213
Hehre, W. J., Ditchfield, R., Radom, R. and Pople, J. A. (1970) *J. Am. chem. Soc.*, **92**, 4796
Hoffmann, R. (1963) *J. chem. Phys.*, **39**, 1397
Horsley, J. A., Jean, Y., Moser, C., Salem, L., Stevens, R. M. and Wright, J. S. (1972) *J. Am. chem. Soc.*, **94**, 279
Hsu, K., Buenker, R. J. and Peyerimhoff, S. D. (1971) *J. Am. chem. Soc.*, **93**, 2117
Hsu, K., Buenker, R. J. and Peyerimhoff, S. D. (1972) *J. Am. chem. Soc.*, **94**, 5639
Klopman, G. (1964) *J. Am. chem. Soc.*, **86**, 4550
Klopman, G. (1965) *J. Am. chem. Soc.*, **87**, 3300
Lifson, S. and Warshel, A. (1968) *J. chem. Phys.*, **49**, 5116
Lifson, S. and Warshel, A. (1970) *J. chem. Phys.*, **53**, 8582
Liskow, D. H., Bender, C. F. and Schaeffer III, H. F. (1972) *J. Am. chem. Soc.*, **94**, 5178
McIver, J. M. and Korminicki, A. (1971) *Chem. Phys. Letters*, **10**, 303
Ohno, K. (1964) *Theoret. Chim Acta*, **2**, 219
Pariser, R. and Parr, R. G. (1953) *J. chem. Phys.*, **23**, 466, 767
Pople, J. A. (1953) *Trans. Faraday Soc.*, **49**, 1375
Pople, J. A. (1972) Lecture at the CECAM Colloquium on *Calculations of Reaction Paths and Reaction Mechanisms*, Paris
Pople, J. A., Beveridge, D. L. and Dobosh, P. A. (1967) *J. chem. Phys.*, **47**, 2026
Pople, J. A., Santry, D. P. and Segal, G. A. (1965) *J. chem. Phys. Suppl.*, **43**, 5129
Pople, J. A. and Segal, G. A. (1966) *J. chem. Phys. Suppl.*, **44**, 3289
Radom, L., Pople, J. A., Buss, V. and Schleyer, P. von R. (1972) *J. Am. chem. Soc.*, **94**, 331
Rauk, A., Allen, L. C. and Clements, G. (1970) *J. chem. Phys.*, **52**, 4133
Roothaan, C. C. J. (1951) *Rev. mod. Phys.*, **23**, 69
Ruedenberg, K. (1962) *Rev. mod. Phys.*, **34**, 326
Salem, L. (1972) Personal communication
Westheimer, F. H. (1956) in *Steric Effects in Organic Chemistry* (edited by Newman, M. S.),
    Wiley, New York
Wolfsberg, N. and Helmholz, L. (1952) *J. chem. Phys.*, **20**, 837
Wolfsberg, M. and Helmholz, L. (1953) *J. chem. Phys.*, **21**, 943
Wolfsberg, M. and Helmholz, L. (1955) *J. chem. Phys.*, **23**, 795

16

The Influence of Wave Mechanics on Organic Chemistry

CHARLES A. COULSON, F.R.S.
Department of Theoretical Chemistry
University of Oxford

1. INTRODUCTION

Schroedinger's wave mechanics was introduced in 1926, and it immediately became evident that a large part of chemistry could in principle be derived from it. The first real evidence that this was so in practice came only one year later, when Heitler and London made their famous excursion into chemical behaviour and explained the formation of molecular hydrogen. This was followed by a short spell devoted to understanding the way in which electron spins could be 'paired' to form a normal bond: and then, in 1931, there began what was later to become an explosion. This burst of interest, increasing exponentially with time, was chiefly concerned with organic molecules. As a result, by 1940, there was a growing body of work attempting to understand the small and the large molecules of organic chemistry. In the preface of his book on *Molecular-Orbital Theory for Organic Chemists* Streitwieser claims that 'a mere handful (approximately 20) in the thirties was followed by approximately 70 papers in the forties, whereas the decade of the fifties has witnessed some 600 papers on this subject'. It is my own belief that all these numbers are rather too small; but they pale before the total of several thousand papers in the sixties, and a continuing increase into the seventies. With the possible exception of the study of natural products, wave mechanics has profoundly influenced all parts of organic chemistry, including several technological processes.

In this chapter we shall be concerned to understand this influence, which is by no means always recognised by the very people affected by it. The influence is shown in two quite distinct ways; both will have to be dealt with. But, in their simplest terms, wave mechanics explains to us those things which we already know by experiment, and it does this at a level which justifies us in the claim that now we really *understand* what is happening;

255

then it predicts for us effects that we do not yet know, and can subsequently seek to discover by experiment.

Claims of this kind need to be substantiated. But first it is worth spending a little while in recognising why the field of organic chemistry provided such excellent ground for the application of wave mechanics. That this certainty is the case can very easily be seen from the fact that far more has been known about benzene and naphthalene than about the permanganate ion or the polyhalogenides. Some of these latter systems are only just yielding to the attack made on them by modern huge computers; but explanations of the high stability of benzene and of the reactive sites of naphthalene have been available for many years.

This difference between organic and inorganic molecules stems from the unique character of carbon, and the existence of many homologous series on which to test any possible theory. The unique character of carbon hardly needs to be stressed. There are other atoms in Group IV of the Mendeleev Periodic Table; but silicon and germanium do not have the versatility of carbon, where chains of atoms can form large molecules, and where single, double and triple bonds occur in great variety. It was a fortunate circumstance that put carbon at the middle of a row of the Periodic Table; for this implied the possibility of covalent bonding with less polar character than with inorganic molecules. As a result, carbon–carbon and carbon–hydrogen bonds could be grouped and classified: both represented almost ideal subjects for relatively simple theoretical treatment.

2. EXPLAINING WHAT WE ALREADY KNOW

Many organic molecules are stable enough, and crystallise regularly enough, for good studies to be made of their structure. As a result we are in possession of excellent information about valence angles and bond lengths. The explanation of valence angles came first. We have grown so used to the concept of hybridisation that we do not always realise how significant it was, when Linus Pauling introduced it in 1931. A ground-state carbon atom has electronic structure $(1s)^2(2s)^2(2p_x)(2p_y)$, and, with two electrons in singly-occupied orbitals, it would be expected to be divalent. Van Vleck and others noticed that if a 2s-electron were promoted to the empty $(2p_z)$ orbital, four bonds could be formed. However, if these bonds were of the Heitler–London type using atomic orbitals for each atom, the four bonds for carbon would not all be equivalent. Pauling's genius here consisted in observing that if we mixed – or hybridised – the four atomic orbitals, then four equivalent hybrids could be formed, leading to four equivalent bonds. Moreover the four hybrids 'pointed' tetrahedrally, giving a nice explanation of the familiar tetrahedral valence angles. Further, a pleasant bonus was added: these hybrids had such excellent overlapping power that (Figure 1 for the case of an aliphatic C–H bond) we recovered enough 'overlap energy' to compensate for the s→p promotion, and so make the tetrahedral valence behaviour of carbon more normal than the divalent character such as is found in CBr_2, derived from a carbon atom without s→p excitation.

The concept of maximum overlapping was crucial in all this. In its simplest terms an atom adjusted its hybridisation so that the total energy was lowest. In a molecule CXYZW there would be small deviations from regular tetrahedral shape because the differing overlap between C and X, and Y, ... would favour slightly different hybridisations in the four bonds. Van Vleck and the present writer were able to discuss this effect, and show that these hybridisation ratios were directly related to the corresponding valence angles.

A particularly interesting situation arose when steric restrictions had to be included. In a molecule such as CXYZW there is usually nothing to prevent the bonds from being straight, since for any chosen hybridisation this maximises the various overlaps. But in a ring molecule such as cyclopropane

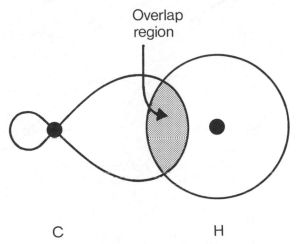

Overlap
region

C   H

Figure 1   Excellent overlap of a carbon tetrahedral hybrid and a hydrogen 1s orbital, as in an aliphatic CH bond

(Figure 2) straight C–C bonds would require two hybrids that pointed in directions making an angle of only 60°. It is easy to show that with s and p atomic orbitals, the orthogonality condition prevents us forming two hybrids at less than 90° to each other. Even then the 'hybrids' are pure p orbitals, with a relatively poor overlapping power. Some compromise therefore has to be achieved. Increasing the overlapping power means that we increase the s-character of the hybrids. But this automatically opens out the angle between them, and so hinders them from overlapping as much as if the bonds could be straight. In his D.Phil. thesis at Oxford the late W. E. Moffitt showed that the compromise occurs when the C–C hybrids make an angle of almost 100°, leading, as in Figure 2, to bent bonds. It also led to an opening-out of the HCH angle, in complete agreement with experiment.

This concept of bent bonds represented a major contribution of wave mechanics. For at one stroke it rationalised the question of strain energy (poorer overlapping implies less binding energy), and rendered obsolete the earlier qualitative account of Baeyer. However, it remained an essentially

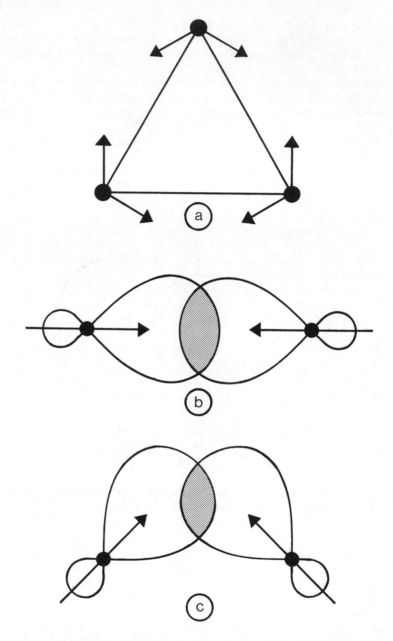

Figure 2 *(a) The carbon atoms in cyclopropane* C_3H_6*: arrows denote directions of hybrids for the C–C bonds. (b) A straight C–C bond; charge density builds up in the overlap region shown shaded. (c) A bent C–C bond, showing charge-density build-up away from line joining the nuclei*

theoretical picture for almost twenty years. To confirm the existence of bent bonds required a very accurate account of the electronic density, to be obtained by X-ray analysis. If the bonds are straight there should be a concentration of charge lying close to the line joining the nuclei: but if the bonds are bent, this concentration should lie off this line, outside the triangle formed by the three carbon atoms. Moffitt's model has recently been triumphantly vindicated by the careful work of Hirshfeld and Hartmann, using the cyano-

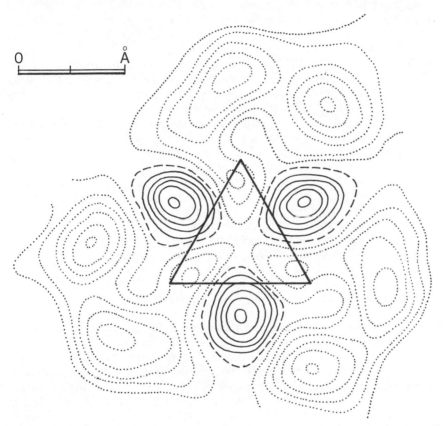

Figure 3   Difference density in the plane of the ring atoms in tricyanocyclopropane (courtesy of Drs. F. L. Hirshfeld and A. Hartmann). Solid lines denote built-up of charge; dotted lines denote decrease of charge density. The vertices of the triangle mark the positions of the carbon atoms

derivative $C_3H_3(CN)_3$. The experimental difference density, shown in Figure 3, reveals an excess of charge in precisely those places where it was to be expected on this theory.

Pauling pointed out that there were many other types of hybridisation than the pure (or nearly pure) tetrahedral kind. If a carbon atom is promoted, as before, to the configuration $(1s)^2(2s)(2p_x)(2p_y)(2p_z)$, and if we keep the $2p_z$-orbital unchanged, we can then mix the s, p_x and p_y orbitals to form sp^2 trigonal hybrids. If the hybrids are equivalent, the valence angle is $120°$, and we have the basis of our understanding of the exceedingly common aromatic bonding, such as occurs in benzene. It may be true that in recent years Pauling has been led to question the usefulness of these trigonal hybrids: but most other theoretical workers still believe them to be important.

In the first place, they lead to the separate discussion of σ and π electrons. The trigonal hybrids are of σ-type, since their wave functions are unchanged by reflection in the xy plane, which is the plane of the aromatic molecule. This distinction into σ and π, which is basic to any full understanding of aromatic systems, depends upon distinguishing between the out-of-plane $2p_z$ atomic orbital on the one hand, and the in-plane 2s, $2p_x$, $2p_y$ orbitals on the other. Recent developments of the photoelectron spectra technique due to D. W. Turner and W. C. Price show conclusively that we must think of the valence electrons in a molecule such as benzene as being in σ or π molecular orbitals, which have their own appropriate energies, and from which they can be ejected by a photon of suitable energy.

In the second place we can see the peculiar importance of the six-membered aromatic ring. Trigonal hybrids point at mutual angles of 120°, the angle appropriate to a regular hexagon. If we either increase or decrease the number of carbon atoms in the ring, we prevent truly effective overlapping of the resulting hybrids, and lower the contribution to the binding energy that comes from the σ bonds. Since this is normally considerably greater than that which comes from the π bonds we can easily see one good reason for the superior stability of the benzene ring.

There is, however, another reason for the special importance of benzene. In the 1920s there developed the concept of an aromatic sextet. The studies of Lapworth, Ingold and Robinson led to the view that ring molecules in which there were precisely six π-electrons would have particular stability. It was not necessary that these should all come from carbon atoms. In pyridine (Figure 4a) one each comes from the carbons and the nitrogen; in pyrrole (Figure 4b) two come from the nitrogen and in furan (Figure 4c) two come from the oxygen, leading in all cases to a total of six.

Such is the experimental situation: what about the explanation? There must be an explanation, and it must be rational; for we are not Pythagoreans, worshipping some magic number such as six or seven, but without rational justification. In our case it is, indeed, quite simple. We suppose that each of the π electrons may be allotted to some orbital – in this situation, a molecular orbital – and we use simple Hückel theory to estimate the energies of these orbitals. The result, for a variety of systems, is shown in Figure 5. Next we remember that most famous of physical principles, the Pauli Exclusion Principle, with its claim that if we are fitting electrons into orbitals, not more than two can be allotted to any such orbital; and even then they must have opposed spins. It is immediately clear that in all the molecules referred to in Figure 5 there is a total of three good bonding orbitals, which together can accommodate exactly six of these π-electrons. If, as in the ring molecule C_7H_7, we have seven, then the seventh is in an antibonding orbital, so that the most stable situation is the cation $C_7H_7^+$, in which this unwanted electron has been lost, or given away. Similarly the five-membered ring C_5H_5 is easily converted to the anion $C_5H_5^-$, since in the neutral system there

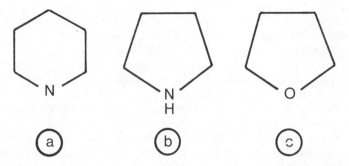

Figure 4 *(a) Pyridine* C_5H_5N; *(b) pyrrole* C_4NH_5; *(c) furan* C_4H_4O

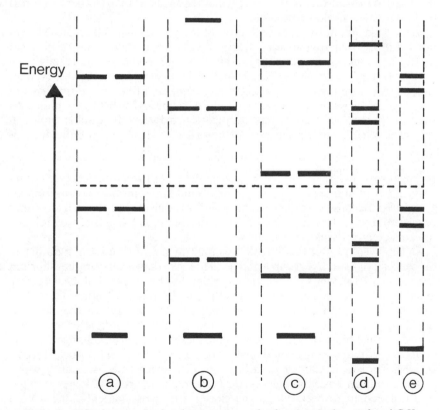

Figure 5 *Orbital energy levels in various ring molecules. (a) Cyclopentadienyl* C_5H_5; *(b) benzene* C_6H_6; *(c) cycloheptatrienyl* C_7H_7; *(d) pyridine* C_5H_5N; *(e) pyrrole* C_4NH_5. *The aromatic sextet. All orbital energies below the dotted line are bonding; all above are anti-bonding*

remains one empty space among the bonding orbitals, with which to welcome any available electron.

This explanation of the aromatic sextet is simple. The computational expert will claim, rightly, that our wave functions should be given much greater flexibility than we have done. He will talk of the failure of π, σ separability, of configuration interaction, and of increasing the size of the basis set. This is largely irrelevant. The great contribution of wave mechanics here, as in so many other situations, is not that it provides a nicely-wrapped-up set of energy values. In this case the energy values of Figure 5 are not even absolutely correct. The really important contribution is that it has given us insight into why Robinson and colleagues could be led to the idea of an aromatic sextet. This insight is indicated by the splendid simplicity of the model, and a sufficiently close numerical agreement (e.g. in energies) for us to feel sure that the model is valid. All this is what the scientist means when he speaks of 'understanding' a phenomenon. Without wave mechanics it is hard to see how in this particular phenomenon, as in many others, any real understanding would be possible.

Let us turn to another example. Almost forty years ago Baker and Nathan noticed that if a methyl group —CH_3 (or some other alkyl group) replaced a hydrogen atom on the periphery of an aromatic molecule, the reactivity of that molecule would be increased. At first it was called the Baker–Nathan effect. But before long R. S. Mulliken had suggested that, in a formal sort of way, we could write the bond structure of a methyl-substituted molecule, such as toluene, as if there was conjugation between the double bonds in the benzene ring and a pseudo-triple-bond —$C{\equiv}H_3$. The symbolic description $\langle\!\!\!\!\bigcirc\!\!\!\!\rangle$ — $C{\equiv}H_3$ suggested that this be called hyperconjugation. Now conjugation – or, as it is better called, delocalisation of electrons – does allow electron migration across the single bond that joins the two unsaturated groups. So this might perhaps be the basis for the electron-releasing character of the methyl group, as inferred from the increased reactivity of toluene towards electron-seeking reactants.

So far, so good. But the trouble was that no reasonable set of parameters for benzene or a methyl group would lead to more than a very tiny equilibrium migration of electrons. Independent estimates by Pople and Dewar gave less than 0.01 for the net transfer into the ring in toluene. Neither the inductive effect, arising among the π-electrons of the ring due to the electron-repelling effect of the methyl substituent, nor the mesomeric effect, represented by the ability of electrons from the methyl group to exchange with the π-electrons of the ring, could give satisfactory numerical values. Hyperconjugation was in danger of falling into disrespect – almost a dirty word.

The situation has now been retrieved, and the concept of hyperconjugation restored to respectability. The clue to the reinstatement came from a study of positive and negative ions. It was clear, for example, that with both the cation and the anion of pyracene (Figure 6), electron-spin measurements could be satisfactorily interpreted only if it could be supposed that the π-electrons of the central naphthalene region could be found with substantial probability on the methylene bridges. In fact, Colpa and de Boer showed that theory and experiment both agree that the odd spin in the positive ion is six times as likely to be on a carbon of the bridges as on a β-position of the naphthalene moiety. This migration onto a methylene group is a form of hyperconjugation. It therefore seems that hyperconjugation is many times more significant for ions than for neutral molecules. So the explanation of the Baker–Nathan effect of increased reactivity of toluene as compared with benzene when attacked by an ion (e.g. NO_2^+ to form nitrotoluene) is that as the ion approaches the neutral molecule it begins to form a bond to it: the formation of this bond requires that the electrons involved in it should come from the toluene molecule, thus forcing it into a state similar to that of a cation. In this condition hyperconjugation can operate effectively, and the greater reactivity of toluene becomes perfectly understandable. But this satisfactory outcome has only been possible because of the strong interplay between experimental organic chemists and the exponents of wave mechanics.

One further example, also from the area of chemical reactivity, must conclude this part of our story. I am thinking of the rules that govern the

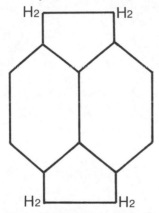

Figure 6   The pyracene molecule

cyclisation of a conjugated molecule such as cis-butadiene. In its chain form (Figure 7a) all the ten atoms lie in a plane. But when the two end carbon atoms are joined by a bond, as in Figure 7b, it is necessary for the four hydrogen atoms a, b, c, d to lie two above and two below the central plane. So the two end $=CH_2$ groups have had to rotate through $90°$. If the two directions of rotation are the same, we call it a conrotatory reaction, whereas if they are opposite, it is disrotatory. Now a few years ago Woodward and Hoffmann noticed that effectively all of these reactions were stereospecific. Thus the cyclisation of butadiene to cyclobutene is always conrotatory if achieved by thermal processes; but disrotatory if achieved photochemically. Similar conclusions were obtained for a host of such reactions, including the opening of the cyclopropane ring and the unwillingness of two ethylene molecules to form a cyclobutane ring.

It was an easy matter to rule out ordinary steric influences, since the sizes of any atoms that replaced a, b, c, d in Figure 7 did not alter the route of the reaction. The explanation must therefore be in the electronic properties of the original butadiene molecule. Moreover, since the absorption of light changes the electronic distribution by lifting one electron from one molecular orbital to another, the explanation must depend upon the different character of the various molecular orbitals. There is no time now to expound this in detail. But one difference between the top-occupied orbital in the ground

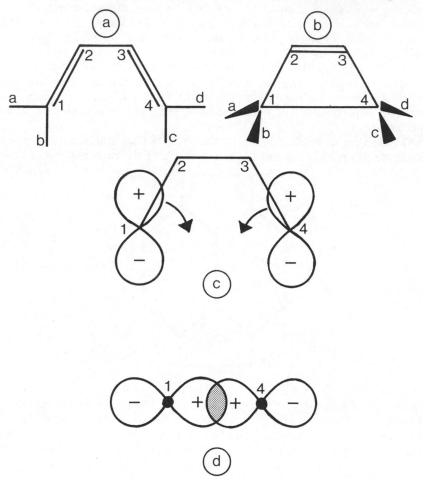

Figure 7   (a) Planar cis-butadiene. (b) Non-planar cyclobutene, in which atoms a, b, c, d lie above or below the plane of the ring. (c) The phase of the first-excited molecular orbital at the two end carbon atoms. (d) The positive overlap that results from dis-rotation in (c)

state, and the newly-occupied orbital in the excited state is that the corresponding de Broglie waves are such that in the former orbital the phases are different at the two ends, whereas in the latter the two phases are the same. Figure 7c shows those parts of the excited orbital which lie at the two end carbon atoms. The similar phase is shown by the allocation of + and − signs.

Now in order to join these two end carbons by a bond, we require to establish a molecular orbital, now of σ-character and not π, in which there is positive overlap between the component atomic orbitals at the two end atoms. If the two —CH_2 groups rotate in opposite directions then, as indicated in Figure 7d there is the basis of a good ring-closure bond. The opposite situation occurs with the molecular orbital from which the electron was photochemically excited. In this way we reproduce the experimental situation by simply asking about the symmetry properties needed to establish the new bond.

Of course a more sophisticated treatment can be given: for example Longuet-Higgins and Abrahamson showed that the same conclusion followed if all the original π-electrons were simultaneously considered. But whether we decide in favour of the more, or the less, sophisticated explanation, it is perfectly clear that our explanation will necessarily be a wave mechanical one. And so the most intriguing and perhaps the most exciting development in organic chemistry in the last ten years gains its real understanding only at this level.

3. PREDICTING WHAT WE DO NOT KNOW

The application of wave mechanics in organic chemistry can be predictive as well as descriptive. Part of this predictive power will lie in its ability to suggest new significant experiments; another part will lie in its introduction of new concepts; and a third part will lie in its ability to settle some unsolved problems. We must say a little about each of these.

Let us begin with cyclobutadiene C_4H_4. A natural assumption would be that this molecule lies in a plane and that the four carbon atoms lie at the corners of a square. But a simple energy-level diagram such as that of Figure 5 indicates that the ground state should then be a triplet. Such an electronic state would be highly reactive. However, closer study showed that if we relaxed the condition of equal bond lengths, and computed the energy for rectangular models, a singlet state could be found lying below the triplet (Figure 8a). In fact it was quite clear that without some additional force the hypothetical molecule would move away from the square configuration and would ultimately dissociate into two acetylene molecules. In this way one could recognise the reasons why traditional synthetic methods had failed to produce a stable cyclobutadiene molecule. However, Orgel and Longuet-Higgins showed that if an appropriate transition element atom (M, Figure 8b) were placed symmetrically between two such cyclobutadiene molecules to form a 'sandwich', then the overlap of certain of the d-orbitals of the central atom with the molecular orbitals of the two π-systems could provide just the necessary force to stabilise the rings. Experiment soon confirmed the prediction.

There is an interesting *sequitur* to this. A few years ago Streitwieser showed that just as the d-orbitals of a transition-metal atom could stabilise a four-membered ring, so the f-orbitals of a rare-earth atom could stabilise the eight-membered ring of cyclo-octatetraene, C_8H_8. In its normal state (Figure 9a) this molecule is not planar, and can be described as a set of four

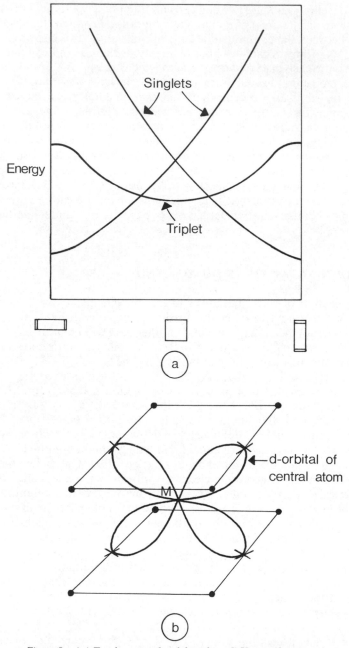

Figure 8   (a) Total energy of cyclobutadiene C_4H_4 as a function of the shape (assumed rectangular) of the four-membered ring. (b) Overlap of a d-orbital of the central metal atom with the two square rings in a cyclobutadiene sandwich molecule. Regions of overlap are marked with a cross weakly interacting ethylenic double bonds, joined by four single bonds. But in the sandwich form (Figure 9b) with a uranium atom in the middle, the planar form is stabilised. Again theory preceded experiment, even though in this case the experiment is very simple.

From these sandwich molecules we turn to some of the radicals so frequently postulated in the mechanisms of organic reactions. One of the most

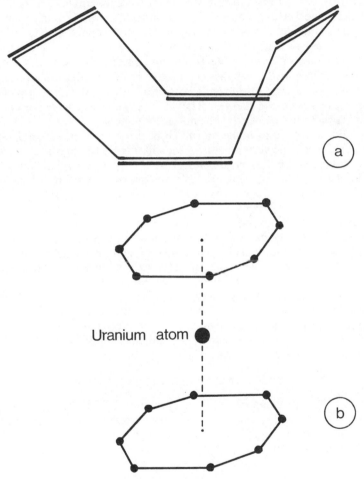

Figure 9   (a) An isolated cyclo-octatetraene molecule C_8H_8. (b) A sandwich molecule with a central uranium atom

famous of these is methylene CH_2. Free radicals of this kind are not easy to measure experimentally; for their lifetimes are usually so short that normal techniques cannot be used. Some experimental measurements on CH_2 radicals had been made by Herzberg, who studied their ultra-violet absorption. It took about twenty years before the experimental techniques showed conclusively that CH_2 molecules were indeed being measured; and even then it was not possible to decide whether the ground state was a singlet or a triplet. If it were a singlet, then, as in water H_2O, one would expect a valence angle of about 105°; but if it were a triplet, the molecule could well be linear. Kinetic studies suggested that under differing circumstances both singlets and triplets were formed, and that they had quite different types of reactivity. If we were to understand these differences it was absolutely necessary to know which was the ground state and what were the angles involved.

At this stage the big computer came into its own (though some more elementary considerations by A. D. Walsh and others gave similar conclusions). For the methylene radical contains only eight electrons, and so much better molecular wave functions and energies can be found for it than for larger molecules. In this case the matter has now effectively been settled by this means. There are indeed two states—a singlet and a triplet— and they do have very close energies. The valence angles are as expected, except that the triplet may be nearly, but not exactly, linear.

It would be wrong to make too much of this sort of situation. Electronic computers—even the very biggest— are still unable to cope adequately with large systems, so that they are more useful to the inorganic than to the organic chemist. But, as this example shows, there are important organic systems where a brute force computation can solve an otherwise unsolved problem.

Nevertheless there are other examples to choose. The trimethylene methyl radical $C(CH_2)_3$ is interesting in this respect. For as long ago as 1948 it was shown by Moffitt and Coulson that in this molecule the central carbon atom attains the highest possible total mobile (i.e. π) bond order*. In fact all calculations of the free-valence numbers of aromatic carbons depend upon the numerical magnitude of this maximum. The molecule itself has frequently been used to test theories of spin distribution. But it was not until 1966 that Paul Dowd obtained an electron-spin-resonance diagram which strongly argued in favour of the presence of the radical. He obtained the radical by irradiation of the pyrazoline system

and found that provided he kept his sample at $-196°C$ it remained stable for several months, but the e.s.r. lines disappeared in a few minutes if the temperature rose to $-150°C$. In the following year, by starting with the cyclobutanone system

$$CH_2 = \text{⟨◇⟩} = O$$

he was able to get an e.s.r. spectrum with the seven lines to be expected for a system with six equivalent protons, as in

An interesting feature was that the splitting between the peaks turned out to be very close to the value previously calculated by valence-bond-theory methods.

* This is for trigonal atoms, for diagonal ones the maximum is a little larger.

We must now turn to the last of the three contributions mentioned at the start of this section – the introduction of new concepts. It is hard to know where to start developing this topic, since many of these concepts, such as covalent–ionic resonance, π- and σ-electrons, hybridisation, lone-pairs, atomic valence states, ring currents in nuclear–magnetic resonance, and the idea of resonance among canonical valence-bond structures, have become part of the everyday language of all sorts of chemists. I have chosen instead to speak of the concept of fractional bond order.

When Mendeleev devised his famous Periodic Table of the elements and set it out in his influential two-volume textbook of chemistry, he related the valence number of an atom to its position in this table. Such numbers were necessarily integral, and they led to the idea of bonds which equally had to be described in integral terms. If G. N. Lewis was right in 1916 to argue that two electrons gave rise to a single bond, four to a double and six to a triple bond, it was inevitable that bonds had integral orders. These bond orders would be associated with particular bond lengths. In carbon–carbon bonds the normal single, double and triple bond lengths are 1.53, 1.34 and 1.20 Å. However, this neat classification failed completely to explain why, in benzene, the bond length was 1.40 Å, intermediate between those of single and double bonds. There were many other examples, chiefly among π-electron molecules, in which almost all lengths appeared possible. The matter was elucidated first by Fox and Martin, and by Pauling. If, as in Figure 10, we put the three basic points on an order-length diagram, we can draw a smooth curve through them. The central idea implicit in this order-length curve is that bonds need not be restricted to be single, double or triple, but can have fractional orders. All that was now necessary was to invent a satisfactory measure of this fractional bond order in terms of the wave functions of the π-electrons. Such a measure was provided, in the valence-bond approximation, by Pauling and Penney; and in the molecular-orbital approximation by the present writer.

An immediate application of this concept will soon show its power. In the molecular-orbital approximation the bonds in benzene have an order 1.67, leading from Figure 10 to a bond length 1.40 Å, in perfect agreement with the experimental value 1.397 Å. Even the infinite graphite layer lattice (Figure 11) can be treated in the same way. The bond order is 1.53, fitting perfectly with the observed internuclear distance 1.42 Å.

This concept of fractional bond order has been widely used, and has proved remarkably effective in predicting, or rationalising, the accurate bond lengths obtained by X-ray methods. So well is it established that if a situation is found where it does not fit, we immediately start looking for some additional effect to complicate the previous simple analysis. One such complication will arise if, for whatever reason, the σ-bond structure which underlies the π-electron distribution, is not as expected. M. J. S. Dewar and the author showed that one such example would be found if we compared bonds in which the hybridisation was not regular trigonal. Instead of the simple curve in Figure 10 a series of curves has to be drawn, one for each pair of hybridisations at the atoms joined by the bond. This at once cleared up a difficulty in butadiene $= - =$, where the curve of Figure 10 suggested a

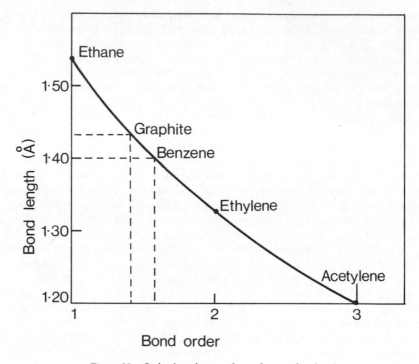

Figure 10 Order–length curve for carbon–carbon bonds

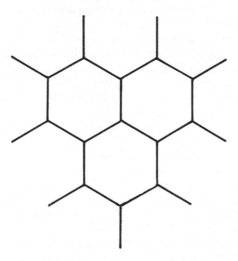

Figure 11 Part of a planar layer in graphite

greater degree of conjugation than was likely on other grounds. Another example is the benzyne system, C_6H_4,

in which the near-acetylenic bond C_1C_2 (length 1.22 Å) prevents the molecule retaining 120° ring angles. This, as Marshall Newton and H. A. Fraenkel have shown, results in hybridisation changes at $C_3, \ldots, C_6$, and corresponding changes in the bond lengths. It is particularly interesting that in the final analysis, the C_4C_5 bond is slightly longer (1.42 Å) than in benzene, but the C_2C_3 and C_3C_4 bonds are scarcely changed from 1.40 Å. Evidently the bond diagram above does not really do justice to the true situation.

It would not be difficult to make a case that the purely theoretical development of the concept of fractional bond order has been as liberating an influence in our understanding of the structure of organic molecules as the three-centre bond in understanding the structure of electron-deficient molecules such as the boron hydrides.

4. CONCLUSION

It should be evident from what has been written above that, although some of the influences of wave mechanics on organic chemistry have depended on elaborate calculations of wave functions and energies, this has not been the case for the greater number of them. As electronic computers increase in size and power, it may become true that more and more of these heavy calculations are made, and, being made, settle issues that previously were uncertain. But I do not believe that this aspect of wave mechanics will ever surpass in importance the new understanding provided by some of the concepts that have been described in this chapter. Organic chemistry really is different now because of the work of De Broglie, Schroedinger, Linus Pauling, Mulliken and others. It is only right that, at this time, this difference should be celebrated.

5. SUMMARY

The influence of wave mechanics on organic chemistry is shown in two ways; first, it provides an explanation, at the deepest level, of chemical behaviour that is already known to us, and second, it enables us to predict new chemical behaviour not already known. Examples of both situations are discussed to illustrate the nature of these contributions. In the first category we consider the varied valence angles in carbon compounds, the nature of the aromatic sextet, the concept of hyperconjugation, the charge and spin distribution in molecules and radicals, and the Woodward–Hoffmann rules for chemical reactions. In the second category we consider the stability of the cyclobutadiene molecule, the structure and shape of the methylene radical CH_2, and the concept of a bond of fractional order. Without wave mechanics it is doubtful whether any of these situations would be properly understood.

17

Quantum Biochemistry

ALBERTE PULLMAN and **BERNARD PULLMAN**
Institut de Biologie Physico-Chimique, Fondation Edmond de
Rothschild, Paris

1. INTRODUCTION

Quantum biochemistry is the youngest offspring on the genealogical tree
of quantum molecular sciences. In so far as it is bound to represent a systematic application of quantum mechanical methods to structures and problems
of biochemistry, its birthdate may be traced to the late 1950s. Its second
distinctive feature consists of dealing with large molecular systems, probably
the largest to which the methods of quantum mechanics have been applied.

Nevertheless, in spite of the relatively short period of existence of quantum
biochemistry, two broad periods may already be distinguished in its
development. The division is based mainly on the type of methodology
utilised but, as it occurs frequently, different methodologies imply also, at
least to some extent, the coverage of different problems.

In the first period which, broadly speaking, extended from 1957 to 1963
the method employed was essentially the simple Hückel theory. The fortunate
fact that a number of biochemical units of fundamental importance are
conjugated heterocycles has encouraged the first theoretical unravelling of
problems connected with their electronic structure in the framework of the
π-electron approximation. Treated in this way were, e.g. the purine and
pyrimidine bases and base pairs, the 'energy rich' phosphates, biological
pteridines, porphyrine, flavins, polyenes, quinones and all the principal
coenzymes. In spite of the simplicity of the procedure utilised, a careful
analysis of the results has allowed the interpretation at the electronic level
of a considerable body of experimental facts and a number of predictions
later confirmed by subsequent experimentation. Some of the predictions, e.g.
those of the values and directions of the dipole moments of the purine and
pyrimidine bases or of the values of their ionisation potentials, made many
years before their experimental determination (some of these predictions,

such as those relevant to the dipole moments of guanine or cytosine still await experimental confirmation) have been of great help in the understanding of such essential problems as the mechanism of the interaction of the bases and their complementarity.

Practically all essential results obtained in this approximation have survived the test of the successive refinements of the π-electron theories, as carried out, e.g. in the framework of the self-consistent field methods and in due time have been complemented by the introduction of a simple representation of the σ-framework.

The first stage in the short but already rich history of quantum biochemistry has been summed up in the book *Quantum Biochemistry* (Pullman, B. and Pullman, A., 1963).

The second stage in the development of quantum biochemistry is intimately linked to the parallel development in the early 1960s of (1) more refined theoretical methods of studying molecular structures, (2) electronic computers, which introduced a new dimension in the practical vistas opened to theoreticians. In particular, the possibility of treating the σ- and π-electrons simultaneously on an equal footing, through the development of the so-called all-valence electrons procedures [the most prominent among which are the Extended Hückel Theory (EHT), the Complete Neglect of Differential Overlap (CNDO/2) method and the Perturbative Configuration Interaction using Localised Orbitals (PCILO) method] or even the non-empirical, so-called *ab initio* SCF method, taking care of all the electrons of the system, enabled one to establish the theory on a firmer basis and to obtain deeper insight into some fine features of electronic structure which were otherwise inaccessible.

Although the first outcome of the penetration of the all-valence and all-electrons methods into biochemistry has been to deepen and refine the studies obtained in simpler approximations, another and more essential benefit of this penetration resided in the fact that it opened also new fields of investigation: among those are in particular the domain of saturated or partially saturated biomolecules and the whole, large area of *conformational problems*, which is of central importance not only for the understanding of the behaviour of large polymers, in particular proteins and nucleic acids, but also for questions related to the mechanism of action of compounds of pharmaceutical and medicinal interest. In fact recent developments in this last field are at the origin of a new direction in quantum-molecular investigations leading to a *quantum pharmacology* (Kier, 1971; Pullman and Courrière, 1973). Last, but not least, the field of *molecular associations* governed by hydrogen-bonding, charge-transfer, Van der Waals–London 'forces', etc., which have until recently been the preserve of semi-empirical perturbative approaches, became amenable to non-truncated calculations, in what is sometimes called the 'super-molecular model'. New possibilities appear also in the treatment of biochemical reactivity.

No general reviews are available yet which describe satisfactorily the achievements of this second stage in the history of quantum biochemistry. Partial accounts are presented in Pullman, A. (1969, 1970); Pullman, B. (1971); Pullman, A. and Pullman, B. (1968, 1972); Pullman, B. and

Pullman, A. (1969, 1971, 1973).

It is, of course, impossible to summarise a whole branch of science in a few pages. Instead of attempting such an impossible task, we shall rather illustrate on a few examples some of its typical aspects.

2. ELECTRONIC PROPERTIES OF BIOMOLECULES

Electronic energy levels and charge distributions have been evaluated for practically all biomolecules of significance. The same concerns a large number of their physico-chemical characteristics, frequently unknown or difficult to establish experimentally, but whose values determine the biological role and fate of the compounds. Thus, as already mentioned, the theoretical predictions of the values of the dipole moments and ionisation potentials of the purine and pyrimidine bases, a long time before their experimental determination, have been instrumental in the comprehension of the mechanism of association of these compounds. In fact, dipole moments are advantageously replaced in these types of studies by the distribution of monopoles (electronic charges), which are more suitable for the evaluation of intermolecular interactions at distances relatively short with respect to the dimensions of the interacting species. Now, quantum mechanical computations are the only reliable procedure for the determination of such charge distributions, which in the most recent achievements by *ab initio* computations have been replaced by isodensity contours as illustrated in Figure 1 on the specific example of adenine.

Illustrating the broad utility of these data is the theoretical exploration of the factors responsible for the exclusiveness of hydrogen-bonded pairing between the bases complementary in the Watson–Crick sense, namely between adenine and thymine (uracil) or guanine and cytosine, as demonstrated by studies on the co-crystallisations of derivatives of the nucleic acids bases carrying a substituent at their glycosidic nitrogen (so as to mimic the situation prevailing in these acids) and on their interaction in non-aqueous solvents. No interactions occur in these circumstances between guanine and thymine or adenine and cytosine, or between cytosine and thymine or guanine and adenine, a result which seems surprising at first sight, because it is easily seen that hydrogen bonds can *in principle* be established in all these cases.

The computations show that all these associations, whether complementary or non-complementary in the Watson–Crick sense, should in fact involve considerable stabilisation energies. A distinct difference appears, however, between the two classes of base pairs when the interaction energies of the cross-associations are compared with those of the auto-associations of the constituent bases. Thus, it is observed that the stabilisation of the A–T association, although relatively moderate, is nevertheless greater than that of the A–A or T–T auto-associations. Similarly, although the stabilisation of the G–G and C–C auto-associations is relatively strong, that of the G–C association is still stronger. On the contrary, all the remaining cross-associations appear always weaker than the auto-association of one of their

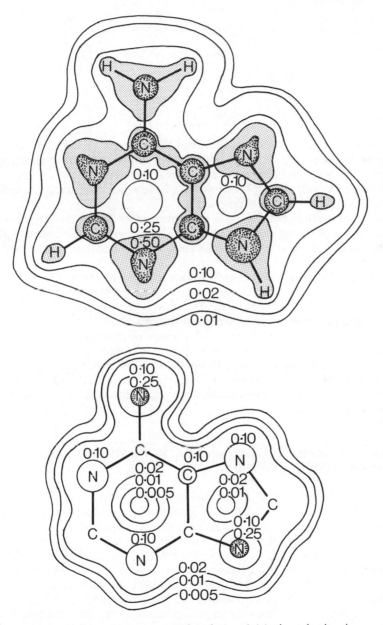

*Figure 1   Global isodensity contours for adenine: (a) in the molecular plane,
(b) above the molecular plane roughly at the position of the radial maximum
of a nitrogen p_z atomic orbital (unit: e/a_0^3)*

constituents, G or C. The general situation may be schematically represented in the following way:

A–T	>	A–A
		T–T
G–C	>	G–G
		C–C
A–C	<	C–C
G–T	<	G–G
C–T	<	C–C
A–G	<	G–G

It points to the conclusion that it is the strength of the G–G and C–C auto-interactions which may possibly be responsible for preventing the in-plane association of the non-complementary bases.

A large amount of experimental data have since confirmed the different facets of this theoretical proposition.

The computations in this field have been extended to the large problem of the in-plane *versus* stacking energy contributions to the overall stability of the nucleic acids, with the general conclusions of an approximate equivalence of the two.

The computation of ionisation potentials and electron affinities of biomolecules, carried out for a very large number of them, while very few of these data are known experimentally, led to a general evaluation of electron donor and acceptor properties of biomolecules which had a strong stimulating influence on the elaboration of a series of proposals related to intermolecule interactions in biochemistry and pharmacology (see e.g. Szent-Gyorgyi, 1960; Pullman, A. and Pullman, B., 1966; Slifkin, 1971) and enabled a critical study of the controversial problem of semiconductivity in biological polymers: proteins and nucleic acids (see e.g. Pullman A., 1965; Pullman, A. and Pullman, B., 1967).

3. THE SUPERMOLECULAR APPROACH TO MOLECULAR ASSOCIATIONS

The most striking development, however, which occurred recently in the field of the theoretical investigation of molecular associations, so primordial in biochemical structures and mechanisms, is related to the advent of 'the supermolecular approach' in which the complex is treated quantum mechanically as a single large molecular entity, the results obtained for the 'supermolecule' then being compared to those obtained for the separate partners, treated in the same approximation. A particularly interesting example is offered by recent studies on hydrogen bonding, a type of interaction of general great importance in biochemistry (nucleic acid base pairing, codon–anticodon recognition, secondary structure of proteins, solvent effects, etc.). Carried out by *ab initio* methods for appropriate models of biochemical situations, this type of computation produced illuminating results on the energetical and structural fine features of the interaction, solving the long debated problem of the respective roles and weights of different 'contributions' to the energy and the none less disputed problem of the nature of the electron displacements accompanying hydrogen-bond formation. Figures 2 and 3 illustrate the type of the results obtained in the case of the 'anti-parallel' dimer of formamide, representing hydrogen bonding between

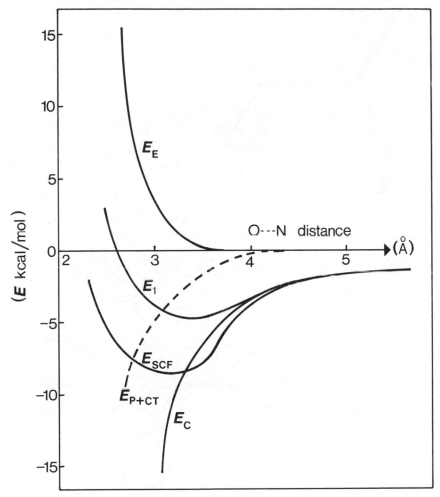

Figure 2 Variation of the energy and its components for a linear approach of two formamides (see text for details)

peptide units of proteins (for details see Pullman A., 1972). Figure 2 shows the variation of the total self-consistent field energy (E_{SCF}) and of its components [first order energy E_1, exchange energy due to antisymmetrisation (E_E) and polarisation and charge transfer energy (E_{p+CT})] and Figure 3 the isodensity differential map at equilibrium, indicating the positions of space which lost

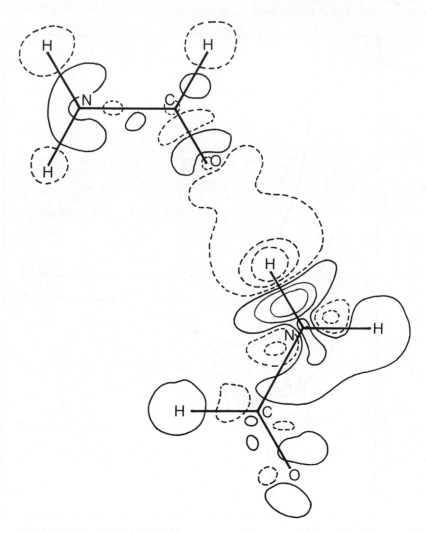

Figure 3   Isodensity differential map at equilibrium for formamide dimer: difference between SCF wave function and isolated monomers. Full lines: gain of electrons; broken lines: loss of electrons ($\pm 10^{-3}$ e/a_0^3, outer curve, to $\pm 2.10^{-2}$ e/a_0^3, inner curve)

or gained electrons upon the bridging. The most striking result hardly accountable for by older techniques is the clear-cut demonstration in Figure 3 of the fact that hydrogen bonding perturbs the distribution of the electronic cloud in the whole system and not only, as frequently thought, in the region limited to the bonded atoms and also the observation of the piling up of negative charges on the H-bonded O and N atoms, in particular the N atom, together with a relative discharge of the intermediate hydrogen.

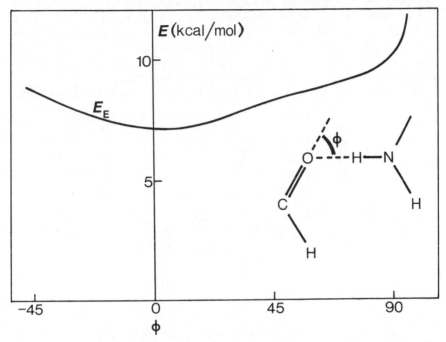

Figure 4   Variation of energy for an in-plane rotation of one monomer with respect to the other

A related most significant result, shown in Figure 4, indicates the flatness of the energy curve for in-plane rotation of one monomer with respect to the other in the region $-45° < \Phi < 75°$, indicating the flexibility of the hydrogen bond, a phenomenon of far-reaching consequence for protein structure.

4. CONFORMATIONAL PROPERTIES OF BIOMOLECULES

This last result brings us to the important contribution made recently by the quantum mechanical approach to the study of conformational properties of biological molecules and polymers. The full value of this contribution may perhaps best be illustrated by comparing its achievements with the previous situation in this field which was dominated by the result of the so-called 'empirical' studies. These consist of partitioning the potential energy of the system into several discrete contributions, such as non-bonded and electrostatic interactions, barriers to internal rotations, hydrogen-bonding, etc., which are then evaluated with the help of empirical formulae deduced from studies on model compounds of small molecular weight (for general reviews of such works, see Ramachandran and Sasisekharen, 1968; Scheraga, 1968).

Praiseworthy as such attempts are, they suffer from two obvious drawbacks. In the first place, whatever the practical justification for the partitioning of the total potential energy into a series of components, the procedure involves necessarily an element of arbitrariness and, possibly, incompleteness. Secondly, the fundamental formulae and parameters used to define the various components are far from being well established and differ, often appreciably, from one author to another. A more rigorous deal may be expected from a quantum mechanical approach which is able to evaluate the total molecular energy corresponding to any given configuration of the constituent atoms and thus to choose the preferred ones.

This expectation was fully satisfied as can be illustrated on two particularly

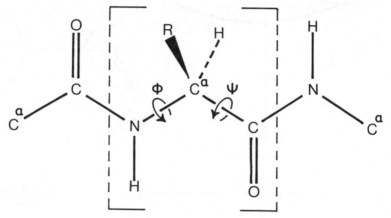

Figure 5   A 'dipeptide' and standard conventions for the study of polypeptide conformations. ⌈ ⌉ limits of a residue

striking examples, referring to the two fundamental biopolymers: proteins and nucleic acids.

One of the principal problems in the study of the conformation of proteins is the determination of the conformational possibilities of their twenty different constituent amino-acid residues. This is done, in general, with the help of the so-called 'dipeptide' model (Figure 5), in which the conformational freedom and restrictions of the residue R are determined by the values of the torsion angles Φ and Ψ. (For details see Ramachandran and Sasisekharen, 1968; Pullman, B. and Pullman, A. 1973.) We have indicated in Figures 6 and 7 the results of typical quantum mechanical and empirical computations on the predicted conformationally stable zones of the two fundamental, glycyl (R = H) and alanyl (R = CH_3), residues. The figures indicate also the experimental conformations of these residues in a large number of globular proteins as determined by high resolution X-ray crystallography. Under these conditions the confrontation leads immediately to two essential conclusions.

(1) It is seen that in both cases the quantum mechanical calculations impose less restrictions on the allowed or preferred conformational space than do the empirical ones.
(2) The agreement between theory and experiment is much better with the quantum mechanical calculations than with the empirical ones, many

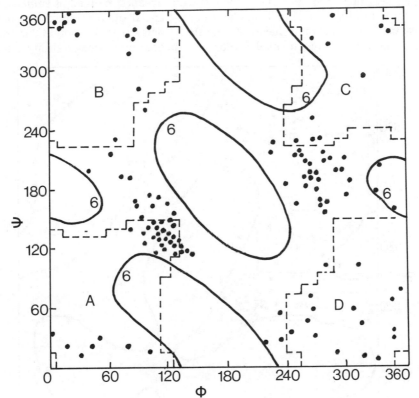

Figure 6   *A simplified conformational energy map for the glycyl residue.* ---- *limits of conformationally stable states in typical empirical computations.* ——— *limits of conformationally stable states in PCILO computations, within 6* kcal/mol *above the global minimum (the number 6 is placed on the side of increasing energies).* ● *conformations of the glycyl residues in globular proteins as indicated by high resolution X-ray crystallography*

representative points lying in the space forbidden by the latter but allowed by the former.

This situation has an important conceptional consequence: it indicates that following the quantum mechanical computations, the 'protein effect' apparently does not create 'extraordinary' conformations, which would correspond to high energy regions of the calculations for the 'dipeptides'. It just operates within the conformational stable zone of the individual residue. A different conclusion would have been drawn from the results of the empirical computations.

Similar results have by now been obtained for *all* the amino-acid residues of proteins and they all lead to similar individual conclusions. They indicate also, however, some general features which enable a substantial extension of the application of the procedure. Thus, one of the basic conclusions which may be drawn from these results is that the general allowed conformational space, within a fixed limit above the individual deepest minima, is similar in all these residues and similar to the general contour obtained for the alanyl residue. This situation is due to the fact that atoms situated beyond C^{β} of the side chain have a much smaller effect on the conformational stability than

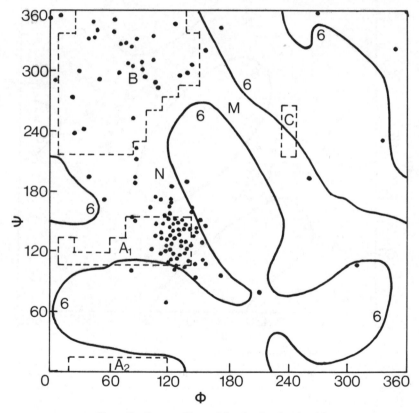

Figure 7   Same as Figure 6 but for the alanyl residue

does the introduction of C^{β}. Naturally, there are differences among the individual residues but they all more or less conform to the pattern obtained for alanyl. What essentially distinguishes these residues from alanyl (and among themselves) is the fine structure of the conformational space and the location of the different energy minima. Under these conditions it seems reasonable to assume that the conformational map of alanyl residue may be considered as representing, at least at first approximation, the conformational map of all C^β-containing residues, i.e. of all residues with the exception of glycine. In order to check the validity of this proposition we may compare the theoretical predictions based on this assumption with the available experimental data on all the amino-acid residues in any of the experimentally determined proteins. Such a confrontation is shown in Figure 8 for the illus-

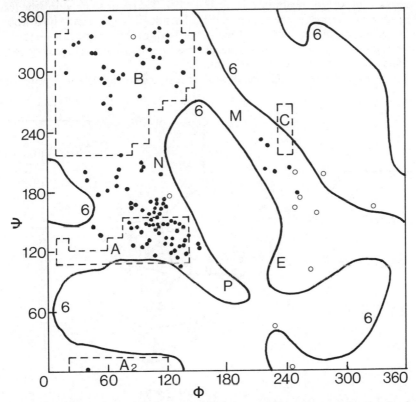

Figure 8   The general significance of the conformational energy map for the alanyl residue. ● *conformations of all the amino acid residues in lysozyme.* ○ *conformations of the glycyl residues*

trative case of lysozyme. The figure is self-explanatory and demonstrates the obvious success of the proposition.

The case of the nucleic acids presents us with a different challenge. The conformational properties of polynucleotide chains depend on a series of torsion angles (Figure 9) which may be divided into three groups.

(1) The glycosidic torsion angle χ, defining the relative orientation of the purine and pyrimidine bases with respect to the sugar.
(2) The torsion angles of the backbone Φ', ω', ω, Φ, Ψ, Ψ'.
(3) the torsion angles about the bonds of the sugars: $\tau_0 - \tau_4$, defining the pucker of this constituent.

This represents a large number of degrees of freedom. Even if we put aside the torsions of the bonds of the sugars and adopt the usual division of the puckering of the sugar into the classical four principal types: C(3′)-endo, C(2′)-endo, C(3′)-exo and C(2′)-exo, we are still left with six essential torsion angles in the mononucleotide unit. If we admit *a priori* that each of these torsions can adopt three preferred values (a reasonable estimation), the number of possible combinations is $3^6 = 729$. Taking into account the four principal puckerings of the sugars, a dinucleoside phosphate may have about 3000 acceptable conformations. One of the goals of the theoretical work is to operate a selection of the most stable and most probable among these possible conformations so as to reduce this number.

The work has been highly successful. It resulted in the determination of a

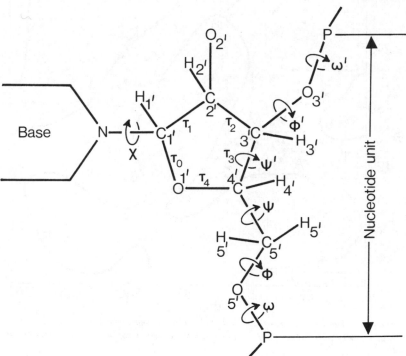

Figure 9   The nucleoside unit

relatively very restricted number of preferred conformations pointing to the relative rigidity of the nucleotide unit and enabling thus consideration of a successful extension of the studies to the determination of the conformational preferences of polynucleotide chains. The superiority of the quantum mechanical treatment with respect to empirical computations is in this field also quite evident (Berthod and Pullman, 1972; Pullman, Perahia and Saran, 1972).

Another field in which the penetration of the quantum mechanical approach combining conformational analysis with the determination of electronic properties has been of outstanding importance, inaugurating practically a new development, is that of molecular pharmacology. Systematic investigations have been carried out or are under way on the conformational and electronic aspects of acetylcholine, nicotine, muscarine, serotonine, histamine, a large series of phenethylamines (ephedrine, norephedrine, dopamine, tyramine, noradrenaline, amphetamines), a number of barbiturates, a number of drugs deriving from hydrazine, penicillins, anti-histamines, sympatholytics, local anaesthetics (procaine, lidocaine, etc.), phenothiazines, opiate narcotic analgesics (morphine, codeine, heroin) etc. (e.g. Kier, 1971; Pullman and Courrière, 1973). In a number of cases the studies have been extended to the examination of drug-receptor interactions, with the ambition of delineating the essential receptor features. Altogether there is a promise here of a most useful guideline in an otherwise very complex field in which progress was based till now essentially on empiricism and intuition.

5. NEW APPROACHES TO CHEMICAL AND BIOCHEMICAL REACTIVITY

It has been customary in the past to utilise, in the study of reactivity, indices based on the knowledge of the molecular electronic structure. There is no difficulty, in principle, in defining in all-valence or all-electrons methods, reactivity indices whether 'static' or 'dynamic' similar to those used, e.g. earlier in the π-electron theory. More fruitful is, however, the exploration of new approaches developed under the impulse of the new methods. One seemingly very promising line is the study of the *molecular potential* seen by an approaching reagent: the knowledge of a molecular wave function allows the accurate calculation of the electrostatic potential $V(r_i)$ created in the neighbouring space by the nuclear charges and the electronic distribution. For a first order density function $\rho(1)$, the average value of the potential $V(r_i)$ at a given point i of space is (Bonaccorsi, Scrocco and Tomasi, 1970)

$$V(r_i) = - \int \frac{\rho(1)}{r_{1i}} dr_1 + \sum_\alpha \frac{Z_\alpha}{r_{\alpha i}}$$

where Z_α is the nuclear charge of nucleus α. The interaction energy between the molecular distribution (considered unperturbed) and, say, an external point charged q placed at point i is $qV(r_i)$, which is rigorously the first order perturbation energy of the molecule in the field of the charge q. Its very definition makes it an appropriate index for studying chemical reactivity, at least in the early phase of approach of an external reagent. Taking q as a *unit positive charge* the interaction potential can then be used for studying proton affinities (basicities) and hopefully electrophilic attacks. In fact, a recent detailed investigation performed for nucleic acid bases for this very problem has given extremely promising results. Such molecules present a particular interest since each of them possesses more than one possible site for protonation or electrophilic attack, thus providing a good test as to the possibilities of isopotential curves to distinguish among different positions of attack, or among different molecules. The more so as molecules with very similar charge distributions on the protonation sites offer frequently very different behaviours: thus, e.g. while in both adenine and guanine, N_7 is the most charged among the pyridine-type nitrogens, alkylation reactions occur preferentially at N_7 and O_6 of guanine but at N_1 or N_3 of adenine. Isopotential maps for an approaching unit positive charge in the molecular plane account for this state of affairs remarkably and they are the first to do so unambiguously. Figure 10 presents such a map for adenine as constructed

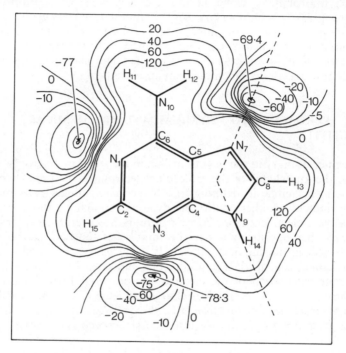

Figure 10   Isopotential map for a unit positive charge approaching adenine in the molecular plane (unit: kcal/mol*)*

with the *ab initio* wave function for this base. It clearly shows that the strongest attraction for an external positive charge occurs towards N_1 and N_3. *Ab initio* results are not yet available for the case of guanine, but the maps obtained for the two bases by the CNDO/2 procedure, although less reliable on the absolute scale, show without any possible doubt the difference in behaviour between the two compounds (Figure 11). This appears again as another very promising line of research, which may potentially be extended to the study of more complex problems.

6. THE PROBLEM OF EXCITED STATES

The refined methods of quantum mechanics reach presently the stage in which, if cautiously handled, they may be applied with equal success to the

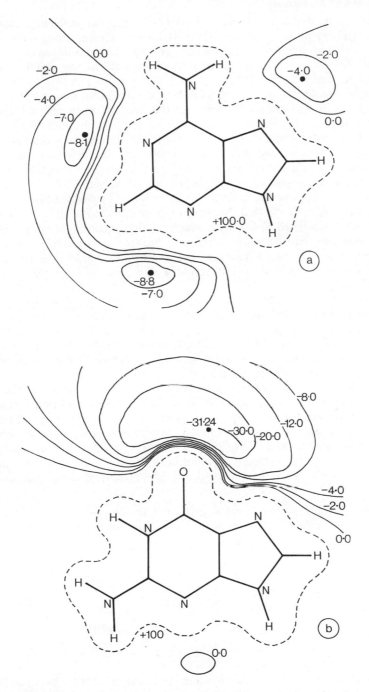

Figure 11 Isopotential maps (CNDO/2 approximation) for the approach of a unit positive charge in the planes of (a) adenine, (b) guanine

excited, at least the lowest excited, as to the ground molecular states. A particularly striking example of their usefulness in unravelling some of the mysteries connected with spectroscopic properties of biomolecules is offered by the curious example of the 'bathochromic shift' in the visual pigments (Mantione and Pullman, 1971).

The problem is the following: while retinal, the chromophore of the visual pigments absorbs around 380 nm and the protein opsin around 280 nm; rhodopsin, the combination of the two, absorbs at much longer wavelength, generally between 480 and 560 nm, depending on the animal species involved. This large bathochromic shift, so drastically responsible for our perception of the outside world, is considered to be one of the most important unexplained problems in the physical chemistry of the visual pigments.

In fact, a part of this shift seems to be properly accounted for at present. Thus, it seems probable that the chromophore is bound to the protein in the form of a *protonated Schiff base* which absorbs around 440 nm. The unaccounted for bathochromic shift, attributable to residual interactions between this protonated base and the apoprotein would thus have to be measured from this limit. It still remains an unusually large shift of about 40–120 nm.

Numerous theories have been proposed in order to explain the origin of this spectral displacement (for a review see Mantione and Pullman, 1971). The most plausible among them is the *point charge perturbation theory*, which postulates that the residual bathochromic shift is brought about by interaction of the protonated Schiff base of 11-*cis* retinal with some reactive groups on opsin placed in its vicinity.

Explicit computations substantiate this general viewpoint and at the same time formulate it in the appropriate form. The procedure utilised consists of the evaluation of the differential stabilisation of the ground state and of the first excited state of the protonated retinylidene iminium ion through interactions with an external point charge located at different positions along the periphery of the cationic chromophore. Both the electrostatic and the polarisation effects are being considered. The procedure thus involves three steps.

(1) Evaluation of the charge distribution in the ground and in the first excited state of the 11-cis retinylidene iminium ion (*I*).

I.   11-cis retinylidene iminium ion

(2) Calculation of the electrostatic interaction energy between these charge distributions in the cationic chromophore and an ion $X^{\delta-}$, placed at different positions around its periphery or in its vicinity.

(3) Calculation of the increment of interaction energy due to the polarisation of retinylidene iminium ion by the $X^{\delta-}$ ion.

The results of the overall treatment are represented schematically in Figure
12 which indicates the difference in the electrostatic stabilisation energy
between the ground and the excited states. (The contribution of the polarisation effect is negligible.) It is observed that as an over-all result this difference
is in favour of the ground state when $X^{\delta-}$ is in the vicinity of the N^+ terminal
of the chromophore (between C_{13} and N_{16}), and in favour of the excited
state for positions of $X^{\delta-}$ between C_1 and C_{12}, and in particular for position

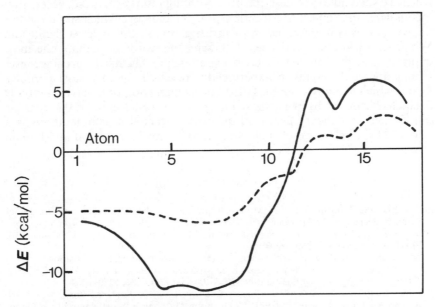

*Figure 12   Differential stabilisation of the ground and the excited states of I through
electrostatic interaction with $X^{-0.6}$ as a function of the position of the external ion along the
periphery. —— close contact approximation, ———— loose contact approximation*

of $X^{\delta-}$ in the vicinity of the region C_5-C_9. Thus, a bathochromic shift of the
spectrum of the retinylidene iminium ion may be expected to occur, as a
result of the electrostatic factor, when the $X^{\delta-}$ group of opsin is located in
the vicinity of these atoms.

A quantitative evaluation shows that the appropriate magnitude of the
shift is obtained provided reasonable assumptions are adopted for the
distance between an appropriate charge on the opsin and the chromophore.

7. A PARTICULAR PROBLEM: CANCER

Finally mention must be made in this short review about a particular
problem, which although it does not belong exclusively to biochemistry,

involves a lot of it and which has been at the forefront of quantum mechanical studies of large molecules for more than a quarter of a century. This is the problem of the possible relationship between the electronic features of different classes of large organic molecules and their carcinogenic potency. The problem was investigated in particular in relation to the polycyclic aromatic hydrocarbons. A theory was elaborated in the early 1940s and continuously developed since (Pullman, A. and Pullman, B., 1955a, 1955b, 1969) which in fact links the appearance of tumours to the existence of certain specific electronic distributions (and hence reactional potentialities) in this type of molecule. This is the so-called K–L theory of carcinogenesis, which we shall not try to describe here. Although thirty years old, necessarily incomplete, and subject to exceptions it is the only one which stood the test of time and is considered by most investigators in this field as significant. Whatever its future, it has been, following the words of perhaps the most eminent specialist in the field of experimental chemical carcinogenesis, stimulating and helpful in experimental research. In this most complex field in which so many attempts failed, and so many proposals have essentially been characterised by their short life-span, the persistence of the possible significance of a theory based on quantum mechanical computations is a symbol of the relevance of such computations for biochemical and human problems.

REFERENCES

Berthod, H. and Pullman, B. (1972) 'Molecular Orbital Calculations on the Conformation of Purine Nucleosides', *The Purines—Theory and Experiment. Proc. 4th Jerusalem Symposium on Quantum Chemistry and Biochemistry* (edited by Bergmann, E. D. and Pullman, B.), 30–47, Academic Press, New York

Bonnaccorsi, R., Scrocco, E. and Tomasi, J. (1970) 'Molecular SCF Calculations for the Ground State of some Three-membered Ring Molecules. $(CH_2)_3$; $(CH_2)_2NH$, $(CH_2)_2NH_2^+$, $(CH_2)_2O$, $(CH_2)_2S$, $(CH_2)_2CH_2$ and N_2CH_2', *J. chem. Phys.*, **52**, 5270–5284

Kier, L. B. (1971) *Molecular Orbital Theory in Drug Research*, Academic Press, New York

Mantione, M. J. and Pullman, B. (1971) 'A Quantum Mechanical Investigation of the Bathochromic Shift in Visual Pigments', *Int. J. quantum Chem.*, **5**, 349–360

Pullman, A. (1965) 'Hydrogen Bonding and Energy Bands in Proteins', *Modern Quantum Chemistry* (edited by Sinanoglu, O.), 284–312, Part 3, Academic Press, New York

Pullman, A. (1969) 'The Electronic Structure of Purines and Pyrimidines', *Ann. N. Y. Acad. Sci.*, **158**, 65–85

Pullman, A. (1970) 'The Present Image of Hetero-aromatics in Quantum-chemistry. Revolution or Evolution', *Quantum Aspects of Heterocyclic Compounds, Proc. 2nd Jerusalem Symposium on Quantum Chemistry and Biochemistry* (edited by Bergmann, E. D. and Pullman, B.), 9–31, Academic Press, New York

Pullman, A. (1972) 'Quantum Biochemistry at the All—or Quasi-all-electron Level', *Topics in Current Chemistry*, Springer, Berlin, **31**, 45–103

Pullman, A. and Pullman, B. (1955b) 'Electronic Structure and Carcinogenic Activity of *Moléculaire*, Masson Ed., Paris

Pullman, A. and Pullman, B. (1955b) 'Electronic Structure and Carcinogenesis Activity of Aromatic Molecules. New Developments', *Adv. Cancer Res.*, **3**, 117–169

Pullman, A. and Pullman, B. (1966) 'Charge Transfer Complexes in Biochemistry', *Quantum Theory of Atoms, Molecules and the Solid State* (edited by Löwdin, P. O.), 345–359, Academic Press, New York

Pullman, A. and Pullman, B. (1967) 'Quantum Biochemistry', *Comprehensive Biochemistry* (edited by Florkin, M. and Stotz, E. H.), Vol. 22, 1–60, Elsevier, Amsterdam

Pullman, A. and Pullman, B. (1968) 'Aspects of the Electronic Structure of the Purine and Pyrimidine Bases of the Nucleic Acids and of their Interactions', *Adv. Quantum Chem.*, **4**, 267–325

Pullman, A. and Pullman, B. (1969) 'A Quantum-Chemist's Approach to the Mechanism of Chemical Carcinogenesis', *Physico-chemical Mechanisms of Carcinogenesis, Proc. 1st Jerusalem Symposium on Quantum Chemistry and Biochemistry* (edited by Bergmann, E. D. and Pullman, B.), 9–24, Academic Press, New York

Pullman, A. and Pullman, B. (1972) 'The Quantum Theory of Purines', *The Purines–Theory and Experiment, Proc. 4th Jerusalem Symposium* (edited by Bergmann, E. D. and Pullman, B.), 1–20, Academic Press, New York

Pullman, B. (1971) 'Etudes Conformationnelles on Biochimie Quantique', *Aspects de la Chimie Quantique Contemporaine* (edited by Daudel, A. and Pullman, A.), 261–300, Editions du C.N.R.S.

Pullman, B. and Courrière, Ph. (1973) 'Molecular Orbital Studies on the Conformation of Pharmacological and Medicinal Compounds', *Conformation of Biological Molecules and Polymers, Proc. 5th Jerusalem Symposium in Quantum Chemistry and Biochemistry* (edited by Bergmann, E. D. and Pullman, B.), Academic Press, New York, in press

Pullman, B., Perahia, D. and Saran, A. (1972) 'Molecular Orbital Calculations on the Conformation of Nucleic Acids and their Constituents. III Backbone Structure of Di- and Polynucleotides, *Biochim. Biophys. Acta*, **269**, 1–14

Pullman, B. and Pullman, A. (1963) *Quantum Biochemistry*, Wiley-Interscience, New York

Pullman, B. and Pullman, A. (1969) 'Quantum-mechanical Investigations of the Electronic Structure of Nucleic Acids and their Constituents', *Progr. nucl. Acid Res. Molec. Biology*, **9**, 327–402

Pullman, B. and Pullman, A. (1971) 'Electronic Aspects of Purine Tautomerism', *Adv. heterocycl. Chem.* **13**, 77

Pullman, B. and Pullman, A. (1973) 'Molecular Orbital Calculations on the Conformation of Amino-acid Residues of Proteins', *Adv. Protein Res.*, in press

Ramachandran, G. N. and Sasisekharan, V. (1968) 'Conformation of Polypeptides and Proteins', *Adv. Protein Chem.*, **23**, 243–437

Scheraga, H. A. (1968) 'Calculations of Conformations of Polypeptides', *Adv. phys. org. Chem.*, **6**, 103–184

Slifkin, M. A. (1971) *Charge Transfer Interactions of Biomolecules*, Academic Press, New York

Szent-Gyorgyi, A. (1960) *Introduction to a Submolecular Biology*, Academic Press, New York

18

Application of Wave Mechanics to the Study of Natural Optical Activity in Organic Molecules

RUDOLF E. GEIGER and GEORGES H. WAGNIÈRE
Institute of Physical Chemistry, the University of Zurich, Zurich

1. GENERAL INTRODUCTION

1(a) OPTICAL ACTIVITY AND STEREOCHEMISTRY

The study of natural optical activity has been of fundamental importance in the development of stereochemistry. One may well say that it has introduced the third spatial dimension into chemical thinking. A more detailed outline of this development than can be presented here has been, for instance, given by Kuhn (1933), or by Mathieu (1957).

Arago in 1811 discovered that crystalline quartz rotates the plane of linearly polarised light. Biot and Seebeck showed a few years later that this effect is not restricted to the solid state, but may be observed in liquids and vapours as well. This implied that in such cases optical activity is to be viewed as a molecular property, and that the existence of the effect is not tied to particular intermolecular interactions. Pasteur in 1848 found that a substance which is optically active may always occur in two oppositely rotating modifications or enantiomers. The search for an explanation of this fact led him around 1860 to propose that a molecule is optically active when it cannot be superposed onto its mirror image. This implies a definite three-dimensional arrangement of the atoms in the molecule and the absence of a plane of symmetry, of a centre of inversion or of a rotation–reflection axis (see also Section 2e). Van't Hoff's and Le Bel's theory of the tetrahedrally bound carbon atom was a further direct consequence of such ideas.

In an optically active medium the index of refraction for right circularly polarised light n_r is different from the one for left circularly polarised light n_l. The angle of rotation per unit length ϕ in radians/cm of a linearly polarised beam is related to the difference

$$\Delta n = n_l - n_r \text{ as } \phi = \pi \, \Delta n/\lambda \tag{1}$$

The measurement of Δn as a function of wavelength is called optical rotatory dispersion (ORD). This effect is inseparably connected to circular dichroism (CD), the difference between the absorption coefficients for left and right circularly polarised light in regions of absorption. The combined phenomena of ORD and CD of a given transition is called the Cotton effect. Incident linearly polarised light emerges as elliptically polarised if it has travelled through a circular dichroic medium. If E_l and E_r are the amplitudes of the left and right circularly polarised beams respectively, the ellipticity ψ is defined as

$$\tan \psi = E_r - E_l/E_r + E_l \tag{2}$$

A practically measured quantity, the ellipticity per unit length θ, is related to the extinction coefficient ε as

$$\theta = \frac{2.303}{4} \cdot C \cdot (\varepsilon_l - \varepsilon_r) \tag{3}$$

C being the concentration of optically active molecules in mol/l.

The problem which confronts the chemist is to deduce the absolute configuration of a molecule from measured ORD and CD spectra. The present paper is in no way a complete review of this field. We will try and outline the search for simple rules and show how an elementary theoretical approach can serve this purpose. Also here the advent of wave mechanics has led to a breakthrough.

1(b) THE CLASSICAL THEORY OF OPTICAL ACTIVITY

From an historic point of view it is of interest that chronologically the development of the classical description of optical activity reached its peak *after* the first attempts to apply quantum mechanics to this effect. Based on prior investigations by Born (1915, 1918), Oseen (1915) and other researchers (Gans, 1923; de Malleman, 1930), Kuhn (1929) studied possibly the simplest optically active model. It consists of two coupled, perpendicular harmonic oscillators separated by a distance d. The kinetic and potential energy are respectively

$$T = \frac{m_1}{2}\dot{x}_1^2 + \frac{m_2}{2}\dot{y}_2^2, \quad V = \frac{k_1}{2}x_1^2 + \frac{k_2}{2}y_2^2 + k_{12}x_1x_2 \tag{4}$$

Let the two oscillators have effective masses m_1 and m_2 and effective charges e_1 and e_2 respectively and let them interact alternately with the electric vector of a left and right circularly polarised beam

$$\vec{E}_{l/r} = E_0[\vec{i} \cos(\omega t - kz) \pm \vec{j} \sin(\omega t - kz)] \tag{5}$$

$\vec{i}$ and $\vec{j}$ being unit vectors in x and y directions, $\omega = 2\pi\nu$, $k = 2\pi/\lambda$. It may be shown in a straightforward way that the amplitude of forced oscillation of one normal mode is larger with a left circularly polarised beam of light than with a right circularly polarised one. For the other normal mode the situation is exactly reversed. The oscillating system of charges emits a secondary wave, and the superposition of this secondary wave on the primary wave leads to a resulting wave, the phase velocity of which gives the index of refraction. It is shown that the secondary wave emitted by a pair of oscillators, as described, is in general elliptically polarised. A manifold of N such pairs per unit volume, evenly distributed around the Z axis, will tend to give a circularly polarised secondary wave. Kuhn obtains the following expression for the rotatory dispersion of such a system, assuming the coupling parameter k_{12} within each pair to be small:

$$\phi = \frac{N}{2\pi c^2} \frac{e_1 e_2}{m_1 m_2} \frac{k_{12}d}{(v_1^2 - v_2^2)} \left[\frac{v^2}{v_1^2 - v^2} - \frac{v^2}{v_2^2 - v^2} \right] \tag{6}$$

v_1 and v_2 designate the frequency of each normal mode and c the light velocity. It is seen that the occurrence of optical activity is dependent upon k_{12} and d being non-zero. For very small and very large frequencies ϕ vanishes. To treat circular dichroism one solves the equations of motion by adding a damping term. Otherwise the amplitude of oscillation would become infinite at resonance. The index of refraction now becomes complex, and the imaginary part of it is proportional to the absorption coefficient. In this way, with the same approximations as for ϕ, Kuhn and Braun (1930) obtain

$$\theta = \frac{N}{2\pi c^2} \frac{e_1 e_2}{m_1 m_2} \frac{k_{12}d}{(v_1^2 - v_2^2)} \left[\frac{v\Gamma}{4(v_1 - v)^2 + \Gamma^2} - \frac{v\Gamma}{4(v_2 - v)^2 + \Gamma^2} \right] \tag{7}$$

One notices the contributions of the two normal modes (equations 6 and 7) to be opposite to each other in sign (see also Section 2c).

The essence of the application of the classical method to actual molecules consists in generalising the above treatment to a large number of coupled oscillators and in identifying every absorption in the actual spectrum with a normal mode. Although Kuhn's approach is a significant and most appealing contribution to the interpretation of ORD and CD spectra, its limitation rests in the failure of classical mechanics to satisfactorily describe the interaction of electrons with each other and with atomic nuclei in a molecule.

1(c) THE QUANTUM MECHANICAL APPROACH

From the point of view of electrodynamics a medium is optically active when the electric polarisation induced by the electromagnetic field also depends on the time derivative of the magnetic field.* To assess the contribution of a given molecule to optical activity the induced electric moment has to be calculated, including the corresponding higher terms. Rosenfeld (1929) appears to be the first to apply quantum mechanics to the problem, using a matrix formalism.

A wave mechanical formulation was given by Condon, Altar and Eyring

---

* Concomitantly the magnetisation depends on the time derivative of the electric field.

(1937). These treatments are semi-classical in the sense that the electromagnetic field is not quantised and represented by a vector potential of the form

$$\vec{A}(\vec{r},t) = \frac{1}{2}\left[\vec{A}_0^0 e^{i\omega t} + \vec{A}_0^0{}^* e^{-i\omega t}\right] \cdot e^{-i\vec{k}\cdot\vec{r}} = \vec{A}_0(t) e^{-i\vec{k}\cdot\vec{r}} \qquad (8)$$

The Hamiltonian for a given electron in the molecule subjected to the field is written

$$\mathcal{H} = \mathcal{H}_0 + \mathcal{H}_1 = \frac{1}{2m}\left[\vec{p} - \frac{e}{c}\vec{A}(\vec{r},t)\right]^2 + V(\vec{r}) \qquad (9)$$

$V(\vec{r})$ stands for the electrostatic interactions of the electron with all other electrons and all nuclei. Within the Born–Oppenheimer approximation the nuclei are assumed fixed at equilibrium positions. Upon expansion one obtains

$$\mathcal{H}_1 = i\hbar\frac{e}{mc}\vec{A}\cdot\vec{\nabla} + i\hbar\frac{e}{2mc}\vec{\nabla}\cdot\vec{A} + \frac{e^2}{2mc^2}A^2 \qquad (10)$$

in the Coulomb gauge $\vec{\nabla}\cdot\vec{A} = 0$. In numerous texts (see, for instance, Eyring, Walter and Kimball, 1964) it is customary to neglect $(e^2/2mc^2)\,A^2$ right away, leaving only the first term on the right-hand side of (10) to be considered. In it the vector potential is expanded in a Taylor's series around a point in the molecule – assuming the wavelength λ to be much larger than the molecular dimensions – to obtain in the Hamiltonian an electric dipole term in the dipole velocity form, a magnetic dipole term, an electric quadrupole term, etc. A more general procedure consists in subjecting $\mathcal{H}$ to a canonical transformation (Richards, 1948; Power and Zienau, 1958; Fiutak, 1963; Wallace, 1966) of the general form

$$\mathcal{H}' = e^\beta \mathcal{H} e^{-\beta} + i\hbar\frac{\partial\beta}{\partial t} \qquad (11)$$

with

$$\beta = -\frac{i}{\hbar}\frac{e}{c}\vec{A}\cdot\vec{r}$$

by which, using the relations

$$\vec{E} = -\frac{1}{c}\frac{\partial\vec{A}}{\partial t}$$

and

$$\vec{H} = \vec{\nabla}\times\vec{A}$$

a multipole expansion of the Hamiltonian is more directly achieved:*

$$\mathcal{H}' = \mathcal{H}_0 - e\vec{r}\cdot\vec{E} - \frac{e}{2mc}\vec{l}\cdot\vec{H} + \dots \qquad (12)$$

* For a rigorous and complete treatment of this question see Power and Zienau (1958) and Fiutak (1963).

The interplay of the electric dipole and magnetic dipole terms are important for the optical activity. The derivation of the measurable quantities ϕ and θ, while somewhat cumbersome, is straightforward and described in detail by Condon (1937) or by Eyring, Walter and Kimball (1964). We here just state the well-known results. For simplicity it is assumed that the N identical molecules per unit volume are all in their ground state a, and that the mean index of refraction of the medium is equal to unity. Designating the excited states collectively by b,

$$\phi = \frac{16\pi^2 N}{3hc} \sum_b \frac{v^2 \text{Im}[\langle a|\vec{R}|b\rangle\langle b|\vec{M}|a\rangle]}{v_{ba}^2 - v^2} \tag{13}$$

$$\vec{R} \equiv \sum_v \vec{\mu}_v \equiv \sum_v e\vec{r}_v; \quad \vec{M} \equiv \sum_v \vec{m}_v \equiv \sum_v \frac{e}{2mc}\vec{l}_v$$

The factor $1/3$ comes from averaging over the randomly oriented molecules, v stands for the frequency of incident light and v_{ba} for the transition frequency. The rotatory strength $\text{Im}[\langle a|\vec{R}|b\rangle\langle b|\vec{M}|a\rangle]$ is, basically, a tensor quantity, but may be considered as a pseudo-scalar when isotropic averaging takes place (Tinoco, 1962).

The contribution of a given transition $a \to b$ to the ellipticity, $\theta_{ab}(v)$, is related to the rotatory strength of that transition by (Condon, Altar and Eyring, 1937; Moscowitz, 1962):

$$\text{Im}[\langle a|\vec{R}|b\rangle\langle b|\vec{M}|a\rangle] = \frac{3hc}{8\pi^3 N} \int \frac{\theta_{ab}(v)}{v} dv \tag{14}$$

where the integration on the right has formally to be carried out from $v = 0$ to $v = \infty$; physically it goes over all populated substates of a and all substates of b.

Equations 13 and 14 are strictly valid only when the dimensions of the individual molecules are small compared with the wavelength λ. If not, the vector potential (8) cannot be expanded into a rapidly converging series. Higher terms in (12) may conceivably have to be taken into account (see also Section 3b).

As between ordinary dispersion and absorption, or, in general, between the real and imaginary part of any complex polarisability, the rotation ϕ and the ellipticity θ are related by transforms of the Kronig–Kramers type (Moffitt and Moscowitz, 1959: Moscowitz, 1962):

$$\phi(v) = \frac{2v^2}{\pi} \int_0^\infty \frac{\theta(v')}{v'(v'^2 - v^2)} dv' \tag{15a}$$

$$\theta(v) = -\frac{2v^3}{\pi} \int_0^\infty \frac{\phi(v')}{v'^2(v'^2 - v^2)} dv' \tag{15b}$$

For a treatment of optical activity considering the quantisation of the electromagnetic field see Gervais (1969) and Philpott (1972). The results do not basically differ from the ones obtained by the semi-classical approach.

2. SIMPLE MODELS TO INTERPRET OPTICAL ACTIVITY

2(a) THE ONE-ELECTRON MODEL

After what we have just seen in Section 1 the main problem confronting the theoretical chemist interested in optical activity consists in evaluating the rotatory strength for the different transitions in actual molecules. In view of the well known difficulties in obtaining accurate molecular wave functions, the problem, though more of an 'engineering' type than of a fundamental nature, is by no means easy or without interest. The rotatory strength $R_{ab} \equiv \text{Im} [\langle a | \vec{R} | b \rangle \langle b | \vec{M} | a \rangle]$ is the product of a polar and an axial vector. As such it is not only sensitive to the overall symmetry, but also to local symmetry properties of a molecule. It is thus conceivable that even crude wave functions, which correctly reflect these symmetry properties, may be adequate to deduce rules of thumb that are of direct use to the practical chemist. It may of course easily be shown that R_{ab} vanishes in a system containing, for instance, a plane of symmetry or a centre of inversion.

Under the influence of Kuhn's work (1933) it appeared that rotatory power requires the existence of coupled oscillators for its explanation. Condon, Altar and Eyring (1937) showed quantum mechanically that a single electron moving in a potential field of the form

$$V = \tfrac{1}{2} k_1 x^2 + \tfrac{1}{2} k_2 y^2 + \tfrac{1}{2} k_3 z^2 + k' xyz \qquad (16)$$

is optically active. This potential is invariant under rotations of 180° around the x, y, z axes and is therefore of symmetry D_2 (Caldwell and Eyring, 1971). The fact that the coupling between different electrons is not a prerequisite to the occurrence of optical activity does not, however, imply that electron interaction effects are unimportant, as we shall see in Section 3a. As a matter of fact, it might actually be of interest to study a many-electron system in the above-mentioned potential, in which the $e^2/r_{\mu\nu}$-terms are explicitly taken into account.

The one-electron approach was pursued by Kauzmann, Walter and Eyring (1940) and applied to an asymmetrically substituted cyclic ketone. Although these authors' description of the electronic structure of the carbonyl group does not coincide with the picture considered today as valid, this paper well marks the beginning of a development which culminated in the statement of the so-called sector rules twenty years later.

Saturated ketones are ideally suited for the one-electron model. They show a generally weak, but spectroscopically easily accessible $n \rightarrow \pi^*$ transition around 300 nm, which is strongly localised on the carbonyl group. It is followed by further transitions only in the vacuum ultra-violet region, among which one also finds a relatively strongly localised $\pi \rightarrow \pi^*$ transition around 160 nm. In symmetric ketones, where the atoms $\underset{\text{C}}{\overset{\text{C}}{>}}$C=O lie in a plane of symmetry, the former transition is magnetic dipole allowed and becomes weakly electrically allowed only by vibronic interaction. The

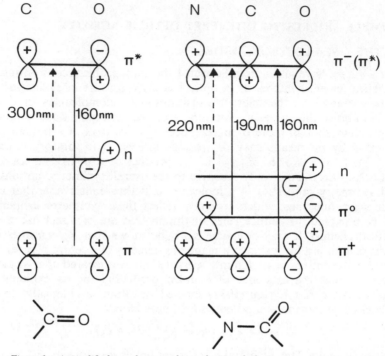

Figure 1   A simplified one-electron scheme showing the longest-wavelength transitions in the carbonyl and amide chromophores. If the environment of the carbonyl chromophore maintains the C_{2v} symmetry, then the $n \to \pi^*$ transition is magnetic dipole allowed in the x-direction (see Figure 2), electric dipole forbidden. The reverse holds for the $\pi \to \pi^*$ transition. In the amide chromophore of symmetry C_s the $n \to \pi^-$ transition is magnetic dipole allowed in the x, y-plane, electric dipole forbidden. The situation is analogous to the carbonyl case

$\pi \to \pi^*$ transition is electric dipole allowed (see Figure 1). One thus finds in this symmetric case with $n \approx p_{yo}$,

$$\pi \approx \frac{1}{\sqrt{2}}(p_{zc}+p_{zo}), \pi^* \approx \frac{1}{\sqrt{2}}(p_{zc}-p_{zo}):$$

$$\langle n|\vec{r}|\pi^*\rangle = 0; \quad \langle n|\vec{l}|\pi^*\rangle \neq 0; \quad R_{n\pi^*} = 0 \qquad (17a)$$

$$\langle \pi|\vec{r}|\pi^*\rangle \neq 0; \quad \langle \pi|\vec{l}|\pi^*\rangle = 0; \quad R_{\pi\pi^*} = 0 \qquad (17b)$$

The effect of an asymmetric substituent S may be viewed as a perturbing potential V_S. It will mix the orbitals n and π

$$n' = n+\lambda\pi, \qquad \lambda = \frac{\langle n|V_S|\pi\rangle}{\Delta\varepsilon_{n\pi}} \qquad (18a, b)$$

leading correspondingly to

$$R_{n\pi^*} = \lambda \langle \pi \,|\, \vec{\mu} \,|\, \pi^* \rangle \, \langle \pi^* \,|\, \vec{m} \,|\, n \rangle \neq 0 \qquad (19)$$

that is, inducing optical activity. Within the frame of this admittedly very crude approximation, only λ depends on the nature and position of the substituent. If V_S is everywhere positive in space, disappearing at infinity, then the relative sign of the contribution of the substituent S to $R_{n\pi^*}$ will be given by the product of the functions n and π, or as the sign of $y \cdot z$, at the position of S. Space around the C—O axis is thus divided into four sectors $+$, $-$, $+$, $-$, and this type of dependence is termed a 'quadrant rule' (see Figure 2) (Schellman and Oriel, 1962; Wagnière, 1966). As $\Delta\varepsilon_{n\pi}$ will be of the order of possibly 3–4 eV and consequently λ, as defined above, may be small, it is entirely conceivable that other mechanisms like the one described are mainly responsible for making $R_{n\pi_*}$ non-vanishing. Moscowitz (1962) has considered the admixture of $3d_{xy}$ on both C and O to π^*. Because of the

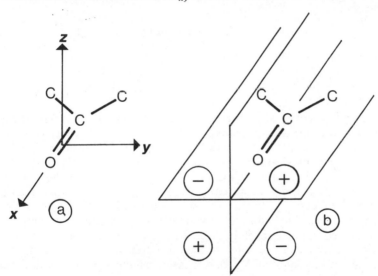

Figure 2   (a) Adopted co-ordinates to describe the electronic structure of the carbonyl group. (b) Illustration of the quadrant rule

node in the π^* orbital bisecting the C—O bond, the relative sign of the contribution of S to $R_{n\pi_*}$ now varies as $x_S \cdot y_S \cdot z_S$, corresponding to an 'octant rule'. In fact, the octant rule for ketones was the first sector rule to be explicitly stated (Moffitt *et al.*, 1961). Whether in reality the quadrant rule, or the octant rule, holds in ketones, is hard to decide because of the difficulty of synthesising compounds with asymmetric substituents in front octants (Kirk, Klyne and Mose, 1972). The impossibility of attaching substituents directly to the oxygen atom may, in the long run, even make the problem rather academic. It can be shown that from a group-theoretical point of view the quadrant rule for ketones is more fundamental (Schellman, 1966).

2(b) GENERALISED SECTOR RULES

If in (16) the asymmetric potential $k'xyz$ is assumed to arise from a point charge Q located at x_Q, y_Q, z_Q, one finds, upon Taylor expansion of V_Q around the origin of the co-ordinate system, that k' is proportional to the product $x_Q \cdot y_Q \cdot z_Q$ (Caldwell and Eyring, 1971). As, on the other hand, the rotatory strength for the different transitions in the model is always proportional to k', an octant rule for the optical activity is obtained with respect to the position of Q.

Schellman (1966) has systematically studied chromophores of different symmetry made optically active by an external, asymmetric potential V. Let us assume that the ground state a of such an unperturbed chromophore transforms like the totally symmetric representation Γ_A of the particular point group. We consider as an example a particular excited state b to which the transition from a is electric dipole allowed, magnetic dipole forbidden. In the asymmetrically perturbed system a goes over into a', b into b' and one has

$$R_{ab} = 0, \qquad R_{a'b'} = i\langle a | \vec{R} | b \rangle \cdot (\lambda_1 + \lambda_2) \qquad (20a)$$

where by first-order perturbation theory

$$\lambda_1 = \sum_{c \neq a} \frac{\langle c | V | a \rangle}{E_c - E_a} \cdot \langle b | \vec{M} | c \rangle, \qquad \lambda_2 = \sum_{d \neq b} \frac{\langle b | V | d \rangle}{E_d - E_b} \cdot \langle d | \vec{M} | a \rangle \qquad (20b)$$

and where c, d cover unperturbed excited chromophore states. Both the state b and the operator $\vec{R}$ transform like Γ_e, say, and $\vec{M}$ transforms like Γ_m. In λ_1, for $\langle b | \vec{M} | c \rangle$ not to vanish, the state c must transform like $\Gamma_e \cdot \Gamma_m$. Consequently V must also transform like $\Gamma_e \cdot \Gamma_m$ for $\langle c | V | a \rangle$ to be non-vanishing. Similar arguments hold for λ_2 and imply that to be effective in inducing optical activity the potential V must contain a part that transforms like a pseudo-scalar, V_{ps}. In the opposite case, in which the transition $a \to b$ is magnetic dipole allowed in the unperturbed chromophore and an electric dipole allowed component is induced by the perturber — as for the $n \to \pi^*$ transition in ketones — one is led to identical conclusions concerning V.

The part of the potential V_{ps} that transforms like a pseudo-scalar may be obtained from V by a projection operator

$$V_{ps} = \frac{1}{n} \sum_R \chi_{ps}(R) \cdot P_R V \qquad (21)$$

where n is the number of operations in the symmetry group, χ_{ps} the character of $\Gamma_{ps} = \Gamma_e \cdot \Gamma_m$, belonging to the group element R, and $P_R V$ is the result of applying the operation R to V. In the idealised case where the perturbing potential is Coulombic, V_{ps} may be interpreted in terms of a multipole expansion. The explicit form of V_{ps} determines the dependence of $\langle c | V | a \rangle$, $\langle b | V | d \rangle$ on the location of the perturber and consequently also the sector rules for $R_{a'b'}$. Schellman obtains the following results. For an unperturbed chromophore of symmetry C_s the simplest pseudo-scalar function has the form z, leading to a planar rule. Asymmetrically substituted amides, peptides, carboxylic acids should show this behaviour. For the point group C_{2v} as for ketones, or nitro compounds, the result is yz, or a quadrant rule. For C_{2h} one finds a planar rule, but for D_{2h} an octant rule. For chromophores of higher symmetry, for groups containing also degenerate representations, the results become somewhat more complicated.

For some applications of these sector rules see Velluz, Legrand and Grosjean (1965), Djerassi (1967), Snatzke (1968). For a somewhat different approach to the problem see Ruch and Schönhofer (1970).

2(c) THE COUPLED-OSCILLATOR MODEL

Suppose one has a molecule composed of two identical monomers 1 and 2. These symmetric monomers are weakly coupled to form an asymmetric

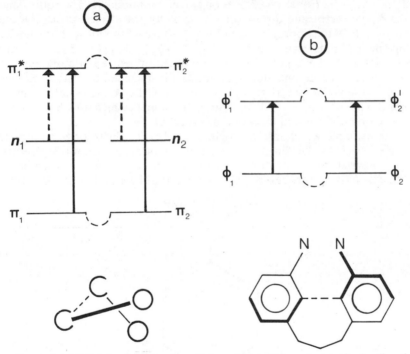

Figure 3   (a) Schematic illustration of the coupling of two electric dipole allowed transitions within a dimer composed of two identical monomers (see also Figure 4). (b) In the dicarbonyl system the longest-wavelength transitions are electric dipole forbidden in the isolated monomer. The coupled-oscillator model in its simplest form cannot here be applied. Both molecular systems shown have a right-handed chirality

dimer. Each monomer has, in the spectral region of interest, a single, electric dipole allowed transition (Figure 3). The coupling in the dimer between these degenerate transitions is purely electrostatic, exchange effects being negligible. The ground state a and the excited states $b_{\pm}$ of the composite system are written in terms of one-electron functions from which ϕ_1 and ϕ_2, and to which, ϕ_1' and ϕ_2', electrons are excited:

$$a = \phi_1\phi_2, \qquad b_\pm = \frac{1}{\sqrt{2}}\{\phi_1\phi_2' \pm \phi_2\phi_1'\} \tag{22}$$

The magnitude of the exciton splitting between b_+ and b_- is given by the dipole–dipole interaction and the rotatory strengths as (Kirkwood, 1937; Tinoco, 1962; Mason, 1963; Schellman, 1968):

$$R_{ab_\pm} = \pm C\vec{R}_{12} \cdot (\vec{\mu}_2 \times \vec{\mu}_1) \tag{23}$$

where C is a positive constant, $\vec{R}_{12} = \vec{R}_2 - \vec{R}_1$ is the distance vector between the monomers, and $\vec{\mu}_i = \langle \phi_i | \vec{\mu} | \phi_i' \rangle$. The direction (as distinguished from the orientation) of $\vec{\mu}_1$ is only defined relative to $\vec{\mu}_2$. There are two possible combinations, corresponding to a resulting electric moment $\vec{\mu}_+ = \vec{\mu}_1 + \vec{\mu}_2$ and $\vec{\mu}_- = \vec{\mu}_1 - \vec{\mu}_2$. Equation 23 rests on two assumptions: (1) The equivalence of the dipole vector and dipole velocity form for the electric transition moment, in spite of the approximate nature of the wavefunctions; (2) the neglect of magnetic dipole terms such as $\langle \phi_1 | \vec{l}^{(1)} | \phi_1' \rangle$, $\langle \phi_2 | \vec{l}^{(2)} | \phi_2' \rangle$, in which $\vec{l}^{(1)}, \vec{l}^{(2)}$ designate the angular momentum operator centred in monomer 1 or 2, respectively. In the uncoupled monomer the magnetic transition moment is indeed assumed to be strictly zero. In the coupled system a (weak) magnetic moment will be induced in the individual monomers which, to a higher order of approximation, should be taken into account.

As an example we choose 2,2'-diaminobiphenyl, which for our purposes may be viewed as the dimer of two aniline molecules. In aniline itself the longest-wavelength absorption around 300 nm is polarised perpendicular to the axis passing through the C—N bond (Labhart and Wagnière, 1963).

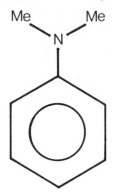

The composite system is forced to acquire a non-planar conformation of overall symmetry C_2, because of the steric interference between the substituents (see Figures 3 and 4). The molecule, as drawn, has a definite chirality or sense of screw. The transition moments are assumed to be point-dipoles centred in the individual monomers. From dipole–dipole interaction and equation 23 one predicts the resulting transition $(a \rightarrow b_+)$ of symmetry B, that is polarised perpendicular to the twofold symmetry axis, to be of lower energy and to have a positive rotatory strength. The transition $(a \rightarrow b_-)$ of symmetry A, polarised parallel to the twofold axis, leads to a negative sign for R. The picture which one obtains thus strikingly resembles the one given by the asymmetric pair of classical oscillators (see Section 1b). The predictions also agree with the

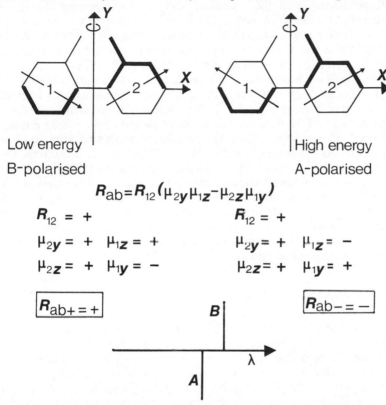

Low energy

B-polarised

High energy

A-polarised

$$R_{ab} = R_{12} \left(\mu_{2y}\mu_{1z} - \mu_{2z}\mu_{1y} \right)$$

$R_{12} = +$	$R_{12} = +$
$\mu_{2y} = + \quad \mu_{1z} = +$	$\mu_{2y} = + \quad \mu_{1z} = -$
$\mu_{2z} = + \quad \mu_{1y} = -$	$\mu_{2z} = + \quad \mu_{1y} = +$
$\boxed{R_{ab+} = +}$	$\boxed{R_{ab-} = -}$

Figure 4 Application of the coupled oscillator model (equation 23) to the longest-wavelength transitions in 2,2'-diaminobiphenyl. The chirality of the system is right-handed. The heavily marked lines symbolise bonds lifted above the plane of the drawing

observed longest-wavelength Cotton effects (Mislow *et al.*, 1963). If the molecule is twisted around the bond connecting both halves into a near-*trans* conformation, maintaining the chirality (see also Section 2e) the correspondence between the symmetry of the transitions and the sign of the rotatory strength remains the same, but the energetic sequence of the transitions is then reversed (see also Mason and Vane, 1965).

2(d) INHERENTLY DISSYMMETRIC CHROMOPHORES

Consider the case of two asymmetrically coupled carbonyl groups (see Figure 3). The one-electron method cannot well be applied here. Although the first group acts as an asymmetric environment for the second one, it also has transitions at the same energies. The idea of the unpolarisable asymmetric surroundings does not hold. If we are mainly interested in the two longest-wavelength transitions, which in the uncoupled chromophores are magnetic dipole allowed and electric dipole forbidden, an attempt to use the exciton model, as described in the preceding section, also appears illusory. We conclude that the optically active dicarbonyl chromophore is better treated as inherently dissymmetric. Explicit, if crude, wave functions are to be obtained that extend over the whole optically active system, and the rotatory strength will be evaluated therewith (Hug and Wagnière, 1970; 1971).

A posteriori one then finds for the two longest-wavelength transitions and for a right-handed chirality (see Section 2e) a situation somewhat analogous to the coupled oscillator model. In the close-to-*cis* conformation the two longest-wavelength transitions are, in the sequence of increasing energy, of symmetry *B* followed by *A*. The corresponding rotatory strengths are positive and negative. As the angle of twist of the two C=O groups around the C—C bond is increased to about 90°, a degeneracy between the two transitions occurs as well as a cancellation of the Cotton effects. In the close-to-*trans* conformation the sequence is reversed to *A* negative followed by *B* positive. The absolute value of the rotatory strength of the transition occurring at longest wavelength is here always greater than the corresponding value for the next transition. This contrasts with the situation encountered in the coupled-oscillator model (see equation 23 and Figure 4).

2(e) CHROMOPHORES OF SYMMETRY C_2

In the two preceding sections we have merely discussed systems of overall symmetry C_2. This is indeed a limitation if one remembers that all point groups containing neither a centre of inversion, nor planes of reflection, nor rotation-reflection axes lead to optical activity (Caldwell and Eyring, 1971). On the other hand the group C_2, being the point group of a perfectly regular helix, or a Moebius strip, plays possibly a fundamental role. In search for working rules, also for inherently dissymmetric chromophores, the following regularities for systems of symmetry C_2 and well-defined chirality are of interest: if the chirality is right-handed, transitions of symmetry *A* will in general tend to show negative Cotton effects in the long-wavelength part of the spectrum, transitions of symmetry *B* positive ones (Wagnière and Hug, 1970; Hug and Wagnière, 1972). To determine the chirality, a chromophore is visualised in the energetically most easily accessible planar conformation of effective C_{2v} symmetry. The C_2-axis lies in this plane and divides the molecule into a right part and a left part. Each half, as viewed by the observer, is further subdivided into a front (close to the observer) and a back. The molecule is now brought into its real conformation. If this is achieved by lifting the back of the right half above the plane and lowering the back of the left half below the plane, then the chirality is right-handed (see also Figures 3–5). It is questionable if the C_2-regularity may be applied to helical molecules of many turns. In a perfectly regular helix transitions polarised parallel to the helical axis are of symmetry $B_{\parallel}$. Transitions polarised perpendicularly to the helical axis may either be of symmetry $A_{\perp}$ or $B_{\perp}$. The distinction between these two latter types of transitions is, however, determined by the relative positions of the endpoints of the helix. As end-effects should decrease in importance with an increase in length of the helix, the fundamental distinction between these transitions should also gradually fade away.

3. THE CALCULATION OF THE OPTICAL ACTIVITY OF LARGE MOLECULES

3(a) ON THE POSSIBILITY AND THE NEED TO IMPROVE WAVE FUNCTIONS

Kirkwood's polarisability theory (1937) was a first attempt to bring equation 13 into a tractable form, without the need to explicitly know the wave functions of a large system. To this purpose a molecule is subdivided into interacting units, between which electronic exchange may be neglected. The individual units are considered to be optically inactive, and the coupling between the magnetic transition moments of a given unit with the electric transition moments of the other, or vice versa, are also considered unimportant. The remaining term, corresponding to equation 23 in Section 2c for the case of the dimer with only one transition per single unit, may be transformed in such a way as to be expressible by the polarisabilities of the individual groups. The strength of the polarisability theory resides in the possibility of relating the optical activity of a molecule to other measurable quantities. Its weakness is to be sought in the fact that in its ultimate formulation it is only applicable outside of absorption regions. It also fails to consider electric dipole forbidden transitions, a realm where the one-electron theory comes into play. For a detailed criticism see Tinoco (1962).

The development of electronic computers has, within the last fifteen years, opened up great possibilities of obtaining explicit molecular wave functions of increasing degrees of accuracy. Although in large molecular systems one still has to be content with quite approximate solutions, if one wants to avoid immense computational expense, the different steps of improvement will indeed provide added insight. We mention here a few examples where semi-empirical wave functions are used.

Hexahelicene (see Figures 5 and 6), a helical molecule composed of six fused benzene rings, has over the years been the subject of a number of theoretical investigations, in an attempt to correlate the absolute configuration with the sign of the long-wavelength optical activity (Fitts and Kirkwood, 1955; Moscowitz, 1962; Tinoco and Woody, 1964; Weigang, Turner and Trouard, 1966). As has been recently shown (Wagnière, 1971; Brickell et al., 1971), electron interaction effects play a decisive role in this connection and have to be taken into account explicitly. Suppose, for simplicity, that the energetically higher-lying molecular orbitals ϕ_i are linear combinations of the carbon 2p atomic functions χ_p with axes parallel to the helical axis: $\phi_i = \Sigma_p c_{ip} \chi_p$. The ϕ_i have been made self-consistent, within the frame of a given semi-empirical procedure (Pariser and Parr, 1953; Pople, 1953), and for the ground state one consequently may write

$$a = |\phi_1 \bar{\phi}_1 \ldots \phi_i \bar{\phi}_i \ldots \phi_{N/2} \bar{\phi}_{N/2}| \qquad (24a)$$

Singly excited singlet configurations are constructed (higher excitations are neglected):

$$\Phi_i^j = \frac{1}{\sqrt{2}} \left\{ |\phi_1 \bar{\phi}_1 \ldots \phi_i \bar{\phi}_j \ldots \phi_{N/2} \bar{\phi}_{N/2}| + |\phi_1 \bar{\phi}_1 \ldots \phi_j \bar{\phi}_i \ldots \phi_{N/2} \bar{\phi}_{N/2}| \right\}$$

$$(24b)$$

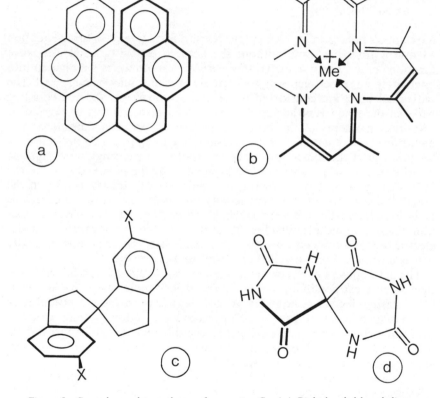

Figure 5   Some large chromophores of symmetry C_2. (a) Right-handed hexahelicene. (b) The corrin-chromophore, assumed left-handed. (c) A disubstituted spiroindane, left-handed. (d) Spirobihydantoin, left-handed

To describe excited states:

$$b = \sum_{ij} B_{ij} \Phi_i^j$$

the B_{ij} being configuration interaction coefficients and assumed real. The rotatory strength then is expressed as

$$R_{ab} = 2\text{Im}\left\{ \sum_{ij} \sum_{i'j'} B_{ij} B_{i'j'} \langle \phi_i | \vec{\mu} | \phi_j \rangle \langle \phi_{j'} | \vec{m} | \phi_{i'} \rangle \right\} \qquad (25)$$

One notices that not only do the scalar products of $\vec{\mu}$ and $\vec{m}$ for every one-electron transition appear in equation 25, but also all the possible crossterms, weighted by the products of B_{ij} and $B_{i'j'}$. Configuration interaction leads to a

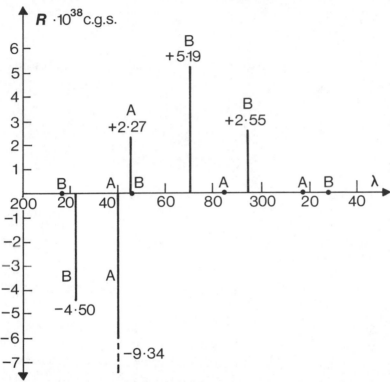

Figure 6   The rotatory strength of the long-wavelength transitions in right-handed hexahelicene, as obtained by the PPP-method including the interaction of singly excited configurations. The system of strong positive Cotton effects, mainly of symmetry B, at longest wavelength probably determines the sign of the optical activity at the Na-D-line. The C_2-regularity is approximately obeyed. States having very small values for the rotatory strength are indicated by dots

'scrambling' of the transitions and their rotatory strengths (see Figure 6). The sequence of strong positive Cotton effects in the long-wavelength part of the ultra-violet spectrum of right-handed hexahelicene allows the conclusion that the optical activity beyond the region of absorption (Na-D-line, say) is, in all likelihood, positive, in agreement with experiment (Groen *et al.*, 1970; Lightner *et al.*, 1972).

To describe the Cotton effects of n → π* transitions in carbonyl compounds within the frame of improved semi-empirical molecular orbital procedures, one must consider both (pseudo-) σ and π electrons simultaneously (Pao and

Table 1. Comparison of calculated (CNDO–CI) and experimental rotatory strengths of the $n \rightarrow \pi^*$-transition of optically active pyrrolidones. From the position of the substituents with respect to the carbonyl group it appears as if a quadrant rule were obeyed (see Sections 2a and 2b)

Compound	Calculated rotatory strength of the $n \rightarrow \pi^* -$ transition (c.g.s. units)	Sign of the experimental CD-band	References for the experimental CD-spectra
	$-\ 3 \cdot 10^{-40}$	Negative	Urry (1968)
	$-20 \cdot 10^{-40}$	Negative	Greenfield and Fasmar (1970)*
	$+16 \cdot 10^{-40}$	Positive	Urry (1968) Greenfield and Fasman (1969)
	$+\ 9 \cdot 10^{-40}$	Positive ($+2.10^{-40}$ c.g.s. units)	Measured by the authors†

*Also private communication from these authors.
†Sample from C. H. Eugster. (see Matsumoto et al., 1969).

Santry, 1966). Configuration interaction effects may also prove to be important here (Hug and Wagnière, 1970; 1971). Some recent results on cyclic amides are given in Table 1, obtained by one of the authors (R.G.) with the SCF–CI procedure, within the CNDO (complete neglect of differential overlap) approximation. Rotational strengths are computed as previously described by Hug and Wagnière (1970).

Theoretical studies of optical activity not of electronic, but of vibrational origin have been carried out by Deutsche and Moscowitz (1968, 1970), and by Holzwarth and Chabay (1972).

3(b) THE OPTICAL ACTIVITY OF BIOPOLYMERS

Biopolymers, such as proteins and nucleic acids, show a multitude of structural modifications. As they always occur in optically active form, they are ideally suited for investigations by the methods of optical rotatory dispersion and circular dichroism. Small changes in the molecular geometry may show up very significantly in the ORD and CD spectra. Since the discovery of the α-helix in proteins and of the double helix in nucleic acids, these molecules have also inspired theoretical studies of their electronic properties. The methods of calculation and adopted models have their roots in solid state physics, on one hand, and in traditional quantum chemistry on the other.

We limit ourselves here to a brief consideration of the amide chromophore in polypeptides, referring the reader to particular reviews (Deutsche et al., 1969; Pino, Ciardelli and Zandomeneghi, 1970) which treat nucleic acids and other stereo-regular polymers. Polypeptides have the repeating structure as shown.

$$
\left[
\begin{array}{c}
R \quad\quad H \\
| \quad\quad\; | \\
CH \!\!-\!\! \overset{\displaystyle}{\underset{\displaystyle O}{C}} \!\!-\!\! N \\
\end{array}
\right]_n
$$

It is generally recognised that the origin of the longest-wavelength ultraviolet and CD bands above 185 nm, accessible today with standard spectroscopic equipment, lies in the $n \to \pi^-$ (or $n \to \pi^*$) and $\pi^\circ \to \pi^-$ (or $\pi \to \pi^*$) transitions of the monomers (see Figure 1). From the discussion in Sections 2c and 2d we conclude: the ($\pi \to \pi^*$) band system of the polymer may conceivably be interpreted by a coupled-oscillator, or exciton-type approach. On the other hand it is doubtful if the ($n \to \pi^*$) band may satisfactorily be interpreted either by the exciton model, or by the one-electron theory.

Inspired by Kirkwood's pioneering work (1937), Moffitt (1956), and Fitts and Kirkwood (1956) developed a quantitative theory of the optical rotatory dispersion of helical polymers. Moffitt's main results, based on an exciton approach, may be stated as follows. An electric dipole allowed monomer transition gives rise to an exciton band in the helical polymer. To this exciton band there are two allowed transitions, namely a simple one, polarised parallel to the helical axis and a doubly degenerate one, polarised perpendicularly. One may define a partial rotation per residue [R'], which far away from the absorption frequencies follows the law (Yang, 1967; Blout, Carver and Shechter, 1965)

$$[R'] = \frac{a_0 \lambda_0^2}{\lambda^2 - \lambda_0^2} + \frac{b_0 \lambda_0^4}{(\lambda^2 - \lambda_0^2)^2} \tag{26}$$

where the first term depends on the nature of the monomer and the second term is a measure of the helicity. The best experimental fit between (26) and the observed rotations in the region 300–600 nm corresponds to values of $\lambda_0 \approx 210$ nm and $b_0 \approx -700$ for a completely right-handed α-helical polypeptide. In a random polypeptide the value of b_0 is small and positive. Equation 26 therefore has proven to be of practical use in identifying helical protein systems in solution and to determine semi-quantitatively the partial helical content. The splitting of the $(\pi \rightarrow \pi^*)$ band predicted by Moffitt was later confirmed both by polarised ultra-violet (Gratzer, Holzwarth and Doty, 1961) and by CD (Holzwarth, Gratzer and Doty, 1962; Holzwarth and Doty, 1965) measurements.

Spectroscopic penetration of the regions of absorption also confirmed the existence of the $(n \rightarrow \pi^*)$ band, which was neglected in the treatment by Moffitt, Fitts and Kirkwood (1957). In the CD spectrum this band shows a definite sensitivity to conformation, namely a decrease of the absolute value of the rotatory strength upon helix-coil transition (Beychok, 1967). It is therefore to be concluded that the $(n \rightarrow \pi^*)$ band also influences the optical activity of the helix far away from the regions of absorption. This casts a shadow on the theoretical foundations of the Moffitt–Yang formula (26). A study of the $(n \rightarrow \pi^*)$ band using the one-electron theory has been carried out by Schellman and Oriel (1962). For an attempt to interpret simultaneously the $(n \rightarrow \pi^*)$ and $(\pi \rightarrow \pi^*)$ bands see, for instance, Woody (1968).

In recent times molecular orbital methods have been used to describe the electronic structure of crystals and polymers, invoking different types of approximations that have proven justifiable in the study of smaller systems (Ladik and Appel, 1964; Imamura, 1970; Morokuma, 1971; André, Kapsomenos and Leroy, 1971; Bacon and Santry, 1972). Instead of viewing a polymer as an entity of relatively weakly coupled units, the conceptual point of departure are one-electron functions that extend over the whole molecule. In spite of the unavoidable simplifications inherent to this approach it is hoped by the authors that in polypeptides the $(n \rightarrow \pi^*)$ and the $(\pi \rightarrow \pi^*)$ transitions may then be treated on an equal footing, which may also lead the way to a more differentiated understanding of the optical activity of these bands.

The question of the applicability of periodic boundary conditions has attracted recent attention. These conditions are strictly justified only in cyclic or in infinite systems of repeating units. Equations 13 and 14, however, are derived under the assumption that the dimensions of a molecule are small compared to the wavelength of incident light [see Section 1(c)]. Moffitt (1956) used concurrently equation 13 and periodic boundary conditions. He obtained results that were at variance with those of Fitts and Kirkwood (1956). It was then shown that this discrepancy was due to the incorrect use of periodic boundary conditions (Moffitt, Fitts and Kirkwood, 1957). Recent investigations (Loxsom, 1969; Deutsche, 1970; Hansen and Avery, 1972) show that periodic boundary conditions may indeed be used, but that a more general formulation for the rotation ϕ is required than equation 13, such as is for instance given by Stephen (1958); see also Tinoco (1962). Also Philpott (1972) treats the optical activity of an infinite single-stranded helical polymer using an exciton approach with periodic boundary conditions. He applies the multipole Hamiltonian (12) in its field-quantised formulation, including higher-order terms.

4. SUMMARY

The study of natural optical activity has been of fundamental importance for the development of stereochemistry. Attempts to explain the effect of optical activity by the classical mechanics of oscillating point charges, though not unsuccessful, nevertheless fail to give a differentiated insight into molecules. Also here, as in many other fields of physical chemistry, has wave mechanics lead to a breakthrough.

In this chapter a brief review is given of how an elementary quantum mechanical approach can be of direct practical use in interpreting the optical activity of different kinds of molecules. Even without the knowledge of accurate molecular wave functions it is often possible to make predictions on the sign and relative magnitudes of Cotton effects. The use of perturbation theory and group theoretical considerations lead to the sector rules, the coupled-oscillator model is based on a simple exciton-type formalism. Chromophores of symmetry C_2 show characteristic regularities that may be relatively easily interpreted. Some CD and ORD spectra cannot be understood without gradually refining the wave functions, however. The explicit inclusion of electron interaction in the form of configuration interaction proves to be important.

The investigation of the optical activity of large helical polymers, such as proteins and nucleic acids, is of significance for biochemistry. These systems have also been extensively investigated theoretically. One question of current interest concerns the applicability of periodic boundary conditions and the sensitivity of the results to different approximations.

REFERENCES

André, J.-M., Kapsomenos, G. S. and Leroy, G. (1971) 'A Comparison of ab initio, Extended Hückel and CNDO Band Structures for Polyene and Polyethylene', *Chem. Phys. Lett.*, **8**, 195–197

Bacon, J. and Santry, D. P. (1972) 'Molecular Orbital Theory for Infinite Systems: Hydrogen-bonded Molecular Systems, *J. chem. Phys.*, **56**, 2011–2016

Beychok, S. (1967) 'Circular Dichroism of Poly-α-amino Acids and Proteins', *Poly-α-amino Acids,* (edited by Fasman, G. D.), 293–337, Dekker, New York

Blout, E. R., Carver, J. P. and Shechter, E. (1965) 'Optical Rotatory Dispersion of Polypeptides and Proteins', *Optical Rotatory Dispersion and Circular Dichroism in Organic Chemistry*, (edited by Snatzke, G.), 224–300, Heyden, London

Born, M. (1915) 'Ueber die Natürliche Optische Aktivität von Flüssigkeiten und Gasen', *Phys. Z.*, **16**, 251–258

Born, M. (1918) 'Elektronentheorie des Natürlichen Optischen Drehungsvermögens Isotroper und Anisotroper Flüssigkeiten', *Annln Phys.*, **55**, 177–240

Brickell, W. S., Brown, A., Kemp, C. M. and Mason, S. F. (1971) 'π-Electron Absorption and Circular Dichroism Spectra of [6]- and [7]-Helicene', *J. chem. Soc. A*, 1971, 756–760

Caldwell, D. J. and Eyring, H. (1971) *The Theory of Optical Activity*, Wiley-Interscience, New York

Condon, E. U. (1937) 'Theories of Optical Rotatory Power', *Rev. mod. Phys.*, **9**, 432–457

Condon, E. U., Altar, W. and Eyring, H. (1937) 'One-electron Rotatory Power, *J. chem. Phys.*, **5**, 753–775

de Malleman, R. (1930) 'The Molecular Theory and the Calculation of Natural Rotatory Power', *Trans. Faraday Soc.*, **26**, 281–292

Deutsche, C. W. (1970) 'Theory of Optical Activity of Crystalline Benzil', *J. chem. Phys.*, **53**, 1134–1149

Deutsche, C. W., Lightner, D. A., Woody, R. W. and Moscowitz, A. (1969) 'Optical Activity', *A. Rev. phys. Chem.*, **20**, 407–448

Deutsche, C. W. and Moscowitz, A. (1968) 'Optical Activity of Vibrational Origin. I. A Model Helical Polymer', *J. chem. Phys.*, **49**, 3257–3272

Deutsche, C. W. and Moscowitz, A. (1970) 'Optical Activity of Vibrational Origin. II. Consequences of Polymer Conformation, *J. chem. Phys.*, **53**, 2630–2644

Djerassi, C. (1967) 'The Role of Optical Rotatory Dispersion and Circular Dichroism in Organic Chemistry—Origin, Present Development and Future Prospects', *Optical Rotatory Dispersion and Circular Dichroism in Organic Chemistry*, (edited by Snatzke, G.), 16–40, Heyden, London

Eyring, H., Walter, J. and Kimball, G. E. (1964) *Quantum Chemistry*, Wiley, New York, London, Sydney

Fitts, D. D. and Kirkwood, J. G. (1955) 'The Theoretical Optical Rotation of Phenanthro [3,4-c] Phenanthrene', *J. Am. chem. Soc.*, **77**, 4940–4941

Fitts, D. D. and Kirkwood, J. G. (1956) 'The Optical Rotatory Power of Helical Molecules', *Proc. natn Acad. Sci. U.S.A.*, **42**, 33–36

Fiutak, J. (1963) 'The Multipole Expansion in Quantum Theory', *Can. J. Phys.*, **41**, 12–20

Gans, R. (1923) 'Das Tyndallphänomen in Flüssigkeiten', *Z. Phys.*, 17, 353–397

Gervais, H. P. (1969) 'Spectres de Transition Electronique de Groupements Chromophores. I. Activité Optique d'une Molécule Libre, *J. Chim. phys.*, **66**, 1292–1301

Gratzer, W. B., Holzwarth, G. M. and Doty, P. (1961) 'Polarization of the Ultraviolet Absorption Bands in α-helical Polypeptides, *Proc. natn Acad. Sci.*, *U.S.A.*, **47**, 1785–1791

Greenfield, N. J. and Fasman, G. D. (1969) 'Optical Activity of Simple Cyclic Amides in Solution', *Biopolymers*, **7**, 595–610

Greenfield, N. J. and Fasman, G. D. (1970) 'The Circular Dichroism of 3-methylpyrrolidin-2-one', *J. Am. chem. Soc.*, **92**, 177–181

Groen, M. B. and Wynberg, H. (1971) 'Optical Properties of some Heterohelicenes. Absolute Configuration', *J. Am. chem. Soc.*, **93**, 2968–2974

Hansen, A. E. and Avery, J. (1972) 'A Fully Retarded Expression for the Rotatory Strength', *Chem. Phys. Lett.*, **13**, 396–398

Holzwarth, G. and Chabay, I. (1972) 'Optical Activity of Vibrational Transitions: A Coupled Oscillator Model', *J. chem. Phys.*, **57**, 1632–1635

Holzwarth, G. and Doty, P. (1965) 'The Ultraviolet Circular Dichroism of Polypeptides', *J. Am. chem. Soc.*, **87**, 218–228

Holzwarth, G., Gratzer, W. B. and Doty, P. (1962) 'The Optical Activity of Polypeptides in the Far Ultraviolet', *J. Am. chem. Soc.*, **84**, 3194–3196

Hug, W. and Wagnière, G. (1970) 'Molecular Orbital Calculations of Rotational Strengths: a Study of Skewed Diketones', *Theoret. Acta*, **18**, 57–66

Hug, W. and Wagnière, G. (1971) 'Die Optische Aktivität von Chiralen Dienen, Enonen und α-Diketonen', *Helv. chim. Acta*, **54**, 633–649

Hug, W. and Wagnière, G. (1972) 'The Optical Activity of Chromophores of Symmetry C_2, *Tetrahedron*, 1972, 1241–1248

Imamura, A. (1970) 'Electronic Structures of Polymers Using the Tight-Binding Approximation. I. Polyethylene by the Extended Hückel Method', *J. chem. Phys.*, **52**, 3168–3173

Kauzmann, W. J., Walter, J. E. and Eyring, H. (1940) 'Theories of Optical Rotatory Power', *Chem. Rev.*, **26**, 339–407

Kirk, D. N., Klyne, W. and Mose, W. P. (1972) 'Front Octant Effects in the Circular Dichroism of Ketones, *Tetrahedron Lett.*, **1972**, 1315–1318

Kirkwood, J. G. (1937) 'On the Theory of Optical Rotatory Power, *J. chem. Phys.*, **5**, 479–491

Kuhn, W. (1929) 'Quantitative Verhältnisse und Beziehungen bei der Natürlichen Optischen Aktivität', *Z. phys. Chem.*, **B4**, 14–36

Kuhn, W. (1933) 'Theorie und Grundsetze der Optischen Aktivität', *Stereochemie*, (edited by Freudenberg, K.), 317–434, Deuticke, Leipzig und Wien

Kuhn, W. and Braun, E. (1930) 'Messung und Deutung der Rotationsdispersion einfacher Stoffe', *Z. phys. Chem.*, **B8**, 281–313

Labhart, H. and Wagnière, G. (1963) 'Experimentelle und Theoretische Untersuchung der Angeregten Elektronenzustände einiger Substituierter Benzole', *Helv. chim. Acta*, **46**, 1314–1326

Ladik, J. and Appel, K. (1964) 'Energy Band Structure of Polynucleotides in the Hückel Approximation', *J. chem. Phys.*, **40**, 2470–2473

Lightner, D. A., Hefelfinger, D. T., Powers, T. W., Frank, G. W. and Trueblood, K. N. (1972) 'Hexahelicene. The Absolute Configuration', *J. Am. chem. Soc.*, **94**, 3492–3497

Loxsom, F. M. (1969) 'Optical Rotation of Helical Polymers: Periodic Boundary Conditions', *J. chem. Phys.*, **51**, 4899–4905

Mason, S. F. (1963) 'Optical Rotatory Power', *Q. Rev. chem. Soc.*, **42**, 20–66

Mason, S. F. and Vane, G. W. (1965) 'The Absolute Configuration of (+)-1,3-diphenylallene from the Electronic Absorption and Circular Dichroism Spectra', *Tetrahedron Lett.*, **1965**, 1593–1597

Matsumoto, T., Trueb, W., Gwinner, R. and Eugster, C. H. (1969) Isolierung von (−)-R-4-Hydroxy-pyrrolidon-(2) und einigen Weiteren Verbindungen aus Amanita Muscaria', *Helv. chim. Acta*, **52**, 716–720

Mislow, K., Bunnenberg, E., Records, R., Wellman, K. and Djerassi, C. (1963) 'Inherently Dissymmetric Chromophores and Circular Dichroism', II. *J. Am. chem. Soc.*, **85**, 1342–1349

Moffitt, W. (1956) 'Optical Rotatory Dispersion of Helical Polymers', *J. chem. Phys.*, **25**, 467–478

Moffitt, W., Fitts, D. D. and Kirkwood, J. G. (1957) 'Critique of the Theory of Optical Activity of Helical Polymers', *Proc. natn Acad. Sci. U.S.A.*, **43**, 723–730

Moffitt, W. and Moscowitz, A. (1959) 'Optical Activity in Absorbing Media', *J. chem. Phys.*, **30**, 648–660

Moffitt, W., Woodward, R. B., Moscowitz, A., Klyne, W. and Djerassi, C. (1961) 'Structure and the Optical Rotatory Dispersion of Saturated Ketones', *J. Am. chem. Soc.*, **83**, 4013–4018

Morokuma, K. (1971) 'Electronic Structures of Linear Polymers. CNDO/2 Calculation for Polyglycine', *Chem. Phys. Lett.*, **9**, 129–132

Moscowitz, A. (1962) 'Theoretical Aspects of Optical Activity. Part one: Small Molecules', *Adv. chem. Phys.*, **4**, 67–112

Oseen, C. W. (1915) 'Ueber die Wechselwirkung zwischen zwei Elektrischen Dipolen und über die Drehung der Polarisationsebene in Kristallen und Flüssigkeiten', *Annln Phys.*, **48**, 1–56

Pao, Y.-H. and Santry, D. P. (1966) 'A Molecular Orbital Theory of Optical Rotatory Strengths of Molecules', *J. Am. chem. Soc.*, **88**, 4157–4163

Pariser, R. and Parr, R. G. (1953) 'A Semi-empirical Theory of the Electronic Spectra and Electronic Structure of Complex Unsaturated Molecules', *J. chem. Phys.*, **21**, 466–471, 767–776

Philpott, M. R. (1972) 'Theory of Optical Rotation by Helical Polymers', *J. chem. Phys.*, **56**, 683–688

Pino, P., Ciardelli, F. and Zandomeneghi, M. (1970) 'Optical Activity in Stereoregular Synthetic Polymers', *A. Rev. phys. Chem.*, **21**, 561–608

Pople, J. A. (1953) 'Electron Interaction in Unsaturated Hydrocarbons', *Trans. Faraday Soc.*, **49**, 1375–1385

Power, E. A. and Zienau, S. (1958) 'Coulomb Gauge in Non-relativistic Quantum Electrodynamics and the Shape of Spectral Lines', *Phil. Trans. R. Soc. A*, **251**, 427–454

Richards, P. I. (1948) 'On the Hamiltonian for a Particle in an Electromagnetic Field', *Phys. Rev.*, **73**, 254

Rosenfeld, L. (1929) 'Quantenmechanische Theorie der Natürlichen Optischen Aktivität von Flüssigkeiten und Gasen', *Z. Phys.*, **52**, 161–174

314   Wave Mechanics

Ruch, E. and Schönhofer, A. (1970) 'Theorie der Chiralitätsfunktionen', *Theoret. chim. Acta,* **19**, 225–287

Schellman, J. A. (1966) 'Symmetry Rules for Optical Rotation', *J. chem. Phys.,* **44**, 55–63

Schellman, J. A. (1968) 'Symmetry Rules for Optical Rotation', *Accts. chem. Research,* **1**, 144–151

Schellman, J. A. and Oriel, P. (1962) 'Origin of the Cotton Effect of Helical Polypeptides', *J. chem. Phys.,* **37**, 2114–2124

Snatzke, G. (1968) ,Circulardichroismus und Optische Rotations dispersion—Grundlagen und Anwendung auf die Untersuchung der Stereochemie von Naturstoffen', *Angew. chem.,* **80**, 15–26

Stephen, M. J. (1958) 'Double Refraction Phenomena in Quantum Field Theory, *Proc. Camb. phil. Soc. math. phys. Sci.,* **54**, 81–88

Tinoco, I. (1962) 'Theoretical Aspects of Optical Activity. Part two: Polymers', *Adv. chem. Phys.,* **4**, 113–160

Tinoco, I. and Woody, R. W. (1964) 'Optical Rotation of Oriented Helices. IV. A Free Electron on a Helix', *J. chem. Phys.,* **40**, 160–165

Tinoco, I., Woody, R. W. and Bradley, D. F. (1963) 'Absorption and Rotation of Light by Helical Polymers: the Effect of Chain Length', *J. chem. Phys.,* **38**, 1317–1325

Urry, D. W. (1968) 'Optical Rotation', *A. Rev. phys. Chem.,* **19**, 477–530

Velluz, L., Legrand, M. and Grosjean, M. (1965) *Optical Circular Dichroism,* Verlag Chemie GmbH., Weinheim

Wagnière, G. (1966) 'On the Optical Activity of Ketones Perturbed in Front Octants', *J. Am. chem. Soc.,* **88**, 3937–3940

Wagnière, G. (1971) 'On the Optical Activity of some Aromatic Systems: Hexahelicene, Heptahelicene, and [n,n]-Vespirene', *Aromaticity, Pseudo-Aromaticity, Anti-Aromaticity,* The Jerusalem Symposia on Quantum Chemistry and Biochemistry, III. The Israel Academy of Sciences and Humanities, Jerusalem, 127–139

Wagnière, G. and Hug, W. (1970) 'Polarization and the Sign of the Long-wavelength Cotton Effects in Chromophores of Symmetry C_2, *Tetrahedron Lett.,* **1970**, 4765–4768

Wallace, R. (1966) 'Diagrammatic Perturbation Theory of Multiphoton Transitions, *Molec. Phys.,* **11**, 457–470

Weigang, O. E., Turner, J. A. and Trouard, P. A. (1966) 'Emission Polarization and Circular Dichroism of Hexahelicene', *J. chem. Phys.,* **45**, 1126–1134

Woody, R. W. (1968) 'Improved Calculation of the $n \rightarrow \pi^*$-rotational Strength in Polypeptides', *J. chem. Phys.,* **49**, 4797–4806

Yang, J. T. (1967) 'Optical Rotatory Dispersion', *Poly-α-amino Acids,* (edited by Fasman, G. D.), 239–291, M. Dekker, New York

19

Photoelectron Spectroscopy and the Electronic Structure of Matter

WILLIAM C. PRICE, F.R.S.
Department of Physics, King's College, London

Photoelectron spectroscopy is concerned with the ejection of electrons from atomic systems according to the Einstein photoelectric equation

$$hv = E + \tfrac{1}{2}mv^2 \tag{1}$$

E is the initial binding energy (ionisation energy) of the electron and $\tfrac{1}{2}mv^2$ the kinetic energy it acquires when ejected by a photon of frequency v. Ordinary (photon) spectroscopy is governed by the equation

$$hv = E' - E \tag{2}$$

where both E' and E are bound states and E can only be obtained by extrapolating the excited states E' to zero energy, that is when the electron becomes free. To use equation (2) it is only necessary to measure the photon frequency, which of course is done with high accuracy by conventional spectroscopy. However, to use equation (1) it is necessary to measure the kinetic energy of the photoelectron ejected by a photon of known frequency, with a precision comparable with that of optical spectroscopy. This has been possible only in recent years as a result of the development of sensitive electron detectors and improved electron energy analysers (both electrostatic and electromagnetic) which are capable of giving high accuracy and good resolution. The reward for these developments has been that by using the methods of photoelectron spectroscopy it has become possible to pick out each electron from a molecule or a solid whether it is an inner (core) or outer (valence shell) electron and to determine accurately the energy with which it is bound into the system. Many other features which reveal the function of the electron in the electronic structure of the system also appear in the photoelectron spectrum. It was previously not possible to obtain this information for inner electrons by optical spectroscopy except in a few simple cases because of the interaction of their excited states E' with the ionisation continua of the outer electrons. By using a high energy photon it was found that the inner electron

315

could be ejected so rapidly past the outer electrons that this interaction was largely eliminated. For example by conventional spectroscopy it had been possible to study only the outer two non-bonding electrons in the water molecule and it was not until the advent of photoelectron spectroscopy that any detailed information became available on the other eight electrons in this relatively simple and vitally important molecule.

Photoelectron spectroscopy is conveniently divided into the two major fields in which it has been developed over the last decade. The first uses X-ray irradiation · and is largely concerned with core electrons. It was pioneered by K. Siegbahn and his coworkers at Uppsala who used their experience on β-ray spectrometry to make the necessary technical development. Extensive details of the work of this school are given in two major publications (Siegbahn *et al.*, 1967, 1970). The second field employs ultraviolet radiation, mostly utilising the resonance line of helium at 58.4 nm 21.22 eV). It was developed in London by D. W. Turner and his associates

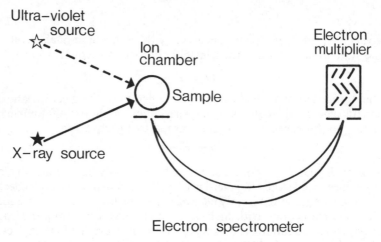

Figure 1   Schematic diagram of ultra-violet X-ray photoelectron spectrometer

who used it to study the outer valence shells of molecules. Their results are collected in Turner *et al.* (1970). The range was later extended by W. C. Price and collaborators who employed the shorter wavelength line of ionised helium (30.4 nm, 40.83 eV) as a photon source which gave deeper penetration into the valence shell (Price, 1970; Potts *et al.*, 1970). The X-ray and ultraviolet irradiation techniques are conveniently abbreviated as XPS and UPS respectively. While the former gives information on all electrons the latter gives much more highly resolved spectra for the more loosely bound outer electrons since the helium linewidth (about 0.001 eV) is less than that of the X-ray lines (about 0.5 eV) used by Siegbahn. Further because the photoelectron velocity is much lower in UPS than XPS the precision of its determination is correspondingly greater and more highly resolved spectra often showing detailed vibrational structure are obtained when ultra-violet photons are used.

INSTRUMENTATION

A schematic diagram of the equipment used in photoelectron spectroscopy is shown in Figure 1. The energy spectrum of the photoelectrons emitted by the sample is obtained by plotting the count rate of electrons arriving at the detector against a continuously increasing energy setting of the electron spectrometer. Thus photoelectron energy increases from left to right along the abscissae and binding energy from right to left. This presentation is used in all published XPS spectra and in some UPS spectra but there is an increasing tendency in the latter case to plot binding energy (ionisation potential) increasing from left to right, since the energy scale is always calibrated against gases of spectroscopically known ionisation potential because the photoelectron energy is affected by contact potentials and other factors and its absolute value cannot be determined accurately.

X-RAY PHOTOELECTRON SPECTROSCOPY

Most X-ray photoelectron spectroscopy has used the soft $K\alpha$ X-ray lines of Mg or Al (1254 and 1487 eV respectively) because these are narrower than for example the lines of Cu $K\alpha$ and they give photoelectrons whose energies are lower and can therefore be measured more accurately. These lines permit examination of the electronic structure down to 1000 eV which covers the 1s shells of the light atoms. The XPS lines from the K shells of the atoms in the second period of the Periodic Table are shown in Figure 2. Their binding energies increase from lithium (55 eV) to neon (867 eV), the differences between successive elements gradually increasing from 56 eV (Li–Be) to 181 eV (F–Ne). The photoelectron lines are thus well separated from one another and can be used as a method of identifying the elements present in the sample. For the heavier atoms the binding energies of electrons in L, M and higher inner shells can be used for this purpose and thus all the elements of the Periodic Table can be identified by X-ray photoelectron spectroscopy. It was for this reason that Siegbahn called his method ESCA, the letters standing for 'electron spectroscopy for chemical analysis'. However, this rather restrictive title is now being dropped.

Although the binding energy of an electron in an inner shell is mainly controlled by the charge Z within the shell it is also affected to a lesser extent by the cloud of valence electrons lying above it. The XPS line of an element is found to vary over a few electron volts according to the state of chemical combination of the atom and the nature of its immediate neighbours. This is illustrated in the spectra of acetone $(CH_3)_2CO$, ethyl trifluoroacetate $CF_3CO_2C_2H_5$ and sodium azide shown in Figure 3. The C(1s) line of the two methyl carbons in acetone is 2–3 eV lower in binding energy than that of the more positive carbonyl carbon and is, as might be expected from the number of C atoms, twice as strong. Similar remarks apply to the two outer and the inner N (1s) electrons of the azide ion. The binding energies of the four different carbons in ethyl trifluoroacetate similarly reflect the different environments of the four carbon atoms in this molecule (note that the formula in the diagram is placed so that the lines lie vertically below the carbon atoms with

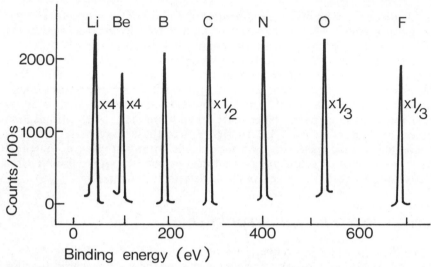

Figure 2   *Photoelectron lines from the* K *shells of elements of the second period when irradiated with Al* Kα *plotted aginst binding energy (i.e. 1487* eV *minus the photoelectron energy)*

which they are associated). These 'chemical shifts' have been related to the electronegativity difference of the atom from its neighbours and provide a means of estimating the effective charge on the core of each atom. The method has been extended to complicated molecules such as vitamin B_{12} where, using Al (Kα), the photoelectron line of one cobalt atom could be studied in the presence of the lines of 180 other atoms. Similarly both the sulphur and the iron electron line in the enzyme cytochrome c can be recorded. It is clear that the technique is a powerful direct method of studying chemical electronic structure.

In addition to the work on mainly covalent materials, ionic and metallic solids have been studied. For metals the conduction band structure can be found and the photoelectron energy distribution curves compared with calculated densities of states. Because of the small depth of penetration of the radiation and the necessity for the photoelectrons to emerge from the material the method is clearly a surface effect. Depths down to at most 10 nm (100 Å) are involved in the process and ultra high vacuum techniques must be employed to eliminate the effects of surface contamination. Surface properties such as are important in catalytic and adsorption processes have received considerable attention from workers in the field. It should be mentioned that although XPS can produce valuable information about the electrons in the valence shell the cross sections for this radiation of valence shell orbitals are almost an order of magnitude less than those of core orbitals. For

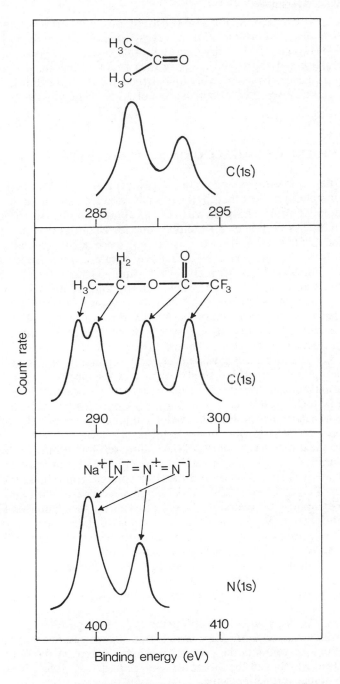

Figure 3 XPS *spectra of* C(1s) *shells of acetone and ethyl trifluoro-acetate and the* N (1s) *shell of sodium azide. Note that the* C *atoms in the formula are arranged to be above the lines to which they correspond*

example in hydrocarbons the cross-sections of C (2s) orbitals are about a tenth those of C (1s) orbitals and the cross-sections of orbitals built from C (2p) and H (1s) are still lower by almost another order of magnitude. This is a limitation not suffered by UPS where the cross-sections particularly of 2p orbitals are generally appreciably higher than those of 2s orbitals. This fact is likely to seriously affect the prospects of XPS for detailed outer electron studies.

ULTRA-VIOLET PHOTOELECTRON SPECTROSCOPY

Ultra-violet photoelectron spectroscopy using line sources obtained from discharges in helium or the other inert gases has produced an enormous amount of new detailed information on virtually all types of molecule and chemical species. This has been such an exciting time for chemists that nearly all the effort has so far been absorbed in studying these substances as gases. Its application to the solid phase which involves greater experimental and interpretative difficulties has scarcely begun though there have been some interesting studies by UPS on the conduction bands of metals such as gold in the solid and liquid phases (Eastman 1971). Apart from the fact that the inner valence electron orbitals of diatomic and polyatomic molecules which cannot be studied by conventional spectroscopy appear directly as features in the photoelectron spectrum, an additional attraction of the technique is that the UPS record itself reveals in a very convincing and readily understandable manner the orbital shell structure of the electrons in a molecule.

For monatomic gases only electronic energy is involved in photoionisation and the photoelectron spectrum of atomic hydrogen (obtained by passing the products of a discharge in H_2 into the ion chamber) gives a single sharp peak corresponding to group of photoelectrons of energy 7.62 eV. This corresponds to the difference in energy of the helium resonance line (21.22 eV) and the ionisation energy of the H (1s) electron (13.60 eV). For argon the outer electrons are in p type orbitals and the $(3p)^5$ configuration of the ion has the two states $^2P_{3/2}$ and $^2P_{\frac{1}{2}}$ with statistical weights in the ratio 2 to 1 and energies differing by spin–orbit coupling. The photoelectron spectrum (see Figure 4) corresponding to this $(3p)^{-1}$ ionised state is thus a doublet, one component of which is twice as strong as the other. Similar doublets are found for the other inert gases, the spin–orbit coupling increasing with their atomic weight.

For the ejection of an electron from an orbital in a molecule we have the equation

$$I_0 + \Delta E_{\text{vib}} + \Delta E_{\text{rot}} = h\nu - \tfrac{1}{2}mv^2$$

where I_0 is the 'adiabatic' ionisation energy (i.e. the pure electronic energy change) and ΔE_{vib} and ΔE_{rot} are the changes in vibrational and rotational energy which accompany the photoionisation. To explain the nature of the information which can be obtained we shall discuss the photoelectron spectra of some simple molecules.

Figure 5b gives the potential energy curves of H_2 and H_2^+. When an electron is removed from the neutral molecule the nuclei find themselves suddenly in the potential field appropriate to the H_2^+ ion but still separated by the distance characteristic of the neutral molecule. The most probable change is thus a transition on the potential energy diagram from the internuclear separation of the ground state to a point on the potential energy curve of the ion vertically above this. This is the Franck–Condon principle and it deter-

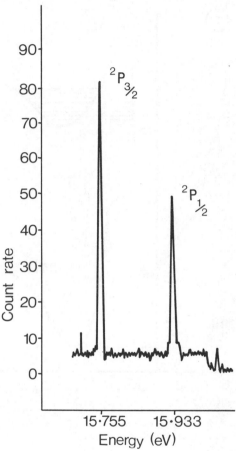

Figure 4   Photoelectron spectrum of argon ionised by Ne (736 Å)

mines to which vibrational level of the ion the most probable transition (strongest band) occurs. The energy corresponding to this change is called the vertical ionisation potential I_{vert}. Transitions to vibrational levels on either side of I_{vert} are weaker, the one of lowest energy corresponding to the vibrationless state of the ion corresponding to I_{adiab}. The photoelectron spectra of H_2 and D_2 are given in Figure 5a and that of H_2 is also plotted along the ordinate of Figure 5b.

From the intensity distribution of the bands in the photoelectron spectrum it is possible to calculate the change in internuclear distance on ionisation.

Clearly when a bonding electron is removed, the part of the photoelectron spectrum corresponding to this will show wide vibrational structure with a frequency separation which is reduced from that of the ground state vibration. The removal of relatively non-bonding electrons on the other hand will give rise to photoelectron spectra rather similar to those of the monoatomic

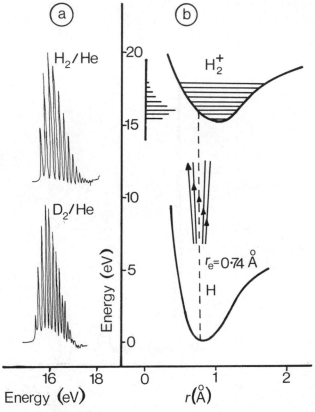

Figure 5a   Photoelectron spectra of H_2 *and* D_2. *b Potential energy curves showing spectra plotted along ordinate*

gases and little if any vibrational structure will accompany the main electronic band. The type of vibration associated with the pattern obtained when a bonding electron is removed can usually be identified as either a bending or a stretching mode and this can throw light on the function of the electron in the structure of the molecule, that is either as angle forming or distance determining. From the band pattern it is frequently possible to calculate values of the changes in angle as well as changes in internuclear distance on ionisation and so the geometry of the ionic states can be found if that of the neutral molecule is known. Changes in bond lengths on angles can be calculated with about 10% accuracy for photoelectron band systems which show structure by using the semi-classical formula $(\Delta r)^2 = l^2(\Delta\theta)^2 = 0.543$

$[I_{\text{vert}} - I_{\text{adiab}}]\mu^{-1}\omega^{-2}$, where r and l are in Å, θ in radians, I in eV, μ in atomic units and ω the mean progression spacing in $1000\ \text{cm}^{-1}$.

PHOTOELECTRON SPECTRA OF HYDRIDES ISOELECTRONIC WITH THE INERT GASES

As examples of the features mentioned above and also to illustrate how the orbitals of isoelectronic systems are related to one another we shall now discuss the photoelectron spectra of those simple hydrides for which the united atoms are the inert gases Ne, Ar, Kr and Xe, that is the atoms obtained by condensing all the nuclei into one central positive charge.

Hydrides such as HF, H_2O, NH_3 and CH_4 can be thought of as being formed from Ne by successively partitioning off protons from its nucleus. The solutions of the wave equation – the wave functions or orbitals – arising from the progressive subdivision of the positive charge pass continuously from the atomic to the molecular cases with a gradual splitting of the p^6 orbital degeneracy as the internal molecular fields are set up. Schematic diagrams of the orbital changes are indicated in Figure 6. The photoelectron

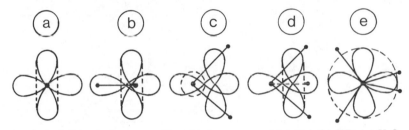

Figure 6   Schematic diagrams of 2p orbital structures of (a) Ne, (b) HF, (c) H_2O, (d) NH_3, (e) CH_4

spectra shown in Figures 7a, b, c and d reveal this orbital subdivision in a striking way. Figure 7a shows the spectra of the halogen acids which is the first stage in this process. The spectra of the corresponding inert gas (i.e. the united atom) is inserted on the records for comparison. The triply degenerate p^6 shell is split into a doubly degenerate π^4 shell and a singly degenerate σ^2 shell. The π^4 shell is non-bonding and represented by two bands in the photoelectron spectrum which are of equal intensity and show little vibrational structure. The σ shell is formed from that p orbital along which the proton is extracted and forms a negative cloud which binds the proton to the residual positive charge. It therefore gives rise in the photoelectron spectrum to a simple progression of bands with a separation corresponding to the vibration frequency of the $^2\Sigma$ state of the ion except where the structure is lost by predissociation.

The second stage of partition leads to the molecules H_2O, H_2S, H_2Se and H_2Te. If two protons were removed colinearly from the united atom nucleus the linear triatomic molecule so formed would not have maximum stability since only the two electrons in the p orbital lying along the line will then be

324

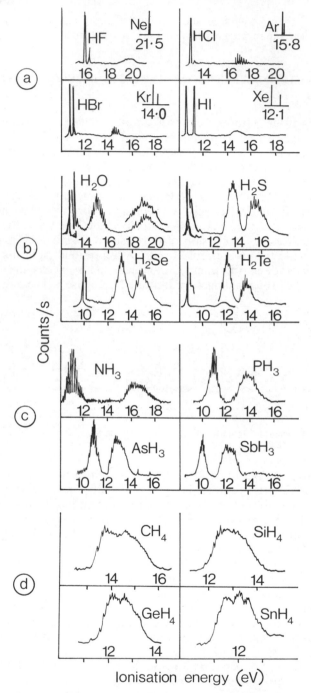

Figure 7   Photoelectron spectra of the hydrides isoelectronic with
Ne, Ar, Kr and Xe obtained with He (21.22 eV) radiation

effective in shielding the protons from the repulsion of the core. By moving off this line, the protons can acquire additional shelter from the central charge and from themselves through shielding by one lobe of a perpendicular p orbital (see Figure 6c). The electrons in this orbital then become 'angle determining' as distinct from the two previously mentioned p electrons which are mainly effective in determining the bond separations of the hydrogen atoms from the central atom. The remaining p orbital, because of its perpendicular orientation to the plane of the bent H_2X molecule, can affect neither the bonding nor the angle, that is its electrons are non-bonding. These expectations are borne out by the photoelectron spectra shown in Figure 7b. These spectra show how in all H_2X molecules the triple degeneracy of the p^6 shell of the united atom is completely split into three mutually perpendicular orbitals of different ionisation energies. The lowest ionisation band is sharp with little vibrational structure. The second has a wide vibrational pattern which turns out to be the bending vibration of the molecular ion. The third also has a wide vibrational pattern with band separations greater than those of the second band. These separations can be identified as the symmetrical bond stretching vibrations of the ion. The geometries of these three ionised states can be calculated from the band envelopes and pattern spacings. Only small changes are associated with the band of lowest IP. Large changes of angle accompany ionisation in the second band which in the case of H_2O causes the equilibrium configuration of this ionised state to be linear. The third ionised state is one in which the internuclear distances are increased but little change occurs in the bond angle. The integrated intensities of all three bands are roughly equal indicating that each originates from the ionisation of a single orbital.

The partitioning of three protons from the nucleus of an inert gas molecule leads to a pyramidal XH_3 molecule. One p orbital is directed along the axis of symmetry and provides the shielding which causes the molecule to have a non-planar geometry, that is it is angle determining. The other two p orbitals are degenerate and provide an annular cloud of negative charge passing through the three XH bonds and thus mainly determine the bond distances. The structure pattern on the first band of the photoelectron spectrum (Figure 7c) can be assigned to bending vibrations and indicates that the molecule flies to a symmetrical planar configuration without much change in the bond distances when an electron is ionised from this orbital. The second band in the photoelectron spectrum shows Jahn–Teller splitting which is consistent with it being doubly degenerate. Although it has limited structure such structure as can be observed is consistent with changes in bond stretching without much change in bond angle.

Finally the photoelectron spectra of the XH_4 type molecules show that the p^6 shell of the united atom has changed to another triply degenerate shell with orbitals of tetrahedral symmetry. The contour shows the presence of Jahn–Teller splitting and the structure on the low energy side shows that on ionisation the molecule moves by contraction along one side of the enveloping cube towards a square coplanar configuration. The structure on the high energy side indicates movement in the opposite direction towards two mutually perpendicular configurations in which opposite XH_2 angles are roughly

90°. In the heavier molecules, for example SnH_4, the large spin–orbit splitting of the heavy atom influences the structure. Further details on the spectra of these hydride molecules are given by Potts and Price (1972a, 1972b).

It should be mentioned that in the process of partitioning off protons from the neon nucleus to form the second row hydrides the 2s orbitals are distorted though to a lesser extent than the 2p orbitals. They are deformed from spherical symmetry in the direction of the extracted proton. They thus have 'bumps' in the bond directions and in this way provide an 's' contribution to the bond. This is evident in the vibrational structure of the photoelectron bands associated with these orbitals. The deformation corresponds to the distortion of the $1s^2$ orbitals of He in forming molecular hydrogen by dividing the central charge.

PHOTOELECTRON SPECTRA OF 'MULTIPLE' BONDED MOLECULES

To discuss 'multiple' bonded systems in which molecular orbitals are formed by p atomic orbitals combining in the 'broadside on' as well as in the 'end on' configuration we shall take as examples the molecules N_2, NO, O_2 and F_2. Their photoelectron spectra are shown in Figure 8. The orbitals upon which their electronic structures are built are the in phase and out of phase combinations of the appropriate 2s and 2p orbitals. These are given schematically in the insert. For N_2 the electronic configuration is $(\sigma_g 2s)^2(\sigma_u 2s)^2(\pi_u 2p)^4(\sigma_g 2p)^2$, the orbitals being in order of decreasing ionisation energy. With the exception of $(\sigma_u 2s)$, all of these might be expected to provide excess negative charge density between the two nuclei and thus to account for the strength of the N_2 'triple' bond. The additional electron of NO has to go into a $\pi^* 2p$ orbital which is largely outside the nuclei and therefore anti-bonding, and so in a loose analogy N_2 is the inert gas of diatomic molecules and NO the corresponding alkali metal, since it contains one electron outside a closed shell of bonding electrons. An inspection of the photoelectron spectrum of N_2 in which the ionised states corresponding to removal of the different orbital electrons are marked, shows that by far the most vibration accompanies the removal of the π_u electron. Thus at the internuclear distance of neutral N_2, the nuclei are held mainly by the negative cloud of the $(\pi_u 2p)^4$ electrons. The short-distance bonding character of these π electrons results in the nuclei being pulled through the $(\sigma 2p)^2$ cloud so that this σ orbital is as much outside the nuclei as between them and therefore supplies no bonding at the N–N equilibrium separation. It can be understood readily from the geometry of their overlap that the bonding of p electrons in the broadside on (π) arrangement optimises at shorter internuclear distances than that in the end on (σ) position. In N_2 the nuclei are in fact on the inside of the minimum of the partial potential energy curve associated with the $(\sigma 2p)^2$ electrons. This orbital is thus in compression and its electrons have both their bonding and their binding (ionisation) energies reduced. In NO, O_2 and F_2 the presence of the additional electrons in antibonding $\pi 2p$ orbitals causes the internuclear distances to be relatively longer than they are in N_2 and it can be seen from Figure 8 that the $\sigma 2p$ bands of these molecules have progressively more

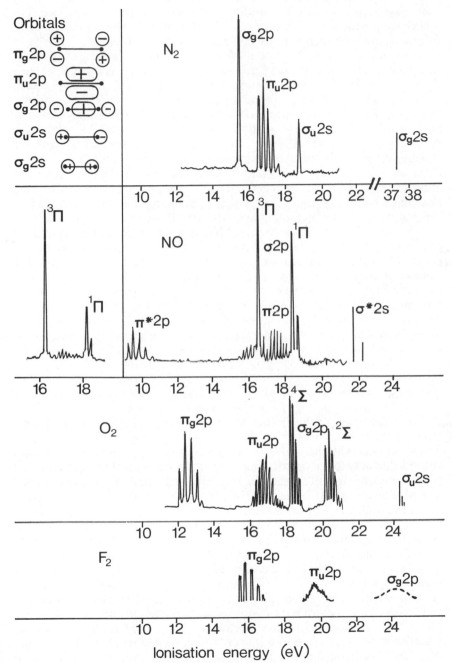

Figure 8   Photoelectron spectra of N_2, NO, O_2 and F_2

vibrational structure as the internuclear distance approaches more closely that separation for which the bonding of the $(\sigma 2p)^2$ orbital is optimised. The associated σ bands move through the $\pi 2p$ systems to higher ionisation energies in accord with their increased effective bonding power.

Other interesting points illustrated by inspection of Figure 8 are that the spacing in the $\pi 2p$ antibonding bands in NO and O_2 (first systems) are larger than those of the $\pi 2p$ bonding bands (second systems). This is to be expected since the removal of an antibonding electron increases the vibration frequency while that of a bonding electron decreases this frequency. Another interesting feature is that the separation of the first and second bands, which reflects the overlap between the out of phase and in phase $(\pi 2p)$ orbital, rapidly decreases with increase in internuclear distance. For NO, O_2 and F_2 these separations are 7.5, 4.5 and 3.0 eV. This indicates the reduction in 'multiple' (π) bonding as the effect of the increasing number of π antibonding electrons reduces and annuls the bonding of π bonding electrons by increasing the interatomic distance and reducing the orbital overlap.

The insert of the 16–18 eV region of NO in Figure 8 shows the $^3\Pi$ and $^1\Pi$ bands with an intensity ratio of 3:1 and thus illustrates how closely the intensities follow the statistical weights of the ionised states, agreeing with the number of channels of escape open to the electrons. Similar remarks apply to the $^4\Sigma$ and $^2\Sigma$ bands of O_2, the integrated intensities of which are in the ratio of 2:1. A further interesting feature which these multiplets illustrate is the greater bonding of the states of lower multiplicity. Since the lower multiplicity corresponds to the states of antisymmetric spin function it is associated with the symmetric (summed) space co-ordinate wave function of the orbitals between which the spin interaction is occurring. The additional overlap to which this gives rise results in greater bonding relative to states of higher multiplicity where the total space co-ordinate wave function is obtained by subtracting those of the interacting states.

The 'building up' of the electronic structures of molecules by adding electrons to the unfilled orbitals as an atom is changed to one with greater Z follows very closely the 'aufbau' process in atoms. Although we have considered this only for diatomic molecules it can be applied to triatomic and polyatomic systems and extends the shell concept to even the valence electrons of molecules. It might be asked whether we should now replace the bond structures of chemists by the molecular orbital description. No spectroscopist for example has ever seen an electron in a hybridised sp^3 bond orbital, a concept the chemist uses constantly. However, no real conflict exists except perhaps in chemiluminescence or photosynthesis where single electrons are specifically concerned. The situation is rather similar to that in infra-red spectroscopy where to interpret spectra it is necessary to describe the motion of the atoms in terms of normal vibrations. It certainly would not be convenient to do this to describe the distortion of molecules in chemical reactions. The orbital description of the electronic structure of molecules can be regarded as its description in terms of the irreducible representations of the structure. The bond pictures may be considered as combinations of these fundamentals which are more suitable in most cases for the description of chemical reactions. Such reorganisations involve energy changes of electrons in many different orbitals even though the major change may be largely localised in one orbital.

THE INTENSITIES OF PHOTOELECTRON SPECTRA

Apart from the quantitative data concerning orbital energies which are to be obtained by photoelectron spectroscopy, the intensities of the bands obviously contain much information about the nature and extent of the orbitals themselves. In this connection we must consider the transition moment of an electron from the initial orbital ϕ to the state ε in the ionisation continuum into which it is ejected by the incident photon. This is illustrated in Figure 9 where it can be seen that the integral representing the transition moment depends upon the spread and nodal character of the wave function of the initial state and the wavelength and phase of the photoelectron (i.e. the final state). For the purposes of this illustration the wave functions are taken as those of a one-dimensional oscillation of an electron in a $V(1/r)$ potential field, that is pseudo Rydberg bound states. Plane wave functions are used for the continuum states ($\lambda[1 \text{ eV}] \sim 12 \text{ Å}$). The simplest fact that this illustrates is the fall-off of the photoionisation cross-section with higher energy (shorter wavelength) photoelectrons. This explains the relative transparency of electrons in valence shells to X irradiation where the area of the graph for the transition moment is divided into small compensating positive and negative regions. It can be appreciated that the cross-section appropriate to any orbital will vary with the dimensions and nodal character of this orbital and the wavelength of the photoelectron. As a rough generalisation it might be expected that the photoionisation cross-section would maximise when $\lambda/4$ of the photoelectron is not less than the orbital dimensions. This would explain experimental observations that near the threshold of ionisation the cross-sections of molecular orbitals built from p atomic orbitals are greater than those arising from s orbitals. This situation is reversed for high photoelectron energies where because of the smaller radial spread of s orbitals the transition moment integral maximises with shorter wavelength photoelectrons. The study of the variation of the relative intensity of the photoelectron spectra of different orbitals with varying irradiating wavelengths is clearly going to yield much information on the nature of orbital wave functions. A summary of the information available for gases on this topic has been given by Price, Potts and Streets (1972). Photon sources of different energies derived from synchrotron radiation have already been used by Eastman and Grobman (1972) to study the band structure of gold and the density of intrinsic surface states in Si, Ge and GeAs.

Many interesting new developments in the field are reported in the Asilomar Conference Proceedings (1972) and a new journal entitled *Electron Spectroscopy* has been founded in which photoelectron spectroscopy is a major topic.

Studies have been made by photoelectron spectroscopy of atoms and transient molecular species produced by electrical discharges as well as the photodetachment of negative ions of atoms and molecules. The angular

330

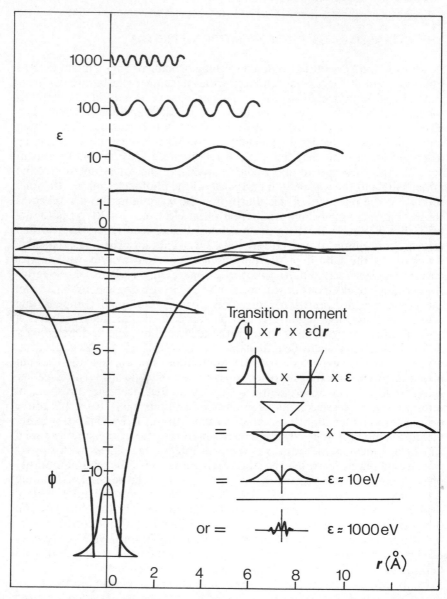

Figure 9 *Diagrammatic illustration of features of importance in the photoionisation of an electron from a bound state* φ *to a continuum state* ε

distribution of the photoelectron emission is another subject of considerable theoretical interest on which experimental information is being accumulated. It is certain that exciting revelations on the electron structure of matter will be forthcoming as this field is further developed.

REFERENCES

Asilomar Conference Proceedings, *Electron Spectroscopy* (1972) (edited by Shirley, D. A.), North-Holland, Amsterdam
Eastman, D. E. (1971) *Phys. Rev. Lett.*, **26,** 1108
Eastman, D. E. and Grobman, W. D. (1972) *Phys. Rev. Lett.*, **28,** 1327 and 1379
Potts, A. W., Lempka, H. J., Streets, D. G. and Price, W. C. (1970) *Phil. Trans. R. Soc.*, **A268,** 59
Potts, A. W. and Price, W. C. (1972a) *Proc. R. Soc. Lond.*, A, **326,** 165
Potts, A. W. and Price, W. C. (1972b) *Proc. R. Soc. Lond.*, A, **326,** 181
Price, W. C. (1970) *Molecular Spectroscopy IV* (edited by Hepple, P. W.), 221, London; Inst. Petroleum, distrib. Elsevier
Price, W. C., Potts, A. W. and Streets, D. G. (1972) Asilomar Conference Proceedings, *Electron Spectroscopy* (edited by Shirley, D. A.), 187, North-Holland, Amsterdam
Siegbahn, K. *et al.* (1967) 'Electron Spectroscopy for Chemical Analysis', *Nova Acta Reg. Soc. Sci. Uppsala* Ser. IV **20**
Siegbahn, K. *et al.* (1970) *ESCA Applied to Free Molecules*, North-Holland, Amsterdam
Turner, D. W. *et al.* (1970) *Molecular Photoelectron Spectroscopy*, Wiley-Interscience, London

20

Determination of the Electronic Structures of Complex Molecules and Solids by the SCF-Xα Scattered-Wave Cluster Method*

KEITH H. JOHNSON

Department of Metallurgy and Materials Science and
Center for Materials Science and Engineering,
Massachusetts Institute of Technology,
Cambridge, Massachusetts

1. INTRODUCTION

In this chapter, we review the self-consistent-field Xα scattered-wave (SCF-Xα-SW) cluster method of calculating the electronic structures of complex molecules and solids. Recent applications of the method, as well as newer ones still in progress, are described. The basic elements of this first-principles technique, also known as the 'multiple-scattering' method, were originally formulated by Slater and the author in 1965 and 1966, respectively (Slater, 1965; Johnson, 1966). The theoretical formalism and computational procedure have been developed so far since then (see, for example, Slater and Johnson, 1972; Johnson and Smith, 1972), that the method can now be implemented accurately for a wide range of polyatomic molecules and complex crystals, but requires only moderate amounts of computer time. Several other review papers devoted to the Xα statistical exchange-correlation theory and multiple-scattered-wave formalism, computational procedure, and range of applications have recently been published or are near publication (Slater and Johnson, 1972; Slater, 1972a; Johnson, 1972; Johnson, Norman and Conolly, 1972).

The method is based, first of all, on the arbitrary division of matter into

* Research sponsored by the Air Force Office of Scientific Research, United States Air Force (AFSC), Contract No. F44620-69-C-0054 and by the National Science Foundation, Grants No. GP-21312 and No. GH-33635 (through the Center for Materials Science and Engineering, M.I.T.).

component clusters of atoms. A cluster may be an isolated polyatomic molecule in the gaseous phase, in a crystalline environment, or in aqueous solution. It may also be a finite group of atoms in a bulk crystal or at a crystalline surface, or it may be part of a macromolecule. The cluster, in turn, is partitioned into three fundamental types of regions: (I) *atomic* — the regions within non-overlapping spheres surrounding the constituent atoms; (II) *interatomic* — the region between the atomic spheres and an outer boundary (generally spherical) surrounding the entire cluster; (III) *extramolecular* — the region outside the cluster. The one-electron Schroedinger equation is numerically integrated within each region in the partial-wave representation for spherically averaged and volume averaged potentials which include Slater's statistical approximation (Slater, Wilson and Wood, 1969; Slater *et al.*, 1969; Slater and Wood, 1971; Slater, 1972a) to exchange correlation. The wave functions and their first derivatives are joined continuously throughout the various regions of the cluster via multiple-scattered-wave formalism (Johnson, 1966; Johnson and Smith, 1972; Johnson, 1972). The effects of a particular environment on the cluster are described by boundary conditions, e.g. in the case of a symmetrical polyatomic molecule, the matching of the solutions of Schroedinger's equation in the extramolecular region to those within the molecule at an artificial spherical boundary surrounding the system. This procedure leads to a set of rapidly convergent secular equations which are solved numerically for the spin-orbital energies and wave functions. The matrix elements of these equations are very simple to evaluate in comparison with those characteristics of *ab. initio* Hartree–Fock LCAO methods. For example, there are no multicentre integrals to evaluate. The electronic charge density throughout the space of the cluster can then be determined from the occupied spin orbitals. This entire procedure is repeated, using the charge density of the preceding iterations in Poisson's equation of classical electrostatics to generate a new potential, until self consistency is attained.

The SCF-Xα-SW technique uses only a small fraction of the computer time required by *ab initio* Hartree–Fock LCAO methods. Furthermore, this technique yields molecular orbitals which are in significantly better quantitative agreement with experiment than do either *ab initio* or semi-empirical LCAO methods. For example, in conjunction with the 'transition-state' concept (Slater, 1972a; Slater and Johnson, 1972), the SCF-Xα-SW method leads to an accurate description of the optical properties of molecules and solids, including the effects of orbital relaxation. The method is practicable, moreover, on molecules and solids where *ab initio* LCAO methods are too difficult and costly to implement. The total energy as a function of stereochemical geometry can also be determined by the scattered-wave approach, using the expression for the Xα statistical total energy (Slater and Wood, 1971; Slater, 1972a; Slater and Johnson, 1972). There are several important consequences of the form of this statistical total energy expression which make the SCF-Xα-SW technique generally superior to the more traditional Hartree–Fock method, and not just an approximation to the latter. First of all the statistical total energy for a molecule or cluster goes to the proper separated-atom limit as the internuclear distances are increased to infinity (i.e. as the system dissociates). This is not true for the Hartree–Fock total energy. Secondly, the SCF-Xα-SW theory satisfies Fermi statistics, thereby ensuring the proper ordering and occupation of electronic energy levels, while the Hartree–Fock theory does not. Finally, Slater (1972b) has shown that under the proper conditions the statistical total energy satisfies both the virial and Hellmann–Feynman theorems, which facilitates the calculation of equilibrium cohesive properties for molecules and solids.

For the electronic structure of an ordered complex crystal, i.e. one with several or more atoms per unit cell (e.g. a 'molecular crystal'), one can assume the periodic cell to be the unit polyatomic cluster. The boundary condition on the cluster orbitals is then just the Bloch condition, and the theoretical model reduces exactly to band theory (Johnson and Smith, 1972). Because it is possible to adopt different boundary conditions and practical to handle reasonably large clusters of atoms, the theoretical formalism can be extended easily to a variety of problems in the electronic structure of complex crystals. For example, since the cluster model is not dependent on the assumption of long-range order, problems such as the bonding of impurities and defects in crystals and the electronic structures of disordered or amorphous solids are readily within the scope of the theory.

2. SUMMARY OF THE SCF-Xα SCATTERED-WAVE THEORY

The principal objective is to solve the one-electron Schroedinger equations (written here in Rydberg units)

$$[-\nabla_1^2 + V_C(1) + V_{X\alpha}(1)]u_i(1) = \varepsilon_i u_i(1) \tag{1}$$

for the spin orbitals $u_i(1)$ and energy eigenvalues ε_i of the cluster, subject to the boundary conditions on the cluster. Each spin orbital can be associated with either the spin-up or spin-down Pauli function. Thus the electronic charge density

$$\rho(1) = \Sigma(i)n_i u_i^*(1)u_i(1) \tag{2}$$

where n_i is the occupation number of the ith spin orbital, can be separated into spin-up $\rho\uparrow(1)$ and spin-down $\rho\downarrow(1)$ parts. In expression (1), $V_C(1)$ is the electrostatic potential energy at position 1 due to the total electronic and nuclear charge, determined classically. The quantity $V_{X\alpha}(1)$ is Slater's Xα statistical approximation to exchange correlation, namely

$$V_{X\alpha}\uparrow(1) = -6\alpha\left[\frac{3}{4\pi}\rho\uparrow(1)\right]^{1/3} \tag{3}$$

with a similar expression for spin down. This potential is dependent only on the local electronic charge density $\rho\uparrow(1)$ and on the scaling parameter α. The value $\alpha = 1$ yields the exchange potential originally derived by Slater (1951). The value $\alpha = 2/3$ leads to the exchange approximation derived independently by Gaspar (1954) and by Kohn and Sham (1965). Actually a value of α systematically chosen between these two limits is generally better for most applications to atoms, molecules, and solids (see below).

With expressions (2) and (3), the one-electron Schroedinger equations (1) define a self-consistent-field problem which can be solved iteratively, in analogy to the more traditional Hartree–Fock method. It should be emphasised, however, that the SCF-Xα method, in conjunction with the multiple-scattered-wave procedure for solving the one-electron equations, has several advantages over the Hartree–Fock method and is therefore not just an approximation to the latter approach.

The numerical solutions of equations (1) are implemented for a polyatomic cluster by first spherically averaging the potential $V_C(1) + V_{X\alpha}(1)$ inside each of the atomic regions I. The potential is generally assumed to be a constant throughout the interatomic region II, equal to the volume average of $V_C(1) + V_{X\alpha}(1)$ over this region. However, spherical averages can also be carried out in region II inside supplementary non-overlapping 'interstitial' spheres, e.g. in the regions of 'lone-pair' electrons (see below). For localised molecular orbitals, the potential is usually spherically averaged in the extramolecular region III with respect to the defined centre of the cluster. Alternative boundary conditions can also be imposed on region III to describe the effects of different local environments on the cluster. The initial potential is based on a superposition of free-atom or free-ion potentials in the SCF-Xα approximation, using atomic self-consistent-field computer programs of the type originally developed by Herman and Skillman (1963). The scaling parameter α in expression (3) is optimised for each component atom using the procedure suggested by Slater (1972a). This procedure consists of matching the Xα statistical total energy of the atom to the Hartree–Fock total energy in the case of a closed-shell atom, or to the average of the multiplet energies for the ground-state configuration in the case of an open-shell atom. Using this scheme, Schwarz (1972) has recently determined the α parameters for the atoms hydrogen through niobium. These values are then adopted for the corresponding atomic regions I of the molecular cluster. Appropriately weighted averages of the component atomic α values are used in regions II and III. This procedure ensures as accurate a description of the cluster in the separated-atom limit as possible, within the framework of the SCF-Xα-SW theory.

The partitioning of the space of the cluster into local regions of spherically averaged and volume averaged potential permits one to introduce a rapidly convergent composite partial-wave representation of the solutions of equations 1. Inside each atomic sphere j (and each interstitial sphere, if present), the spin orbitals are expanded in the form

$$u_i^I(1) = \Sigma(L)C_L{}^j R_l^j(\varepsilon; r)Y_L(\vec{r}) \tag{4}$$

where $L = (l, m)$ is the partial-wave (angular-momentum) index. The functions $R_l^j(\varepsilon; r)$ are solutions of the radial Schroedinger equation, the functions $Y_L(\vec{r})$ are spherical harmonics, and the coefficients C_L^j are to be determined. The radial functions are generated by outward numerical integration of the radial Schroedinger equation for each partial wave l and each trial energy ε. In the extramolecular region III, the orbitals are expanded with respect to the centre of the cluster in the representation

$$u_i^{III}(1) = \Sigma(L)C_L^0 R_l^0(\varepsilon; r)Y_L(\vec{r}) \tag{5}$$

where the functions $R_l^0(\varepsilon; r)$ are obtained by inward numerical integration of the radial Schroedinger equation. For the intersphere region II, Schroedinger's equation (1) reduces to the ordinary scalar wave equation

$$(\nabla_1^2 + \varepsilon - \bar{V}_{II})u_i^{II}(1) = 0 \tag{6}$$

where $\bar{V}_{II}$ is the volume average of $V_C(1) + V_{X\alpha}(1)$ over that region. The exact solutions of equation 6 for the energy range $\varepsilon < \bar{V}_{II}$ can be expanded in the multicentre representation

$$u_i^{II}(1) = \Sigma(L)i^{-l}A_L^0 j_l(i\kappa r_0)Y_L(\vec{r_0})$$
$$- \Sigma(j)\Sigma(L)i^{-l}A_L^j h_l^{(1)}(i\kappa r_j)Y_L(\vec{r_j}) \tag{7}$$

where

$$\kappa = (\bar{V}_{II} - \varepsilon)^{\frac{1}{2}} \tag{8}$$

is the 'wave propagation constant', $j_l(i\kappa r_0)$ is a spherical Bessel function, and $h_l^{(1)}(i\kappa r_j)$ is a spherical Hankel function of the first kind. For the energy range $\varepsilon > \bar{V}_{II}$, the solutions of equation 6 are

$$u_i^{II}(1) = \Sigma(L)A_L^0 j_l(\kappa r_0)Y_L(\vec{r_0}) + \Sigma(j)\Sigma(L)A_L^j n_l(\kappa r_j)Y_L(\vec{r_j}) \tag{9}$$

where

$$\kappa = (\varepsilon - \bar{V}_{II})^{\frac{1}{2}} \tag{10}$$

and $n_l(\kappa r_j)$ is a spherical Neumann function.

The first term in each of the expressions (7) and (9) may be interpreted as a superposition of 'incoming' spherical waves which have been 'scattered' by the extramolecular region of spherically averaged potential and which are directed towards the centre of the cluster. The second term may be interpreted as a superposition of 'outgoing' spherical waves which have been 'scattered' by the atomic regions of potential. The arguments of the spherical Hankel functions in equation 7 are imaginary. Thus for localised spin orbitals of the cluster in the energy range $\varepsilon < \bar{V}_{II}$, one is not dealing with progressive waves in the usual sense of scattering theory, but rather spherical waves which in region II decay exponentially away from the atoms. The occurrence of spherical Neumann functions with real arguments in equation 9 suggests 'standing waves' in the intersphere region for the energy range $\varepsilon > \bar{V}_{II}$.

The intersphere wave functions (7) and (9) and their first derivatives are required to be continuous with the solutions (4) and (5) of Schroedinger's equation in the atomic and extramolecular regions, respectively. This leads to equations relating the coefficients A_L^j and C_L^j and the coefficients A_L^0 and C_L^0. Thus the cluster spin orbitals undergo a smooth transition from atomic-like behaviour inside each atomic sphere to exponentially decreasing behaviour away from each atom. This is very much like the behaviour of molecular orbitals represented as traditional linear combinations of analytic atomic orbitals (the LCAO method). However, the scattered-wave representation suffers from none of the convergence problems and other computational difficulties associated with the LCAO method.

By relating the waves incident on each region of the cluster to the waves scattered by all the other regions, it is possible for one to derive a set of compatibility relations among the coefficients A_L^i and A_L^0. These relations are in the form of a set of secular equations which have non-vanishing solutions only for certain values of the energy ε, the eigenvalues. The secular equations are similar in form to those found in the Korringa–Kohn–Rostoker (KKR) method (Korringa, 1947; Kohn and Rostoker, 1954) of crystal band theory, in that there are two principal kinds of matrix elements. There are those elements which depend only on the geometrical arrangement of atoms in the cluster (the 'structure factors') and those elements which depend only on the nature of the atoms, through the logarithmic derivatives of the radial wave functions at the various sphere radii separating regions I, II and III (the 'scattering factors'). In the event that the cluster is a repeating or periodic unit cell of a crystal and the boundary condition of wave function localisation in the extramolecular region III is replaced by the Bloch condition of band theory, the secular equations can be transformed exactly to the KKR form (Johnson and Smith, 1972).

The eigenfunctions of the one-electron equations (1) for an initial superposed-atom potential $V_C(1) + V_{X\alpha}(1)$ are the starting point for a complete SCF–Xα–SW calculation. The initial set of spin orbitals $u_i(1)$ is substituted into expression (2) to obtain the electronic charge density throughout the cluster. This charge density is then substituted into Poisson's equation of classical electrostatics to determine a new Coulomb potential $V_C(1)$ and into equation 3 to determine a new exchange-correlation potential. As before, these potentials are spherically averaged in regions I and III and volume averaged in region II. A weighted average of the new potential and the initial potential serves as input for the first iteration. A new set of spin orbitals is calculated by the scattered-wave method, and the entire computational procedure is repeated until self consistency is attained.

From the self-consistent cluster spin orbitals and electronic charge density, one can compute the expectation value of the Xα statistical total energy, namely (Slater and Wood, 1971; Slater, 1972a; Slater and Johnson, 1972)

$$
\begin{aligned}
\langle E_{X\alpha}\rangle = {}& \Sigma(i)n_i\!\int u_i^*(1)f_1 u_i(1)\,\mathrm{d}v_1 \\
&+\tfrac{1}{2}\!\int\rho\!\uparrow\!(1)[\int\rho(2)g_{12}\,\mathrm{d}v_2 + \tfrac{3}{2}V_{X\alpha}\!\uparrow\!(1)]\,\mathrm{d}v_1 \\
&+\tfrac{1}{2}\!\int\rho\!\downarrow\!(1)[\int\rho(2)g_{12}\,\mathrm{d}v_2 + \tfrac{3}{2}V_{X\alpha}\!\downarrow\!(1)]\,\mathrm{d}v_1
\end{aligned}
\tag{11}
$$

The term containing the one-electron operator f_1 represents the average of the kinetic and potential energies of the electrons in the electrostatic field of the nuclei. The term involving the two-electron operator g_{12} represents the average Coulomb interaction of the electrons with the electron cloud and includes the interaction of an electron with itself. The term in g_{12} also includes the average internuclear Coulomb repulsion which, for fixed nuclear positions, is a constant to be added to the total electronic energy at the end of the calculation. The terms involving $V_{X\alpha}\!\uparrow\!(1)$ and $V_{X\alpha}\!\downarrow\!(1)$ (see equation 3) remove the electron self interaction and account for exchange effects. These quantities are different for spin-up and spin-down electrons, so that the two terms are written separately.

The eigenvalues of the one-electron equations (1) can be shown (Slater,

1972a) to be related to the statistical total energy expression (11) as the first derivatives of the latter quantity with respect to occupation numbers, i.e.

$$\varepsilon_{iX\alpha} = \frac{\partial \langle E_{X\alpha} \rangle}{\partial n_i} \tag{12}$$

This is in distinct contrast to the Hartree–Fock method, where the one-electron energies are given by the difference between the total energies calculated respectively when the ith spin orbital is occupied and when it is empty, i.e.

$$\varepsilon_{iHF} = \langle E_{HF}(n_i = 1) \rangle - \langle E_{HF}(n_i = 0) \rangle \tag{13}$$

In general, $\langle E_{X\alpha} \rangle$ is not a linear function of n_i, so that the values of $\varepsilon_{iX\alpha}$ will usually be different from the values of ε_{iHF}. This is true even if the spin-orbital eigenfunctions and the total energies of the two methods are identical.

One very important consequence of expressions (11) and (12) is that they facilitate the calculation of ionisation energies and optical excitations. In the Hartree–Fock method, the ionisation energies of an atom, molecule, or solid are usually calculated on the basis of Koopmans' theorem. This theorem states that the ionisation energy associated with the excitation of an electron from a particular spin orbital is equal to the negative of the orbital energy. Unfortunately, Koopmans' theorem ignores the orbital relaxation which accompanies ionisation, and thus is generally a poor approximation for most systems. In order to include this relaxation within the Hartree–Fock framework, it is necessary for one to perform separate total energy calculations for the neutral system and the ion, subtracting the two total energies to obtain the ionisation energy. This is too costly a computational procedure to implement in general, even for relatively small or medium sized molecules of the type listed below.

On the other hand, in the SCF-Xα-SW method the difference between the statistical total energies of the neutral system and the ion is equal, to very good approximation, to the energy of a spin orbital from which one-half a unit of electronic charge has been removed (Slater, 1972a; Slater and Johnson, 1972). This modified spin orbital, calculated self-consistently, is called a transition state, since it represents a one-electron state 'halfway' between the initial and final states. The transition states automatically include the effects of spin-orbital relaxation, and the negative values of their energies may be identified with the ionisation energies.

The SCF-Xα-SW transition-state procedure has been applied to many molecules (Johnson, 1972; Johnson, Norman and Conolly, 1972). Results are consistently in better agreement with photoelectron spectroscopic (ESCA) data than the results of both *ab initio* and semi-empirical LCAO calculations, yet they have required relatively small amounts of computer time. The reader is referred to the references listed above for details of these applications.

Another advantage of the SCF-Xα-SW method over the Hartree–Fock LCAO method is that the former approach leads to a reasonably accurate description of the unoccupied or virtual spin orbitals to which electrons in the ground-state orbitals may be excited by optical or other means. Hartree–

Fock theory typically leads to virtual orbitals of positive energy which have no quantitative physical connection with the final states of an optical excitation. This is due to the fact that in the Hartree–Fock theory the occupied and unoccupied orbitals are treated differently. An electron in an occupied orbital moves in the average potential field of $N-1$ electrons, as it should, but an electron in a virtual orbital moves in the potential of all N electrons.

In the SCF-Xα-SW method the two types of electrons move in the same local potential field of $N-1$ electrons, so that the unoccupied spin orbitals are treated as accurately as the occupied ones. In calculating the excitation energy involved in promoting an electron from an occupied orbital to an empty one it is, of course, important for one to use the transition-state procedure, which accounts for the effects of spin-orbital relaxation. Thus it is necessary to calculate the difference between the energy of an initial spin orbital whose electron occupation number has been reduced by one half and the energy of a virtual orbital with occupation number equal to one half. In general, for optical excitations from occupied orbitals just below the Fermi level to empty orbitals immediately above the Fermi level, the relaxation effects are considerably less than those which accompany ionisation.

A part of the multiplet structure can be determined by the SCF-Xα-SW method if the transition-state calculations are carried out in spin-unrestricted (spin-polarised) form. In this case it is possible to perform separate averages over multiplet states of the same multiplicity.

More detailed information about the multiplet structure of the excited states of a molecule or cluster, within the framework of the SCF-Xα-SW theory, can be obtained only by using the spin-orbital wave functions as a basis for a full multiplet calculation. That is, just as in atomic theory, one will have to use the SCF-Xα-SW spin orbitals to construct determinantal functions and linear combinations of them, to describe the multiplets. This has not yet been implemented, although plans are under way to do so. The scattered-wave representation of the spin orbitals is very rapidly convergent and constitutes a complete solution of the Schroedinger equation throughout the space of the molecule. Therefore, it is reasonable to expect that multiplet calculations by this approach will be practical, perhaps more easily implemented than multiplet calculations based on Hartree–Fock SCF-LCAO molecular orbitals.

3. RECENT APPLICATIONS OF THE SCF–Xα–SW METHOD

(A) APPLICATIONS TO METAL COMPLEXES

Much of the inorganic chemistry, organic chemistry, and solid-state physics is concerned with the behaviour of metal atoms in complex molecular and crystalline environments. Among the many examples are:

(1) Isolated metal complexes in the vapour phase (e.g. $TiCl_4$).
(2) Metal–metal bonded molecules [e.g. $Mn_2(CO)_{10}$].
(3) Heavy-metal complexes (e.g. rare-earth complexes).

(4) Complex metal anions in ionic crystals and aqueous solutions (e.g. MnO_4^- in $KMnO_4$).

(5) Antiferromagnetic insulating crystals (e.g. NiO).

(6) 'One dimensional' crystals with metal–metal bonded columnar sequences [e.g. Magnus' green salt $Pt(NH_3)_4PtCl_4$].

(7) Metal–atom clusters (e.g. $Mo_6Cl_8^{4+}$).

(8) Metal impurities in semiconductors (e.g. 'luminescent' impurities in II–VI compounds).

(9) Transition-metal complexes with 'σ-π donor-acceptor' ligands [e.g. $Ni(CO)_4$ and $Fe(CN)_6^{4-}$].

(10) Transition-metal reaction mechanisms (e.g. ligand exchange).

(11) Organometallic complexes, crystals, and polymers (e.g. metal phthalocyanines and porphyrins; metallo-enzymes).

It has been shown (Johnson and Smith, 1972; Johnson, 1972; Johnson, Norman and Conolly, 1972) that with the SCF-Xα-SW method, it is possible for one to calculate with reasonable accuracy the electronic energy levels, charge distributions, electronic excitations, and spin polarisation of poly-atomic molecules and clusters, using only moderate amounts of computer time. Because the computational effort does not increase inordinately with the number of electrons per atom, this technique is therefore ideally suited for the description of electronic structure in metal complexes such as those listed above. Applications to systems like those cited in categories (1), (4), (8), (9) and (11) have already been described in the literature (Johnson and Smith, 1972; Johnson, 1972; Johnson, Norman and Conolly, 1972). More recently, calculations have been carried out on a number of transition-metal carbonyls.

The nature of the electronic structures of transition-metal carbonyls has been a subject of considerable controversy in inorganic chemistry (Cotton and Wilkinson, 1966). It has long been argued that the stability of these systems is due to the combined effects of the donation of electrons from the 5σ carbon 'lone-pair' orbitals of the CO ligands to the unfilled d-shell of the transition-metal atom and the 'back-donation' of d-electrons into the empty 'acceptor' 2π orbitals of CO. Infra-red spectroscopy has shown that there is a gradual decrease of the CO stretching vibrational frequencies and force constants through the isoelectronic sequences of tetrahedral and octa-hedral first-row transition-metal carbonyls, $Ni(CO)_4$, $Co(CO)_4^-$, $Fe(CO)_4^{2-}$ and $Mn(CO)_6^+$, $Cr(CO)_6$, $V(CO)_6^-$ (Jones, 1958, 1963; Edgell, Wilson and Summitt, 1963).

This has been cited as direct evidence for an increasing amount of 2π back-bonding as we progress from nickel carbonyl to vanadium carbonyl, strengthening the metal–carbon bond at the expense of some C—O bond strength (Cotton, 1964). This argument also seems consistent with thermo-dynamic data which indicate increasing metal–CO bond dissociation energies through the isoelectronic sequences (Cotton, Fisher and Wilkinson, 1956, 1959; Fischer, Cotton and Wilkinson, 1957).

Several approximate and semi-empirical LCAO molecular-orbital calcula-tions have been carried out for transition-metal carbonyls (Nieuwpoort, 1965;

Caulton and Fenske, 1968; Carroll and McGlynn, 1968; Schreiner and Brown, 1968; Beach and Gray, 1968; Hillier, 1970). However, the results of these investigations are at considerable variance with respect to orderings and magnitudes of orbital energies and with respect to charge distributions. Thus they have not been able to resolve unambiguously the relative amounts of ligand-to-metal 5σ electron donation and metal-to-ligand 2π back-donation. Furthermore, the LCAO work is not in quantitative agreement with measured optical properties, namely the occurrence of strong optical absorption in the near ultra-violet (Garratt and Thompson, 1934; Callear, 1961; Beach and Gray, 1968).

We have begun, therefore, a systematic investigation of the electronic structures of transition-metal carbonyls by the first-principles SCF-Xα-SW method, with the hope that we shall arrive at a better understanding of the nature of their chemical bonding, optical spectra, and other physical properties. Calculations for nickel tetracarbonyl [$Ni(CO)_4$], iron penta-carbonyl [$Fe(CO)_5$], and chromium hexacarbonyl [$Cr(CO)_6$] have already been completed (Noodleman and Johnson, 1972). For example, the valence orbital energy levels which are responsible for the bonding and optical properties of $Ni(CO)_4$ are illustrated in Figure 1. Also shown for comparison are the SCF-Xα energy levels of a free Ni atom in the $3d^8 4s^2$ configuration and the SCF-Xα-SW orbital energies of a free CO molecule. $Ni(CO)_4$ is a closed-shell system. The highest occupied level is $9t_2$. This level and the $2e$ level immediately below correspond principally to Ni 3d-like orbitals, split in energy by the ligand-field interactions. These levels are completely filled with ten 3d-like electrons (in contrast to the $3d^8$ configuration of the nickel atom), primarily because of the slight overlap of CO 5σ lone-pair electrons with Ni 3d orbitals. The stability of the molecule is the collective result of this covalency, the concomitant formation of the $7a_1$, $6t_2$, $8a_1$, and $7t_2$ orbitals, and some π-type bonding. The $7a_1$ and $6t_2$ orbitals are covalent admixtures of predominantly Ni 4s-like and 4p-like partial waves, respectively, with the 5σ carbon lone-pair orbitals (suggestive of sp^3 hybridisation), while the $7a_1$ and $7t_2$ orbitals together primarily involve Ni 4s-like and 3d-like covalency with the 5σ orbitals (suggestive of sd^3 hybridisation). There is a small degree of covalency of Ni 4s electrons with CO 4σ electrons present in the $8a_1$ orbital. The $8t_2$ orbital is essentially a non-bonding one due to the CO 4σ electrons. The π-type bonding is very polar and occurs mainly through the interaction of Ni 3d electrons with CO 1π electrons in the $1e$ and $2e$ orbitals. The $1t_1$ orbital is a purely non-bonding CO 1π level. Thus it is possible to explain the metal–ligand interaction in $Ni(CO)_4$ without having to introduce the concept of strong net metal-to-ligand $3d$-2π charge transfer, i.e. 'back-bonding'. The fact that the $6a_1$ and $5t_2$ levels are accidentally degenerate and only slightly less negative in energy than the energy of the 3σ bonding orbital of free CO suggests that the C—O bond strength is only slightly reduced in the $Ni(CO)_4$ molecule. This agrees with the observation that the infra-red stretching-mode vibrational frequencies of CO are only slightly reduced from the free CO values.

The cluster of unoccupied energy levels labelled $10t_2$, $3e$, $9a_1$, and $2t_1$ are localised principally on the ligands and correspond closely to the empty CO

2π orbitals. The onset of optical absorption in $Ni(CO)_4$ is at a wavelength of approximately 3000 Å or energy 4 eV. There are principal absorption maxima near 5 eV and 6 eV, followed by continuous absorption far into the ultra-violet. By applying the transition-state concept described in Section 2 to the SCF-Xα-SW orbitals of $Ni(CO)_4$, we can attribute the absorption between 4 eV and 5 eV to electronic transitions between the highest occupied

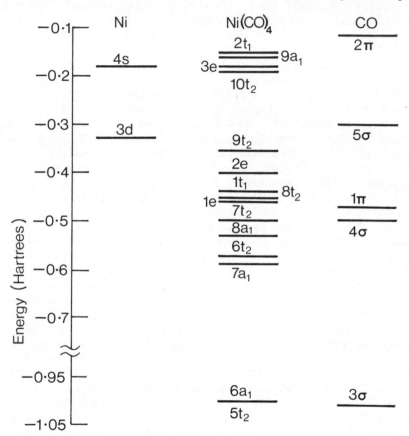

Figure 1 SCF-Xα-SW *electronic energy levels of the nickel tetracarbonyl molecule. The energies are labelled according to the irreducible representations of the tetrahedral* (T_d) *symmetry group. The highest occupied level in the ground state is* $9t_2$. *Also shown are the corresponding* SCF-Xα *energy levels of a free nickel atom in the* $3d^8 4s^2$ *configuration and the* SCF-Xα-SW *orbital energies of a free carbon monoxide molecule*

level $9t_2$ and the unoccupied levels $10t_2$ to $2t_1$. The absorption between 5 eV and 6 eV can be assigned to transitions to the latter levels originating from the next-to-highest occupied level $2e$. Because the initial states $9t_2$ and $2e$ correspond closely to Ni 3d orbitals and the final states correspond to $CO2\pi$ orbitals, the optical absorption is the result of 'metal-to-ligand charge transfer'. This may be contrasted with the situation in MnO_4^-, where we have shown (Johnson and Smith, 1972) that the principal optical absorption involves 'ligand-to-metal charge transfer'. The continuous absorption observed far into the ultra-violet is probably due to photodissociation of the $Ni(CO)_4$ molecule. Current research on transition-metal carbonyls, in particular those involving metal–metal bonding [e.g. $Mn_2(CO)_{10}$], is described in Section 4A.

(B) CALCULATION OF MOLECULAR TOTAL ENERGIES AND CONFORMATIONS

The success of the SCF-$X\alpha$-SW method in accurately predicting the spectral behaviour of complex molecules and crystals, including the effects of spin-orbital relaxation, suggests that one consider applications of the method to other chemical problems. Until recently, all calculations have been carried out for the experimental geometry of the molecule under consideration, and it has not been known how accurately the method will predict potential surfaces and the related geometries and force constants. Thus, it is important to develop an expression for the total energy of a molecule in this approxi-mation, and to test the method on experimentally and theoretically well described systems in order to determine the reliability of the method for these properties. Consequently, the formalism for determining the $X\alpha$ statistical total energy has been developed, and the computer programs have been modified to implement this formalism. In particular, the traditional problems of the internal rotation barriers in C_2H_6 and H_2O_2 and the inversion barrier in NH_3, as well as the total energies and force constants of molecules such as CH_4, H_2O, and P_4, serve as critical tests of the scattered-wave method and $X\alpha$ statistical total energy formulation. Work has already been completed on the barriers in C_2H_6, NH_3, and H_2O_2 (Wahlgren and Johnson, 1972; Wahlgren, 1972). For example, we have calculated the internal rotation barrier between the staggered and eclipsed conformations of C_2H_6 to be 2.9 kcal/mol, which agrees very closely with the experimental barrier of 2.93 kcal/mol. For accurate calculations of the barriers in H_2O_2 and NH_3, it has been necessary to introduce supplementary regions of spherically averaged $X\alpha$ potential in the interatomic regions of the 'lone-pair' electrons. Calculations of the total energies of H_2O (without the introduction of 'lone-pair' spheres) and CH_4 as a function of bond distance have recently been carried out successfully by Connolly and Sabin (1972) and Danese (1972).

Another very interesting problem which has been under investigation is the determination of the stereochemical stability of the tetrahedral phos-phorus molecule (P_4) relative to that of the hypothetical cubic form (P_8). This is a case where there is significant interplay between molecular and solid-state concepts. It is well known that the most chemically reactive and least stable crystalline form of phosphorus is white phosphorus, in which the atoms have local tetrahedral (P_4) co-ordination. The most stable crystalline form is black phosphorus, which consists of corrugated sheets, each phos-phorus atom being bonded to three neighbours approximately at $101°$ angles. Because the phosphorus atom has three 3p electrons, it should be energetically favourable for the atom to form right-angle (p_x, p_y, p_z) bonds in the molecule as well as in the solid, suggesting that a P_8 molecule should be more stable than a P_4 molecule (Cotton, 1972). Indeed, the so-called 'strain energy' and possible occurrence of 'bent bonds' in the tetrahedral P_4 molecule have been long-standing issues of controversy (Pauling and Simonetta, 1959). Thus it is worth while to apply the SCF-Xα-SW method to both the P_4 and P_8 molecules, and to determine the difference between their statistical total energies as a measure of their relative stabilities.

This has been carried out (Norman, Johnson and Cotton, 1972), and it has been found that the equilibrium statistical total energy of P_4, namely $\langle E_{X\alpha} \rangle = -1362.272$ Hartrees, agrees reasonably well with the value of $\langle E_{HF} \rangle = -1362.004$ Hartrees determined by Brundle *et al.* (1972), using an *ab init.* SCF-LCAO method. The P_4 ionisation energies, calculated by the transition-state procedure, are also in excellent agreement with photo-electron data. Of most significance is the fact that the statistical total energy of the P_8 molecule, namely $\langle E_{X\alpha} \rangle = -2724.787$ Hartrees, is lower than the sum of the statistical total energies of two P_4 molecules, i.e. $2\langle E_{X\alpha} \rangle = -2724.544$ Hartrees, suggesting that P_8 is indeed more stable than P_4. If this difference in energy (0.243 Hartree) is divided among the twelve possible P—P near-neighbour bonds, P_8 turns out to be more stable than P_4 by approximately 13 kcal per bond. This figure is significantly greater than the ~ 6 kcal estimate for the P—P bond 'strain energy' in P_4 (Pauling and Simonetta, 1959). There is very little discussion in the literature about the possible occurrence of the P_8 molecule in nature. The fact that the black crystalline form of phosphorus is very difficult to prepare (in comparison with the white form), requiring several days of high pressure and elevated temperature plus a catalyst, suggests that the chemical kinetics may also be unfavourable for the preparation of P_8 in molecular (vapour) form. An attempt is now being made to determine the experimental conditions which are essential for the preparation of P_8 (Cotton, 1972).

(C) APPLICATIONS TO COMPLEX SOLIDS

In explaining the principal features of the optical and magnetic properties of permanganate crystals (Johnson and Smith, 1972), we found it necessary only to calculate the detailed electronic structure and optical excitations (via the transition-state procedure) for a single MnO_4^- cluster in the average stabilising electrostatic field of the crystal. There are many other examples of problems involving localised states in complex materials which are accessible with the SCF-Xα-SW method. Among these are:

(1) Localised crystal defects (e.g. 'deep-level' impurities and vacancies).
(2) Localised optical excitations in crystals (e.g. excitons).
(3) Localised magnetic properties (e.g. Heisenberg exchange integral; local moments).
(4) Chemisorption (e.g. chemisorption of atoms and small molecules at catalytic interfaces).

Applications of the SCF-Xα-SW method to all these areas are in progress.

For example, the exciton problem and a procedure for calculating the Heisenberg exchange integral have been discussed by Slater and Johnson (1972). Calculations of the deep levels associated with 'luminescent' impurities in II–VI compounds (e.g. Mn in ZnS) have been described by Johnson and Smith (1972) and by Johnson (1972). A determination of vacancy and interstitial levels in lead telluride (PbTe), including relativistic effects, is being carried out in co-operation with Professor G. W. Pratt, Jr. of the Electrical Engineering Department at M.I.T. Also under investigation are the electronic structure of interstitial carbon in iron lattices and the chemisorption of CO, NO, H_2, and O on transition-metal and transition-metal-oxide surfaces.

It is sufficient to summarise here that a certain amount of useful information about the effects of a localised perturbation (such as a defect or local excitation) on the electronic structure of an otherwise perfect crystal or interface can be extracted from only a knowledge of the electronic structure of a sufficiently large but finite cluster of atoms centred on the perturbation. The chief difficulty in placing confidence on the results of such calculations is the proper treatment of the termination effects or boundary conditions on the cluster. Within the framework of the SCF-Xα-SW method, there are three principal ways to account for these effects, to first approximation. For reasonably ionic crystals, one can use the same procedure as described by Johnson and Smith (1972) for a MnO_4^- cluster in a crystalline environment, namely introducing a spherical shell of charge around the cluster to approximate the average electrostatic field (Madelung constant) of the rest of the lattice. Reasonably small clusters generally suffice in this situation. For predominantly covalent solids, larger clusters should be used, and the 'saturating effects' of the remainder of the solid should be accounted for, unless 'surface states' due to 'dangling bonds' are to be tolerated. This can be accomplished by saturating the atoms at the periphery of the cluster with hydrogen-like atoms or small molecules. Finally, it may be meaningful under certain circumstances to view the cluster as a periodic entity (i.e. a complex unit cell) in the bulk solid or at the interface. Then it is possible to replace the cluster boundary condition of wave function localisation by the Bloch condition. It has been shown (Johnson and Smith, 1972) that in this limit, the SCF-Xα-SW theory reduces to a band-structure formalism.

4. APPLICATIONS TO SYSTEMS WITH METAL–METAL BONDS

In addition to continuing applications of the SCF-Xα-SW method to the areas described in the preceding section, we have more recently begun several applications of this technique to complex molecules and solids involving metal–metal bonding. The occurrence of metal–metal bonding in isolated polyatomic molecules and in crystals (other than elemental metals and alloys) once considered to be primarily a laboratory curiosity, is now recognised to be widespread and very important (Cotton, 1966, 1967, 1969; Penfold, 1968). Our current research efforts are directed principally to metal–metal bonding in the following systems: (1) binuclear transition-metal carbonyls; (2) metal–metal bonded square-planar complexes in crystals;

and (3) polynuclear metal–atom clusters. We now describe each of these areas in somewhat more detail.

(A) BINUCLEAR TRANSITION-METAL CARBONYLS

Calculations are in progress on binuclear transition-metal carbonyls of the type illustrated in Figure 2, namely $Mn_2(CO)_{10}$, $Co_2(CO)_8$, and $Fe_2(CO)_9$. From the observed molecular structure of $Mn_2(CO)_{10}$, the manganese atoms appear to be joined by a direct metal–metal bond, the exact nature of which is still a mystery. Metal–metal bonding is likewise important in $Co_2(CO)_8$, but there are also bridging carbonyl groups which contribute to the bonding of the dimer. $Fe_2(CO)_9$ has one more bridging CO group than does $Co_2(CO)_8$, and although no metal–metal bond is indicated in Figure 2c, the shortness of the Fe–Fe internuclear distance and the measured diamagnetism (i.e. no unpaired electron spins) suggest the presence of some direct metal–metal covalency. There is as yet no suitable explanation of why a bridged structure for $Mn_2(CO)_{10}$ [e.g. $(OC)_4Mn(CO)_2Mn(CO)_4$, including a Mn–Mn bond] should not be stable, or why a non-bridged version of $Co_2(CO)_8$ should be unstable. Although there have been a few attempts to account for these structural differences in terms of estimated bond energies or qualitative ligand-field theory (Orgel, 1960) the present level of understanding is far from satisfactory.

Having successfully applied the SCF-Xα-SW method to mononuclear transition-metal carbonyls (as described earlier in Section 3A) we think that it is possible to develop, for the first time, a truly quantitative understanding of the electronic structures of the binuclear carbonyls. Contour maps of the electronic charge density [similar to those generated for MnO_4^- by Johnson and Smith (1972)] will be generated, showing the nature of the metal–metal bonds. We hope to establish the relative importance of direct metal–metal covalency and carbonyl bridging to the stereochemical stabilities of these systems.

(B) METAL-METAL BONDED SQUARE-PLANAR COMPLEXES IN CRYSTALS

Another very important and interesting class of materials are crystals based on square-planar metal complexes (typically involving platinum as the central metal atom) which are stacked one on top of the other, producing direct interaction between the metal atoms perpendicular to the plane of each complex. A typical example is illustrated in Figure 3 for a chain of $Pt(CN)_4^{2-}$ complexes such as those which are stable in a tetragonal $K_2Pt(CN)_4Br_{0.30} \cdot 2.3H_2O$ crystal (Krogmann, 1969). Another classic example is Magnus' green salt $Pt(NH_3)_4PtCl_4$, in which $PtCl_4^{2-}$ and $Pt(NH_3)_4^{2+}$ square-planar complexes are joined by Pt–Pt bonds in stacked columnar arrays (Atoji, Richardson and Rundle, 1957).

Recently, there has been a growing amount of interest in the fundamental nature of the electronic structures (particularly the metal–metal bonds) of

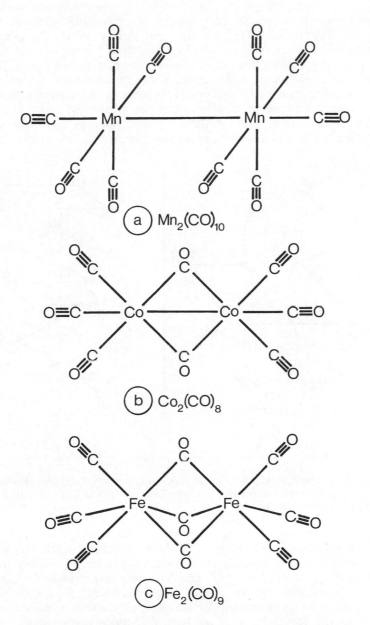

Figure 2   The molecular structures of several binuclear transition-metal carbonyls (see Orgel, 1960)

these crystals, because of their unusually high electrical conductivity (as a function of temperature and pressure), striking optical properties, and other measured physical properties (magnetic, transport, etc.) (Krogmann, 1969; Minot and Pearlstein, 1971; Interrante and Bundy, 1971). The electrical conductivity is, as are most of the other measured properties, very anisotropic, i.e. it is high only along the direction of the metal–metal bonded chains. One plausible qualitative explanation of the nature of the metal–metal bonds and transport effects is that, out of the plane of each complex there is a directed metal 'd_{z^2}-type' molecular orbital (see Figure 3a) which contains two electrons and which overlaps the d_{z^2} orbitals pointing out of the planes of neighbouring complexes in the chain (see Figure 3b). This overlap could

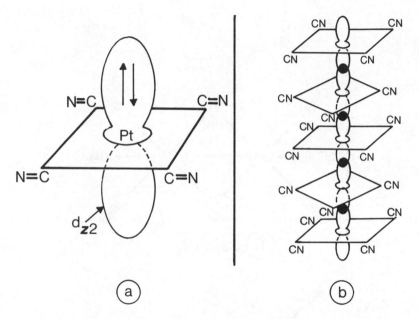

Figure 3   (a) The d_{z^2} orbital of the square-planar complex ion Pt $(CN)_4^{2-}$. (b) Overlap of d_{z^2} orbitals between stacked complex Pt$(CN)_4^{2-}$ ions in a K_2Pt$(CN)_4$Br$_{0.30}$ · 2.3H$_2$O crystal (see Minot and Pearlstein, 1971)

lead to a band of electronic energy levels which is filled and which is separated by an energy gap from a higher unoccupied band of levels (possibly arising from the overlap of metal 'p_z-type' orbitals). Electrons thermally or optically excited from the filled 'valence' bands across the energy gap to the empty 'conduction' band could account for the observed electrical conductivity, in analogy to the conductivity of an intrinsic semiconductor material. Of course, it is also possible that the conductivity is extrinsic, i.e. due to the presence of defect or impurity 'donor' and 'acceptor' levels lying within the band gap. Finally, the electronic energy band description may not be appropriate at all, and it may be better to adopt a purely localised chemical bond description, accounting for observed transport effects in terms of localised electronic charge transfer.

There has been at least one major attempt to calculate the electronic structure of Magnus' green salt [$Pt(NH_3)_4PtCl_4$], using the traditional semi-empirical extended-Hückel LCAO approach (Interrante and Messmer, 1971). The results of these calculations are only of qualitative significance, but they do seem to indicate the importance of intermolecular interactions in Magnus' green salt and the manner in which these interactions depend upon the important structural parameters in the crystal. As we have described in recent publications (Johnson and Smith, 1972; Johnson, 1972), semi-empirical LCAO methods are not generally reliable for transition-metal complexes (even ones as simple as MnO_4^-), so we should not be surprised if they do not lead to quantitative agreement with experiment when they are applied to systems as complicated as Magnus' green salt. On the other hand, it has been demonstrated that the SCF-Xα-SW method leads to accurate quantitative descriptions of transition-metal systems of increasing complexity, with only modest computational effort. Therefore, it is likely that one will arrive at a more complete understanding of the electronic structures and related properties of metal–metal bonded square-planar complexes, using the SCF-Xα-SW method.

SCF-Xα-SW calculations for the isolated $PtCl_4^{2-}$ square-planar complex have already been completed, using the transition-state procedure described in Section 2 to determine the optical properties (Messmer, Wahlgren and Johnson, 1972). A comparison of theory and experiment is summarised in Table 1 for 'd → d type' optical transitions.

By adding neighbouring complexes such as $Pt(NH_3)_4^{2+}$, one can systematically build up the electronic structure of the metal–metal bonded system.

Table 1. Comparison of theoretical and experimental $d \rightarrow d$ optical transition energies (in eV) for $PtCl_4^{2-}$

Transition	SCF–Xα–SW Transition energies	Experiment* (multiplet average)
$e_g(d_{xz}, d_{yz}) \rightarrow b_{1g}(d_{x^2-y^2})$	2.4	2.5(^3E_g, ^1E_g)
$b_{2g}(d_{xy}) \rightarrow b_{1g}(d_{x^2-y^2})$	2.8	2.8($^3A_{2g}$, $^1A_{2g}$)
$a_{1g}(d_{z^2}) \rightarrow b_{1g}(d_{x^2-y^2})$	3.3	3.4($^3B_{1g}$, $^1B_{1g}$)

*See Basch and Gray (1967).

Calculations on crystals involving $Pt(CN)_4^{2-}$ complexes like those illustrated in Figure 3 are also in progress. In particular, we hope to establish how well the localised electronic structure of a single dimer like [$Pt(NH_3)_4PtCl_4$] or [$Pt(CN)_4$]$_2^{4-}$, in the stabilising electrostatic field of the crystalline environment, determines the behaviour of the complete chain. The consequences of adopting periodic boundary conditions on the dimer, utilising the band-theoretical formalism developed by Johnson and Smith (1972), are also being investigated. Finally, we hope to determine the relative contributions of metal–metal covalency and ionic bonding to the stability of the polymeric chains.

(C) POLYNUCLEAR METAL–ATOM CLUSTERS

The third new area in which applications of the SCF-Xα-SW method are in progress is the electronic structure of polynuclear metal–atom clusters, i.e. systems in which the dominant structural feature is a cluster of metal atoms in a symmetrical array (often polyhedral), at internuclear distances which suggest reasonably strong metal–metal interactions (Cotton, 1969). Examples of polyhedral clusters are the $Mo_6Cl_8^{4+}$ and $M_6X_{12}^{2+}$ (e.g. M = Nb, Ta; X = Cl, Br) systems shown in Figures 4a and 4b, respectively. An accurate calculation of the electronic structures of these clusters, particularly the nature of the metal–metal bonds, is of considerable interest, and the results should be of some relevance to the fundamental issue of how metal atoms bond in bulk solids. Indeed, these clusters may be viewed very crudely as finite metal 'lattices' whose 'surfaces' are chemically saturated by halogen atoms. Speculations about the electronic structures of these clusters have been made, using both qualitative valence-bond and molecular-orbital theories (Cotton and Haas, 1964), although the latter approach seems to be most appropriate. With the SCF-Xα-SW method, it should be possible to provide an accurate quantitative description of the molecular orbitals and to establish the amount of covalent interaction among the metal atoms.

Another very interesting and important metal–atom cluster is the iron carbonyl carbide ion $Fe_6(CO)_{16}C^{2-}$ (Churchill et al., 1971), illustrated in Figure 5. Here we have 6 iron atoms in an approximately octahedral array, with 13 terminal carbonyls [two per Fe atom, except for Fe(4) which has three], and 3 bridging carbonyls. At the centre of cluster, approximately equidistant from the 6 Fe atoms, is a single carbon atom. The exact chemical role of the central carbon atom is uncertain, although it is suspected that its valence electrons contribute to the stability of the cluster by allowing the occupied molecular orbitals to assume a closed-shell configuration. It is hoped that a calculation of the detailed electronic structure of this system by the SCF-Xα-SW method will lead to a quantitative understanding of the relative importance of Fe–Fe, Fe–CO, and Fe–C bonding to the stereo-chemical stability. Again this application has some relevance to solid-state chemistry, since the iron carbonyl carbide may be viewed as a molecular analogue of interstitial carbon in an iron lattice, e.g. steel.

5. IMPROVEMENTS IN THE THEORETICAL MODEL

The principal source of error in the SCF-Xα-SW method is the assumption of spherically averaged and volume averaged potentials in the atomic and interatomic regions, respectively, of the cluster. This is the so-called 'muffin-tin' representation of the potential, an approximation which has long been used successfully in the band theory of crystals (Slater, 1968). It would not be

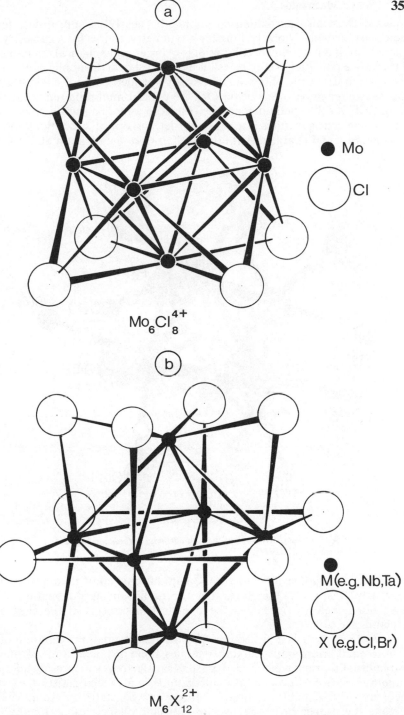

Figure 4   Examples of metal–atom clusters with strong homonuclear metal–metal bonds (see Cotton, 1969)

expected that this approximation should be nearly as appropriate for a molecule, particularly one lacking high symmetry, as it is for a cubic crystal where there is a natural cancellation of low-order non-spherical components of the lattice potential. Nevertheless, it has been shown (Connolly *et al.*, 1972) that even in cases of highly non-spherically-symmetrical molecules, like the linear system carbon suboxide (C_3O_2), the present muffin-tin representation of the SCF-Xα-SW method generally results in a more accurate description of the one-electron features (e.g. orbital orderings, optical excitations, ionisation energies, and charge distributions) than does the *ab initio* SCF-LCAO

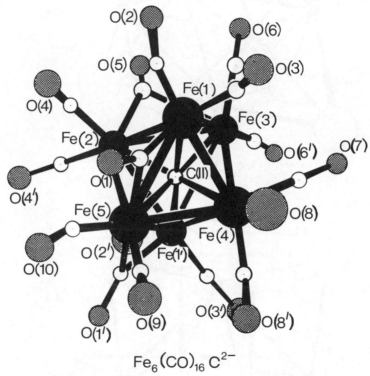

Figure 5   *Molecular structure of the iron carbonyl carbide ion* $Fe_6(CO)_{16}C^{2-}$
(*see Churchill et al., 1971*)

technique, and with significantly less computational effort than in the latter approach. It appears, therefore, that the muffin-tin approximation is an asset, not a liability, as far as the one-electron aspects of the electronic structure are concerned.

As we have described in Section 3(B), the muffin-tin approximation often leads to equilibrium statistical total energies, bond distances, force constants, and internal barriers which are at least as satisfactory as the results of more elaborate Hartree–Fock LCAO calculations, and which satisfy the virial theorem to reasonable accuracy. Moreover, the statistical total energy exhibits the proper behaviour as the molecule dissociates, as long as the atomic sphere radii are increased in proportion to the internuclear distances. In general, however, to obtain a reasonable minimum equilibrium total energy as a function of molecular conformation (e.g. bond angle), it is necessary for one to include at least the first-order interatomic non-muffin-tin corrections to the charge density. This can frequently be accomplished, without any modifications to the computer programs, through the introduction of supplementary interatomic regions of spherically averaged potential (e.g. in the regions of lone-pair electrons).

Preliminary results for the statistical total energies of certain diatomic molecules, such as carbon (C_2) and nitrogen (N_2), are not as satisfactory, although the one-electron transition-state energies are in excellent agreement with ESCA ionisation energies (Connolly, 1972). This is probably due to the fact that the total energy expression (11) is presently being calculated on the basis of the muffin-tin approximation to the electronic charge distribution. In other words, the computed SCF-Xα-SW charge distribution, which is *not* muffin-tin like at the final stage of the SCF procedure (e.g. see the wave function and charge density contour maps generated by Johnson and Smith, 1972) is spherically averaged within the atomic and extramolecular regions and volume averaged, i.e. constant, within the interatomic region. This is, of course, a much more stringent assumption for calculating the total energy than it is for the one-electron energies, especially for molecules like C_2 and N_2 where the presence of p orbitals on both atoms leads to a non-spherical charge distribution along the bond. Thus it is important to calculate the statistical total energy from the complete non-muffin-tin charge distribution which is automatically obtained in the SCF-Xα-SW computational procedure. Modifications of the existing computer programs to accomplish this, without significantly sacrificing the computational advantages of the method, have already been applied successfully to molecules such as C_2, yielding much improved total energies (Connolly, 1972).

A further improvement of the theoretical model includes important low-order non-spherical components in the spherical harmonic expansions of the potential in the atomic, interatomic, and extramolecular regions. This leads to coupled radial Schroedinger equations which can be solved numerically for the radial parts (which now depend on both the quantum numbers l and m) of the wave function expansions (4) and (5). First-order perturbation theory can also be used to correct for the non-spherical components of the potential in the interatomic regions. The SCF-Xα-SW computational procedure is now being modified to permit the implementation of these options.

Finally, there are applications where it will be important to go beyond the one-electron description provided by the SCF-Xα-SW theory. For example, in the analysis of optical excitations, ionisation energies, and crystal-field splittings for open-shell molecules and clusters, the transition-state procedure yields at most the differences between the average energies of multiplets of differing multiplicity. For more detailed information about multiplet structure and oscillator strengths, it will be necessary to determine linear combinations of determinantal functions of the calculated spin-orbital wave functions. Since the method leads to a rapidly convergent complete set of spin orbitals, the latter are ideally suited for the construction of determinantal functions.

6. ACKNOWLEDGEMENTS

I am very grateful to Professor J. C. Slater for originally suggesting the SCF-Xα-SW method, for his collaboration in the development of the theory, and for his encouragement throughout the course of this research. I am grateful also to Dr. F. C. Smith, Jr., for his expertise in originally writing and continuing to improve the computer programs used in this work. Professor J. W. D. Connolly of the University of Florida has carried out many important applications of the method and also deserves much credit for recent improvements in the computational procedure. I wish to thank Professor F. A. Cotton of Texas A and M University and Professor J. G. Norman, Jr., of the University of Washington for suggesting and collaborating in many interesting applications of the method, including the problems of cubic molecular phosphorus and metal–metal bonding. My thanks also go out to Dr. R. P. Messmer of the General Electric Research and Development Centre and to Dr. U. Wahlgren of the University of Stockholm who have contributed significantly to results reported in this review. I am pleased to have had the opportunity to contribute a chapter to this volume on wave mechanics dedicated to Professor de Broglie.

7. SUMMARY

A spin-unrestricted self-consistent-field cluster method has been developed for calculating, from first principles, the electronic structures of complex molecules and localised states in solids. Based on the combined use of Slater's Xα statistical theory of exchange correlation and multiple-scattered-wave formalism, this method permits the accurate calculation of the electronic structures and related physical properties (optical, magnetic, cohesive, etc.) of many-electron polyatomic systems, but requires only moderate amounts of computer time. The following applications are described: metal complexes; molecular total energies and conformations; localised crystal defects; localised optical excitations in solids; localised magnetic properties of solids; chemisorption at catalytic interfaces; metal–metal bonded molecules and crystals; polynuclear metal–atom clusters. Possible improvements in the theoretical model and computational procedure are also described.

REFERENCES

Atoji, M., Richardson, J. W. and Rundle, R. E. (1957) 'On the Crystal Structure of the Magnus Salts, Pt(NH$_3$)$_4$PtCl$_4$', *J. Am. chem. Soc.*, **79**, 3017–3020
Basch, H. and Gray, H. B. (1967) 'Molecular Orbital Theory for Square-planar Metal Halide Complexes', *Inorg. Chem.*, **6**, 365–369
Beach, N. A. and Gray, H. B. (1968) 'Electronic Structures of Metal Hexacarbonyls', *J. Am. chem. Soc.*, **90**, 5713–5721

Brundle, C. R., Kuebler, N. A., Robin, M. B. and Basch, H. (1972) 'Ionization Potentials of the Tetraphosphorus Molecule', *Inorg. Chem.*, **11**, 20–25

Callear, A. B. (1961) 'The Decomposition of Nickel Carbonyl Studied by Flash Photolysis', *Proc. R. Soc. (London)*, **A265**, 71–87

Carroll, D. G. and McGlynn, S. P. (1968) 'Semiempirical Molecular Orbital Calculations. V. The Electronic Structure of Arenechromium Tricarbonyls and Chromium Hexacarbonyl', *Inorg. Chem.*, **7**, 1285–1293

Caulton, K. G. and Fenske, R. F. (1968) 'Electronic Structure and Bonding in $V(CO)_6^-$, $Cr(CO)_6$, and $Mn(CO)_6^+$', *Inorg. Chem.*, **7**, 1273–1284

Churchill, M. R., Wormwald, J., Knight, J. and Mays, M. J. (1971) 'Synthesis and Crystallographic Characterization of $[Me_4N^+]_2[Fe_6(CO)_{16}C^{2-}]$, a Hexanuclear Carbodocarbonyl Derivative of Iron', *J. Am. chem. Soc.*, **93**, 3073–3074

Connolly, J. W. D. (1972) Private communication

Connolly, J. W. D. and Sabin, J. R. (1972) 'Total Energy in the Multiple Scattering Formalism: Application to the Water Molecule', *J. chem. Phys.*, **56**, 5529–5533

Connolly, J. W. D., Nordling, C., Gelius, U. and Siegbahn, H. (1972) Unpublished work

Cotton, F. A. (1964) 'Vibrational Spectra and Bonding in Metal Carbonyls. III. Force Constants and Assignments of CO Stretching Modes in Various Molecules; Evaluation of CO Bond Orders', *Inorg. Chem.*, **3**, 702–711

Cotton, F. A. (1966) 'Transition-metal Compounds Containing Clusters of Metal Atoms', *Quart. Rev. (London)*, **20**, 389–401

Cotton, F. A. (1967) 'Quadruple Bonds and other Multiple Metal-to-metal Bonds', *Rev. pure appl. Chem.*, **17**, 25–40

Cotton, F. A. (1969) 'Strong Homonuclear Metal–Metal Bonds', *Accounts chem. Res.*, **2**, 240–247

Cotton, F. A. (1972) Private communication

Cotton, F. A. and Haas, T. E. (1964) 'A Molecular Orbital Treatment of the Bonding in Certain Metal Atom Clusters', *Inorg. Chem.*, **3**, 10–17

Cotton, F. A. and Wilkinson, G. (1966) *Advanced Inorganic Chemistry* (2nd Edn), p. 720–743, Interscience, New York

Cotton, F. A., Fischer, A. K. and Wilkinson, G. (1956) 'Heats of Combustion and Formation of Metal Carbonyls. I. Chromium, Molybdenum and Tungsten Hexacarbonyls', *J. Am. chem. Soc.*, **78**, 5168–5171

Cotton, F. A., Fischer, A. K. and Wilkinson, G. (1959) 'Heats of Combustion and Formation of Metal Carbonyls. III. Iron Pentacarbonyl; the Nature of the Bonding in Metal Carbonyls', *J. Am. chem. Soc.*, **81**, 800–803

Danese, J. B. (1972) 'Methane Calculations Using the Multiple-scattering Technique with Xα Exchange', *Int. J. quantum Chem.* (in press)

Edgell, W. F., Wilson, W. E. and Summit, R. (1963) 'The Infrared Spectrum and Vibrational Assignment for $Fe(CO)_5$', *Spectrochim. Acta.*, **19**, 863–872

Fischer, A. K., Cotton, F. A. and Wilkinson, G. (1957) 'Heats of Combustion and Formation of Metal Carbonyls. II. Nickel Carbonyl', *J. Am. chem. Soc.*, **79**, 2044–2046

Garratt, A. P. and Thompson, H. W. (1934) 'The Spectra and Photochemical Decomposition of Metal Carbonyls. Part I. Spectral Data', *J. chem. Soc.*, **1934**, 524–528

Gaspar, R. (1954) 'Uber eine Approximation des Hartree–Fockschen Potentials durch eine Universelle Potentialsfunktion', *Acta Phys. Hung.*, **3**, 263–285

Herman, F. and Skillman, S. (1963) *Atomic Structure Calculations*, Prentice-Hall, Englewood Cliffs, N.J.

Hillier, I. H. (1970) 'Approximate Molecular Calculations for the Transition-metal Carbonyls, $Ni(CO)_4$, $Co(CO)_4^-$, $Fe(CO)_4^{--}$, $Fe(CO)_5$, and $Cr(CO)_6$', *J. chem. Phys.*, **52**, 1948–1951

Interrante, L. V. and Bundy, F. P. (1971) 'Studies of Intermolecular Interactions in Square-planar, d^8 Metal Complexes. I. The Physical Properties of Magnus' Green Salt and some Related Complexes at High Pressure', *Inorg. Chem.*, **10**, 1169–1174

Interrante, L. V. and Messmer, R. P. (1971) 'Studies of Intermolecular Interactions in Square-planar d^8 Metal Complexes. II. A Molecular Orbital Study of Magnus' Green Salt', *Inorg. Chem.*, **10**, 1174–1180

Johnson, K. H. (1966) '"Multiple-scattering" Model for Polyatomic Molecules', *J. chem. Phys.*, **45**, 3085–3095

356 Wave Mechanics

Johnson, K. H. (1972) 'Scattered-wave Theory of the Chemical Bond', *Advances in Quantum Chemistry* (edited by Löwdin, P. O.), Vol. 7, Academic Press, New York

Johnson, K. H. and Smith, F. C., Jr. (1972) 'Chemical Bonding of a Molecular Transition-metal Ion in a Crystalline Environment', *Phys. Rev.,* **B5,** 831–843

Johnson, K. H., Norman, J. G., Jr. and Connolly, J. W. D. (1972) 'The SCF-Xα Scattered-wave Method', *Computational Methods for Large Molecules and Localized States in Solids* (edited by Herman, F., McLean, A. D. and Nesbet, R. K.) Plenum Press, New York

Jones, L. H. (1958) 'Vibrational Spectrum of Nickel Carbonyl, *J. chem. Phys.,* **28,** 1215–1219

Jones, L. H. (1963) 'Vibrational Spectra and Force Constants of the Hexacarbonyls of Chromium, Molybdenum, and Tungsten, *Spectrochim. Acta,* **19,** 329–338

Kohn, W. and Rostoker, N. (1954) 'Solution of the Schroedinger Equation in Periodic Lattices with an Application to Metallic Lithium', *Phys. Rev.,* **94,** 1111–1120

Kohn, W. and Sham, L. J. (1965) 'Self-consistent Equations Including Exchange and Correlation Effects', *Phys. Rev.,* **140,** A1133–A1138

Korringa, J. (1947) 'On the Calculation of the Energy of a Bloch Wave in a Metal', *Physica,* **13,** 392–400

Krogmann, M. (1969) 'Planar Complexes Containing Metal–Metal Bonds', *Angew. Chem. Int. Ed. Engl.,* **8,** 35–42

Messmer, R. P., Wahlgren, U. and Johnson, K. H. (1972) Unpublished work

Minot, M. J. and Perlstein, J. H. (1971) 'Mixed-valence Square Planar Complexes. A New Class of Solids with High Electrical Conductivity in one Dimension', *Phys. Rev. Letters,* **26,** 371–373

Nieuwpoort, W. C. (1965) 'Charge Distribution and Chemical Bonding in the Metal Carbonyls $Ni(CO)_4$, $[Co(CO)_4]^-$, and $[Fe(CO)_4]^{--}$. A Non-empirical Approach', *Phillips Res. Rept.,* Suppl. No. 6, 1–104

Noodleman, L. and Johnson, K. H. (1972) Unpublished work

Norman, J. G., Johnson, K. H. and Cotton, F. A. (1972) Unpublished work

Orgel, L. E. (1960) *An Introduction to Transition-Metal Chemistry,* p. 145–147, Methuen, London

Pauling, L. and Simonetta, M. (1959) 'Bond Orbitals and Bond Energy in Elementary Phosphorus', *J. chem. Phys.,* **20,** 29–34

Penfold, B. R. (1968) 'Stereochemistry of Metal Cluster Compounds', *Perspectives in Structural Chemistry* (edited by Dunitz, J. D. and Ibers, J. A.), Vol. II, p. 71–149, Wiley, New York

Schreiner, A. F. and Brown, T. L. (1968) 'A Semiempirical Molecular Orbital Model for $Cr(CO)_6$, $Fe(CO)_5$, and $Ni(CO)_4$', *J. Am. chem. Soc.,* **90,** 3366–3374

Schwarz, K. (1972) 'Optimization of the Statistical Exchange Parameter α for the Free Atoms H Through Nb', *Phys. Rev.,* **B5,** 2466–2468

Slater, J. C. (1951) 'A Simplification of the Hartree–Fock Method', *Phys. Rev.,* **81,** 385–390

Slater, J. C. (1965) 'Suggestions from Solid-state Theory Regarding Molecular Calculations', *J. chem. Phys.,* **43,** S228

Slater, J. C. (1968) *Quantum Theory of Matter* (2nd Edn), p. 610–639, McGraw-Hill, New York

Slater, J. C. (1972a) 'Statistical Exchange-correlation Correction in the Self-consistent Field', *Advances in Quantum Chemistry* (edited by Löwdin, P. O.), Vol. 6, p. 1–92, Academic Press, New York

Slater, J. C. (1972b) 'The Hellmann–Feynman and Virial Theorems in the Xα Method', *J. chem. Phys.* (in press)

Slater, J. C. and Johnson, K. H. (1972) 'Self-consistent-field Xα Cluster Method for Polyatomic Molecules and Solids', *Phys. Rev.,* **B5,** 844–853

Slater, J. C. and Wood, J. H. (1971) 'Statistical Exchange and the Total Energy of a Crystal', *Int. J. quantum Chem.,* **4S,** 3–34

Slater, J. C., Wilson, T. M. and Wood, J. H. (1969) 'Comparison of Several Exchange Potentials for Electrons in the Cu^+ ion', *Phys. Rev.,* **179,** 28–38

Slater, J. C., Mann, J. B., Wilson, T. M. and Wood, J. H. (1969) 'Nonintegral Occupation Numbers in Transition Atoms in Crystals', *Phys. Rev.,* **184,** 672–694

Wahlgren, U. (1972) Unpublished work

Wahlgren, U. and Johnson, K. H. (1972) 'Determination of the Internal Rotation Barrier in Ethane by the SCF-Xα Scattered-Wave Method', *J. chem. Phys.,* **56,** 3715–3716

21

Density Matrix Methods in X-Ray Scattering and Momentum Space Calculations

ROBERT BENESCH

Department of Applied Analysis and Computer Science, Faculty of Mathematics, University of Waterloo, Ontario

and

VEDENE H. SMITH, Jr.

Department of Chemistry, Queen's University, Kingston, Ontario

INTRODUCTION

In the development of physical science in this century, the concept of wave–particle duality has been of wide-reaching significance. During the fifty years since the de Broglie hypothesis of the duality for matter, electron scattering and diffraction have proven to be powerful tools for the investigation of the structure of matter while wave (quantum) mechanics has developed as the theoretical basis of our understanding of the nature of matter.

In the present chapter, we discuss the use of density matrix methods for the calculation of properties of electronic systems which are measurable by the use of X-ray scattering and equivalently by the use of electron scattering in the first Born approximation. We confine our discussion of applications to light atomic systems.

This chapter is dedicated to Professor de Broglie and his outstanding ideas upon which this field of research is based.

1. X-RAY SCATTERING BY ATOMS

In this section we shall summarise the basic results of the quantum mechanical theory of X-ray scattering as developed by Waller and Hartree (1929), and indicate briefly the problems encountered when applying the theory to atoms characterised by correlated wave functions.

Under the assumption that the incident X-ray is plane polarised, Waller and Hartree (WH) showed that the expression for the total intensity (the sum of the elastic and inelastic components) of radiation scattered by an N-electron atom is approximated by

$$I_t/I_{cl} = \sum_m^{E_m - E_g < h\nu} (v_{gm}/v)^3 \, |D_{gm}|^2 \qquad (1)$$

where I_{cl} is the intensity scattered according to the classical expression of Thomson (1906). The terms D_{gm} are given as

$$D_{gm} = \langle \Psi_g(X) | \sum_{j=1}^N \exp\left[2\pi i/c(v_{gm}\vec{S} - v\vec{S}_0)\cdot\vec{r}_j\right] | \Psi_m(X)\rangle \qquad (2)$$

The initial and final states of the target atom are characterised by the normalised wave functions $\Psi_g(X)$ and $\Psi_m(X)$, X denotes the combined space and spin co-ordinates of all N electrons, c is the speed of light, and $\vec{S}_0, \vec{S}$ are unit vectors in the incident and scattered directions, respectively. In the derivation of equation 1, WH assumed that the incident X-ray frequency v is greater than the atomic K-shell absorption frequency. This assumption is noted by the summation restriction in (1) where E_g and E_m denote the energies of the initial and final states of the atom.

With the additional assumption that all components yielding an appreciable intensity in the total scattered radiation have a frequency $v_{gm} \sim v$, WH dropped the summation restriction in (1) to obtain

$$I_t/I_{cl} = \sum_m \langle \Psi_g | P | \Psi_m \rangle \langle \Psi_m | P^* | \Psi_g \rangle \qquad (3)$$

where the scattering operator P has the definition

$$P = \sum_{j=1}^N \exp(i\vec{\mu}\cdot\vec{r}_j) \qquad (4)$$

In terms of the scattering vector $\vec{S} = \vec{S}_0 - \vec{S}$ and the propagation constant $K = 2\pi/\lambda$ of the X-ray, $\vec{\mu} = K\vec{S}$. The magnitude of $\vec{S}$ is related to the scattering angle ω by $|\vec{S}| = 2\sin(\omega/2)$.

Using the closure relation for a complete set of eigenstates $\{\Psi_m\}$, WH obtain the expression

$$I_t/I_{cl} = \langle \Psi_g | PP^* | \Psi_g \rangle \qquad (5)$$

for the total scattered intensity, while from (1) WH obtain the coherent intensity directly as

$$I_c/I_{cl} = |\langle \Psi_g | P | \Psi_g \rangle|^2 \qquad (6)$$

The usual notation for the quantities introduced above is

$$I_t(\vec{\mu})\,|\,I_{cl} = |F(\vec{\mu})|^2 + S(\vec{\mu}) = \langle \Psi_g | PP^* | \Psi_g \rangle$$
$$I_c(\vec{\mu})/I_{cl} = |F(\vec{\mu})|^2 = |\langle \Psi_g | P | \Psi_g \rangle|^2 \qquad (7)$$
$$S(\vec{\mu}) = I_t(\vec{\mu})/I_{cl} - |F(\vec{\mu})|^2$$

Here, $F(\vec{\mu})$ and $S(\vec{\mu})$ are the atomic form factor (X-ray scattering factor) and the incoherent scattering function, respectively. Corrections to these expressions have been given by Bonham (1965).

By approximating the ground state wave function $\Psi_g(X)$ by a single anti-symmetrised product $\Phi(X)$ of N orthonormal spin orbitals $\phi_i(x_j)$, WH obtained the well-known expressions

$$I_t(\vec{\mu})/I_{cl} = N + |\sum_{j=1}^{N} f_{jj}|^2 - \sum_{j=1}^{N} |f_{jj}|^2 - \sum_{j=1}^{N} \sum_{k=1}^{N} |f_{jk}|^2 \qquad (8a)$$

and

$$F(\vec{\mu}) = \sum_{j=1}^{N} f_{jj} \qquad (8b)$$

Denoting the combined space and spin co-ordinates of electron i by $x_i = (\vec{r}_i, \eta_i)$, individual matrix elements are defined as

$$f_{jk}(\vec{\mu}) = \int \phi_j^*(\vec{r}) \overset{\cdot}{\vec{i\mu}} e^{\overset{\cdot}{\vec{i\mu}}\,r} \phi_k(\vec{r}) \, d\vec{r} \delta(\eta_j, \eta_k) \qquad (9)$$

We thus see that the computation of X-ray scattering factors and total scattered X-ray intensities is relatively straightforward for single determinant wave functions constructed from orthonormal spin orbitals $\phi_i(x_j)$.* If correlated wave functions are used in (5) and (6), the computations become tedious, especially for large configuration interaction (CI) wave functions. For Hylleraas-type (HY) wave functions, the evaluation of matrix elements involves products and powers of interelectron co-ordinates, the r_{ij}. These can be expanded, however, via Laplace-like expansions developed by Sack (1964).

In order to study the effects of electron correlation on the total intensity and the scattering factor, Benesch and Smith (1970a) developed a density matrix formalism of X-ray scattering. The applicability of the density matrix formalism to the electron scattering problem was pointed out in that paper. Since the use of density matrices offer computational advantages over the direct use of the total wave function and also leads to a simple description of the X-ray scattering problem in terms of distribution functions, the density matrix method is described in the next section.

2. DISTRIBUTION FUNCTIONS AND THE DENSITY MATRIX FORMALISM

The quantum mechanical expression for the total scattered X-ray intensity is, according to equation 5, the expectation value of the two-electron operator

$$PP^* = \sum_{j=1}^{N} \sum_{k=1}^{N} \exp[i\vec{\mu}\cdot(\vec{r}_j - \vec{r}_k)]$$

$$= N + 2 \sum_{j=1}^{N} \sum_{k>j}^{N} \exp(i\vec{\mu}\cdot\vec{r}_{jk}) \qquad (10)$$

with $\vec{r}_{jk} = \vec{r}_j - \vec{r}_k$. On the other hand, the coherent scattering factor is the

---

*However, until the work of Freeman (1959), the exchange term (i.e. the double summation) in (8a) was omitted in calculations of Restricted Hartree–Fock (RHF) incoherent scattering functions.

expectation value of the one-electron operator defined in (4). As shown previously (Benesch and Smith, 1970a, hereafter referred to as I (5) and (6) can be cast in the form

$$I_t(\vec{\mu})/I_{cl} = N + 2\int \exp\left[i\vec{\mu}\cdot(\vec{r}_1 - \vec{r}_2)\right]\Gamma(x_1, x_2 \mid x'_1, x'_2)\,dx_1 dx_2 \tag{11}$$

and

$$F(\vec{\mu}) = \int \exp(i\vec{\mu}\cdot\vec{r})\gamma(x \mid x')\,dx \tag{12}$$

The two-particle reduced density matrix (two-matrix) is defined, according to Löwdin's (1955) normalisation convention, as

$$\Gamma(x_1, x_2 \mid x'_1, x'_2) = \binom{N}{2}\int \Psi_g(x_1, x_2, x_3, \dots, x_N),$$

$$\Psi_g^*(x'_1, x'_2, x_3, \dots, x_N)\,(dx)_3 \tag{13}$$

While the one-particle reduced density matrix (or one matrix) is defined by

$$\gamma(x \mid x') = N\int \Psi_g(x_1, x_2, \dots, x_N)\Psi_g^*(x'_1, x_2, x_3, \dots, x_N)\,(dx)_2 \tag{14}$$

where Ψ_g represents a normalised N-electron wave function for the atomic state and $x_i = (\vec{r}_i, \eta_i)$ again denotes the combined space and spin co-ordinates of electron i. The notation $(dx)_k$, $k = 2, 3$ represents the combined operations of integration over spatial co-ordinates and summation over spin variables of electrons $k, k+1, \dots, N$.

If the eigenfunction expansion of the one-matrix is employed, equation 12 yields an expression which resembles the WH expression given in equation 8b. On the other hand, insertion of the eigenfunction expansion of the two-matrix into equation 11 yields an expression bearing little resemblance to the WH expression for the total intensity. For the special case where Ψ_g is a single Slater determinant constructed from orthonormal spin orbitals, the density matrix expressions have been shown (see I) to reduce to the WH expressions given in (8) for both the total intensity and for the X-ray form factor.

In actual computations, it was pointed out in I that either the natural spin geminals (NSG) or the natural geminals (NG), which are the respective eigenfunctions of the two-matrix and of its spin-free counterpart, can be used for evaluating total scattered X-ray intensities. Similarly, either the natural spin orbitals (NSO), or the natural orbitals (NO), eigenfunctions of the one-matrix and of its spin-free analogue, respectively, can be used to evaluate scattering factors. Symmetry properties of the reduced density matrices, and the practical implications of these symmetry properties for the computational aspects of the X-ray scattering problem, are also discussed in I.

To establish the connection between X-ray scattering and distribution functions, it is only necessary to note that the one- and two-electron scattering operators are spin-independent and operate on the one- and two-matrices only in a multiplicative sense. We thus dispense with the primed 'book-keeping' notation in equation 12, sum over the spin variables and denote the resulting diagonal, spin-free one-matrix (or charge density matrix) by $\bar{\gamma}(\vec{r})$, thereby obtaining

$$F(\vec{\mu}) = \int \exp(i\vec{\mu} \cdot \vec{r}) \, \bar{\gamma}(\vec{r}) \, d\vec{r} \tag{15}$$

similarly, (11) becomes

$$I_t(\vec{\mu})/I_{cl} = N + 2 \int \exp(i\vec{\mu} \cdot \vec{r}_{12}) P(\vec{r}_1, \vec{r}_2) \, d\vec{r}_1 \, d\vec{r}_2 \tag{16}$$

where $\vec{r}_{12} = \vec{r}_1 - \vec{r}_2$ and $P(\vec{r}_1, \vec{r}_2)$ is used to indicate the diagonal spin-free two-matrix (or two-electron charge density matrix).

If the atomic state is spherically symmetric, or if one averages over random orientations (α, β) of the scattering vector $\vec{S}$ for fixed values of its magnitude, then equation 16 becomes

$$I_t(\mu)/I_{cl} = \int_0^{2\pi} \int_0^\pi I_t(\vec{\mu})/I_{cl} \sin \alpha \, d\alpha d\beta$$

$$= N + 2 \int P(\vec{r}_1, \vec{r}_2) j_0(\mu r_{12}) \, d\vec{r}_1 \, d\vec{r}_2 \tag{17}$$

$$= N + 2 \int_0^\infty P_0(r_{12}) j_0(\mu r_{12}) \, dr_{12}$$

while the spherical average of $I_c(\mu)/I_{cl}$ leads to

$$F(\mu) = \int_0^\infty D_0(r) j_0(\mu r) \, dr \tag{18}$$

as shown previously (Benesch and Smith, 1970b, hereafter referred to as II). In (17) and (18), $j_0(x)$ is the zero-order spherical Bessel function, $j_0(x) = \sin x / x$.

The radial electron–electron distribution function $P_0(r_{12})$ is the generalisation to N-electron atoms of the pair distribution function $f(r_{12})$ introduced by Coulson and Neilson (1961) in their study of electron correlation in the He atom ground state. Although Bartell and Gavin (1964) introduced pair distribution functions $P(r_{ij})$ defined in terms of the total wave function Ψ_g, their definition tended to confuse particle and orbital indices. The electron–electron or pair distribution function was therefore defined (Benesch and Smith, 1970b, 1970c) in terms of the diagonal spin-free two-matrix as

$$P_0(r_{12}) = \int P(\vec{r}_1, \vec{r}_2) \frac{d\vec{r}_1 d\vec{r}_2}{dr_{12}} \tag{19}$$

In (19) the integration is to be performed over all co-ordinates of electrons 1 and 2 *except* for the interelectron co-ordinate r_{12}. As pointed out by Coulson and Neilson (1961) for $f(r_{12})$, $P_0(r_{12}) \, dr_{12}$ is a measure of the probability that the distance separating electrons 1 and 2 lies between r_{12} and $r_{12} + dr_{12}$. It is normalised to the number of electron pairs.

The radial electron-nuclear distribution function $D_0(r)$ employed in (18) was defined in II via the charge density matrix $\bar{\gamma}(\vec{r})$ as

$$D_0(r) = r^2 \int_0^{2\pi} \int_0^{\pi} \bar{\gamma}(\vec{r}) \sin \theta \, d\theta d\phi \qquad (20)$$

The probability of locating an electron at a distance r from the nucleus within a spherical shell of thickness dr is given by $D_0(r) \, dr$; $D_0(r)$ is normalised to the number of electrons N. The normalisation conditions for $P_0(r_{12})$ and $D_0(r)$ are, of course, consequences of using Löwdin's (1955) normalisation convention for the one- and two-matrices.

3. COMPUTATIONAL ASPECTS

It was mentioned in Section 1 that the density matrix formalism represents a simplification of the computational aspects of the scattering problem. This is especially true for CI wavefunctions if the configurations are constructed from non-orthogonal basis orbitals. For this case, one must keep track of numerous overlap integrals in addition to the matrix elements. The use of NSG's (or NG's) and NSO's (or NO's) not only eliminates the overlap integral problem, but simplifies the programming details of matrix element evaluation.

The computation of scattering factors via the density matrix formalism is particularly straightforward. Having averaged over random orientations of the scattering vector $\vec{S}$, we see from equations 18 and 20 that only the totally symmetric component (under spatial rotations) of the charge density matrix is required. This means that X-ray scattering factors can be computed from its eigenfunctions, the symmetric component natural orbitals (SCNO) [see Smith and Harriman, 1970. They are also called by some authors (Bingel and Kutzelnigg, 1968) the symmetry adapted natural orbitals (SANO)].

Similarly the evaluation of the total intensities may be made using the symmetric component natural geminals [(SCNG) sometimes called symmetry adapted natural geminals (SANG)] which are the eigenfunctions of the totally symmetric component of the spin-free two-matrix.

For Hylleraas-type (HY) wave functions, the form factor calculation is only as good as the NSO or NO expansion of the relevant one-matrix or charge density matrix. In order to fully represent the r_{ij} terms, a complete set of basis functions is required for the eigenfunction expansion. In practice one must settle for a truncated basis set. A similar observation holds for total intensity values calculated via eigenfunction expansions of HY two-matrices.

Although the distribution functions $D_0(r)$ and $P_0(r_{12})$ have a natural expansion in terms of the eigenfunctions of the one- and two-matrices (Benesch and Smith, 1971a), equations 17 and 18 indicate that only the distribution functions themselves are required. Equations 19 and 20 have thus been used (Benesch, 1971) to obtain $P_0(r_{12})$ and $D_0(r)$ in algebraic form from twenty-term HY functions for several members of the He iso-electronic sequence. These particular distributions have an exceedingly simple analytic form and facilitate the evaluation of various expectation values in addition to those of the one- and two-electron scattering operators.

The algebraic formulation of $P_0(r_{12})$ for N-electron atoms has been considered in some detail (Benesch, 1972). However, the necessary radial integrals become quite cumbersome for total wave functions constructed from a spin-orbital basis. On the other hand, the radial integrals encountered in the algebraic evaluation of $P_0(r_{12})$ from a wave function of the HY type are, for the most part, quite straightforward. Details relating to the algebraic evaluation of $P_0(r_{12})$ and $D_0(r)$, including various expressions for radial integrals, are given elsewhere (Benesch, 1971, 1972).

4. ASYMPTOTIC AND LIMITING BEHAVIOUR OF X-RAY FUNCTIONS

The density matrix formalism provides a convenient starting point for obtaining some useful relations between X-ray values and atomic properties. We shall consider these relations as a motivation for the examination of the large and small μ behaviour of $F(\mu)$ and $I_t(\mu)/I_{cl}$.

The determination of charge density and pair distribution functions by Fourier inversion of equations 18 and 17, namely

$$D_0(r)/r = 2/\pi \int_0^\infty \mu F(\mu) \sin \mu r \, d\mu \tag{21}$$

and

$$P_0(r_{12})/r_{12} = 1/\pi \int_0^\infty \mu \sin \mu r_{12}[I_t(\mu)/I_{cl} - N]d\mu \tag{22}$$

has experimental limitations. Specifically, the magnitude of μ is limited to the range of obtainable X-ray wavelengths. Experimental verification of the sum rule (Silverman and Obata, 1963) for the expectation value of $1/r$,

$$\langle 1/r \rangle = 2/\pi \int_0^\infty F(\mu) \, d\mu \tag{23}$$

and the sum rule (Tavard and Roux, 1965) for the expectation value of $1/r_{12}$,

$$\langle 1/r_{12} \rangle = 1/\pi \int_0^\infty (I_t(\mu)/I_{cl} - N) \, d\mu \tag{24}$$

is limited in the same fashion.

In order to use (21) through (24) with experimental data, the behaviour of $F(\mu)$ and $I_t(\mu)/I_{cl}$ must be investigated for large values of μ. The small μ behaviour is easily established by expanding the spherical Bessel function in (17) and (18). The small μ behaviour of the total scattered X-ray intensity is readily seen to be

$$I_t(\mu)/I_{cl} \sim N + 2\int_0^\infty P_0(r_{12})\,dr_{12} - \frac{\mu^2}{3}\int_0^\infty r_{12}^2 P_0(r_{12})\,dr_{12} + 0(\mu^4)$$

$$\sim N^2 - \frac{\mu^2}{3}\langle r_{12}^2 \rangle + 0(\mu^4) \tag{25}$$

while the small μ behaviour of the form factor is approximated by

$$F(\mu) \sim N - \frac{\mu^2}{6}\langle r^2 \rangle + O(\mu^4) \tag{26}$$

Equations 25 and 26 make rigorous the notion that for small values of μ, one is probing the *outer* regions of the pair-distribution and charge density, respectively.

The asymptotic behaviour of $F(\mu)$ has been studied by Goscinski and Lindner (1970) while Smith (1970) studied the large μ behaviour of the incoherent scattering function $S(\mu)$. The asymptotic behaviour of the total intensity is

$$I_t(\mu)/I_{cl} - N \sim 2g^{(2)}(0)\mu^{-4} + O(\mu^{-6}) \tag{27}$$

where $g(r_{12}) = P_0(r_{12})/r_{12}$. The notation $g^{(2)}(0)$ indicates the second derivative of the reduced pair distribution function $g(r_{12})$ evaluated at $r_{12} = 0$. Smith (1970) also obtained an integral relation for $g^{(2)}(0)$ which involved the electron–electron cusp condition (Kato, 1957; Bingel, 1963; Steiner, 1963).

The asymptotic behaviour of $F(\mu)$ can be obtained in an analogous manner. Repeated integration of (18) by parts yields

$$F(\mu) \sim -2\left(\frac{\partial \rho_0(r)}{\partial r}\right)_{r=0} \mu^{-4} + O(\mu^{-6}) \tag{28}$$

where $\rho_0(r) = r^2 D_0(r)$. Equation 27 can be simplified for those atomic wave functions which obey the Kato (1957) electron–nucleus cusp condition,

$$\left(\frac{\partial \rho_0(r)}{\partial r}\right)_{r=0} = -2Z\rho_0(0) \tag{29}$$

thereby yielding

$$F(\mu) \sim 4Z\rho_0(0)\mu^{-4} + O(\mu^{-6}) \tag{30}$$

as previously shown by Goscinski and Lindner (1970).

Equation 27 has been tested (Benesch and Smith, 1971b) for He and He-like ions; the asymptotic results are in excellent agreement with values computed from (17). Again, (26) and (27) make rigorous the notion that for large values of the scattering variable, one is probing the *inner* regions of the charge density and pair distribution function, respectively.

5. THE EFFECT OF ELECTRON CORRELATION ON X-RAY VALUES

Computations of Restricted Hartree–Fock (RHF) form factors $F(\mu)$ and incoherent scattering functions $S(\mu)$, both for electron and X-ray scattering, have been made by several authors with a view to providing fundamental atomic data. However, the effect of electron correlation on these quantities has received comparatively little attention. This is due, no doubt, to the unavailability of correlated wave functions except for some of the lightest atomic systems. For those systems where computations of correlated $F(\mu)$'s and $S(\mu)$'s have been made, the consensus is that the correlated $F(\mu)$'s generally agree with their RHF counterparts to approximately $\pm 1\%$ over the entire range of the scattering variable. On the other hand, correlated and RHF $S(\mu)$'s have been found to differ appreciably, especially for small values of μ. For large values of μ, the agreement is usually quite good.

Computations by Kim and Inokuti (1968), and those of Bartell and Gavin (1964, 1965) and Kohl and Bonham (1967) indicate a discrepancy of 6% between the correlated and RHF incoherent scattering functions $S(\mu)$ for He. The correlated $S(\mu)$'s are smaller for a fixed value of μ, than those computed from RHF wave functions. This behaviour has also been verified for Be in II, where differences of the order of 25% and larger for $S(\mu)$ were observed for values of $K \leq 0.1$ (Å) $^{-1}$; $\mu = 4\pi K$. Similar behaviour has been noted by Brown (1970) for correlated $S(\mu)$ values obtained for 3- and 4-electron atoms and ions. Calculations by Brown (1972) for atomic carbon have turned up discrepancies of the order of 22%, while computations of $S(\mu)$ by Tanaka and Sasaki (1971) using CI wave functions incorporating only L-shell correlation effects indicate maximum differences of the order of 15% for C, 12% for N, 11% for O, 10% for F and 10% for Ne. For the latter system, Peixoto, Bunge and Bonham (1969) have found differences of the order of 12%.

It is interesting to investigate the direction of the deviations between correlated and RHF values for both $S(\mu)$ and $F(\mu)$. For values of K, $K \sim 0.2(\text{Å})^{-1}$ or less, the maximal value of the differences for the systems listed above

$$\Delta S(\mu) = S_{\text{corr}}(\mu) - S_{\text{RHF}}(\mu) \tag{31}$$

is negative. Although correlated and RHF form factors agree to approximately 1%, the differences

$$\Delta F(\mu) = F_{\text{corr}}(\mu) - F_{\text{RHF}}(\mu) \tag{32}$$

for Be (see Figure 3 of II) and for C (see Table V of Tanaka and Sasaki, 1971) are usually positive, rather than negative. Examination of total intensity differences

$$\Delta I_t(\mu) = I_t^{\text{corr}}(\mu) - I_t^{\text{RHF}}(\mu) \tag{33}$$

for Be (Figure 1 of II) and for Be through Ne (Figure 3 of Tanaka and Sasaki) show that these values are negative over the entire range of the scattering variable. Again, the total intensity difference attain their maximal values for small K, $K \sim 0.2\,(\text{Å})^{-1}$

For small values of K, $\mu = 4\pi K$, use of (25) in (33) gives, neglecting terms of $O(\mu^4)$,

$$\Delta I_t = \frac{\mu^2}{3}\left[\langle r_{12}^2\rangle_{\text{RHF}} - \langle r_{12}^2\rangle_{\text{corr}}\right] \tag{34}$$

Since $\Delta I_t(\mu)$ is strongly negative for small μ, it immediately follows that

$$\langle r_{12}^2\rangle_{\text{RHF}} < \langle r_{12}^2\rangle_{\text{corr}} \tag{35}$$

which implies that the correlated pair distribution function $P_0(r_{12})$ must be

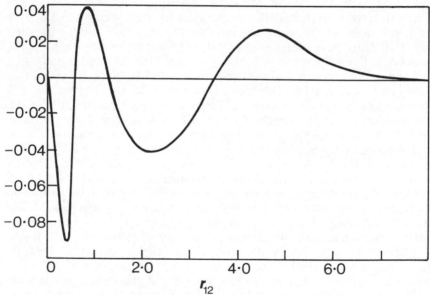

Figure 1   Approximate Coulomb Hole Function $\Delta(r_{12})$ for the ^1S ground state of beryllium (atomic units)

more diffuse and approaches zero more slowly than its RHF counterpart, provided that both pair distributions are of the same general shape.

Examination of the Coulomb hole function (Coulson and Neilson, 1961),

$$\Delta r_{12} = P_0^{\text{exact}}(r_{12}) - P_0^{\text{RHF}}(r_{12}) \tag{36}$$

which is shown in Figure 1 for Be and in Figure 2 for Ne (Peixoto, Bunge and Bonham, 1969) indicates that correlated pair distributions are indeed more diffuse than RHF distributions.

In a similar fashion, use of (26) in (32) for small values of μ leads to the observation that

$$\langle r^2\rangle_{\text{RHF}} > \langle r^2\rangle_{\text{corr}} \tag{37}$$

provided that $\Delta F(\mu)$ values are positive for small μ. The implication of equation 37 is that the charge density obtained from a correlated wave function is contracted relative to the RHF charge density, a result which would be intuitively expected from arguments based on shielding considera-

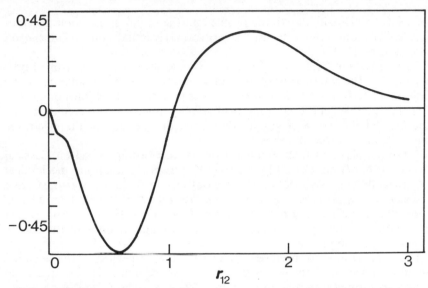

Figure 2   Approximate Coulomb Hole Function $\Delta(r_{12})$ for the ^1S ground state of neon (atomic units)

tions. Finally, we note that $\Delta S(\mu)$ should be negative for small μ, again as a result of the Coulomb hole. Since $\Delta S(\mu)$ is given by

$$\Delta S(\mu) = \Delta I_+(\mu) - \Delta F^2(\mu) \tag{38}$$

we can relate $\Delta S(\mu)$ to differences between $\langle r_{12}^2 \rangle$ and $\langle r^2 \rangle$, namely

$$\Delta S(\mu) = \frac{\mu^2}{3}(\langle r_{12}^2 \rangle_{\text{RHF}} - \langle r_{12}^2 \rangle_{\text{corr}}) - \frac{N\mu^2}{3}(\langle r^2 \rangle_{\text{RHF}} - \langle r^2 \rangle_{\text{corr}}) + O(\mu^4) \tag{39}$$

6. THE COMPTON PROFILE

If an X-ray is scattered by an unbound electron initially at rest, it undergoes a shift in wavelength given by the Compton expression

$$\Delta\lambda = \lambda_c - \lambda = 2h \sin^2(\omega/2)/mc \tag{40}$$

where λ and λ_c are the incident and shifted wavelengths, h is Planck's constant, c is the speed of light, ω is the scattering angle and m is the electron mass. If the X-ray is inelastically scattered by an atom, a distribution of shifted wavelengths centred about the Compton wavelength is observed.

As pointed out by DuMond (1929, 1930), the broadening of the Compton line into a profile can be understood in terms of the Doppler effect, due to the motion of electrons in the atom and hence proportional to their velocities.

This idea was developed by DuMond who treated the electrons classically and obtained the relation

$$J(q) = \tfrac{1}{2} \int_{|q|}^{\infty} \frac{I_0(p)}{p} \, dp \tag{41}$$

between the intensity $J(q)$ of the profile and the radial momentum distribution $I_0(p)$ of the atomic electrons. The form of (41) is actually due to Duncanson and Coulson (1945).

As pointed out by Bloch (1934), the average speed of an electron is proportional to the square root of its binding energy and the broadening of the Compton line is also proportional to this square root. Bloch thus points out that the Doppler effect can only explain a symmetrical broadening of the line around its centre of gravity (40) since the average of the electronic velocities in the atom vanishes.

Although (41) was first obtained from considerations based on classical physics, it has been derived by Kilby (1965) from the quantum mechanical scattering theory of Waller and Hartree outlined in Section 1. Starting from a one-electron approximation, Kilby assumes that the one-electron excited states $\Psi_m(x)$ in (2) can be represented by plane waves. He also assumes that the binding energy of the electron is small in comparison with the energy transfer from the proton.

Equation 41 has also been obtained within the framework of the so-called impulse approximation, which we shall now examine. The scattering of X-rays by atomic electrons in the non-relativistic region, $h\omega \ll m_0 c^2$, where ω is the X-ray frequency and m_0 is the electron rest mass, is treated properly in the Waller–Hartree (1929) approximation. The quantity of interest for us is the differential scattering cross-section per unit solid angle $d\Omega$ and per unit of outgoing photon energy, $d\omega$, namely $d^2\sigma/d\Omega d\omega$, which in turn is proportional to the dynamical structure factor

$$S(\vec{k}, \omega) = \sum_f |\langle f| \sum_j e^{i\vec{k} \cdot \vec{r}_j} |0\rangle|^2 \, \delta(E_f - E_0 - \omega) \tag{42}$$

evaluated for the energy transfers $\omega = \omega_1 - \omega_2$ and momentum transfers $\vec{k} = \vec{k}_1 - \vec{k}_2$. The subscripts 1 and 2 refer to the incoming and scattered waves, respectively. The constant of proportionality between $d^2\sigma/d\Omega d\omega$ and $S(\vec{k}, \omega)$ is $r_0^2 \, (\omega_2/\omega_1) \, (\vec{e}_1 \cdot \vec{e}_2)$ where r_0 is the classical electron radius and $\vec{e}_1 \cdot \vec{e}_2$ describe the polarisation of the incident and outgoing photons.

Use of the integral relation

$$\delta(x) = \frac{1}{2\pi} \int_{-\infty}^{\infty} e^{ixt} \, dt \tag{43}$$

for the delta function as well as the relation

$$e^{iHt} |j\rangle = e^{iE_j t} |j\rangle \tag{44}$$

and the closure relation

$$\sum_j |j\rangle\langle j| = 1 \tag{45}$$

for a complete set of eigenstates allows (42) to be rewritten in the form

$$S(\vec{k}, \omega) = \frac{1}{2\pi} \int_{-\infty}^{\infty} dt \; e^{i\omega t} \langle 0| e^{iHt} \sum_m e^{i\vec{k} \cdot \vec{r_m}} e^{-iHt} \sum_n e^{-i\vec{k} \cdot \vec{r_n}} |0\rangle \qquad (46)$$

Now $H = H_0 + V$, $H_0 = \sum_j P_j^2/2$ and we can expand (Wilcox, 1967) the Hamiltonian e^{iHt} as

$$e^{iHt} = e^{iH_0 t} e^{iVt} e^{-[H_0, V]t^2/2} \dots \qquad (47)$$

As Eisenberger and Platzman (1970) have noted, the important times in the integral are those of $O(1/\omega)$. Hence if we perform our scattering experiment at energy transfers ω which are much larger than the characteristic energies of the scattering system but not so large as to invalidate this non-relativistic treatment, then

$$e^{-[H_0, V]t^2/2} \sim 1 \qquad (48)$$

This relation is the basis of the so-called impulse approximation. Then

$$S^I(\vec{k}, \omega) = \frac{1}{2\pi} \int_{-\infty}^{\infty} dt \; e^{i\omega t} \langle 0| e^{iH_0 t} \sum_m e^{i\vec{k} \cdot \vec{r_m}} e^{-iH_0 t} \sum_n e^{-i\vec{k} \cdot \vec{r_n}} |0\rangle \qquad (49)$$

as $V(r)$ commutes with r, implying that the potential term has vanished from the approximation 47 for the cross-section. The implications of the apparent vanishing of the potential term are discussed by Eisenberger and Platzman (1970). We rewrite (49) as

$$S^I(\vec{k}, \omega) = S_1^I(\vec{k}, \omega) + S_2^I(\vec{k}, \omega)$$

where

$$S_1^I(\vec{k}, \omega) = \frac{N}{2\pi} \int_{-\infty}^{\infty} dt \; e^{i\omega t} \langle 0| e^{iH_0 t} e^{i\vec{k} \cdot \vec{r_1}} e^{-iH_0 t} e^{-i\vec{k} \cdot \vec{r_1}} |0\rangle \qquad (50)$$

$$S_2^I(\vec{k}, \omega) = \frac{N(N-1)}{2\pi} \int_{-\infty}^{\infty} dt \; e^{i\omega t} \langle 0| e^{iH_0 t} e^{i\vec{k} \cdot \vec{r_1}} e^{-iH_0 t} e^{-i\vec{k} \cdot \vec{r_2}} |0\rangle \qquad (51)$$

Equation (50) may be evaluated by employing a complete set of eigenstates of H_0, namely the plane wave momentum eigenfunctions $e^{i\vec{p} \cdot \vec{r_j}}$. Then equation (50) yields

$$S_1^I(k, \omega) = \frac{1}{2\pi} \int dt \; d\vec{p} \, e^{i\omega t} \gamma(\vec{p} \,|\, \vec{p}) e^{-i(k^2/2 + \vec{k} \cdot \vec{p})t} \qquad (52)$$

which from (43) gives

$$S_1^I(k, \omega) = \int d\vec{p} \, \gamma(\vec{p} \,|\, \vec{p}) \delta(\omega - \frac{k^2}{2} - \vec{k} \cdot \vec{p}) \qquad (53)$$

Similarly, (51)

$$S_2^I(k, \omega) = 2 \int d\vec{p_1} \, d\vec{p_2} \, \Gamma(\vec{k} + \vec{p_1}, \vec{p_2} - \vec{k} \,|\, \vec{p_1}, \vec{p_2}) \delta(\omega - k^2/2 - \vec{k} \cdot \vec{p_1}) \qquad (54)$$

In the above, $\gamma(\vec{p} \,|\, \vec{p})$ and $\Gamma(\vec{p_1}, \vec{p_2} \,|\, \vec{p'_1}, \vec{p'_2})$ are the momentum space analogues of the spin-free one- and two-matrices in position space.

The term $S_2^I(\vec{k}, \omega)$ was overlooked by Eisenberger and Platzman in their derivation of the impulse approximation. For the case of electron electron scattering, a similar term has been obtained by Tavard and Bonham (1969). It can be shown (Smith, 1972) to have a negligible contribution for sufficiently large momentum transfer $\vec{k}$ (i.e. the usual experimental condition). This term should be taken into account when one integrates a theoretical $d^2\sigma/d\Omega d\omega$ curve to obtain the total scattered intensity. We note the recent study by Currat, DeCicco and Weiss (1971) who integrated over the $d^2\sigma/d\Omega d\omega$ arising only from $S_1^I(\vec{k},\omega)$ and found significant discrepancies at small scattering angles.

We direct our attention now to $S_1^I(\vec{k},\omega)$. The delta function energy conservation condition shows that the scattered photon is shifted in frequency by the momentum transfer $k^2/2$ and the Doppler shift $\vec{k} \cdot \vec{p}$. Furthermore we see that differential scattering cross-section $d^2\sigma/d\Omega d\omega$ is simply related to a one-dimensional projection of the momentum space charge distribution function.

The isotropic component of $S(\vec{k},\omega)$, $S_0(k,\omega)$, under the conditions of sufficiently large energy and momentum transfer, becomes from (53)

$$S_0^I(k,\omega) = J(q) = 1/2 \int_{|q|}^{\infty} p^{-1} I_0(p) \, dp \tag{55}$$

where $q = \vec{k} \cdot \vec{p}/|\vec{k}|$ is the projection of the electron's momentum on the scattering vector k. $I_0(p)$ is the radial momentum distribution which is discussed in detail in the next section.

7. DENSITY MATRIX FORMALISM OF THE MOMENTUM DISTRIBUTION

There are three possible methods for determining the distribution of electrons in momentum space. The first method involves transforming the Schroedinger equation into an integral equation and solving iteratively for the momentum space wave function. Except for the early work of McWeeny and Coulson (1949) for He and McWeeny (1949) for H_2^+, this method has received little or no attention in recent years. The second method involves taking the total wave function in position space, computing its $3N$-dimensional Fourier transform, squaring the transform, and then integrating out the momentum space co-ordinates of N-1 electrons. It is noted that some workers have actually advocated that this laborious procedure should be the only method for computing the momentum distribution. The third method, the density matrix method, developed by the authors in a series of papers, and described in this section is an elegant and computationally attractive method that allows momentum distributions to be obtained from the most complicated wave functions wherein electron correlation is adequately represented.

The radial distribution of electrons in momentum space, $I_0(p)$, in terms of which the Compton profile $J(q)$ was defined in (41), can be obtained very easily from the eigenfunction expansion of either the one-matrix or the charge density matrix. It has been defined (Smith and Benesch, 1968) as

$$I_0(p) = \int p^2 \hat{\gamma}(\vec{p} \,|\, \vec{p}) \, d\Omega \tag{56}$$

where $\vec{p} = (p,\alpha,\beta)$, $d\Omega = \sin \alpha \, d\alpha d\beta$ is an element of solid angle in momentum space and $\hat{\gamma}(\vec{p}\,|\,\vec{p})$ is the momentum space counterpart of the position space charge density matrix,

$$\hat{\gamma}(\vec{p}\,|\,\vec{p}) = (2\pi)^{-3} \int \hat{\gamma}(\vec{r}\,|\,\vec{r}')\,e^{-i\vec{p}\cdot\vec{r}}\,e^{i\vec{p}'\cdot\vec{r}'}\,d\vec{r}'\,d\vec{r} \tag{57}$$

The zero subscript on $I_0(p)$ again serves as a reminder that only the totally symmetric component [under spatial rotations, compare with $D_0(r)$] of $\hat{\gamma}(\vec{p}\,|\,\vec{p})$ is involved in its evaluation*. As in position space, $\hat{\gamma}(\vec{p}\,|\,\vec{p}')$ has an expansion in terms of its eigenfunctions, the momentum space NO's $\hat{\chi}_i(\vec{p})$,

$$\hat{\gamma}(\vec{p}|\vec{p}') = \sum_i \lambda_i \hat{\chi}_i(\vec{p})\hat{\chi}_i^*(\vec{p}') \tag{58}$$

where λ_i is the occupation number. Given the NO (or NSO) $\chi_i(\vec{r})$ in position space, $I_0(p)$ for correlated system is very easily evaluated by computing the Fourier transform

$$\hat{\chi}_j(\vec{p}) = (2\pi)^{-3/2} \int e^{-i\vec{p}\cdot\vec{r}} \chi_j(\vec{r})\,d\vec{r} \tag{59}$$

Evaluation of (59) is discussed elsewhere (Benesch and Smith, 1971c) for NO's constructed from STO basis functions. Since $I_0(p)$ is related to the totally symmetric component of $\hat{\gamma}(\vec{p}|\vec{p})$, symmetric component natural orbitals (SCNO's) can be used for its evaluation. The reader will recall from Section 3 that SCNO's, NO's or NSO's can also be used for evaluating the X-ray form factor.

It is worth noting the normalisation conditions

$$N = \int \hat{\gamma}(\vec{p}|\vec{p})\,d\vec{p}$$
$$= \int_0^\infty I_0(p)\,dp \tag{60}$$

and

$$\int_0^\infty J(q)\,dq = 1/2 \int_{-\infty}^\infty J(q)\,dq = N/2 \tag{61}$$

which serve as useful checks for computed $I_0(p)$ and $J(q)$ values. These normalisation conditions are again a consequence of using Löwdin's (1955) normalisation convention for the position space one-matrix. We also have the useful relation

$$J(0) = 1/2\langle p^{-1}\rangle \tag{62}$$

which follows immediately from (41), while (twice) the kinetic energy can be obtained from

$$\langle p^2\rangle \equiv \langle -\nabla^2\rangle = \int_0^\infty p^2 I_0(p)\,dp \tag{63}$$

8. LIMITING AND ASYMPTOTIC BEHAVIOUR OF $I_0(p)$ AND $J(q)$

In order to establish sum rules for $J(q)$, it is necessary to investigate the behaviour of $I_0(p)$ and $J(q)$ for both large and small values of their respective

---

*Note that $\hat{\gamma}(\vec{p}\,|\,\vec{p})$ is not the Fourier transform of $\gamma(\vec{r}\,|\,\vec{r})$. The relationship between $F(\vec{\mu})$ and $\gamma(\vec{p}\,|\,\vec{p})$ has been pointed out elsewhere (Benesch, Singh and Smith, 1971).

arguments. The cusp condition (Smith, 1971) for individual NO's can be employed to obtain the asymptotic behaviour of both $I_0(p)$ and $J(q)$. Since the radial momentum distribution defined by (56) involves only the totally symmetric component of the momentum space one-matrix, this means that we need consider only position spaces NO's which are already eigenfunctions of the angular momentum operator l^2. That is, we can restrict our attention to NO's constructed for S states, which are of s-type, p-type, and so forth, or else we can use SCNO's which by definition are eigenfunctions of l^2.

For s-type NO's, we obtain from (59) after integration over the angular co-ordinates,

$$\hat{\chi}_s(\vec{p}) = (2/\pi)^{1/2} \int_0^\infty \chi_s(r) j_0(pr) r^2 dr \tag{64}$$

which yields for large p

$$\hat{\chi}_s(\vec{p}) = -(8/\pi)^{1/2} \left[\left(\frac{\partial \chi_s(r)}{\partial r} \right)_{r=0} p^{-4} + O(p^{-6}) \right] \tag{65}$$

As orbitals of higher angular momentum give leading contributions of $O(p^{-5}), O(p^{-6}), \ldots$, for p-type, d-type, $\ldots$, orbitals respectively, we shall confine our attention to s-type orbitals. If the s-type orbitals obey the cusp condition (Smith, 1971)

$$\left(\frac{\partial \chi_s(r)}{\partial r} \right)_{r=0} = -Z\chi_s(0) \tag{66}$$

then equation 65 simplifies to

$$\chi_s(\vec{p}) \sim (8/\pi)^{1/2} \{ Z\chi_s(0)p^{-4} + O(p^{-6}) \} \tag{67}$$

and the leading component of the asymptotic form for $I_0(p)$ is therefore given as

$$I_0(p) \sim \frac{32Z^2}{\pi} \sum_s \lambda_s \chi_s^2(0) p^{-6} \tag{68}$$

by (58) and the definition (56) of $I_0(p)$. However, the summation term in (68) is just

$$\sum_s \lambda_s \chi_s^2(0) = \rho_s(0) = \rho_0(0) \tag{69}$$

since $\rho_p(0)$, $\rho_d(0)$, etc., vanish by definition. In terms of the totally symmetric component $P_0(r)$ of the one-matrix, the asymptotic behaviour is given by

$$I_0(p) \sim \frac{32}{\pi} Z^2 \rho_0(0) p^{-6} + O(p^{-8}) \tag{70}$$

while by (41) the behaviour of $J(q)$ for large q is obtained as

$$J(q) \sim \frac{8Z^2}{3} \rho_0(0) q^{-6} + O(q^{-8}) \tag{71}$$

To obtain the dominant term contributing to the small p behaviour of

$I_0(p)$, we shall again consider only s-type NO's. Expansion of the spherical Bessel function in (64) and subsequent integration gives

$$\chi_s(\vec{p}) \sim (2/\pi)^{\frac{1}{2}}[(\chi_s) - \frac{p^2}{6}(r^2\chi_s) + O(p^4)] \tag{72}$$

where

$$(r^k\chi_s) = \int_0^\infty r^{k+2}\chi_s(r)\, dr \tag{73}$$

It should be noted that expansion of the spherical Bessel function of order n, $n = 1, 2, 3, \ldots$, for orbitals of p-type, d-type, f-type, $\ldots$, symmetry, gives leading contributions of $O(p)$, $O(p^2)$, $O(p^3)$, $\ldots$. For this reason, the small p behaviour of $I_0(p)$ is restricted to a consideration of only s-type NO's.

From the density matrix definition of $I_0(p)$ and from equation 72 we obtain

$$I_0(p) \sim 8(\rho_s)p^2 + O(p^4) \tag{74}$$

for small p while the limiting behaviour of $J(q)$ becomes

$$J(q) \sim J(0) - 2q^2(\rho_s) + O(q^4)$$

where

$$(\rho_s) = \sum_s \lambda_s(\chi_s)^2 \tag{75}$$

The limiting and asymptotic behaviour of $J(q)$ can now be used to obtain some interesting sum rules. Consider the integral

$$I_n = \int_0^\infty q^n J(q)\, dq \tag{76}$$

which is integrated by parts to give

$$I_n = \frac{q^{n+1}}{n+1} J(q)\Big|_0^\infty - \frac{1}{n+1}\int_0^\infty q^{n+1} J'(q)\, dq \tag{77}$$

For large q, we have established that $J(q)$ approaches zero as q^{-6}. We thus observe that for $0 \le n \le 4$, the first term of (77) certainly vanishes. Observing from (41) that

$$J'(q) = (-1/2)q^{-1}I_0(q)$$

for positive q, equation 77 reduces to

$$I_n = \frac{1}{2(n+1)}\int_0^\infty q^n I_0(q)\, dq$$

$$= \frac{1}{2(n+1)}\langle p^n \rangle \tag{78}$$

thus yielding sum rules valid for $0 \le n \le 4$. To the best of our knowledge, these sum rules have been noted only previously for the normalisation of $J(q)$, i.e. equation 61.

9. THE EFFECT OF ELECTRON CORRELATION ON MOMENTUM DISTRIBUTIONS AND COMPTON PROFILES

As is the case for X-ray form factors and incoherent scattering functions, there is a paucity of reported calculations of correlated momentum distributions and Compton profiles. A number of experimental measurements of Compton profiles have been made, however, and comparison of experiment with theory has been for the most part with RHF values (such as tabulated

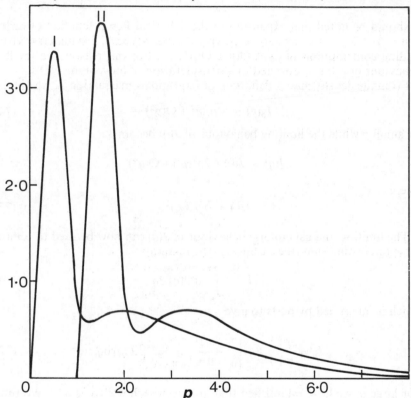

Figure 3   Comparison of Correlated (Curve I) and RHF (Curve II) radial momentum distributions for the ^1S ground state of beryllium (atomic units). The origin for the RHF distribution is translated one unit to the right of that for the correlated distribution

by Weiss, Harvey and Phillips (1968). For a recent survery of experimental work, the reader is referred to Cooper (1971).

The recent Compton profile measurements for helium by Eisenberger and Platzmann (1970), analysed within the framework of the impulse approximation, provide a momentum distribution in excellent agreement with the RHF distribution. As well, correlated wave functions of various quality for He have been found (Smith and Benesch, 1968) to agree well with the RHF distribution.

Calculations of correlated $I_0(p)$ values for Li (Benesch and Smith, 1970d) indicated discrepancies of the order of 2–3% from the RHF values, while correlated $I_0(p)$'s for Be (Benesch and Smith, 1972) unearthed discrepancies of the order of 25% in the valence shell portion of $I_0(p)$. These differences were observed for small values of p in the valence shell portion of the distribution (see Figure 3 for Be).

As we also see in this figure, the valence shell portion of the $I_0(p)$ curve is sharply peaked while the core or K-shell region is comparatively shallow and broad. In lithium and beryllium, one can observe a maximum in the latter region as well. For boron through neon this K-shell distribution is almost masked by the valence shell contribution.

The calculated discrepancy around the valence shell peak is less than 1% in lithium, 10% in beryllium, 4.5% in boron (Smith, Brown and Benesch, 1972) and 0.5% in neon (Brown and Smith, 1972). In the K-shell regions, the calculated RHF and correlated values are always in good agreement (better than 1%) for all these systems. Correlation effects on $J(q)$ are observed to be smaller than those for the $I_0(p)$ from which it is derived, and may be explained by the averaging out of fluctuations due to the integration process. However, they are still experimentally significant, being of the order of 7% for the Be $J(0)$ (Benesch and Smith, 1970b).

Some authors have claimed on the basis of the previously cited experimental and theoretical data for He that correlation effects are unimportant in discussions of $I_0(p)$ and $J(q)$. However, the cited calculations for Li, Be and B indicate that electron correlation is quite important for the *valence* shell electrons. This is important because the Compton profile method is expected to be most accurate when probing this portion of the electron distribution. In contrast to these non-closed shell systems, for well-closed shell systems such as He or Ne, or for the core shells of other systems, correlation effects are quite small.

It should be noted that the impulse approximation (IA) is expected to hold for the valence shell electrons for the small systems mentioned above. Thus, effects of electron correlation on the valence shell $J(q)$ can not be explained on the basis of any failure of the IA. However, the recent measurements on neon (Eisenberger, 1972; Eisenberger, Henneker and Cade, 1972) show that in the core region, corrections to the IA are important. In several forthcoming publications (Smith, Brown and Benesch, 1972; Smith and Brown, 1972) correlation effects will be discussed and explained on the basis of a rigorous decomposition (Sinanoğlu, 1971) of the atomic wave function into terms describing semi-internal, polarisation, internal, and external contributions beyond RHF. For non-closed shell states, a wave function which carefully includes all of these contributions beyond RHF except the external ones, should account for the principal correlation effects on Compton profiles and momentum distributions.

ACKNOWLEDGEMENTS

We wish to acknowledge the support of this research by the National Research Council of Canada. One of us (V.H.S.) wishes to express his gratitude to the Deutsche Forschungsgemeinschaft for the Richard–Merton Professorship during 1971–72 and Professor G. L. Hofacker for his hospitality in Munich, where a portion of this article was written. Helpful comments by Professor R. A. Bonham on this manuscript are gratefully acknowledged.

REFERENCES

Bartell, L. S. and Gavin, R. M. (1964) 'Effects of Electron Correlation in X-ray and Electron Diffraction', *J. Am. chem. Soc.*, **86**, 3493–3498

Bartell, L. S. and Gavin, R. M. (1965) 'Effects of Electron Correlation in X-ray and Electron Diffraction. II. Influence of Nuclear Charge in Two-Electron Systems', *J. chem. Phys.*, **43**, 856–861

Benesch, R. (1971) 'Algebraic Determination of Electron–Nuclear and Electron–Electron Distribution Functions from Hylleraas-type Wavefunctions', *J. Phys. B: Atom. Molec. Phys.*, **4**, 1403–1414

Benesch, R. (1972) 'On the Determination of Radial Electron–Electron Distribution Functions', *Int. J. quantum Chem.*, **6**, 181–192

Benesch, R., Singh, S. R. and Smith, V. H. (1971) 'On the Relationship of the X-ray Form Factor to the 1-Matrix in Momentum Space', *Chem. Phys. Letters*, **10**, 151–153

Benesch, R. and Smith, V. H. (1970a) 'Correlation and X-ray Scattering. I. Density Matrix Formalism', *Acta. Cryst.*, **A26**, 579–586

Benesch, R. and Smith, V. H. (1970b) 'Correlation and X-ray Scattering. II. Atomic Beryllium', *Acta. Cryst.*, **A26**, 586–594

Benesch, R. and Smith, V. H. (1970c) 'Correlation Effects in X-ray and Electron Scattering', *Int. J. quantum Chem.*, **3S**, 413-421

Benesch, R. and Smith, V. H. (1970d) 'Radial Momentum Distributions for the ^{2}S Ground State of the Lithium Atom', *Chem. Phys. Letters*, **5**, 601–604

Benesch, R. and Smith, V. H. (1971a) 'Radial Electron–Electron Distribution Functions and the Coulomb Hole for Be', *J. chem. Phys.*, **55**, 482–488

Benesch, R. and Smith, V. H. (1971b) 'Exact and Asymptotic X-ray Intensity Values from 20-parameter Hylleraas-type Wavefunctions He and He-like ions', *Int. J. quantum Chem.*, **5S**, 35–45

Benesch, R. and Smith, V. H. (1971c) 'Natural Orbitals in Momentum Space and Correlated Radial Momentum Distributions. 1. The ^{1}S Ground State of Li$^+$', *Int. J. quantum Chem.*, **4S**, 131–138

Benesch, R. and Smith, V. H. (1972) 'Radial Momentum Distributions $I_0(p)$ and Compton Profiles $J(q)$. The ^{1}S Be Atom Ground State', *Phys. Rev.*, **5A**, 114–125

Bingel, W. (1963) 'The Behaviour of the First-Order Density Matrix at the Coulomb Singularities of the Schrödinger Equation', *Z. Naturforsch.*, **18a**, 1249–1253

Bingel, W. A. and Kutzelnigg, W. (1968) 'Symmetry Properties of Reduced Density Matrices and Natural p-states', *Adv. quant. Chem.*, **5**, 201–218

Brown, R. T. (1970) 'Coherent and Incoherent X-ray Scattering by Bound Electrons. II. Three- and Four-electron Atoms', *Phys. Rev.*, **A2**, 614–620

Brown, R. T. (1972) 'Incoherent Scattering Function for Atomic Carbon', *Phys. Rev.*, **A5**, 2141–2144

Cooper, M. (1971) 'Compton Scattering and Electron Momentum Distributions', *Adv. Phys.* **20**, 453–491

Coulson, C. A. and Neilson, A. H. (1961) 'Electron Correlation in the Ground State of Helium', *Proc. Phys. Soc. (London)*, **78**, 831–837

Currat, R. DeCicco, P. D. and Weiss, R. J. (1971) 'Impulse Approximation in Compton Scattering', *Phys. Rev.*, **B4**, 4256–4261

DuMond, J. W. H. (1929) 'Compton Modified Line Structure and its Relation to the Electron Theory of Solid Bodies', *Phys. Rev.*, **33**, 643–658

DuMond, J. W. H. (1930) 'Breadth of Modified Compton Line', *Phys. Rev.*, **36**, 146–147

Duncanson, W. E. and Coulson, C. A. (1945) 'Theoretical Shape of the Compton Profile for Atoms from H to Ne', *Proc. Phys. Soc. (London)*, **57**, 190–198

Eisenberger, P. (1970) 'Electron Momentum Density of He and H$_2$; Compton X-ray Scattering', *Phys. Rev.*, **A2**, 1678–1686

Eisenberger, P. (1972) 'Compton Profile Measurements of N$_2$, O$_2$ and Ne using silver and molybdenum X-rays', *Phys. Rev.*, **A5**, 628–635

Eisenberger, P. and Platzman, P. M. (1970) 'Compton Scattering of X-rays from Bound Electrons', *Phys. Rev.*, **A2**, 415–423

Eisenberger, P., Henneker, W. H. and Cade, P. E. (1972) 'Compton Scattering of X-rays from Ne, N_2 and O_2: A Comparison of Theory and Experiment', *J. chem. Phys.*, **56**, 1207–1209

Freeman, A. J. (1959) 'Compton Scattering of X-rays from Aluminum', *Phys. Rev.*, **113**, 176–178

Goscinski, O. and Lindner, P. (1970) 'Asymptotic Properties of Atomic Form Factors and Incoherent Scattering Functions', *J. chem. Phys.*, **52**, 2539–2540

Kato, T. (1957) 'On the Eigenfunctions of Many-particle Systems in Quantum Mechanics', *Commun. pure appl. Math.*, **10**, 151–177

Kilby, G. E. (1965) 'A Wave-mechanical Derivation of the Intensity Distribution of the Compton Line', *Proc. Phys. Soc. (London)*, **86**, 1037–1040

Kim, Y.-K. and Inokuti, M. (1968) 'Atomic Form Factor and Incoherent Scattering Function of the Helium Atom', *Phys. Rev.*, **165**, 39–43

Kohl, D. A. and Bonham, R. A. (1967) 'Effect of Bond Formation on Electron Scattering Cross-sections for Molecules', *J. chem. Phys.*, **47**, 1634–1646

Löwdin, P.-O. (1955) 'Quantum Theory of Many-particle Systems. I. Physical Interpretations by Means of Density Matrices, Natural Spin Orbitals and Convergence Problems in the Method of Configurational Interaction', *Phys. Rev.*, **97**, 1474–1489

McWeeny, R. (1949) 'The Computation of Wavefunctions in Momentum Space. I. The Hydrogen Molecule Ion', *Proc. Phys. Soc. (London)*, **A62**, 519–528

McWeeny, R. and Coulson, C. A. (1949) 'The Computation of Wavefunctions in Momentum Space. I. The Helium Atom', *Proc. Phys. Soc. (London)*, **A62**, 509–518

Peixoto, E. M. A., Buinge, C. F. and Bonham, R. A. (1969) 'Elastic and Inelastic Scattering by He and Ne Atoms in their Ground States', *Phys. Rev.*, **181**, 322-328

Sack, R. A. (1964) 'Generalization of Laplace's Expansion to Arbitrary Powers and Functions of the Distance between Two Points', *J. math. Phys.*, **5**, 245–251

Silverman, J. N. and Obata, Y. (1963) 'Sum Rules Relating Coherent X-ray Scattering Data to the Diamagnetic Nuclear Shielding Constant and to the Self-energy of the Charge Distribution of the Scatterer', *J. chem. Phys.*, **38**, 1254–1255

Sinanoğlu, O. (1971) 'New Atomic Structure Theory and Beam Foil Spectroscopy', *Rev. Mex. Fís.*, **20**, 143–160

Smith, V. H. (1970) 'Asymptotic Behaviour of Incoherent Scattering Functions', *Chem. Phys. Letters*, **7**, 226–228

Smith, V. H. (1971) 'Cusp Conditions for Natural Functions', *Chem. Phys. Letters*, **9**, 365–368

Smith, V. H. (1972) 'On the Impulse Approximation in Compton Scattering from Bound Electrons' (Unpublished paper in the Proceedings of the International Symposium on Quantum Chemistry and Solid-State Physics, Beito, Norway)

Smith, V. H. and Benesch, R. (1968) 'Atomic He in Momentum Space', Unpublished paper in the *Proceedings of the International Symposium on the physics of One- and Two-Electron Atoms* (Munich, Germany).

Smith, V. H. and Brown, R. E. (1972) 'Influence of Electron Correlation on the Compton Profile and Momentum Distribution of Atomic Neon' (unpublished)

Smith, V. H. and Harriman, J. E. (1970) *A Proposal Concerning Terminology in Reduced Density Matrix Theory*, University of Wisconsin Technical Report WIS-TCI-379

Smith, V. H., Brown, R. E. and Benesch, R. (1972) 'Momentum Space and Compton Profile Calculations for the Ground State of Atomic Boron' (to be published)

Steiner, E. (1963) 'Charge Densities in Atoms', *J. chem. Phys.*, **39**, 2365–2366

Tanaka, K. and Sasaki, F. (1971) 'Configuration Interaction Study of X-ray and Fast Electron Scattering Factors for Light Atomic Systems', *Int. J. quantum Chem.*, **5**, 157–175

Tavard, C. and Bonham, R. A. (1969) 'Quantum Mechanical Formulas for the Inelastic Scattering of Fast Electrons and their Compton Line Shape. Non-relativistic Approximation', *J. chem. Phys.*, **50**, 1736–1747

Tavard, C. and Roux, M. (1965) 'Cacul des intensités de Diffraction des Rayons X et des Electrons par les Molecules', *C.r. hebd. Séanc. Acad. Sci., Paris*, **260**, 4933–4936

Thomson, J. J. (1906) *Conduction of Electricity Through Gases'*, Cambridge University Press

Waller, I. and Hartree, D. R. (1929) 'On the Intensity of Total Scattering of X-rays', *Proc. R. Soc.*, **A124**, 119–142

Weiss, R. J., Harvey, A. E. and Phillips, W. C. (1968) 'Compton Line Shapes for Hartree-Fock Wavefunctions', *Phil. Mag.*, **17**, 241–254

Wilcox, R. M. (1967) 'Exponential Operators and Parameter Differentiation in Quantum Physics', *J. math. Phys.*, **8**, 962–982

22

An Approach to the Teaching of Quantum Theory

The Science Foundation Course Team,
Open University, Bletchley, Bucks.

Foreword

PROFESSOR F. R. STANNARD

The foundation Course in Science at the Open University is a broadly based course written by a team of physicists, chemists, biologists and earth scientists. It ranges over such diverse topics as quantum theory, the manufacture of polymers, the role of DNA, sea-floor spreading and the financing of big scientific projects such as the building of proton synchrotrons. All Open University science students, as well as many non-science students, take the course in their first year. The enrolment is about 5000 per year.

Not only does the course have to be of interest to both scientists and non-scientists, but it also has to be accessible to students of different backgrounds and initial abilities. This is because the Open University has no formal entry requirements based on previous academic attainment. On the one hand there may be a student struggling to recall Newton's laws and the meaning of 'sine' and 'cosine'; on the other there is the man who already holds a previous degree.

With such a heterogeneous intake it was clear to the Course Team that we would have to start from a low baseline. And yet for the course to be recognised as being of university standard, it had to develop rapidly to the point where it became intellectually challenging even to the brightest student. For this reason we chose subjects that were *conceptually* demanding. Topics such as quantum theory, relativity, atomic and nuclear physics are particularly suitable in this respect in that, even in the absence of mathematics, they can be taught in such a way as to stretch a student's capabilities.

The decision to include an introduction to quantum theory presented us with interesting problems. Because we could assume little background knowledge in physics, it would have been difficult for us to adopt the traditional historical approach to the subject through a study of black body radiation. We think the approach we finally adopted is more straightforward and potentially less confusing to the student. Then again, because we could assume only minimal skills in mathematics, we were encouraged to dwell on the more qualitative and philosophical aspects of the subject; perhaps this might also be regarded as a desirable feature, especially as we are here concerned only with laying a basic foundation. (A quantitative treatment of quantum theory is given later in a higher level course – one that specifies certain mathematical courses as prerequisites.)

There follows an amended reprint of the part of the foundation course concerned with introducing quantum theory. In reading it you should bear several points in mind. It is Unit 29 of a 34 Unit course (a unit representing approximately twelve hours work for the student). The correspondence material reproduced here represents only a part of the learning experience of the Open University student. Integrated with this text is a television and a radio programme. In the television programme, for example, the student participates in experiments involving electron diffraction and the photoelectric effect. Having seen the experiments performed, he has to analyse the data for himself and verify the de Broglie relationship between wavelength and momentum and evaluate Planck's constant.

The whole of Unit 29 was far too long to be reproduced here in full. In its original form the Unit contains a list of objectives which spell out to the student exactly what he should be able to do having studied the material. There are self-assessment questions and answers to help him assess how well he has met the various objectives. The text itself is punctuated at regular intervals with questions aimed at prompting responses from the student rather in the manner of programmed learning. There are numerous summaries as well as marginal comments highlighting important terms and concepts, as well as appendices containing remedial material.

Regrettably, these educational devices have had to be omitted because the space available in this volume is limited. However, the main line of the argument as developed in the original text is fully covered in this abridged version.

INTRODUCTION

The ideas you will be introduced to this week will undoubtedly appear very strange. Physicists worked for the whole of the first quarter of this century unravelling the mystery of the quantum. Because of the conceptual difficulties involved in this subject it has long been customary to reserve it for the later stages of a degree course in physics. In recent years, however, many universities have experimented by bringing quantum theory forward in the curriculum and we shall follow their example.

Usually the subject is presented in a manner closely following its historical development. We shall be leaving aside this historical approach in favour of one that, to us at least, appears more straightforward and economical. The Unit as a whole should be regarded as a continuously developing line of argument. We shall ask certain questions of Nature, describe the answers given by experimental observation, and show how these observations logically lead to certain remarkable conclusions. As the argument unfolds, follow closely each step. In the process we shall challenge some ideas which at present appear to you to be plain common sense. Remember Einstein's description of common sense as 'that layer of prejudices laid down in the mind prior to the age of eighteen'. This week we think you are in for some surprises – the world never seems quite the same once one knows about its quantum behaviour.

HOW ENERGY AND MOMENTUM ARE PROPAGATED

The laws of reflection and refraction indicate the directions in which light will travel after it has struck a boundary between two media. The state of polarisation of the incident beam has a bearing on how much light is reflected and how much refracted at the surface, i.e. how much is turned back and how much goes forward. The theories of diffraction and interference are concerned with calculating how much light travels in a given direction after passing through or round obstacles.

These various phenomena are concerned with a single general type of question, namely: how is light propagated? Although this question is sometimes answered by invoking interference, or perhaps reflection or refraction, there is only one general answer to the question, and that is: *light is propagated as a wave*. We shall now broaden this investigation by asking: *how are other forms of energy propagated*?

ELECTRONS

We begin by studying the behaviour of a beam of electrons. There is only one way of knowing how electrons are propagated and that is to perform an experiment to find out. Indeed our whole approach in this Unit will be to present you with the experimental observations first, and only then discuss the theoretical ideas that stem from them.

We begin by considering an experiment in which electrons from an electron gun are directed at two parallel slits in a barrier, much in the same way as light was shone onto parallel slits in Young's interference experiment. The electrons are subsequently recorded on a photographic plate placed on the opposite side of the barrier. The rather surprising outcome of this experiment (surprising if you *did* picture an electron as a tiny billiard ball) is shown in Figure 1. As you can see, there is essentially no difference between the intensity distribution of the electrons and that obtained with light in the conventional form of the experiment. Interference phenomena such as this can only be described in terms of a wave theory. The alternating maximum and minimum intensities of the electrons are therefore a clear indication that *in order to determine the distribution of scattered electrons, it is necessary to use the mathematical analysis associated with wave properties.* It is difficult to perform Young's interference experiment using electrons because the wavelength of the beam is found to be very tiny. In order to produce an appreciable separation of the maxima and minima, the distance between the slits has to be very small too.

The experiments by Davisson and Germer and by G. P. Thomson, which first showed the wave behaviour of electrons, were not performed with

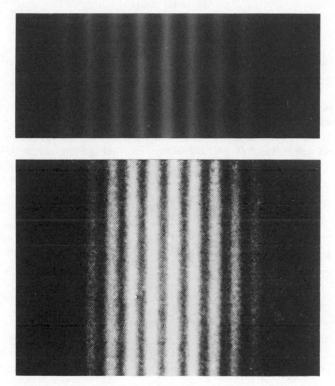

Figure 1 (a) A double-slit interference pattern for electrons. (b) A double-slit pattern for light, shown for comparison. (The electron pattern was obtained by C. Jönsson at the University of Tubingen)

man-made slits, but with a naturally occurring 'diffraction grating'. Crystals consist of rows of regularly spaced atoms. These rows of atoms can be regarded as straight parallel lines from which light may be diffracted. A crystal surface therefore acts like a very finely ruled grating. When an electron beam is directed at such a surface it is preferentially scattered in certain directions giving diffraction maxima and minima.

The wavelength of the light incident on a grating can be determined from the known spacing of the lines and the directions of the diffraction maxima.

The wavelength of an electron beam can be estimated in the same way. In Figure 2, the difference, ΔD, between the lengths of the paths of the interfering rays must be a whole number of wavelengths if there is to be a maximum at angle θ. This angle is given in terms of ΔD and the separation of the atoms, d, by the equation

$$\sin \theta = \frac{\Delta D}{d}$$

For the nth maximum (n being zero for the maximum along the normal to the crystal surface), we have

$$\Delta D = n\lambda, \text{ and}$$

$$\sin \theta = \frac{n\lambda}{d}$$

Therefore
$$\lambda = \frac{d \sin \theta}{n}$$

For a crystal of a given substance, the separation of the atoms is known, so a measurement of θ allows λ to be determined.

(It should be borne in mind that the crystal represents not one diffraction grating but a whole set of plane gratings one behind the other. A complete

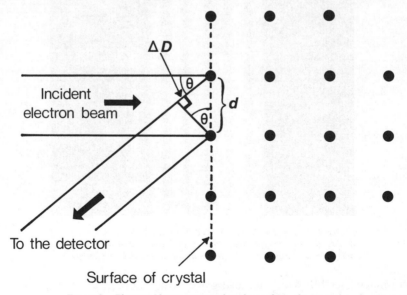

Figure 2   *Electrons being scattered at the surface of a crystal*

description of the observed diffraction pattern is therefore more complicated than the one we have presented. Radiation penetrating to, and diffracting from, these additional layers will combine with that diffracted from the surface layer, and will sometimes lead to destructive interference and a diminishing of the expected maximum.)

What value does this type of experiment give for the wavelength associated with an electron beam? There is no unique answer; the angles of the diffraction maxima, and hence the wavelength, are found to vary as the momentum, p, of the electrons is changed. The momentum, depending as it does on the electron's velocity, may be controlled by the accelerating electric field in the electron 'gun', i.e. by the voltage on the accelerating anode. By observing how λ varies as a result of changes in the voltage, a very simple relationship is found:

$$\lambda = h/p \tag{1}$$

This is an exceedingly important formula. The relationship was originally proposed in 1924 by the French physicist, de Broglie, and so λ is called the *de Broglie wavelength* of the beam. The constant h is called *Planck's constant*. Its value is: $h = 6.62 \times 10^{-34}$ J s. (Note that it is not a dimensionless number — it has the units of energy × time.) This value is so very small that even

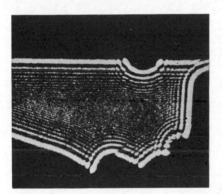 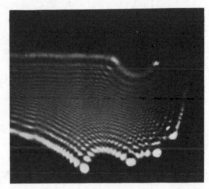

Figure 3 The picture on the right was formed by electrons emitted from a sharp point and passing through an aperture between tiny opaque crystals. For comparison, the other picture shows light being diffracted through an opening cut in a metal plate (Cavendish Laboratory)

though the momentum of an electron is also small, the wavelength of an electron beam is generally very short.

We offer in Figure 3 just one final piece of evidence to convince you of the wave-like characteristics of electron propagation. Here electrons are diffracted at the edges of opaque crystals and, for comparison, light is shown diffracted through a similarly shaped opening; the similarities are striking.

Although earlier we spoke of electrons as particles, the conclusion to be drawn from the experimental evidence here presented is inescapable — in order to describe how electrons are *propagated* it is necessary to use wave theory.

NUCLEI, ATOMS AND MOLECULES

We now move on to other types of radiation. How are beams of nuclei, e.g. protons or deuterons, propagated? Because nuclei are so much heavier than electrons, it is more difficult to produce a beam of them with very low momentum. If the de Broglie formula were to apply to them as well as to electrons, and if the same constant, h, were to be used in the formula, then clearly the wavelength of the beam of nuclei would in general be less than that of an electron beam; this would make it even more difficult to observe diffraction effects. However, diffraction *has* been observed for such beams, as can be seen in Figure 4. Here nitrogen nuclei act as obstacles in the way of a beam of deuterons (i.e. nuclei consisting of one neutron and one proton) and these cause the beam to be diffracted. The large peak in Figure 4 shows

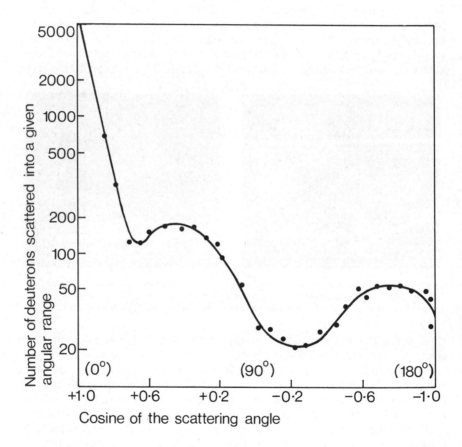

Figure 4   Angular distribution of deuterons elastically scattered by nitrogen (W. M. Gibson and E. E. Thomas)

that most of the deuterons are only slightly affected and are still to be found approximately in the forward direction. There are, however, several lesser peaks indicating other likely directions – these are diffraction maxima. If you compare the curve of Figure 4 with that of Figure 5, showing the diffraction of light, you see the two have remarkably similar features.

The same diffraction phenomena have now also been observed for atoms and even molecules. These latter observations are significant because atoms and molecules are hardly to be considered as elementary quantities—they are groups of particles (electrons and nuclei) bound together. Even so, the group is found to be diffracted *as a complete unit*. Once again it is found that

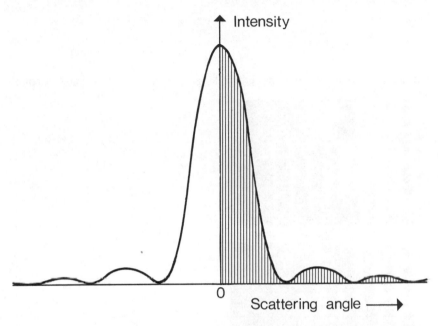

Figure 5   Intensity distribution of light diffracted through a slit, or round a small obstacle. The right half of the symmetrical curve is shown shaded for easier comparison with Figure 4 (where left and right scatterings were superimposed)

the diffraction pattern sizes depend upon momentum; de Broglie's formula applies to all these various types of radiation.

MACROSCOPIC OBJECTS

But what if we extend the investigation still further? Suppose energy is being 'radiated' in the form of a stream of macroscopic objects—for example, a stream of billiard balls? Surely that is taking things too far!

In the simplest case, where no forces act on the billiard balls, they continue moving in a straight line with constant speed. If they pass through a gap in a barrier, they continue moving in a straight line.

In Figure 6, we show water waves passing through gaps of different sizes in a barrier. When the wavelength is smaller than the dimensions of the gap, the wave is seen to continue moving in the forward direction and there is little evidence of any change of direction. Because the de Broglie wavelength of billiard balls would be very small compared with the size of any gap through which the balls could pass, a deviation due to diffraction would not be noticeable.

To stress this point, we take the case of a ball weighing 0.5 kg moving with speed 3 m s^{-1}. Its momentum, p, is given by $mv = 0.5 \times 3 = 1.5 \text{ kg m s}^{-1}$. Applying de Broglie's formula we have:

$$\lambda = \frac{h}{p} = \frac{h}{mv} = \frac{6.62 \times 10^{-34}}{1.5}$$

$$= 4.4 \times 10^{-34} \text{ m}$$

Suppose the ball approaches a gap large enough for it to pass through.

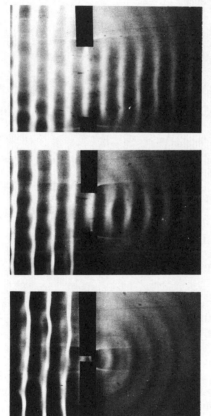

Figure 6   Photographs of water waves passing through a gap in a barrier. Diffraction becomes more apparent the smaller the size of the gap compared with the wavelength

If the width of the gap is 0.1 m, the angle of deviation θ to the first diffraction minimum can be obtained from the expression:

$$\sin \theta = \frac{\lambda}{d} = \frac{4.4 \times 10^{-34}}{0.1} = 4.4 \times 10^{-33}$$

Because θ is obviously very small, we can write $\sin \theta \approx \theta$. Therefore $\theta = 4.4 \times 10^{-33}$ radians.

This means that if the ball is directed at a nearby slit, but then subsequently observed only after it has travelled to the edge of the observable universe (a distance of about 10^{25} m), it would be found to have deviated sideways a few millionths of a centimetre. As we said — hardly noticeable!

Thus the wave theory of propagation could adequately describe the motion of large objects as well as small ones. *Motion in a straight line can be regarded simply as the limiting case of wave motion for small wavelengths.* There is absolutely no reason to introduce a new kind of mathematics to be able to handle this type of motion of billiard balls. (Note that up to this point at least we have not *proved* that the de Broglie formula applies to macroscopic objects; we have simply shown that such an extension of the use of the formula would not lead to any contradiction with known behaviour.)

THE PRINCIPLES GOVERNING LIGHT PROPAGATION AND MECHANICS

But of course macroscopic objects do not always travel in straight lines — under the influence of gravity they move in curved trajectories. When faced with this kind of behaviour, it is customary to rely on Newton's laws of motion and introduce concepts like mass, force, gravitational attraction,

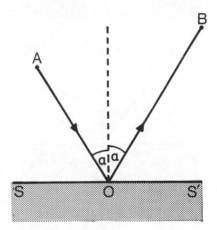

Figure 7   The path of a light ray passing from A to B via reflection at the surface SS'

etc. — a language that appears to have little or no point of contact with that of waves. In this section you will take a fresh look at your customary ways of regarding wave and particle propagation to discover whether they really are as different as they seem.

In unit 22, Huygen's method, we begin with waves to prove the law that the angle of reflection is equal to the angle of incidence. Thus a ray of light going from A to B in Figure 7 and reflecting from the surface SS' takes the path AOB. Let us now point out a remarkable characteristic of this path; *it is impossible for light to go from A to B via a reflection at SS' any quicker than by AOB.* Neighbouring paths such as those illustrated in Figure 8 all take longer.

The same also applies to cases of refraction. The path taken between points A and B in Figure 9 is such that angle r is related to angle i through Snell's law. But another way of looking at the problem is to say that the path AOB is the quickest route for light to travel between A and B. It is even quicker than by going the 'direct route' APB.

The reason is that the wave speed is lower in the second medium, so although the total distance travelled is less, the extra time spent moving

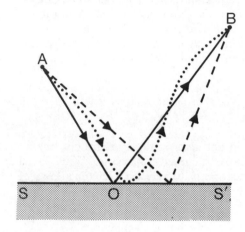

Figure 8   *A selection of possible hypothetical routes from* A *to* B *via a reflection at the surface* SS′

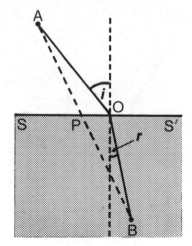

Figure 9   *The path of a light ray passing from* A *to* B *via refraction at the surface* SS′

slowly outweighs the advantage of the shorter distance; if the light remains longer in the medium in which its speed is greater it reaches B quicker.

Exactly the same considerations apply to waves moving in media with gradually changing properties. Light coming from the Sun, for example, follows a curved path as it traverses the Earth's atmosphere. The lower layers are generally denser, so the wave's speed decreases the deeper the penetration. It is quicker for the light to reach its destination on the Earth's surface by following a curved trajectory than for it to take the straight path and find itself spending longer travelling with lower speed in the denser atmosphere.

Thus, light moves between two points by the route that is quicker than any of the other routes nearby; this is *Fermat's Principle of Least Time*. With this powerful principle we are able to derive the laws of reflection and refraction and compute the curved trajectories of light rays moving in media of varying properties.

It is not too difficult to see how Fermat's Principle and Huygen's approach are related. In Huygens's method, each point on a wave front is to be regarded as a source of secondary wavelets. As these wavelets travel, they interfere with each other. Where they interfere constructively the new wave front is formed; elsewhere they interfere destructively. The particular series of wavelets that allows the light to reach its destination by the quickest route obviously cannot be cancelled out by wavelets travelling via some other route, as the latter will take longer to arrive and will be too late to give destructive interference. Therefore the wavelets that keep ahead of 'the rest of the field' are certain to survive. In other words, the path taken up by the wave is the quickest route between any two points—which is Fermat's Principle.

For greater convenience later, let us express this principle in a mathematical form. Imagine the possible paths between A and B in Figure 10 to

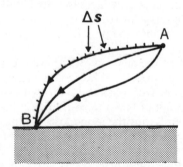

Figure 10   A selection of possible hypothetical routes from A to B through the air. One of them is shown divided into equal segments of length Δs

be divided up into small segments of length, Δs. (We show this for one of the paths.) For the first segment, the speed of light while traversing it is v_1. Because Δs is very small we may assume that v_1 does not vary significantly from one end of the segment to the other: v_1 is the average speed for that segment. The average speed of the light over the second segment is v_2, over the third segment v_3, and so on. These various average speeds will differ from each other if the refractive index of the medium is changing.

The time, Δt_1, taken to traverse the first segment, is given by

$$\Delta t_1 = \frac{\Delta s}{v_1}$$

Similarly, the time taken to traverse the second is

$$\Delta t_2 = \frac{\Delta s}{v_2} \text{ , etc.}$$

The total time taken to go from A to B is t_A^B given by

$$t_A^B = \Delta t_1 + \Delta t_2 + \Delta t_3 + \ldots$$

i.e.
$$t_A^B = \frac{\Delta s}{v_1} + \frac{\Delta s}{v_2} + \frac{\Delta s}{v_3} + \ldots$$

This is a rather cumbersome expression, so we adopt a more economical way of writing it

$$t_A^B = \Sigma_A^B (\Delta s / v) \tag{2}$$

Fermat's Principle of Least Time then states that light follows the path from A to B such that t_A^B, *given by equation 2, is a minimum.*

We turn now to particle propagation. Consider a particle moving in a curved trajectory under the influence of a field that changes slowly from place to place, say the Earth's gravitational field; for example, we could take the case of a small stone thrown from one point on the Earth's surface to another. The trajectory of the stone could, of course, be worked out using Newton's laws of motion, but what is being sought here is a fresh approach to the problem – one that is more closely akin to the application of Fermat's Principle to light paths.

From energy conservation, it is known that as the stone rises and its potential energy increases, its kinetic energy decreases an equivalent amount (neglecting for simplicity the effects of air resistance). Bearing this in mind we ask whether the path of the stone is governed by Fermat's Principle of Least Time.

If all one has to worry about is conserving energy, the quickest route would be the shortest – a horizontal straight line. In actual fact, of course, it is found that the stone has to be thrown upwards at an angle as in Figure 11. In this way, not only is the path longer than the straight line joining the two points, A and B, but the stone loses kinetic energy with height and therefore travels more slowly. Clearly a straightforward application of Fermat's Principle does not work for particles!

However, a principle of sorts *does* exist for the paths of particles. Although not taking the path of shortest time, the stone nevertheless follows a path with a unique characteristic. This is embodied in *Maupertuis's Principle of Least Action.*

From the knowledge that a stone has a certain total energy, that it starts its journey at a given instant and completes it in a given fixed time, all manner of possible paths may be drawn between A and B; we show a selection in Figure 12. From all of these trajectories Maupertuis's Principle singles out the right one. To apply the principle, each path is divided up into segments of length Δs, exactly as was done for the paths of light. The average momentum for the first segment is p_1. This is obtained by noting that the total energy, E, is fixed and equals the sum of the kinetic ($\frac{1}{2} mv^2$) and potential (U) energy:

$$E = \tfrac{1}{2} mv^2 + U$$

The momentum, p, equals mv and can be expressed in terms of U and E.

$$p = mv = \sqrt{[2m(E - U)]}$$

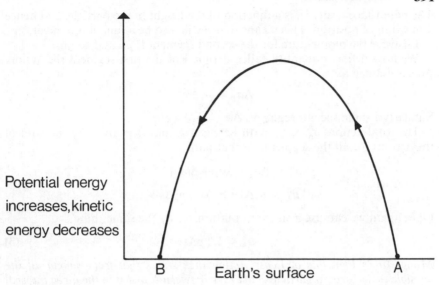

Figure 11   The path taken by a stone thrown from A to B (neglecting the effects of air resistance)

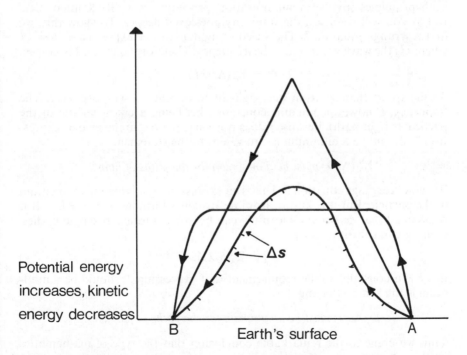

Figure 12   Possible hypothetical paths for a stone passing from A to B. One of them is shown divided into equal segments of length Δs

The potential energy, U, is a function of the height of the particle, and hence a function of position. This means that p will also be a function of position.

Likewise the momentum for the second segment is p_2, and so on.

We now define a quantity called *action*. For the first segment the action, Δa_1, is defined as:

$$\Delta a_1 = p_1 \Delta s$$

Similarly for the second segment, $\Delta a_2 = p_2 \Delta s$, etc.

The total action, a_A^B, for a path between A and B is given by the sum of the actions for all the segments in that path:

$$a_A^B = \Delta a_1 + \Delta a_2 + \Delta a_3 + \ldots$$

$$\therefore a_A^B = p_1 \Delta s + p_2 \Delta s + p_3 \Delta s + \ldots$$

Once again we can use a shorter notation to say the same thing:

$$a_A^B = \Sigma_A^B p \Delta s \tag{3}$$

Maupertuis's Principle of Least Action then states that from among all the possible trajectories a particle could take between A and B in the given interval of time, the actual path is that for which a_A^B *is a minimum.*

So for light there is a principle involving the minimisation of expression 2, and for particles another, in which expression 3 is minimised. The fact that both principles involve a minimisation procedure is in itself interesting. But as you will soon see, the similarity goes even deeper. To show this, we first rearrange equation 2. The speed of light, v, can be expressed as $v = \lambda f$ where λ is the wavelength and f the frequency. Therefore equation 2 becomes:

$$t_A^B = \Sigma_A^B (\Delta s / \lambda f) \tag{2'}$$

As the speed changes from one segment to another, λ will also vary. The frequency, f, however, remains constant (this being a characteristic of the particular light path). Because f does not vary, the requirement that expression 2′ should be a minimum is equivalent to the statement:

$\Sigma_A^B (\Delta s / \lambda)$ *must be a minimum for the paths of light.*

We now rearrange equation 3. This is an expression involving the momentum of the particle. But it is known from de Broglie's formula that $p = h/\lambda$. If it is assumed that de Broglie's formula can be applied to macroscopic bodies, then we can substitute for p:

$$a_A^B = \Sigma_A^B (h/\lambda) \Delta s \tag{3'}$$

But h is a constant, so the requirement that expression 3′ should be a minimum is the same as saying:

$\Sigma_A^B (\Delta s / \lambda)$ *must be a minimum for the paths of particles.*

Thus we come to the remarkable conclusion that the type of mathematics used to describe how particles are propagated is the same as that used to describe how light is propagated.

What is the physical interpretation of this powerful unifying principle? The quantity $(\Delta s/\lambda)$ is the length of a segment of the path expressed as a fraction of the wavelength of the radiation when traversing that segment. Therefore $\Sigma_A^B(\Delta s/\lambda)$ is the total length of the path expressed in wavelengths. *Thus any type of radiation chooses that path between two points that is characterised by the least number of wavelengths.*

This, then, is the fundamental principle that applies equally to light and to particles. Its power is really quite astonishing, e.g. it allows us to compute not only the curved path of the Sun's rays as they pass through the various layers of the Earth's atmosphere, but also the curved path of the Earth itself as it orbits the Sun.

WAVE PACKETS AND WAVE TRAINS

If this idea of describing the propagation of objects in terms of the mathematics of waves is new to you, as it probably is, you may well find it takes a time to assimilate. For example, you may want to argue that macroscopic objects do not *look* like waves. Waves spread out through space, but particles are localised—their centres of gravity, for instance, have well-defined positions.

If this is how you are thinking, remember that there are different kinds of waves. There are continuous *wave trains*, consisting of long sequences of regularly spaced crests and troughs, and there are *wave packets*, consisting of a single irregular hump, like that which passes along a horizontal rope when the end is rapidly jerked upwards.

As we have said, a macroscopic object normally has a position that can be rather well specified. If one wants to use a mathematical wave to specify its position, it is natural to choose a wave that is also localised in space, i.e. a wave packet rather than a long wave train.

This raises a problem. For an infinitely long wave train, the idea of a wavelength takes on a definite meaning—it is the characteristic distance between successive crests or troughs. But for an irregularly shaped packet, consisting perhaps of a single hump, how can one possibly speak of it as having a wavelength?

In order to answer this, we must introduce you to an interesting fact— wave packets of almost any shape can be regarded as the superposition of a selection of infinitely long wave trains. The basic idea is to choose certain infinitely long wave trains, each with constant characteristic wavelength and amplitude, and allow them to interfere with each other. If the right choices of wavelength and amplitude have been made, then constructive interference takes place in a localised region (this gives the wave packet) and complete destructive interference everywhere else. There is a special mathematical procedure (called Fourier analysis) which enables the right selection of component waves to be made. We cannot go into it here, but at least we can give you a feel for how these waves interfere with each other.

In Figure 13 two wave trains of slightly different wavelength are shown repeatedly getting in and out of step. The resultant disturbance consists of a series of short-wavelength fluctuations and superimposed upon them a longer-wavelength variation giving rise to minima and maxima at, for example, X and Y respectively. At the minima the two waves are out of step, and at the maxima in step.

Thus the addition of only two waves has the marked effect of breaking up the regular wave trains into a series of wave packets. The length of the wave packets obviously depends on the distance between two points where the wave trains cancel each other, and this in turn depends upon their wavelengths. By choosing a different pair of waves, we could have produced wave packets of different lengths. The bigger the difference in wavelength, the shorter the distance over which they get out of step and hence the shorter the length of the wave packets.

By adding more wave trains, further modifications can be made. It is possible to add waves that enhance one particular wave packet and get out of step at progressively longer distances. In this way, all the wave packets

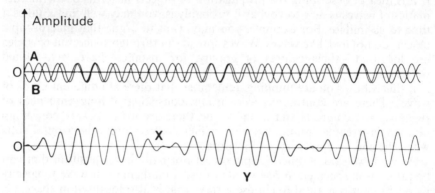

Figure 13 *The two wave trains,* A *and* B, *interfere to produce the resultant disturbance shown above*

can be wiped out except one. As the spread in wavelengths increases, so the wave packet becomes sharper and more pronounced.

It is clear that *a wave packet does not have a precisely defined wavelength; it is a mixture of waves of different wavelengths.* Only if certain waves with approximately the same wavelength dominate all other waves, i.e. if these have much larger amplitudes than the others, is it possible to speak of even an approximate wavelength for the packet. *This property of wave packets assumes a crucial significance later in the Unit.*

HOW ENERGY AND MOMENTUM ARE TRANSFERRED IN INTERACTIONS

Having seen how radiations are propagated in space and time, we now turn to a totally different question—how do radiations behave when they reach their destination and interact with matter?

THE INTERACTIONS OF MACROSCOPIC OBJECTS, NUCLEI AND ELECTRONS

We need say little about macroscopic objects such as pucks on air tables; these have already been dealt with in Units 3 and 4. There you saw that collisions between bodies involve sudden exchanges of energy and momentum and these exchanges are governed by the laws of energy and momentum conservation. The same is true also of collisions between nuclei and between electrons. For example, in Figure 14 an electron enters a bubble-chamber

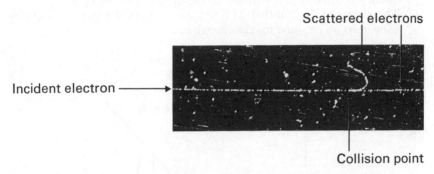

Figure 14 A bubble-chamber photograph showing an electron entering from the left and leaving a trail of bubbles. It strikes one of the electrons belonging to an atom of the liquid. The atomic electron is violently knocked away from its original atom and forms a track of its own

and strikes another electron belonging to an atom of the liquid. The second electron is hit sufficiently hard to be ejected from its parent atom and it then goes on to make its own trail of bubbles. Measurements on the tracks of the recoiling electrons show that momentum and energy are conserved in these collisions just as in the collisions of large objects. (You will learn more about bubble-chamber tracks and the measurements made on them in Unit 32.)

Macroscopic objects, molecules, atoms, nuclei, and electrons are all found to interact in the same way — their interactions can all be explained in terms of a sudden exchange of energy and momentum according to the principles of energy and momentum conservation.

THE INTERACTION OF ELECTROMAGNETIC RADIATIONS

To help you understand how light interacts we describe two experiments. The first involves the *photoelectric effect*.

When an electron leaves a piece of metal, the metal acquires a net positive electric charge (the result of the electron taking away a negative charge); this gives rise to a force of attraction which drags the electron back. If the metal is heated, however, the energies of the electrons are increased and some now acquire the necessary 'escape velocity'.

Heating the metal is only one way of ejecting electrons. It is found that electrons can also be emitted when light is shone upon the surface.

The escaping electrons must be absorbing energy from the incident radiation. Electron emission of this type is called the photoelectric effect. One of its useful features is that it allows light to be converted into a flow of electric charge (i.e. a current) and this property is made the basis of many types of instrument, e.g. a photographic exposure meter, a photomultiplier tube, a colorimeter, etc.

Here are some basic experimental observations on the photoelectric effect.

(1) The number of electrons emitted per second by light of a given frequency is directly proportional to the intensity of the light.
(2) The electrons are emitted with a range of kinetic energies extending from zero up to some maximum value.

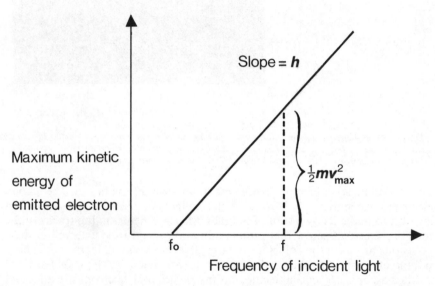

Figure 15   *The maximum kinetic energy of electrons emitted from a metal surface by incident light of varying frequency*

(3) This maximum energy of the emitted electrons is independent of the intensity of the light; it depends instead upon the frequency of the light as shown in Figure 15.
(4) Below the frequency, f_0 (Figure 15), no electrons are emitted, regardless of the intensity of the light. Above f_0 electrons are emitted immediately light arrives at the surface, no matter how weak the intensity.
(5) Electrons emitted from different metals all exhibit similar behaviour to that shown in Figure 15, except that the threshold frequency, f_0, varies from one metal to another. The slope of the line is independent of the nature of the metal, and in fact is found to be equal to h, Planck's Constant — the same constant as appears in the formula for the de Broglie wavelength.

On the basis of this experimental evidence, which of these models most satisfactorily describes the interaction of light with the electrons in the metal?
(A) Light energy falls on the metal plate in small localised packets. Individual electrons absorb these energy packets and are emitted.
The energy in each packet is proportional to the frequency of the light.
(B) As (A) except that the energy in each packet is proportional to the intensity of the light.
(C) Light energy falls diffusely on the whole area of the metal surface. Certain electrons gather the energy from a comparatively wide area and, when they have sufficient, they escape.

Choose your model before reading further.

Model C offers no explanation of the cut-off frequency f_0 noted under observation (4). Likewise model B is inconsistent with (1), (3) and (4). [For example (3) specifically says that the maximum energy of the electrons is independent of the intensity.] Model A fits the data. According to this model the photoelectric effect is explained as follows:
Light transfers energy to individual electrons in localised packets. These packets are called *photons. The intensity of the light governs how many photons arrive on a given area per second.* The electrons, on being struck by the photons, gain energy. Some of this is then expended by the electrons in escaping from the surface, the remainder appearing as the electron's kinetic energy. The least tightly bound electrons lose least energy in escaping and so emerge with the maximum kinetic energy. As we have said, the slope of the graph in Figure 15 is h, so we have the relation

$$h = \frac{\frac{1}{2}mv_{max}^2}{(f-f_0)}$$

$$hf - hf_0 = \tfrac{1}{2}mv_{max}^2$$

$$\therefore \quad hf = hf_0 + \tfrac{1}{2}mv_{max}^2 \tag{4}$$

where $\tfrac{1}{2}mv_{max}^2$ is the maximum kinetic energy of electrons liberated by light of frequency f. From this expression it is concluded that *the energy of the incident photon is* hf. This energy is transferred to the electron on impact. For the least tightly bound electron a part of the energy, hf_0, is needed to overcome the attractive forces appropriate to the particular metal being considered, while the remainder $(hf-hf_0)$ goes into kinetic energy. The value of f_0 will vary from metal to metal (Figure 16) depending upon the strength of these forces. Electrons that are more tightly bound to their parent nuclei, i.e. are in the lower energy atomic states, require more than hf_0 in order to escape and so emerge with less than the maximum kinetic energy.

Now go back and check that all the experimental observations (1) to (5) are in agreement with this explanation.

In Unit 6 it was found that when electrons are transferred from one atomic state to another the light is emitted in the form of discrete packets of energy. These photons were assigned an energy hf. This assignment allowed the energy difference between the initial and final electron states to be determined from a knowledge of the frequency of the spectral line associated with the transition. Thus, whether light is being emitted or absorbed, its interaction with matter may be described by a theory involving the transfer of energy in the form of photons.

We turn now to the second of the two experiments concerned with the

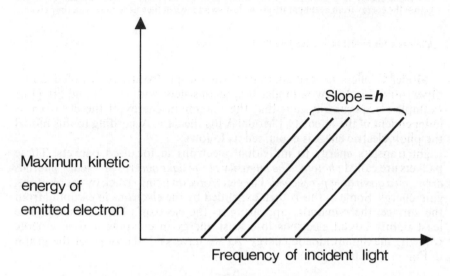

Figure 16   *The same as Figure 15 for four different metal surfaces. Whereas the slopes are identical, the intercepts on the abscissa differ for each metal*

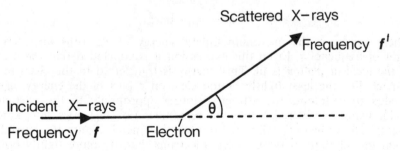

Figure 17   *X-rays scattered from electrons have their frequency reduced, i.e. f′ is less than f. The value of f′ depends upon the scattering angle, θ*

interaction of electromagnetic radiations. How is momentum exchanged in interactions between light and matter?

In 1924, the American physicist Compton studied the scattering of X-rays. In this experiment, it was found that when material is irradiated with a beam of X-rays of a single frequency, f, the radiation scattered from the electrons at any given angle, θ, has a reduced frequency f' (Figure 17). This is called *Compton scattering*. For each angle of scattering, there is a characteristic value of f'. That the frequency should be reduced is not surprising. In the interaction, the photon transfers energy $\frac{1}{2}mv^2$ to the electron, so its own final energy, hf', must be smaller than its initial energy hf. This in turn means f' must be less than f. An electron recoiling with velocity v must also acquire momentum, mv, from the interaction. Compton found this could be done without violating the law of conservation of momentum, if a momentum is assigned to each photon that, like the energy, is directly proportional to

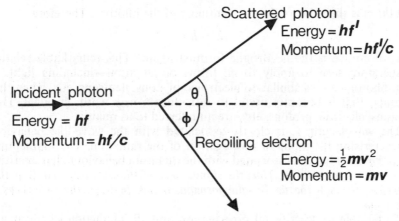

Scattered photon
Energy $= hf'$
Momentum $= hf'/c$

Incident photon
Energy $= hf$
Momentum $= hf/c$

θ
ϕ

Recoiling electron
Energy $= \frac{1}{2}mv^2$
Momentum $= mv$

Figure 18   Diagram for illustrating the collision between a photon and an electron (Compton scattering)

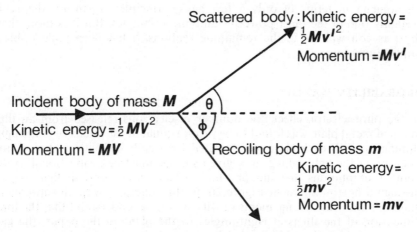

Scattered body : Kinetic energy $=$
$\frac{1}{2}Mv'^2$
Momentum $= Mv'$

Incident body of mass M
Kinetic energy $= \frac{1}{2}MV^2$
Momentum $= MV$

θ
ϕ

Recoiling body of mass m
Kinetic energy $=$
$\frac{1}{2}mv^2$
Momentum $= mv$

Figure 19   Diagram for illustrating the collision between two bodies

the frequency of the radiation. The constant of proportionality is (h/c) where c is the velocity of light. *Thus the momentum of the incident photon is* (hf/c) *and that of the scattered photon* (hf'/c). In this way the collision between the photon and the electron (Figure 18) is analogous to a collision between two bodies (Figure 19), in the ways both energy *and* momentum are exchanged.

The similarity between the behaviour of electromagnetic radiation and other types of radiation, such as beams of electrons and photons, can be shown to be even more striking than has been suggested so far. The wavelength of electromagnetic radiation in a vacuum is given by $\lambda = (c/f)$. Multiplying both the numerator and denominator of this ratio by h we get:

$$\lambda = \frac{hc}{hf} = h\left(\frac{c}{hf}\right)$$

But (hf/c) is the value of the momentum p of the photons. Therefore:

$$\lambda = h/p$$

This of course is the de Broglie formula again! This remarkable relation is therefore seen to apply to *all* forms of radiation — including light. In fact, photons are *so* similar to electrons, protons, neutrons, etc., when they interact, that it becomes useful to introduce a new word, *quantum*. Thus photons, electron, protons, etc., are all referred to as quanta.

The wavelength, λ, is clearly associated with the wave-like behaviour characterising the mode of propagation of the radiation. The momentum, p, on the other hand is associated with the quantum behaviour characterising the mode of interaction. *Thus the significance of Planck's constant, h, is that it forges, through the de Broglie formula, a link between the two types of behaviour.*

To be able to describe all propagation and all interaction so succinctly, you must surely agree, is immensely satisfying. One of the primary aims of any scientist is so to order his observations that he can describe all phenomena in terms of only a few basic principles; quantum theory is certainly one of the most powerful unifying principles. But it is more than that; as you will see in the remaining sections, it has deep philosophical implications too.

PROBABILITY WAVES

In the photoelectric effect, the number of electrons released from an illuminated metal plate was found to be proportional to the intensity of the light shining on it. If each electron is ejected by a collision with a photon, the intensity of the light falling on a given surface must be proportional to the number of photons arriving on that surface. Now suppose light passes through a narrow slit as in Figure 20. In this diagram, as in all subsequent diagrams showing arrangements of slits, it is to be understood that the long dimension of the slit is at right-angles to the plane of the paper; the gap shown represents the width of the slit.

Earlier it was shown that for this experimental arrangement the form of the intensity distribution is that of Figure 5. From this distribution, it can be seen that, for example, the intensity of the central maximum is about 20 times that of the first diffraction maximum. It therefore follows that 20 times the number of photons arrive on the screen in a small region opposite the slit (position A of Figure 20) as will arrive on an area of equal

size situated at the first diffraction peak (position B). Thus, the wave behaviour determines how many photons arrive on each part of the screen
(the quantum behaviour of course determines how the photons interact
once they arrive).

So far so good. But what if the slit is opened and closed very rapidly so
that the average energy transmitted in the time it is open is less than the
energy of a single photon. Does only a fraction of a photon arrive at the
screen? We appeal once again to experiment. It is found that when light of
very low intensity is allowed to pass through an opening for a tiny fraction
of a second (in one particular version of the experiment the time was one

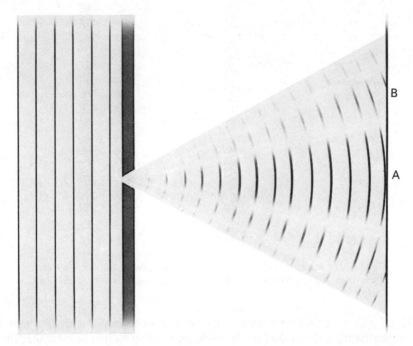

Figure 20   Diagram illustrating the diffraction of light at a slit. There are regions of
maximum intensity on the screen at positions such as A *and* B

nanosecond) either a photon with the whole energy, *hf*, arrives at the screen
or none at all.

Given then that only a single photon's worth of light arrives at the screen,
what happens to the intensity distribution? Does the photon smear itself
out to form a diffraction pattern?

No. The photon does *not* spread itself out — that much is once again
established by experiment. When a single photon arrives, it strikes one, and
only one, position on the screen. Whereabouts on the screen it will strike
cannot be predicted.

All that can be known in advance is that it will arrive at a point on the
screen where the diffraction pattern (*calculated* from its wavelength and

the size of the slit) would be non-zero, i.e. the photon is liable to arrive anywhere on the screen other than at positions lying on a calculated diffraction minimum. (Note the stress on the word 'calculated'. When dealing with a single photon there is no *observable* diffraction pattern; it can only be *calculated* from the known wavelength and the size of the slit.)

Likewise a second photon will go to some position on the calculated diffraction pattern. Subsequent photons do the same, and gradually with time the form of a diffraction pattern materialises as in Figure 21.

Thus, the calculated diffraction pattern describes the *probability* of a photon arriving at various parts of the screen. Where exactly any individual

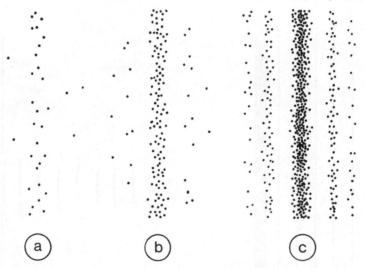

Figure 21   As more and more photons pass through the slit and reach the screen the diffraction pattern gradually builds up and becomes recognisable as such

photon will go cannot be predicted—we can only hope, with the help of wave mathematics, to calculate the odds on it going to one place rather than another.

To talk of being able to predict only the odds on the outcome of an experiment may well strike you as very strange. After all, you have probably been told in the past that physics is an exact science. If a physicist repeats an experiment he has often performed before, he is expected to be able to specify in advance what the outcome will be. To a large extent, of course, this remains true; if he takes a quantity of gas and halves the volume, keeping the temperature constant, he knows this will double the pressure; if he doubles the temperature difference across the faces of a sheet of material, he knows this will double the heat transmitted through it, etc. In these, as in many other examples you can no doubt think of for yourself, the outcome is known—at least to a high degree of precision. However, you should note that these examples deal only in gross features; they each involve the average effects produced by an enormous number of atoms—in one case, the momentum imparted by many gas atoms to the walls of a container averaged over a period of time; in the other, the energy transferred between large numbers of vibrating atoms as they jostle each other. In this respect, they are like the experiment in which a substantial amount of light is allowed to pass through the slit — in that case also the physicist is able to predict the outcome of the experiment, i.e. he knows in advance that, given enough photons, he will get a diffraction pattern of a certain size and intensity distribution, such as will allow him to predict, for example, that 20 times as many photons will arrive at region A as at region B (Figure 20). But, when he is asked to predict the behaviour of a *single* quantum, he is unable to do it.

We repeat: *there is no way of predicting where exactly a particular quantum will arrive.*

That is a statement we do not expect you to accept lying down! So let us try and anticipate your objections. Consider the following argument:

The nature of the experiment you have described has not been specified sufficiently well; all you have told me is that the light passes somewhere through the slit. Now, if I were allowed to make a really thorough study of the initial photon, I could determine very precisely its original trajectory and could then work out which atom (or atoms) in the rim of the slit it would interact with. I could then (in theory at any rate) calculate precisely how the photon would scatter from this atom, and hence determine its final direction. That would tell me exactly where it would go on the screen.

This incidentally is an example of a *thought experiment*; in these imaginary experiments *the question of whether the experiment is practicable or not is ignored. All equipment and measurements are assumed to be as close to the ideal as can be imagined.* Thought experiments have always been regarded as particularly useful in clarifying the meaning and significance of quantum theory.

At this point we throw down a challenge to the really bright student!

Think back over the material that has so far been presented in this Unit, and see if you can formulate two or three convincing arguments to show what is wrong with the assumptions or line of reasoning in this thought experiment.

If you think that this is too difficult, most of the Course Team would probably agree with you! But don't give up just yet. Try the three clues below and make a genuine attempt to develop the line of reasoning suggested in each before reading on.

Clue 1
The precise study of the collision between the photon and the atom would presumably involve a detailed knowledge of the struck atom. Have you been told anywhere that the diffraction of light through a slit in a barrier depends on the nature of the material of the barrier?

Clue 2
Is it possible, even in principle, to determine precisely the trajectory of the initial photon? Try designing an experimental arrangement for doing this.

Clue 3
Suppose you opened up a second slit in the barrier very close to the first—what effect would this have on the calculated intensity distribution? Can you reconcile this change of probabilities with a photon whose final direction can be precisely specified by the way it scatters on the rim of only one of the holes?

In a simple theory of a precisely determined photon–atom collision, it is difficult to see why there should be no possibility at all of the photons scattering at certain angles (to the diffraction minima). Moreover, it is natural to assume that such directions of zero probability would in any case depend in some way on the atom involved in the collision – how

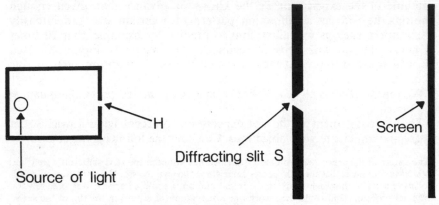

Figure 22   A hole, H, *introduced in front of the source in an attempt to produce a well-defined beam of light*

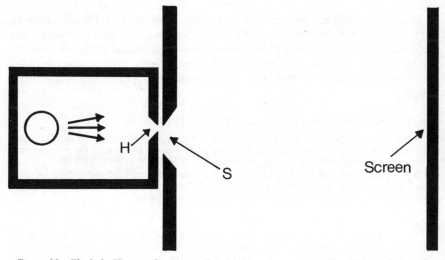

Figure 23   The hole, H, *is made very small and is brought up close to the slit* S. *In this manner the position of the photon at* S *is well defined—but not its direction*

heavy it is, how it is oriented and how strongly it is bound to neighbouring atoms. Yet, if you remember, the directions of the diffraction minima depend *only* on the wavelength of the radiation and the width of the slit, not on the nature of the material surrounding the slit.

Secondly, it is all very well saying that the original trajectory is to be precisely determined, but how could one do this, even in principle? One might think, for example, of placing a tiny hole, H, in front of the source of light as in Figure 22. By bringing up H very very close to S, and by making it exceedingly small, one could learn precisely where the photon was as it passed through S (Figure 23). But in order to know which atom was hit on the rim of S, H would have to be so very small, and the consequent diffraction it causes to the emergent light so large, that the photon could be moving in almost any direction as it emerged from H. Not knowing

Figure 24   The hole, H, is removed a long way from S. In this arrangement the direction of the photon at S is well defined—but not its position

the direction of the photon makes it impossible, of course, to work out the details of its collision with the atom.

In order to fix the direction of the photon, the source and hole H would have to be taken a very long way from S (Figure 24). Now, because the angle subtended by the width of S at the source is so small, the fact that the photon

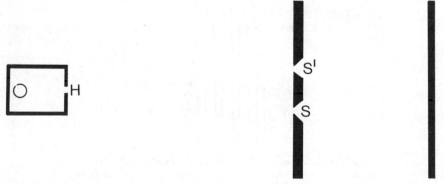

Figure 25. The opening of a second slit S', produces a radical change in the intensity distribution at the screen on the right

is known to pass somewhere between the edges of S defines its direction very satisfactorily.

Unfortunately, in order to get this precise knowledge of the initial direction of the photon, our knowledge of the position of the photon as it passes through S has had to be sacrificed. It is now not known whether the atom is being struck a glancing blow or a direct hit, or indeed even which atom is being struck.

A discussion of this kind soon reveals that it is quite impossible, even in principle, to specify a precisely determined trajectory; an argument that assumes such a determination must therefore be fallacious.

A third and final refutation: suppose a second slit, S′, identical and parallel to the first, is opened close to S as in Figure 25. This is Young's double-slit arrangement. The intensity distribution on the screen is now radically different to what it was when only one slit was open as can be seen by comparing Figure 26 with Figure 5. The combined distribution is not simply the sum of two single distributions (one slightly displaced relative to the other because of the separation of the two slits). Because of interference between the waves from the two slits, *there are now regions of zero intensity* where previously the screen was illuminated. This means that if one slit is open, a photon may arrive at one of these regions, but if *both* are open the photon is unable to do so. How can the opening of the second slit (a slit the photon presumably does not pass through) affect the outcome of the photon–atom collision in the first slit in a way that makes it no longer possible for the photon to go to one of these regions? Clearly anything resembling conventional dynamics could not possibly give such a result.

Thus, on several counts, the attempt to interpret the experiment in terms other than those involving probability has failed. No matter how strange

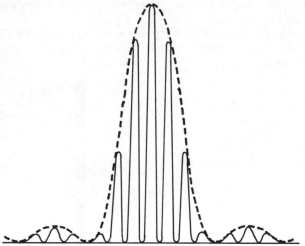

Figure 26   The intensity distribution on the screen resulting from light passing through Young's double-slit arrangement

the idea appears to you at first, the fact remains that it is only possible to predict the probability that a photon (or indeed any other quantum) will arrive at any given region on the screen.

As was said earlier, in order to calculate this probability, the same type of mathematics must be used as is appropriate in describing the progress of, for example, ripples on the surface of water or sound in a gas; it is the mathematics of waves. A quantity can be defined to represent the 'amplitude' of the 'probability wave'. This behaves mathematically in the same way as the amplitudes of water or sound waves, inasmuch as two or more probability wave amplitudes can be added to give constructive or destructive interference. The square of the amplitude of a water or sound wave is a measure of the energy of that wave; the square of the amplitude of the probability wave at any given position is a measure of the probability of finding the quantum at that position.

But what, you may ask, is 'waving' in these probability waves? What is the medium? Here the analogy with water and sound waves breaks down (you have been warned before of the dangers of taking analogies too far!). The latter require a medium — probability waves do not. The similarity is a similarity in mathematics only. *The probability waves say nothing about the state of any medium; instead they tell us of the state of our knowledge.*

HEISENBERG'S UNCERTAINTY RELATION

The situation described in the last section was quite frustrating. When the position of the photon was fixed (as in Figure 23), the direction of the photon was not. When the experimental arrangement was altered to remedy this (as in Figure 24), it was found that the newly-acquired knowledge of the photon's direction had been bought at the expense of losing knowledge of the photon's position.

The trouble stems from the wave behaviour governing propagation and its probabilistic interpretation. Moreover, because wave behaviour is the underlying principle governing *all* propagation, it is only to be expected that this kind of frustration will be encountered again and again; it is not some peculiarity associated only with photons.

In this section, we shall study this fundamental problem of observation in greater depth. We shall investigate to what extent it is possible simultaneously to measure two properties of a quantum; its position and its momentum. For this purpose, we shall again use a slit. The act of measurement is considered to take place at the instant the quantum arrives at the slit. (Although we are sticking to our example of diffraction at a slit, you should realise that what is being discussed has universal significance; we could just as well have chosen some other experimental arrangement with which to illustrate the point we wish to make.)

In Figure 27, a drawing of the experimental arrangement is combined with one of the intensity distribution to remind you of the appearance of the diffraction pattern on the screen when the slit is illuminated with mono-chromatic radiation from a distant source. The y axis has been drawn at right angles to the length of the slit. (Remember that the length of the slit is taken to be perpendicular to the plane of the paper.)

At the instant at which the quantum arrives at the slit, its position along the y axis is uncertain; it is only known that it must lie somewhere between the extreme edges of the slit. If the width of the slit is d, then the uncertainty in the position of the quantum along the y axis — call it Δy — at the instant it arrives at the slit is given by: $\Delta y = d$.

At this instant, the quantum is 'diffracted', i.e. receives a sideways impulse resulting in it subsequently arriving at the far screen at a position L, which may be other than position O directly opposite the slit from the source (Figure 27). How large the impulse will be, no one can say. The photon must therefore be regarded as having an uncertain momentum in the y direction as well as an uncertain position. (Let us emphasise that we are here talking

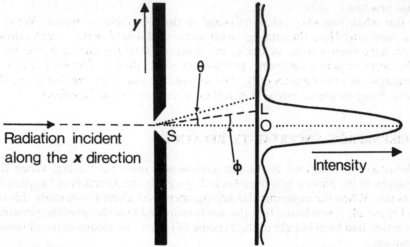

Figure 27   A drawing combining the experimental arrangement and the intensity distribution on the screen appropriate to radiation of a single wavelength

of uncertainty of both position and momentum at the moment the quantum arrives at the slit. Do not be confused into thinking we are referring in any way to uncertainty of position on the screen; the spread of positions of quanta on the screen will only be used as a convenient measure of the uncertainty of the y component of momentum at the slit.)

If the point L is such that the direction SL makes an angle ϕ to the original

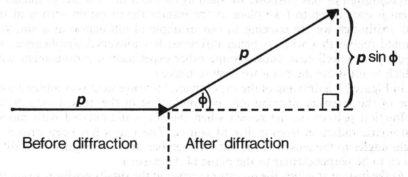

Figure 28   After diffraction, the quantum acquires a sideways momentum $p \sin \phi$

direction of the light, then the y component of momentum would be $p \sin \phi$ (Figure 28), where p is the total momentum of the quantum (both before and after 'diffraction'). But, as we have said, at the time of the measurement there is no way of telling what the value of ϕ will be. All we can do is take a look at the spread of the calculated diffraction pattern and decide on some average 'typical' value for the angle.

What would be a typical value of ϕ? One possibility is to say that most of the radiation would be diffracted to somewhere within the central intensity peak, i.e. ϕ will typically extend to the first diffraction minimum. The direction of this minimum makes an angle θ to the original direction. The 'typical' sideways momentum is therefore $p \sin \theta$, and this we take as a measure of the uncertainty in sideways momentum, Δp_y.

$$\therefore \Delta p_y = p \sin \theta \tag{5}$$

As shown in Unit 28, there is a relation connecting θ, λ and d:

$$\sin \theta = \lambda/d$$

Substituting this value for $\sin \theta$ in equation 5 we have:

$$\Delta p_y = p(\lambda/d)$$

If Δy and Δp_y are multiplied together we get

$$\Delta y \, \Delta p_y = d.p(\lambda/d)$$

$$\Delta y \, \Delta p_y = p\lambda \tag{6}$$

But we have the de Broglie relation connecting λ and p, and from this we can substitute (h/p) for λ in equation 6:

$$\Delta y \, \Delta p_y = p(h/p)$$

$$\Delta y \, \Delta p_y \approx h \tag{7}$$

This is Heisenberg's Uncertainty Relation – or at least one form of it. Note that the 'equals' sign has been changed in the last line to another sign meaning 'of the order of' (i.e. 'roughly within a factor of ten of'). This is because of the rather cavalier way in which a typical angle of diffraction was defined. In other texts you will find h replaced by $(h/2\pi)$ or $(h/4\pi)$; it all depends on how exactly one chooses to define the quantities on the left-hand side of equation 7. We shall not concern ourselves with such details (including the fact that strictly speaking $h/4\pi$ is not 'of the order of' h because it differs from it by a factor greater than ten!). After all, the diffraction pattern in Figure 27 only gradually falls off in intensity as one goes further from the point O, so the product of the uncertainties can on occasion be very much greater than h.

The crucial point you are to grasp is that *the product of the uncertainties in the simultaneous measurements of position and momentum in a given direction is equal to a small but finite quantity of the order of Planck's constant.*

We remind you that in this connection one is completely ignoring the usual sources of error associated with day-to-day experimental work. Throughout this Unit we are concerned only with the fundamental principles of measurement. These show that *even with a perfect idealised experiment,* the measurements of position and momentum are subject to limitations imposed by the uncertainty relation.

Note also that the quantities on the left-hand side of equation 7 refer to measurements of position and momentum component in one direction only (in this case the y direction). Similar relations exist relating uncertainties in position and momentum component along the other two directions at right angles, viz. x and z (the initial direction of the radiation, and the direction perpendicular to the plane of Figure 27, respectively).

$$\Delta x \, \Delta p_x \approx h \tag{7'}$$

$$\Delta z \, \Delta p_z \approx h \tag{7''}$$

In essence, the uncertainty relation does not tell us anything more than is already contained in the material presented in the previous sections of this Unit; it is certainly not a new postulate. It arises directly out of the concept of probability waves. The idea of probability waves is sufficient in itself to tell us that a slit of small width, used to fix precisely the position of a quantum, will produce large diffraction effects and hence large uncertainties in the sideways momentum of the quantum. It also tells us that one can alternatively have very little diffraction if one is prepared to open up the slit wide. The importance of the uncertainty relation is that it sums up in a particularly cogent fashion what one can or cannot hope to measure in any experimental situation — it is a powerful 'rule of thumb'.

In order to specify a trajectory exactly, you would have to be able to specify at a given instant both the exact position of the quantum and the exact values of the components of its momentum. Equation 7 says quite simply that you cannot do it. A precisely determined position (i.e. no uncertainty in position and hence $\Delta y = 0$) means that in order to keep the left-hand side of equation 7 from going to zero, Δp_y would have to be infinite, i.e. there would be no knowledge of the momentum. Alternatively a precisely determined momentum (i.e. $\Delta p_y = 0$) implies no knowledge of position (i.e. Δy is infinite). This is exactly the kind of dilemma encountered in the arrangements of Figures 23 and 24, when these two extremes were approached. Between these extremes, there is a continuous span of intermediate possibilities such that some gain in the precision of one variable may be made at the expense of a loss of precision of the other.

Thus, although it is necessary to know the exact experimental arrangement if one is properly to evaluate from wave theory the probabilities of the various outcomes of the experiment, the uncertainty relation allows one to make strong generalised statements. It is indeed somewhat unfortunate that the relation carries the name 'uncertainty' for this word implies hesitancy, indecision, and a general state of confusion. The student of quantum theory may well feel that if the professional physicist is 'uncertain' as to what he is doing, there is little hope for the student! For this reason the word 'indeterminacy' is sometimes preferred instead of 'uncertainty'. Regardless of what it is called though, you will soon come to recognise that the relation is more than just a semi-quantitative expression of frustration — it is a positive tool for describing many features of Nature.

Perhaps you still feel it goes against common sense to believe that one is unable simultaneously to measure both the position and the momentum of an object to any desired precision given that the measuring apparatus is ideal. Until some such time as the material of this Unit finds its way into the primary school curriculum, it will continue to be regarded as against common sense, and students will spend hours trying to find a way round the uncertainty relation. It is something we all apparently have to get out of our systems at one time or another. Even Einstein devoted long periods to devising ingenious schemes. The diversion of designing experimental arrangements for getting round Heisenberg's relation can be likened in many ways to the old quest for the perpetual motion machine!

The observer, in making an observation, always disturbs the system he is observing. He cannot play a passive role, i.e. obtain his information without actively disturbing the system. This in itself is nothing new; classical ideas of measurement also recognise that an observer interferes with the object under investigation. What *is* new in quantum theory is that the disturbance is unpredictable — the measurement cannot be suitably corrected to take account of the disturbance.

The observer has to make a choice between different courses of action — *either* to concentrate on precision of momentum, *or* precision of position, *or* some kind of compromise. Having decided on a particular kind of measurement, the unpredictable disturbance associated with the act of taking that measurement denies the observer the opportunity of ever gaining the information he *could* have obtained had he taken one of the alternative courses of action. For example, he could consider the alternatives whereby he either uses a small slit to measure the position of a quantum precisely, or a large slit (so keeping the diffraction effects small) to measure the momentum precisely. If he decides to use the small slit, then the very act of measuring the position of the quantum produces a disturbance (a large diffraction) that destroys the information on the quantum's momentum that he *could* have obtained had a large slit been used instead.

This is the essence of *Heisenberg's Uncertainty Principle*. When an observer is faced with a range of alternative procedures for measuring the momentum and position of an object (the precision of these measurements being governed by the uncertainty relation), he can gain only the information appropriate to one of the alternatives.

Indeed, the principle extends further than just to the measurement of the two variables: momentum and position. There are other pairs of variables linked by similar uncertainty relations. Thus, for example, closely associated with the concept of momentum is that of angular momentum. If the momentum of an object is uncertain, then it is only to be expected that the value of its angular momentum about some centre will also be uncertain. An uncertainty relation exists connecting the uncertainty in angular momentum, ΔL, and the uncertainty in angular position, $\Delta \psi$:

$$\Delta L \, \Delta \psi \approx h \tag{8}$$

There is one more uncertainty relation you should know about; it concerns uncertainty in energy. It is so important that we devote the next section to it.

THE ENERGY–TIME UNCERTAINTY RELATION

In the experiment where a slit was used to measure the position and momentum component of a quantum, *when* exactly was the measurement made?

You may immediately answer: At the instant the quantum arrived at the slit. But how would one estimate the instant of arrival?

Perhaps the most direct way is to fix a shutter onto the slit as in Figure 29a. The slit could then be opened for only a limited time (Figure 29b). In order

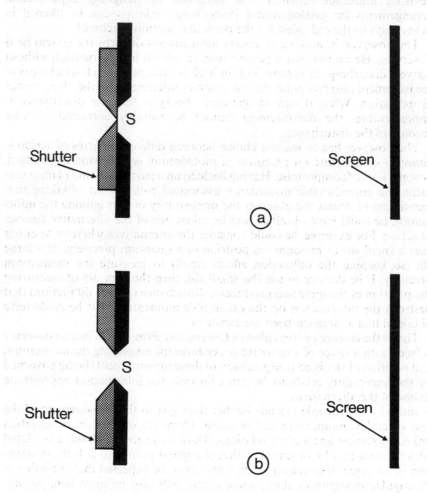

Figure 29 (a) A shutter is now placed over the slit through which the radiation is to pass. (b) The shutter can be opened to admit the radiation

for the y position and momentum of the quantum to be measured, the quantum has to arrive within the time interval for which the slit is open. This interval can be made as small as one likes. In this way, it is possible to fix exactly the instant at which the measurement is made.

However, this procedure has serious repercussions. If the slit is only open for a short time, then the radiation transmitted is no longer in the form of an infinitely long monochromatic wave train – it is in the form of a sharp wave packet. But this is quite contrary to our earlier assumption; the diffraction pattern shown in Figure 27 is derived on the basis of a *single* wavelength. If we are dealing with a wave packet, this is the same thing as saying our radiation has many different wavelengths; we are now talking of a *different* experiment.

We can look at it another way. If the instant at which the quantum arrives at the slit is known, it follows that, at that instant, it is known exactly where the quantum is along the incident direction, i.e. along the x axis (it is at the position S). The wave packet describing its position in the x direction is therefore a sharp spike with a spread, Δx, almost zero. But, according to the uncertainty relation for this direction (equation 7′), a small value of Δx implies a large uncertainty in the momentum in the x direction, i.e. in the incident momentum p. In the original version of the experiment (as in Figure 27), it was assumed that p was known precisely. We now see that such precision could only be achieved with a permanently open slit (resulting in Δx becoming infinitely large and Δp zero).

To summarise: in previous sections we were concerned only with making a measurement of the position and momentum of the quantum in the y direction. For this purpose it was *assumed* that we were dealing with a quantum of known momentum, p. If, however, we wish to say something about the time at which the measurement is made, it is necessary to perform a *different* experiment – one involving, for example, the opening and closing of the slit. This modification of the original experimental arrangement, while permitting us to say something about the time at which the measurement is made, affects our knowledge of p.

If our knowledge of p is uncertain, then this is the same as saying our knowledge of the energy, E, is uncertain; corresponding to Δp, there will be a ΔE. The value of ΔE (like Δp) will depend upon the time the slit is open. Let us call the latter Δt; it is the time available for the measurement. There exists a relation connecting ΔE and Δt. This is a perfectly general relation and applies to all experimental arrangements (not just the slit arrangements we have been considering).

The uncertainty in the energy of a system, ΔE, and the time available for measurement, Δt, is given by:

$$\Delta E \, \Delta t \approx h \tag{9}$$

You should note that this third form of uncertainty relation, although in appearance very similar to the previous ones (equations 7, 7′, 7″ and 8), has a different kind of interpretation. The previous relations were all concerned with measurements of two variables (momentum and position, or angular momentum and angular position) *at a given instant of time*; this third one can be understood as a relation connecting the uncertainty in the energy to the length of time available for making that energy measurement.

It means, for example, that an infinitely long time ($\Delta t \rightarrow \infty$) is required if the energy is to be measured precisely ($\Delta E = 0$). How does this come about? Well, in the case of a free electron, say, an exact value of E implies an exact value of p, and this in turn implies an exact value of the wavelength (through de Broglie's relation $p = h/\lambda$). Such an exact wavelength is characteristic only of an infinitely long wave train; strictly speaking the wave form in Figure 30 does *not* have an exactly definable wavelength, even though the

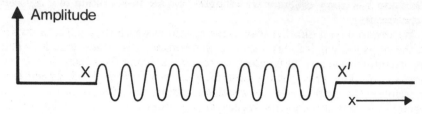

Figure 30   A wave train of finite extension

distances between successive peaks and troughs between the cut-off points X and X′ are equal. This is because many different continuous wave trains must be superimposed in order to produce destructive interference beyond X and X′ and so give the resultant wave train a finite extension.

For a wave form to have only one characteristic wavelength, it is not sufficient that the distance between successive crests and troughs should be equal; the wave train must also extend to infinity. If one has to check that a passing wave train is infinitely long, one needs eternity to complete this measurement!

QUANTUM THEORY—IS IT A COMPLETE THEORY?

Having now presented you with the basic ideas of quantum theory, we are in a position to describe a number of interesting consequences. But before applying the new ideas to a variety of situations, we thought you might like to learn of a famous controversy that split the world of physics in the 1920s and still has repercussions today. Its substance has little to do with your everyday life, or indeed with any scientific work in which you may be engaged. Nevertheless we hope you will find this excursion into philosophy an intriguing one.

An important concept involved in the controversy is that of causality. It is not easy to define this term in a way that would satisfy all philosophers, but we shall content ourselves with the following simple definition used by Planck

> There can be no more incontestable way to prove the causal relationship between any two events than to demonstrate that from the occurrence of one it is always possible to infer in advance the occurrence of the other. . . An occurrence is causally determined if it can be predicted with certainty.

It is important to distinguish between two types of causality. You remember there was the type of causality that related to large statistical effects such as the passage of many photons through a slit, halving the volume of a gas, etc. Causality in this sense holds good and no one denies this. For example, an intense beam of monochromatic light passing through a slit causes a diffraction pattern of a certain size and intensity distribution, and this effect can be accurately predicted in advance. On the other hand, causality in the sense of predicting where an individual photon will go is known as *strict* causality. Quantum theory does not help us to make statements of a strictly causal nature.

THE COPENHAGEN INTERPRETATION

In a paper delivered at a conference in Brussels in 1927, Born and Heisenberg declared,

> Quantum mechanics leads to accurate results concerning average values but gives no information as to the details of an individual event. The determinism which so far has been regarded as the basis of the exact sciences has to be given up. Every additional advance in our understanding of the formulas has shown that a consistent interpretation of the quantum mechanical formula is possible only on the assumption of a fundamental indeterminism. . . . We maintain that quantum mechanics is a complete theory; its basic physical and mathematical hypotheses are not further susceptible of modifications.

This provocative statement in effect declared that quantum theory was to be regarded, in a sense, as the ultimate theory. It did not mean that quantum theory was incapable of improvement, refinement, and general adaptation to meet new situations; but it did express the view that quantum theory would never be replaced by a radically different theory founded on strict causality. According to this claim there was no hope of ever fulfilling Laplace's proud boast

> Give me the initial data on all particles and I will predict the future of the universe.

The uncertainty relations would always deny us the initial data from which to start. This view of quantum theory came to be known as the Copenhagen Interpretation (Copenhagen being the centre where people such as Heisenberg, Born and Bohr worked in the 1920s, while quantum theory was being developed).

Most, if not all, people at first find this view disquieting. It is not easy to abandon strict causality. It is somehow deep-rooted within us to think that the work of science will not be completed until it has established causal relationships which allow us to predict with certainty the outcome of all processes.

The Copenhagen school would argue that you cannot abandon something you have never had. There is no experimental evidence for strict causality and never has been. The only evidence for causality has been that relating to the weaker form of causality involving statistical processes, and the conclusions of quantum theory are in no way at variance with such experiments. To maintain that all physical processes operate in a certain way (i.e. according to strict causality), when there is not the slightest evidence to support the view, is an act of faith strangely uncharacteristic of a scientist.

But, you may argue, if it is agreed that our *observations* of the world are subject to an imprecision imposed by the uncertainty relations, cannot one think of the world *itself* as being governed by causality. Here is Heisenberg's opinion:

In view of the intimate connection between the statistical character of the quantum theory and the imprecision of all perception, it may be suggested that behind the statistical universe of perception there lies hidden a 'real' world ruled by causality. Such speculations seem to us— and this we stress with emphasis—useless and meaningless. For physics has to confine itself to the formal description of the relations among perceptions.

In support of this view he puts forward this argument:

We can say that physics is a part of science and as such aims at a description and understanding of nature. Any kind of understanding, scientific or not, depends on our language, on the communication of ideas. Every description of phenomena, of experiments and their results, rests upon language as the only means of communication. The words of this language represent the concepts of daily life, which in the scientific language of physics may be refined to the concepts of classical physics. These concepts are the only tools for an unambiguous communication about events, about the setting up of experiments and about their results. If, therefore, the atomic physicist is asked to give a description of what really happens in his experiments, the words 'description' and 'really' and 'happens' can only refer to the concepts of daily life or of classical physics. As soon as the physicist gave up this basis he would lose the means of unambiguous communication and could not continue in his science. Therefore any statement about what has 'actually happened' is a statement in terms of the classical concepts and—because of thermodynamics and of the uncertainty relations—by its very nature incomplete with respect to the details of the atomic events involved. Their demand to 'describe what happens' in the quantum-theoretical process between two successive observations is a contradiction *in adjecto*, since the word 'described' refers to the use of the classical concepts, while these concepts cannot be applied to the space between the observations; they can only be applied at the points of observation.

One of the great advantages man has over the other animals is the power of language. Like all animals he observes the world, but having done so he can communicate those observations to other men. Gradually over the ages a common fund of scientific knowledge is accumulated. But, in order to combine the results of the experiences of different persons, a common unambiguous language must be evolved. Such a language can only be ambiguous if it stems from some common experience or observation (perhaps not even then). A word used by a person X can only take on significance for person Y if it is related to some common aspect of X and Y's experience of the world, i.e. some common observation. Thus language has grown out of observation. Heisenberg argues in the passage quoted above that language is therefore only meaningful when applied to the acts of observation.

Thus we have one observation—an electron leaves an electron gun; and a second observation—the electron makes an image on a photographic plate on the other side of the barrier with the two slits in it. *But it is meaningless to argue which slit the electron went through.* The fact that it is difficult to visualise why an electron's collision with an atom in the rim of one of the slits should be affected by whether the second slit is open or closed is, according to the Copenhagen Interpretation, just one example of the trouble encountered when trying to apply concepts evolved *specifically* for describing observations to situations *in between* observations. Heisenberg declared:

> It is a matter of personal belief whether such a calculation concerning the past history of the electron can be ascribed any physical reality or not.

In order to bring home to you just how much the processes of observation are inextricably linked with anything one tries to say about the micro-physical world, consider the following argument:

> I take a sodium lamp and observe it with a counting device that produces an audible click the moment a photon arrives. Knowing how far the counter is from the source, and knowing that the photon travelled with the speed of light, I can then calculate how long it must have taken for the photon to travel from the source to the counter. By noting the time at which the click occurred, I can then calculate the instant at which the photon was emitted from the source. Note that I am now describing something that took place when no observation was being made. Indeed, I can discard the counting device altogether and make the following general statement: at certain instants of time, the atoms in a sodium lamp emit photons even when they are not being observed.

To see whether this statement is reasonable, consider the following behaviour of sodium lamps:

> I take a sodium lamp and carry out an experiment with it involving interference, e.g. I might illuminate a Young's double slit and observe the fringes. This phenomenon I explain in terms of wave trains, one from each slit, travelling different paths HSP and HS′P (Figure 31) and

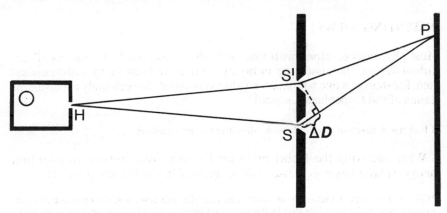

Figure 31 ΔD represents the difference in the path lengths HSP *and* HS′P

then interfering with each other at P. In order for the wave trains to recombine and produce interference at P, their lengths must be at least as long as the difference in the distances they travel, i.e. ΔD in Figure 31. (If they were not, the wave train travelling the shorter distance would have wholly arrived at P before the other had begun to arrive, and so they could not overlap and produce interference.) With such an arrangement, I make the source of light so weak that only the light from a single atom is allowed to pass through the apparatus at any given time. It is found that the phenomenon of interference has still to be taken into account in determining where the light is likely to go. Light emitted by a single atom is therefore in the form of a long wave train. In certain experimental arrangements, it has been found possible to produce interference between wave trains that have travelled paths differing in length by about 1 metre. The wave trains must therefore be at least this length. I finally conclude that for a sodium atom to produce a wave train of 1 metre length it must have emitted light over a period of time equal to 1 metre/velocity of light, i.e. $(1/3 \times 10^8)$ seconds, i.e. 3 nanoseconds. Indeed, I can discard all reference to apparatus and make the following general statement: the atoms in a sodium lamp emit long wave trains of light over extended periods of time even when they are not being observed.

In the first statement, the light was thought of as being emitted at a given instant in a sharp pulse (a photon), whereas in the second the light was emitted continuously over a long period of time (a wave train). You cannot have it both ways. Which mental picture is right? It is impossible to say. *The atoms are not being observed*, so neither viewpoint can be verified.

In this way, you can see that the attempt to say something objective about the 'real' world divorced from the act of observation failed. *Both statements were still inextricably linked to a certain type of observation.*

Whether, therefore, one wishes to continue with mental pictures of what happens in between the observations in the face of such difficulties is a matter of personal preference. Some may find it helpful, others not. Whatever our preference, *science* has nothing to say on the matter. *Science is the study of our observations of the world, not of the world itself.* Moreover, the Copenhagen Interpretation holds that this is no temporary restriction to the activities of the scientist — a restriction likely to disappear with further advances in knowledge — but a fundamental limitation that will apply for all time.

DISSENTING VIEWS

Most physicists go along with this 'orthodox' view, but by no means all do. Indeed among the dissenting minority there have been many distinguished men. Einstein was one who was very dubious about the seemingly extravagant claims of the Copenhagen school:

I look upon quantum mechanics with admiration—and suspicion.

When receiving the Nobel prize for his own contributions to quantum theory, Schroedinger expressed his misgivings in the following words:

Will we have to be permanently satisfied with this? On principle, yes. On principle there is nothing new in the postulate that in the end exact science should aim at nothing more than the description of what can really be observed. The question is only whether from now on we shall have to refrain from tying description to a clear hypothesis about the real nature of the world. There are many who wish to pronounce such abdication even today. But I believe that this means making things a little too easy for oneself.

De Broglie began by trying to salvage strict causality, but was quickly won over to the Copenhagen Interpretation. In 1947 he wrote:

I should also like to explain how, little by little, starting from the classical deterministic conception of physical phenomena which I adopted more from habit of mind than by philosophical conviction, I was led to range myself entirely on the side of the probability and indeterminist interpretation which Bohr and Heisenberg have given to the new mechanics. . . Perhaps this 'confession' will have the further advantage of preventing certain investigators one meets with pretty frequently in these days, forcing themselves to turn back to the past, to question again the new conceptions of the physics of uncertainties . . . it is not out of mere wantonness that I have given up the traditional positions of classical physics but rather was I compelled and constrained to do so; perhaps this statement itself will show the incredulous how much these traditional positions have become impossible to defend.

And yet in spite of this statement, since 1951 he has again had doubts about the completeness of quantum theory:

At the level now reached by research in microphysics it is certain that the methods of measurement do not allow us to determine simultaneously all the magnitudes which would be necessary to obtain a picture of the classical type of corpuscles (this can be deduced from Heisenberg's uncertainty principle), and that the perturbations introduced by the measurement, which are impossible to eliminate, prevent us in general from predicting precisely the result which it will produce and allow only statistical predictions. The construction of purely probabilistic formulae that all theoreticians use today was thus completely justified. However, the majority of them . . . have thought that they could go further and assert that the uncertain and incomplete character of the knowledge that experiment at its present stage gives us about what really happens in microphysics is the result of a real indeterminacy of the physical states and of their evolution. Such an extrapolation does not appear in any way to be justified. It is possible that looking into the future to a deeper level of physical reality we will be able to interpret the laws of probability and quantum physics as being the statistical results of the development of completely determined values of variables which are at present hidden from us. . . One is glad to see that in the last few years there has been a development towards re-examining the basis of the present interpretation of microphysics. . . It seems desirable that in the next few years efforts should continue to be made in this direction. One can, it seems to me, hope that these efforts will be fruitful and will help to rescue quantum physics from the cul-de-sac where it is at the moment.

A leading advocate these days of the dissentient view is David Bohm. His opinion has been expressed as follows:

. . . with regard to the current formulation of the quantum theory, we are led to criticise assumptions such as those of Heisenberg and Bohr, that the indeterminacy principle and the restriction to complementary pairs of concepts will persist no matter how far physics may progress into new domains. It should be clear, however, that in making such criticisms, it is not our intention to imply that the quantum theory is not valid or useful in its own domain. On the contrary, the quantum theory is evidently a brilliant attainment of the highest order of importance, a theory whose value it would be absurd to contest. . . What we wish to stress here is, however, that the brilliant achievements of the quantum mechanics in no way depend on the notion that the features mentioned above (or any other features) of the current theory represent absolute and final limitations on the laws of nature. For all these achievements could equally well be obtained on the basis of the more modest assumption that such features apply within some limited domain and to some limited degree of approximation, the precise extent of which limits remains to be discovered. In this way we avoid the making of arbitrary *a priori* assumptions which evidently could not conceivably be subjected to experimental proof, and we leave the way open for the consideration of basically new kinds of laws that might apply in new domains, laws that cannot be considered if we assume the absolute and final validity of certain features of the theories that are appropriate to the quantum-mechanical domain.

Among the new kinds of laws that one is now permitted to consider if one ceases to assume the absolute and final validity of the indeterminacy principle, a very interesting and suggestive possibility is then that of a sub-quantum mechanical level containing hidden variables.

What is meant by 'hidden variables'? Let us explain with the help of an analogy. You no doubt recall seeing Brownian motion in Unit 5—the tiny smoke particles moving about in an agitated fashion. Suppose you knew nothing of the kinetic theory of gases – how would you explain the motion? You could not the behaviour of the particles would seem strange and unpredictable. However, once the idea of rapidly moving molecules of air is introduced, the motion of the smoke particle is easily explained in terms of its collisions with these air molecules. A movement in any direction can be attributed to a statistical fluctuation in the number of impacts on opposite sides of the smoke particle; the phenomenon appears rational and predictable.

What is being suggested by Bohm is that there may exist hidden variables (analogous to the moving air molecules) at a sub-quantum level, behaving in accordance with strict causality. It is these variables that determine the *apparently* unpredictable behaviour of quanta (analogous to the motion of the smoke particles).

Matter exists at different levels of organisation and of complexity, one of which is occupied by the organism called man. There is a great deal of evidence that ways of describing the behaviour of matter (or our observations of matter) may be adequate and valid at one level but not at another.

Likewise, the uncertainty principle is relevant to the diffraction of photons at a slit, but has little to do with the growth of the rabbit population in Australia. In this Unit great stress has been laid on the fact that physical variables (at least the ones known at present) can only say what happens at an observation and not what happens in between. But at the macroscopical levels of biology, geology and human practice quite the contrary is assumed. Descriptions given by these and other studies are based on the notion that objective processes actually take place 'in between' our observation of them (and, in the case of the Earth sciences often millions of years before there was any man to observe them!).

Thus we find qualitative changes in the language and concepts of science, corresponding to the qualitative changes in the behaviour of matter as we pass from one level to another. *It is near the boundaries between one level and another that the difficulties seem to arise.* In the Brownian motion example, the macroscopic object (smoke particle) is small enough to be affected by sub-macroscopic processes (motion of gas molecules), so it behaves in a 'strange' way.

Is there a level of organisation of matter below that of the nuclear atom? Quantum mechanics describes adequately what happens at the microphysical level (the behaviour of electrons, atoms and molecules). But it is by no means clear that it is adequate when it comes to describing the structure of sub-nuclear matter. There may indeed be a more fundamental sub-microphysical level than that of nuclear and atomic physics, to which different and even more fundamental concepts than those of quantum mechanics may be applicable. The investigation of sub-nuclear matter is still at a very early stage, but even now there are some indications that we are once again near a new kind of threshold.

Classical mechanics has been found to be an adequate approximation at the macrophysical level, but at the microphysical level quantum mechanics has to be used. Is quantum mechanics, in its turn, only an approximation that is adequate at the microphysical level but inadequate at the sub-micro-physical? And, if so, to what is it an approximation?

Even if the idea of hidden variables behaving in accordance with strict causality is not the correct one, a deeper understanding of Nature than that at present provided by quantum theory may yet evolve along some other lines — *assuming steps are taken to seek one.*

As for the Bohr–Heisenberg view that one can never hope to use language evolved in the context of observation to describe what happens in between observation, this has also been challenged by Bohm. Here, his argument is that language can go further than mere vocabulary. Although the words of the basic language may originate in the context of common observation, language can use such words to evolve abstractions that go beyond observation. Such abstractions may one day yield a description of the real world independent of our observation of it.

Such, then, are the arguments used to counter the Copenhagen Interpretation.

BIBLIOGRAPHY

Feynman, R. P. (1963) *Lectures on Physics,* Vol. 1, Chapters 26, 37 and 38, Addison-Wesley, New York
Quanta and Reality: a symposium (1962) British Broadcasting Corporation, Third Programme lectures and discussion, Hutchinson, London

Author Index

Subject Index

429